Syntheses in Limnogeology

The aim of this book series is to focus on syntheses or summaries of modern and/or ancient lake systems worldwide. Individual books will present as much information as is available for a particular lake basin or system of basins to offer readers one distinct reference as a guide to conduct further work in these areas. The books will synthesize the tectonics, basin evolution, paleohydrology, and paleoclimate of these basins and provide unbiased new interpretations or provide information on both sides of controversial issues. In addition, some books in the series will synthesize special topics in limnogeology, such as historical records of pollution in lake sediments and global paleoclimate signatures from lake sediment records.

More information about this series at https://link.springer.com/bookseries/10029

Alexander Orkhonselenge
Munkhjargal Uuganzaya • Tuyagerel Davaagatan

Lakes of Mongolia

Geomorphology, Geochemistry and Paleoclimatology

Springer

Alexander Orkhonselenge
Laboratory of Geochemistry
and Geomorphology
School of Arts and Sciences
National University of Mongolia
Ulaanbaatar, Mongolia

Munkhjargal Uuganzaya
Laboratory of Geochemistry
and Geomorphology
School of Arts and Sciences
National University of Mongolia
Ulaanbaatar, Mongolia

Tuyagerel Davaagatan
Laboratory of Geochemistry and
Geomorphology
School of Arts and Sciences
National University of Mongolia
Ulaanbaatar, Mongolia

Division of Physical Geography
Institute of Geography and Geoecology
Mongolian Academy of Sciences
Ulaanbaatar, Mongolia

ISSN 2211-2731 ISSN 2211-274X (electronic)
Syntheses in Limnogeology
ISBN 978-3-030-99119-7 ISBN 978-3-030-99120-3 (eBook)
https://doi.org/10.1007/978-3-030-99120-3

This Springer imprint is published by the registered company Springer Nature Switzerland AG
The registered company address is: Gewerbestrasse 11, 6330 Cham, Switzerland

A. Orkhonselenge dedicates this book to the honorable memory of her mother, B. Lkhagvasuren (1954–2017), whom she owes everything, including enormous support for her education and spawn love for science.
This book is also dedicated to all the members of the Laboratory of Geochemistry and Geomorphology (LGG), National University of Mongolia (NUM) where Mongolian future geoscientists have been educating and growing with theoretical knowledge, analytical methods, and empirical skills in Earth Science fields since 2015.

Preface

New research data from lakes of Mongolia arise from time to time, which provide a fundamental clue in understanding of surface processes operating in lake basins of Mongolia. Such data in lake study of Mongolia continue to emerge giving with deeper insights into basic principles of the processes. In other words, over the past two decades, lake studies in Mongolia have profoundly challenged us for the need to synthesize valuable implications regarding the surface processes (e.g., lacustrine, fluvial, aeolian, glacial) within lake basins in the context of principles of geomorphology, sedimentology, and geochemistry and of the reconstruction of past and present climate changes.

The following aspects promote to publish this book *Lakes of Mongolia: Geomorphology, Geochemistry,* and *Paleoclimatology*:

1. Recent studies revealing significant vulnerability of Mongolian lakes and their rapid evolution due to climate change have raised wider concerns.
2. Comprehensive study in geomorphology, geochemistry, and paleoclimatology of Mongolian lakes has yet to be conducted with the exception of general descriptions of lakes by Tserensodnom (1971, 2000) who introduced a number of lakes and classified them based on their distributions according to each aimag (meaning province), administration unit of Mongolia.
3. The study of lakes in Mongolia in terms of interdisciplinary fields (e.g., limnogeology, hydrogeomorphology, hydrogeochemistry) needs more leaderships of Mongolian geoscientists, i.e., most of recent results about lakes in Mongolia have been provided or led by foreign scientists including those outside of the former Soviet bloc (e.g., Grunert et al., 2000; Fowell et al., 2003; Kashiwaya et al., 2010; Kang et al., 2015; Mischke et al., 2020).
4. There have emerged powerful advanced methodological progresses in the application of numerical techniques, conceptual models, and mapping tools in the field of lacustrine geomorphology, geochemistry, and paleoclimatology.
5. It is important to let foreigners know the correct physiographic names and their spellings not only of lakes but also of other geographical objects in Mongolia.

All these points demand a new edition in the book form in order to bring readers up to date on them.

In This New Book

- The opportunity to include some older Russian and Mongolian literature, which have never been written in English, is to present in the book in order to review the early stage of lake study in Mongolia.
- New, pertinent references to more recent research article, report, presentation, and book publications help keep this new book up to date for informing readers about the advancement in lake study of Mongolia during the last two decades.
- The book presents statistical data about the lake distribution in five physiographic regions of Mongolia. The regions are holistically determined based on landscape type and landform feature of Mongolia providing a representative categorization.
- The book describes data on expansion and reduction of lake area and change in lake level for representative lakes in each region of Mongolia controlled by the prevailing semiarid to arid climates.
- The book addresses the sedimentation dynamics and geochemical characteristics derived from stratigraphic features of lake sediments for selected lakes.
- The book reflects on the scientific background of modern lakes throughout the country and gives detailed information on their evolution over the geological time scales and paleoclimate changes in lake basins during the Pleistocene, Holocene, and Anthropocene.
- The paleoclimate changes inferred from high-resolution multi-proxy data recorded in lake sediments are newly synthesized. The chronologies are primarily revealed with the radiocarbon isotope (^{14}C) extending to the late Pleistocene and with some optically stimulated luminescence (OSL) records.
- The book introduces recent new data on modern academic, economic, and touristic prospects and concerns of lakes in Mongolia.
- New figures and photos illustrate the topics discussed in the related text and represent current research advances.
- The correct spelling from modern Mongolian Cyrillic alphabet for geographical place names is introduced by transliteration into English.

This book is meant to be not only a scientific summary of Mongolian lakes but also a reference source in the applied fields within Earth Science. We hope and intend that the book contents would be still valid for future multi-decades for lake studies in Mongolia and Eurasia.

Ulaanbaatar, Mongolia

Alexander Orkhonselenge
Munkhjargal Uuganzaya
Tuyagerel Davaagatan

References

Fowell, S. J., Hansen, B. C., Peck, J. A., Khosbayar, P., & Ganbold, E. (2003). Mid to late Holocene climate evolution of the Lake Telmen basin, north central Mongolia, based on palynological data. *Quaternary Research, 59*(3), 353–363.

Grunert, J., Lehmkuhl, F., & Walther, M. (2000). Paleoclimatic evolution of the Uvs Nuur basin and adjacent areas (Western Mongolia). *Quaternary International, 65–66*, 171–192.

Kashiwaya, K., Ochiai, S., Sumino, G., Tsukamoto, T., Szyniszewska, A., Yamamoto, M., Sakaguchi, A., Hasebe, N., Sakai, H., Watanabe, T., & Kawai, T. (2010). Climato-hydrological fluctuations printed in long lacustrine records in Lake Hövsgöl. *Mongolia Quaternary International, 219*(1–2), 178–187.

Kang, S., Lee, G., Togtock, C., & Jang, K. (2015). Characterizing regional precipitation-driven lake area change in Mongolia. *Journal of Arid Land, 7*(2), 146–158.

Komatsu, G., Brantingham, P. J., Olsen, J. W., & Baker, V. R. (2001). Paleoshoreline geomorphology of Boon Tsagaan Nuur, Tsagaan Nuur and Orog Nuur: The Valley of Lakes, Mongolia. *Geomorphology, 39*, 83–98.

Krivonogov, S. K., Sheinkman, V. S., & Mistruykov, A. A. (2005). Stages in the development of the Darhad dammed lake (Northern Mongolia) during the Late Pleistocene and Holocene. *Quaternary International, 136*(1), 83–94.

Lehmkuhl, F., Grunert, J., Hülle, D., Batkhishig, O., & Stauch, G. (2018). Paleolakes in the Gobi region of southern Mongolia. *Quaternary Science Reviews, 179*, 1–23.

Mischke, S., Lee, M. K., & Lee, Y. I. (2020). Climate history of Southern Mongolia Since 17 ka: The Ostracod, Gastropod and Charophyte Record from Lake Ulaan. *Frontiers in Earth Science, 8*(221), 1–15.

Orkhonselenge, A., Krivonogov, S. K., Mino, K., Kashiwaya, K., Safonova, I. Y., Yamamoto, M., Kashima, K., Nakamura, T., & Kim, J. Y. (2013). Holocene sedimentary records from Lake Borsog, eastern shore of Lake Khuvsgul, Mongolia, and their paleoenvironmental implications. *Quaternary International, 290–291*, 95–109.

Peck, J. A., Khosbayar, P., Fowell, S. J., Pearce, R. B., Ariunbileg, S., Hansen, B. C. S., & Soninkhishig, N. (2002). Mid to Late Holocene climate change in north central Mongolia as recorded in the sediments of Lake Telmen. *Palaeogeography, Palaeoclimatology, Palaeoecology, 183*, 135–153.

Prokopenko, A. A., Khursevich, G. K., Bezrukova, E. V., Kuzmin, M. I., Boes, X., Williams, D. F., Fedenya, S. A., Kulagina, N. V., Letunova, P. P., & Abzaeva, A. A. (2007). Paleoenvironmental proxy records from Lake Hovsgol, Mongolia, and a synthesis of Holocene climate change in the Lake Baikal watershed. *Quaternary Research, 68*, 2–17

Tserensodnom, J. (1971). *Lakes of Mongolia*. State Publishing, Ulaanbaatar, 202 p. [In Mongolian].

Tserensodnom, J. (2000). *A catalog of lakes in Mongolia*. Shuvuun Saaral Publishing, 141 p. [In Mongolian].

Wang, W., Ma, Y., Feng, F., Narantsetseg, T., Liu, K., & Zhai, X. (2011). A prolonged dry mid-Holocene climate revealed by pollen and diatom records from Lake Ugii Nuur in central Mongolia. *Quaternary International, 229*, 74–83.

Yu, K., Lehmkuhl, F., Schlütz, F., Diekmann, B., Mischke, S., Grunert, J., Murad, W., Nottebaum, V., Stauch, G., & Zeeden, C. (2019). Late quaternary environments in the Gobi Desert of Mongolia: Vegetation, hydrological, and palaeoclimate evolution. *Palaeogeography, Palaeoclimatology, Palaeoecology, 514*, 77–91.

Acknowledgments

This book *Lakes of Mongolia*: *Geomorphology*, *Geochemistry*, and *Paleoclimatology* is primarily based on the results, concerns, prospects, and insights gained from our own and others' research and also on discussions with our colleagues and graduate students at the Laboratory of Geochemistry and Geomorphology (LGG), National University of Mongolia (NUM). We thank all of them.

We would like to express many thanks to executive editor P.V. Steenbergen for receiving our wishes to publish this book and for making his kind acceptance. We are also very grateful to book editorial director, G.Z. Landolfo and vice president, D. Merkle for having agreement and permission to publish the book *Lakes of Mongolia*: *Geomorphology*, *Geochemistry,* and *Paleoclimatology*, the book project advisor Dr. A. Schiller and coordinators Mr. D.M. Manoharan, Henry Rodgers and P. N. Kala for their continuous technical support, and anonymous reviewers for improvement of the book and precious contribution to the enhancement of the quality of the book. Finally, we would also like to acknowledge the precious suggestions and constant availability of Springer senior publisher and the assistance by the Springer book project coordinators who took care of this book with remarkable dedication and patience, in particular the volume production and quality. It was a real pleasure to work on such an exceptional book describing Mongolian lakes, associated landforms, and paleoclimate change reconstruction of Mongolia.

A. Orkhonselenge expresses her sincere gratitude, in particular, to loving mother B. Lkhagvasuren (1954–2017) and families. She acknowledges the important influence of her supervisors, co-workers, and co-authors Prof. Ch. Gonchigsumlaa (1963–2008), Dr. G. Undral, Prof. K. Kashiwaya, Prof. J. Harbor, Prof. M. Wagreich, Prof. F. Lehmkuhl, Prof. G. Nanson, Prof. F. Zucca, Prof. S. Fowell, and Dr. S. Ariunbileg. She thanks her students N. Amgalan-Erdene, T. Davaagatan, M. Uuganzaya, Ts. Davaakhuu, N. Altansukh, B. Myadagbadam, N. Khishigsuren, D. Gerelsaikhan, O. Bulgan, D. Batzorig, P. Bolor-Erdene, G. Tserendorj, Ts. Bulgantamir, and G. Narmandakh at the LGG, NUM for their enormous assistance in fieldworks, analyses, and maps. She also thanks Dr. G. Komatsu for reading the book manuscript and contributing to its advancement.

M. Uuganzaya expresses her gratitude to all colleagues at the LGG, NUM, especially to her supervisor A. Orkhonselenge, who taught and introduced glacial geomorphology during her graduate school and gave advice during compiling maps for her contribution to this book. She also thanks her families.

T. Davaagatan acknowledges the influence of her supervisor and families. She thanks all colleagues at the LGG, NUM, and Division of Physical Geography, Institute of Geography and Geoecology, Mongolian Academy of Sciences. She expresses her gratitude to her supervisor A. Orkhonselenge, who led her with analytic methods in lacustrine geomorphology and paleoclimatology and gave advice on research publication and opportunity to deliver her contribution to this book.

About the Book

The book *Lakes of Mongolia*: *Geomorphology*, *Geochemistry*, and *Paleoclimatology* aims to systematically provide an upgraded overview of the most spectacular lakes in Mongolia from scientific, economic, and scenic points of view, presenting lake area changes, associated outstanding landforms in lake basins, their sedimentological and geochemical characteristics, valuable economic and geoheritage resources, paleogeographical evolution and paleoclimate change reconstruction in a comprehensive manner.

Understanding geomorphic evolution of lacustrine landscape and investigating changes in lake areas in Mongolia are important because they contribute to obtain an overall view of histories of the paleo- and modern-day climate changes in Mongolia and Eurasia. However, despite some early works on paleoclimate and paleoenvironmental changes recorded in lakes of Mongolia intending to shed light on large-scale regional changes (e.g., Grunert et al., 2000; Komatsu et al., 2001; Peck et al., 2002; Krivonogov et al., 2005; Prokopenko et al., 2007; Wang et al., 2011; Orkhonselenge et al., 2013; Lehmkuhl et al., 2018; Yu et al., 2019), temporal and spatial analyses of modern lakes in Mongolia have not been conducted sufficiently.

This book covers lacustrine geomorphology, geochemistry, and paleoclimatology which best contribute to represent uniqueness and geodiversity of lakes in the country and past and present climate changes in Mongolia and Eurasia. It also introduces these lakes with scientifically essential lacustrine landscapes, economically useful resources, and recreationally fascinating geosites, which are of crucial importance to understanding the geomorphic evolution of lake basins in Mongolia.

In this book, a special attention is given to the lakes from where recent intensive investigations have brought new insights into the spatial and temporal reconstructions of paleoclimate and paleoenvironmental changes within Mongolia and Eurasia. The book not only emphasizes internationally well-known lakes of Mongolia, but it also tends to describe far less popular lakes which have been remained unrecognized of their scientific importance.

This book is the first effort to synthesize the geomorphological, geochemical, sedimentological, paleogeographical, and paleoclimatological implications with present climate change obtained from the lakes of Mongolia. The book is divided into three parts.

Part I introduces an overview of past and present lake studies in Mongolia. Moreover, it is concerned with aspects of formation, evolution, origin, and classification of Mongolian lakes taking account of the great variety of the lakes existing in the country.

Part II introduces the geological, climatological, geomorphological, cryological, and hydrological characteristics of each region in the east, south, west, north, and center of Mongolia. In these regions, 10 lakes that are geomorphologically, sedimentologically, geochemically, and paleoclimatologically valuable are discussed in alphabetical order as Buir, Buun Tsagaan, Khargal, Khoton, Khuvsgul[1], Khukh, Terkhiin Tsagaan, Ugii, Ulaan, and Uvs. In addition, it deals with paleoclimatic and paleoenvironmental reconstructions in Mongolia and Eurasia inferred from the long-term geomorphological, sedimentological, and paleogeographical evolutionary histories of lake basins.

Part III provides readers essential core aspects of academic, economic, and geotouristic significance, especially those valuable lakes in Mongolia, and their implications on the lakes and landscapes. This part highlights an extension based on the scientific values of lake study to applicable fields of higher education, economy, and geopark.

All the chapters of the book begin with an overview clue abstract and ends with a review comprehensive summary. The overall goal of the book is to provide an accessible, highly illustrated, virtually attractive, and well-integrated publication suitable for pure and applied sciences and their interdisciplinary fields.

We hope that the book *Lakes of Mongolia*: *Geomorphology*, *Geochemistry,* and *Paleoclimatology* will be used as a primary source for scholars, researchers, experts, and professionals in the fields of Earth science, Quaternary science, and Environmental science. The observed surface processes in lake basins, resulting in geomorphological features and sedimentological and geochemical characteristics, have a fundamental bearing on our understanding not only of lake evolution over various geological time scales and paleo- and modern-day climate changes in lake basins but also of the present academic issues, future scientific development, and sustainability of economic and geoheritage resources.

This book is the culmination of years of experience in training undergraduate and graduate students in fields of introduction to Earth science, geomorphology, sedimentology, geochemistry, and paleogeography, and in working for research projects over the last two decades. It intends to provide a valuable source material useful for describing and analyzing lake sediments, related landforms and Earth surface processes in lake basins, and regional patterns of paleoclimate changes in large lake basins of Mongolia.

---

[1] *Khuvsgul* has been misspelled as *Hovsgol* and *Khubsugul* in publications. Khuvsgul is the right English transliteration from Mongolian *Хөвсгөл*.

Contents

About the Authors

Alexander Orkhonselenge's research interests focus on glacial, lacustrine, fluvial, and aeolian sedimentological, geochemical, and geomorphological processes, paleoclimate changes, and Quaternary science. Her interest in the galactic formation, evolution, and processes in the Solar System and the Earth was inspired by a fascinating lecture on geochemistry at her freshman. Later on, her continuous learning from eminent professors and researchers expanded in the fields of geomorphology, sedimentology, and paleoclimatology.

Her research has largely focused on lake sedimentations and alpine glaciations in Mongolia and on reconstruction of paleoclimate changes based on their deposits. She also has studied peatland formation and evolution in northeastern Mongolia and how paleoclimate changes have influenced peatlands. She has been studying stratigraphic sequences of outcrops in the Govi[1] region, southern Mongolia. Over the last two decades, she worked in the Mongolian-Russian-Japanese-Korean joint research projects in Lake Khuvsgul[2] (HDP) and paleolake Darkhad (DDP), and the Swedish-American-Chinese-Mongolian joint research project (CAPP) in the Mongolian Altai Mountain Range.

She established the Laboratory of Geochemistry and Geomorphology (LGG) at the National University of Mongolia (NUM) on September 28, 2015 for training undergraduate and graduate students in Earth science and for developing fields in Earth science, especially geomorphology, sedimentology, and geochemistry, in Mongolia. The first LGG's research project was to reconstruct paleoclimate change in southern Mongolia based on lakes in the Valley of Lakes of the Govi region. The research has led her, together with her students, to work in eastern, northern, and central Mongolia. She is an author and a coauthor of about 20 scientific articles published internationally. Laboratory of Geochemistry and Geomorphology,

[1] *Govi* has been often spelled as *Gobi* in the international literature. Govi is the correct English transliteration from Mongolian *Говь*.

[2] *Khuvsgul* has been misspelled as *Hovsgol* and *Khubsugul* in many publications. The Khuvsgul is the right English transliteration from Mongolian *Хөвсгөл*

School of Arts and Sciences, National University of Mongolia, Ulaanbaatar, Mongolia

Munkhjargal Uuganzaya graduated from the National University of Mongolia (NUM) with BSc in Geography in 2011 and MSc in Environmental Science in 2015. Since her graduate school, she has been deeply interested in glacier changes of Mongolia and joined the colleagues at the Laboratory of Geochemistry and Geomorphology (LGG). Her master thesis under the supervisor A. Orkhonselenge entitled *Estimating modern glacier changes of Mongolia using remote sensing: the case of Mt. Ikh Turgen* contributed and upgraded her research in advanced level professionally.

Her research focuses on alpine glaciations and lake area changes in Mongolia. She has specialized in GIS and remote sensing techniques. She participated in research projects at the LGG and published over ten scientific articles related to paleo- and modern glaciers in the Mongolian Altai, Khangai and Khentii. Mountain Ranges and lake sedimentations in Lake Ulaan at national and international peer-reviewed journals. Laboratory of Geochemistry and Geomorphology, School of Arts and Sciences, National University of Mongolia, Ulaanbaatar, Mongolia

Tuyagerel Davaagatan graduated from Mongolian State University of Education with BSc in Geography in 2011 and National University of Mongolia (NUM) with MSc in Geography in 2014. After her graduate school, she joined her colleague at Laboratory of Geochemistry and Geomorphology (LGG), NUM.

Her research focuses on lake sedimentations and past and present climate changes in Mongolia. She participated in the fieldwork for lake sedimentations in the Mongolian-Russian-Japanese-Korean joint international research project in paleolake Darkhad (DDP) with her supervisor A. Orkhonselenge from NUM and professors and graduate students from Kanazawa University, Japan, in 2011 and the national research project *Landscape Structure, Change, Planning and Proper Zonation in eastern Mongolia* led by the Division of Physical Geography, Institute of Geography and Geoecology in 2016. She has specialized in analytic methods to determine physical and chemical properties of lake sediments since her training at Kanazawa University in 2012.

She also participated in some research projects at the LGG and published more than ten scientific articles at national and international peer-reviewed journals. Recently, she published the coauthored books in Mongolian entitled *Landscape Ecological Potential of Mongolia* in 2020 and *Geographical Uniqueness of Mongolia* in 2021. Laboratory of Geochemistry and Geomorphology, School of Arts and Sciences, National University of Mongolia, Ulaanbaatar, Mongolia

Division of Physical Geography, Institute of Geography and Geoecology, Mongolian Academy of Sciences, Ulaanbaatar, Mongolia

Abbreviations

AD	Anno Domini meaning a calendar year, the start of the era
Al	Aluminum
Al_2O_3	Aluminum oxide
&	And
~	Approximately, nearly, about
Ar	Argon
As	Arsenic
a.s.l.	Above sea level
ASTER	Advanced Spaceborne Thermal Emission and Reflection Radiometer
B	Boron
BA	Bølling–Allerød
BP	Before Present
Br	Bromine
BRZ	Baikal Rift Zone
Ca	Calcium
$CaCO_3$	Calcium carbonate
cal.	Calibrated
CaO	Calcium oxide
CAOB	Central Asian Orogenic Belt
carb	Carbonic
^{14}C	Radiocarbon-14
$\delta^{13}C$	Ratio of stable isotopes carbon-13 (^{13}C) : carbon-12 (^{12}C)
CHs	Carbohydrates
°C	Celsius degrees
CRU	Climatic Research Unit
Cl	Chlorine
C/N	Carbon to nitrogen ratio
DDP	Darkhad Drilling Project
DEM	Digital Elevation Map
dm	Decimeter
EASM	East Asian Summer Monsoon

EAWM	East Asian Winter Monsoon
EC	Electrical conductivity
e.g.	For example
et al.	And others
etc.	And so on
Fe	Iron
Fe_2O_3	Hematite
FeOOH	Goethite
Fig.	Figure
GC	Gas chromatography
g/l	Gram per liter
GPa	Gigapascal
HCO_3	Bicarbonate
HDP	Hovsgol (or Khuvsgul) Drilling Project
H_2S	Hydrogen sulfide
i.e.	In other words
IR	Infrared
IRSL	Infrared optically stimulated luminescence
ITM	International Travel Mart
K	Potassium
ka	Kilo annum meaning thousands of years ago
KDP	Khuvsgul Drilling Project
km	Kilometer
km^2	Square kilometer
km^3	Cubic kilometer
K_2O	Potassium oxide
LGG	Laboratory of Geochemistry and Geomorphology
LGM	Last Glacial Maximum
Li	Lithium
LLGM	Local Last Glacial Maximum
LOI	Loss on ignition
l/s	Liter per second
m	Meter
m^2	Square meter
m^3	Cubic meter
Ma	Mega annum meaning millions of years ago
MAS	Mongolian Academy of Sciences
mg/l	Milligram per liter
MgO	Magnesium oxide
MIS	Marine Isotope Stage
MHPS	Mongolian High-Pressure System
MnO	Manganese oxide
mm	Millimeter
MODIS	Moderate Resolution Imaging Spectroradiometer
MPI	Max Planck Institute

m/s	Meters per second
MS	Magnetic susceptibility
Mt	Mountain
N	North
Na	Sodium
NAO	North Atlantic Oscillations
Na_2O	Sodium oxide
NDVI	Normalized Difference Vegetation Index
NDWI	Normalized Difference Water Index
NGO	Non-governmental organizations
NH_4	Ammonium
NMR	Nuclear Magnetic Resonance
NOAA	National Oceanic and Atmospheric Administration
NO_3	Nitrate
NUM	National University of Mongolia
$\delta^{18}O$	Ratio of stable isotopes oxygen-18 (^{18}O)/oxygen-16 (^{16}O)
org	Organic
OSL	Optically stimulated luminescence
PAHs	Polycyclic aromatic hydrocarbons
PCA	Principal Component Analysis
%	Percentage
±	Plus or minus
pH	Potential of hydrogen
ppm	Part per million
P_2O_5	Phosphorus oxide
RAS	Russian Academy of Sciences
S	South
SAR	Synthetic Aperture Radar
SDSN	Sustainable Development Solutions Network
Si	Silicon
SiO_2	Silicon dioxide
SIRM	Saturation Isothermal Remnant Magnetization
SO_4	Sulfate or Sulphate
Sr	Strontium
TC	Total carbon
TDS	Total dissolved solids
Th	Thorium
Ti	Titanium
TIC	Total inorganic carbon
TiO_2	Titanium dioxide
TL	Thermoluminescence
TOC	Total organic carbon
TM	Thematic Mapper
TN	Total nitrogen
TS	Total sulfur

U	Uranium
USSR	Union of Soviet Socialist Republics
vs.	Versus
YD	Younger Dryas
yr	Year

Chapter 1
Introduction

Abstract Mongolia consists of the highly elevated Mongolian Plateau between the Siberian and Chinese Cratons on the Central Asian Orogenic Belt (CAOB) and occupies a transitional region between the Siberian taiga forest in the north and the Govi desert in the south. Mongolia is situated at the region characterized by the highest degree of continentality within Eurasia developing under the interaction of three large-scale climatic systems (the Siberian high- and the Asian low-pressure cells and the westerlies). Because of its unique physiographic condition, landscape of Mongolia is diverse and very peculiar and extraordinary with glaciated high mountains, large freshwater and saline lakes, spacious plains, spectacular Govi deserts, and large rivers. Landscape and lakes of Mongolia represent the natural museum of the northeastern, eastern and southeastern Eurasian continent because of its untouched, wild, and less anthropogenic nature. In this chapter, the whole of Mongolian landscape and lakes and their evolution and feature are introduced in detail, and the scope and worth of the book are highlighted.

Keywords Physical Geography · Geomorphology · Landscapes · Landforms · Lakes · Mongolia

1.1 Overview of Mongolian Physiography

Mongolia, a highly uplifted country in Eurasia, is located in the region characterized by the highest degree of continentality with extreme diurnal, seasonal, and annual air temperature amplitudes. Mongolia lies at the southern limit of both the Siberian taiga forest and permafrost, and the northern limit of both the Govi (or Gobi)[1] desert area and steppe grassland. Because of its unique physiographic condition, landscape of Mongolia is diverse and very peculiar within Eurasia.

[1] *Govi* has been often spelled as *Gobi* in the international literature. Govi is the correct English transliteration from Mongolian *Говь*.

A. Orkhonselenge et al., *Lakes of Mongolia*, Syntheses in Limnogeology,
https://doi.org/10.1007/978-3-030-99120-3_1

Lakes of Mongolia, situating in the eastern part of a great lake zone extending from the Mediterranean Sea to the Lake Baikal[2] (Murzaev, 1952; Tsegmid, 1969), constitute a spectacular scenery among numerous extraordinary attractive landscapes of the country (e.g., glaciated high mountains such as the Altai, Khangai,[3] and Khentii[4] Mountain Ranges; large freshwater and saline lakes such as Lakes Buir, Khuvsgul,[5] and Khyargas; open wide plains such as Dornod, Menen, and Sulin Kheer plains; Govi Desert regions such as Borzon, Uush, and Galba and large rivers such as Orkhon, Kherlen, and Selenge[6] Rivers). An outstanding overview regarding physical geography of Mongolia was published in a book by Russian physical geographer E.M. Murzaev in 1952 (see Chap. 2) with remarkable observations and a series of valuable topographic maps, and this book was translated and published from Russian to Mongolian language by Tsegmid (1969). Landscape diversity in each region of Mongolia is deeply connected with the long-term geological evolution and climate changes over the geological time scale. The Mongolian landscape is diverse and distinctive owing to the country's highly elevated intercontinental location in the Central Asian Orogenic Belt (CAOB), the deep interior of the Eurasian continent occupying a transitional area between the Siberian taiga forest in the north and the Govi Desert in the south.

Landscapes and landforms of Mongolia reflect interactive imprints of the major Central Asian geotectonic setting consisting of the Siberian Craton in the north and the Tarim and Sino-Korean Cratons in the south. The landscapes and landforms also are under the influence of an extremely arid extra-continental climate developing under the interaction of three large-scale climatic systems (the Siberian high- and the Asian low-pressure cells and the westerlies) (Hilbig, 1995; Gong & Ho, 2002; Panagiotopoulos et al., 2005), which are modulated by the North Atlantic Oscillations (NAO) (Visbeck, 2002). The extreme continental climate of Mongolia is reflected in the annual air temperature amplitude of approximately 45 °C and the low annual precipitation with dominant supplies from June to August (Academy of Sciences of Mongolia and Academy of Sciences of USSR, 1990), decreasing from more than 400 mm/year in the north to less than 50 mm/year in the south affecting the latitudinal trends in distributions of soil and vegetation covers from the north to the south. This considerable variation in the climatic condition makes Mongolia sensitive to climate change (Sugita et al., 2007). Data on paleoclimate change help us to understand the responses of various sensitive landscapes, especially lake basins in

[2] *Baigali* meaning *nature* has been misspelled as *Baikal* based on the Russian pronunciation. Baigali is the right English transliteration from Mongolian *Байгаль*.

[3] *Khangai* has been misspelled as *Khangay* (or *Hangay*) and *Hangai* in many publications. Khangai is the right English transliteration from Mongolian *Хангай*.

[4] *Khentii* has been misspelled as *Khentey* (or *Hentey*) and *Khentei* (or *Hentei*) in many publications. Khentii is the right English transliteration from Mongolian *Хэнтий*.

[5] *Khuvsgul* has been misspelled as *Hovsgol* and *Khubsugul* in publications. Khuvsgul is the right English transliteration from Mongolian *Хөвсгөл*.

[6] *Selenge* has been often incorrectly written as *Selenga* internationally by the Russian pronunciation. Selenge is the correct English transliteration from Mongolian *Сэлэнгэ*.

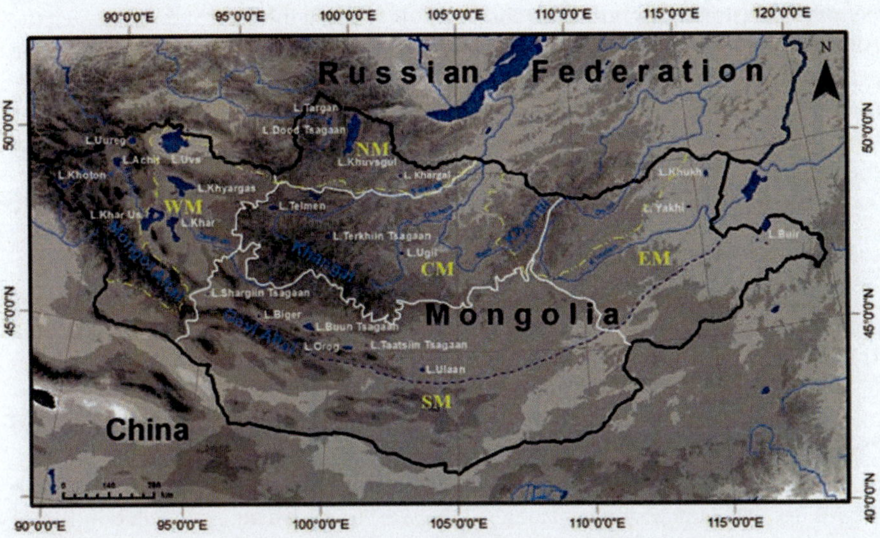

Fig. 1.1 Major lakes representing each region of Mongolia (*EM* eastern Mongolia (Chap. 6); *SM* southern Mongolia (Chap. 9); *WM* western Mongolia (Chap. 12); *NM* northern Mongolia (Chap. 15); and *CM* central Mongolia (Chap. 18), including the ten lakes described in Part II) with the Main Mongolian Lineament (blue dotted line) and a delineation of the Siberian and Central Asian geomorphological regions (yellow dotted line)

Mongolia (Davaagatan et al., 2015), to present and future climate changes (Orkhonselenge et al., 2019a). In addition, the geological structure of Mongolia is considerably manifested in the lake feature and evolution regarding their origin, diversity, and distribution.

The territory of Mongolia is divided by an approximate regional topographic and structural boundary, the Main Mongolian Lineament (Fig. 1.1), into the northern domain consisting of the Precambrian and early Paleozoic rocks and the southern domain of the early to late Paleozoic rocks (Badarch et al., 2002). In the northern domain, the Precambrian and early Paleozoic metamorphic rocks occur with the Neoproterozoic ophiolites, early Paleozoic island arc volcanic and volcaniclastic sediments, the Devonian to Carboniferous sediments, and the Permian volcanic-plutonic belts with marine and nonmarine sediments (Badarch et al., 2002). These geological elements of the basement are overlain by the postglacial landscapes in the Mongolian Altai Mountain Range, periglacial phenomena in the Khangai, Khuvsgul and Khentii Mountain Ranges, vast intermontane valleys and depressions between these ranges, large lakes in the Depression of Great Lakes and the Valley of Lakes, and the dominant river network of the Pacific Ocean and North Arctic Ocean drainage basins. In the southern domain, there is a complex of predominant early to late Paleozoic arc-related volcanic and volcaniclastic rocks with ophiolites and serpentinite mélanges and subdominant Silurian and Devonian fossil-rich reef limestones associated with terrigenous and volcaniclastic rocks in the north and of the

Permian limestones and turbidites and volcanic rocks in the suture zone between the Altaid collage and the Manchuride Orogenic Belt intruded by Mesozoic granite plutons and overlapped by late Jurassic to Cretaceous nonmarine volcanic and sedimentary rocks in the south (Badarch et al., 2002). This complex is characterized by the Govi Altai Mountain Range in the west and the Khyangan[7] Mountain Range in the east, numerous small lakes in the plains, and the Govi Desert.

A possibility to conduct research in geomorphologically rich Mongolia is highly attractive for geoscientists from all around the world. In 1990, Russian (e.g., S.S. Korjuev, D.A. Timofeev, E.V. Telegina, etc.) and Mongolian (e.g., J. Natsag, Sh. Tsegmid, S. Jigj) physical geographers compiled and provided the first geomorphological categorization of the country at the scales of 1:3,000,000 and 1:12,000,000 in the national atlas (Academy of Sciences of Mongolia and Academy of Sciences of USSR, 1990) based on the delineation of morphostructures, geomorphological provinces, and general morphometric properties of the Siberian and Central Asian geomorphological origins (Fig. 1.1).

In terms of the geomorphological view, Mongolia extends over 2405 km from the Mongolian Altai Mountain Range in the west to the Khyangan Mountain Range in the east and about 1263 km from the Khuvsgul Mountain Range in the north to the Govi Desert in the south. The altitude of the Mongolian territory, highly elevated in Eurasia, ranges from 552 m a.s.l. (Lake Khukh in the northern corner of the Dornod plain) in the east up to 4653 m a.s.l. (Tsegmid, 1969; Jigj, 1975) or 4374 m a.s.l. (Academy of Sciences of Mongolia and Academy of Sciences of USSR, 1990) (Mt. Maanit in the Altai Tavan Bogd Mts of the Mongolian Altai Mountain Range) in the west with an average elevation at 1850 m a.s.l., i.e., 84.7% of the total territorial area of 1,569,963 km^2 is located above 1000 m a.s.l., while 15.3% lies below 1000 m a.s.l. (Tsegmid, 1969; Jigj, 1975). Among the diverse landscapes in Mongolia, over 3500 lakes occupy nearly 1% (Tserensodnom, 1971) or 0.4% (Ministry of Environment and Green Development, 2013) of the total area of the country, having origins related to the Hercynian (or Variscan) Orogeny at the end of the Paleozoic Era (Jigj, 1975; see Chap. 3). Large rivers draining high mountain ranges in Mongolia feed other large rivers (e.g., Ob, Yenisei, Amar, Erchis) and lakes (e.g., Baikal, Dalai[8]) in Eurasia because the Mongolian Plateau is one of the major interfluves in the world. The global interfluve passes peaks of the high mountain ranges, and they divide the land of Mongolia into the three drainage basins of the Pacific Ocean, North Arctic Ocean, and Central Asian internal.

Due to the complex geologic, geotectonic, and geomorphic structures, the lakes of Mongolia (see Sect. 1.2) can be classified as mountainous and prairie of first-order importance. The Depression of Great Lakes and the Valley of Lakes in the Govi Desert region include very large lakes (see Chaps. 9 and 12), which are among the less investigated but are important paleoenvironmental archives of the Neogene

[7] *Khyangan* has been misspelled as *Kingan* (or *Hingan*) and *Hyangan* in many publications. Khyangan is the right English transliteration from Mongolian *Хянган*.

[8] *Dalai* meaning *ocean* has been often spelled as *Hulun* in many publications. Dalai is the right English transliteration from Mongolian *Далай*.

and Quaternary climate changes in Mongolia and Eurasia. Although other examples are small lakes in eastern and central Mongolia (see Chaps. 6 and 18), these lakes preserve valuable traces of regional landscape evolution over the geological time scale and regional and local past and present climate changes.

1.2 Overview of Mongolian Lakes

The global trend of lake occurrence is that there are many small lakes and large lakes are fewer (Cael & Seekell, 2016). Lakes of Mongolia also follow this pattern (see Chap. 5). In Mongolia there are an estimated over 3500 lakes (Tsegmid, 1969) or 3061 lakes (Tserensodnom, 2000) with an area greater than 0.1 km^2, and they cover an area of 15,640 km^2 (Tsegmid, 1969) or 16,003.3 km^2 (Tserensodnom, 2000) in total. Among them, nearly 1000 lakes have been previously noted by scientists (Murzaev, 1952). The total number, origin, type, classification, and distribution of lakes in Mongolia remain poorly understood because there are insufficient studies about the lake inventory of the country. Only some aspects of the lakes have been documented, such as a general background information (Murzaev, 1952; Tsegmid, 1969), morphometric values (Tserensodnom, 1971), and brief descriptions for lakes with an area over 0.1 km^2 in each aimag meaning *province* (Tserensodnom, 2000). Despite the major improvements in our scientific knowledge of lakes in Mongolia, there are still gaps for crucial information about spatial and temporal distributions of the lakes (see Part II).

Unlike the previous general information regarding lakes of Mongolia classified according to the each administration unit, this book contributes to present the lakes of Mongolia according to five physiographic regions (Fig. 1.1). Each region of Mongolia hosts different numbers of lakes with diverse origins and temporally and spatially heterogeneous distributions depending on geomorphic features or landscape types, including orogeny, altitude, and landform (see Chaps. 3–5). For instance, in western and northern Mongolia, lakes are generally found in mountainous areas, tectonic rifts, and periglacial regions (see Chaps. 12 and 15), whereas in eastern, southern, and central Mongolia (see Chaps. 6, 9, and 18), some lakes are found in endorheic basins and along the courses of mature rivers (see Chap. 4). Mountainous lakes with an area over 0.1 km^2 constitute 34.1% among the total lakes in Mongolia, and they dominantly occur in the Mongolian Altai, Khuvsgul, and Khangai Mountain Ranges in western, northern, and central Mongolia (see Chaps. 12, 15, and 18). These lakes are mostly deep and oval shaped. Because some of the mountainous lakes were originally related to paleoglaciations or modern glaciations, they are formed in glacial cirques or by moraine dams (see Chap. 4). Prairie lakes are dominantly distributed in eastern, southern, and central Mongolia, and they are divided into plain and Govi lakes (see Chaps. 9, 12, and 18). These prairie lakes are mainly formed by fluvial and aeolian processes in tectonic depressions, thermokarst cavities, and river valleys (see Chap. 4). The prairie lakes with an area over 0.1 km^2 each constitute 65.9% of the total lakes in Mongolia.

The geological history during the Mesozoic and Cenozoic Eras (see Chap. 3) plays an important role in the distribution of lakes in Mongolia. The present-day distribution of lakes throughout Mongolia is affected mainly by orogeny, landform, altitude, geology, river network, and climate (Tsegmid, 1969; Tserensodnom, 1971). A small number of large lakes reside in tectonic rifts or depressions and periglacial intermontane valleys in western and northern Mongolia, whereas a large number of small lakes occupy plains, endorheic basins, and river valleys in eastern, southern, and central Mongolia (see Chap. 5). For example, large lakes are abundantly distributed in the Mongolian Altai Mountain Range in western Mongolia (see Chap. 12) and in the Khuvsgul Mountain Range in northern Mongolia (see Chap. 15) due to their dominant orogenic landscapes, high precipitations, and dense river networks. Moreover, in central Mongolia (see Chap. 18), lakes in the northern Khangai Mountain Range are larger in area because of high precipitations and dense river networks in comparison to lakes in the southern Khangai Mountain Range of southern Mongolia (see Chap. 9) characterized by low precipitations and sparse river networks. Medium-sized lakes occur primarily in forest steppe regions (see Chaps. 5, 15, and 18), while small lakes are occurring throughout eastern, southern, and central Mongolia (see Chaps. 6, 9, and 18). Some present-day lakes of Mongolia are relict lakes of larger paleolakes existed during the Cretaceous and Neogene periods (see Chap. 3), and they are primarily distributed in tectonic rift grabens and postglacial intermontane valleys (see Chaps. 4, 12, and 15). Plains, Govi, and deserts of Mongolia contribute to the emergence of densely distributed lakes because they become terminal basins of inflows. The lake distribution is also affected by the presence of thermokarst, ice wedge, and permafrost (see Chap. 4).

To date, there has been no detailed classification scheme with attributes for the lake distribution in Mongolia. An exception might be the natural zonation used by Tserensodnom (1971) in Table 1.1. However, it is significantly underestimated lakes in the Khuvsgul and Khentii Mountain Ranges. Moreover, the lake distribution in Mongolia has never been temporally and spatially shown in detail with the context of landform features and landscape types. Hence, there is a strong need to present the temporal and spatial distributions of lakes more properly. In this book the lake distribution is temporally and spatially described according to each of the five regions based on the physiographic condition, landform features, and landscape

Table 1.1 Lakes in natural zones of Mongolia. After Tserensodnom (1971)

#	Natural zones	Percentage in total number of lakes	Percentage in total area of lakes
1	Altai Mountain Range	13.3	10.0
2	Khangai Mountain Range	20.8	25.0
3	Dornod plain	29.4	11.0
4	Govi region	36.5	54.0

Table 1.2 Lake distribution in five regions of Mongolia

#	Regions	Landscape types	Number of lake and its share		Lake area and its share		Details
			Number	Percentage	Area (km^2)	Percentage	
1	Eastern Mongolia	Eastern Khentii Mountain Range, Dornod plain, Khyangan Mountain Range, Kherlen River drainage basin, Khalkh River drainage basin	885	28.9	1448.9	9.0	Chapter 6
2	Southern Mongolia	Southern Khangai Mountain Range, Govi Altai Mountain Range, Valley of Lakes, Govi region, Desert region	756	24.7	1246	7.8	Chapter 9
3	Western Mongolia	Mongolian Altai Mountain Range, Depression of Great Lakes	567	18.5	9549.2	59.7	Chapter 12
4	Northern Mongolia	Khuvsgul Mountain Range, Darkhad basin, Eg River drainage basin, eastern Selenge River drainage basin	236	7.7	3297.4	20.6	Chapter 15
5	Central Mongolia	Northern and eastern Khangai Mountain Range, western Khentii Mountain Range, Orkhon River drainage basin, western Selenge River drainage basin	617	20.2	461.8	2.9	Chapter 18
	Total		3061	100.0	16,003.3	100.0	

types (see Part II). This is the first-time effort to provide regionality for the lake distribution in Mongolia (see Table 1.2).

In this book lakes of Mongolia are firstly described in each region characterized by specific landscape types (Fig. 1.1, Table 1.2). Lakes of Mongolia are distributed depending on the physiographic condition, landform features, and landscape types in each region. Eastern Mongolia has the largest number of lakes (28.9%) out of the total number of lakes in the country, whereas the largest lake area (59.7%) is found in western Mongolia (Table 1.2). A total number of 3061 lakes, occupying an area over 0.1 km^2, are distributed as follows as 28.9% in eastern, 24.7% in southern, 18.5% in western, 7.7% in northern, and 20.2% in central Mongolia (Table 1.2).

1.3 Scope and Structure of the Book

The book *Lakes of Mongolia: Geomorphology, Geochemistry and Paleoclimatology* aims to systematically provide an overview of the most spectacular lakes in Mongolia from scientific, economic, and scenic points of view presenting lake areas, associated outstanding landforms in lake basins, their sedimentological and geochemical characteristics, paleogeographical evolution, and paleoclimatological review in a comprehensive way. The book focuses on lake evolution, related geomorphology which best contributes to representing particularity and diversity in lake basins of the country, and paleoclimate and paleoenvironmental changes in Mongolia and Eurasia. The book covers the lakes with scientifically valuable landscapes, economically useful resources, and recreationally spectacular touristic spots, which are of crucial importance for the understanding of impact of geomorphological processes on lake evolution of Mongolia.

Mongolia may be called a country of lake water resources because lakes constitute 84% (Davaa et al., 2007) or 75% (Ministry of Environment and Green Development, 2013) of the total freshwater resources of the country. A total of 3061 lakes possess a surface area greater than 0.1 km^2 for each, including the largest Lake Uvs (3451.6 km^2) and the deepest Lake Khuvsgul (2773.2 km^2) (Table 1.3). Lake Khuvsgul alone constitutes 74.0% (Davaa et al., 2007) or 66.4% (Ministry of Environment and Green Development, 2013) of the total freshwater resources of Mongolia. Small lakes with an area less than 1.0 km^2 constitute 83.7% of the total lake number but only 5.6% of the total lake area amounting to 16,003.3 km^2 (Tserensodnom, 2000). Special attention in this book is given to lakes from which recent intensive investigations and valuable results have brought new insights into reconstructions of paleoclimate and paleoenvironmental changes within Mongolia and Eurasia. The book emphasizes internationally well-known lakes of Mongolia (e.g., Lake Ulaan in Chap. 11; Lake Uvs in Chap. 13; Lake Khuvsgul in Chap. 16), but it also tends to describe far less popular lakes (e.g., Lake Khukh in Chap. 8; Lake Khargal in Chap. 17; Lake Terkhiin Tsagaan in Chap. 19), which have been remained unrecognized for scientific importance (Fig. 1.1).

This book is the first effort to synthesize the geomorphological, sedimentological, geochemical, geochronological, and paleoclimatological implications obtained from the lakes of Mongolia. In other words, this book links hydrological processes in lake basins and sedimentary deposits in lakes, geochemical characteristics of lake sediments, and modern and paleoclimatic reconstructions in lake basins with the presently changing climatic systems, which have not been integrated to date in Mongolia. The book is divided into three parts.

Part I introduces an overview of past and present lake studies in Mongolia (see Chap. 2). The earliest phase of Mongolian lake study in the late nineteenth and twentieth centuries focused on the context of physical geography in the country. In the past two decades, lake studies in Mongolia have been rapidly widened and multiplied in research areas within the Earth Science disciplines (e.g., geomorphology,

Table 1.3 Morphometric components of lakes representing each region of Mongolia

#	Lakes	Altitude (m a.s.l.)[5]	Area of lake basin (km^2)[2]	Area of lake (km^2)[13]	Shoreline (km)[3]	Length (km)[2]	Width (km)[2]		Depth (m)[3]		Volume (km^3)[3]
						Maximum	Average	Maximum	Average	Maximum	
1	Buir	581	20,200[0]	625.4	118.2	40.0[1]	15.4	21.0[1]	6.0	15.0[8]	8.7[7]
2	Buun Tsagaan	1312[3]	33,500	241.12[9]	81.0	24.0	10.7	19.0[3]	9.9	16.0	0.638
3	Khargal	1057[10]	–	13.83[12]	14.8[5]	4.9[5]	–	4.4[5]	–	15.6[5]	0.127[5]
4	Khoton	2084[2]	2670	53.79	48.8	21.5[3]	2.3	4.0[3]	26.6	58.0	1.341
5	Khukh	560	–	56.7	52.6	14.5[3]	7.0	10.5[3]	10.3	14.0	0.26[7]
6	Khuvsgul	1645[2]	5300	2773.2	414.0	139.0[3]	20.3[4]	45.0[3]	139[4]	262.0[2]	383.3[4]
7	Terkhiin Tsagaan	2060	7880	61.99	66.0	16.0[3]	4.0	6.0[3]	10.0	20.0	0.368
8	Ugii	1337	4655	23.8	24.7	7.4[3]	3.4	5.3[3]	6.6	17.0[6]	0.171[7]
9	Ulaan	1024[9]	–	44.16[9]	87.0	45.0	29.0	40.0	0.9[5]	1.6[5]	0.158
10	Uvs	759[2]	71,100	3451.6	425.0	84.0	40.0	79.0	11.9	20.0	39.6

Source in order of publication year: [0] Murzaev (1952); [1] Badarch et al. (1965); [2] Tsegmid (1969); [3] Tserensodnom (1971); [4] Bogoyavlensky et al. (1989); [5] Tserensodnom (2000); [6] Wang et al. (2011); [7] Davaa (2015); [8] Enkhtaivan et al. (2016); Orkhonselenge et al. ([9] 2018, [10] 2019b, [11] 2020); [12] Gerelsaikhan and Orkhonselenge (2019); [13]Orkhonselenge and Uuganzaya (unpublished) sedimentology, geochemistry, geochronology, etc.) with advanced methods and techniques (see Chap. 2).

Part I is also concerned with aspects of formation, evolution, origin, and classification of lakes in Mongolia (see Chaps. 3–5), considering the great variety of lakes existing in Mongolia. Emergence and development histories of lakes in Mongolia are described, showing their formation and evolution over geological time scales (see Chap. 3). Outstanding lacustrine geomorphological features of different origins of lakes are described, showing their high geodiversities in the country, which include tectonic, volcanic, landslide, glacial, fluvial, karst, and aeolian lakes (see Chap. 4). Seven types of classification of lakes in Mongolia depending on geomorphic features, elevation above sea level, occupying area, holding volume, water level, salinity, outflow, and inflow are presented (see Chap. 5).

Part II is the core of the book, and it includes 16 chapters. These chapters present ten lakes of Mongolia that are geomorphologically, sedimentologically, geochemically, and paleoclimatologically valuable (Table 1.3; Fig. 1.1), beginning from the region in the east (see Chaps. 6–8), then south (see Chaps. 9–11), west (see Chaps. 12–14), north (see Chaps. 15–17), and concluding in the center (see Chaps. 18–20). The geomorphological, sedimentological, geochemical, and paleoclimatological characteristics of the ten lakes, in alphabetical order as Lakes Buir, Buun Tsagaan, Khargal, Khoton, Khuvsgul, Khukh, Terkhiin Tsagaan, Ugii, Ulaan, and Uvs (Table 1.3), are introduced and discussed in detail in Part II. From the geomorphological and hydrological points of view, which are linked to geological and climatological factors, these lake basins belong to the diverse physiographic conditions of each region in Mongolia. Therefore, each region of the country is introduced and described with geological, climatological, geomorphological, glaciological, cryological, hydrological, and lacustrine aspects (see Chaps. 6, 9, 12, 15, and 18). The presence of unique landforms such as paleoshorelines and paleoterraces associated with the lakes of Mongolia indicates some of the most valuable imprints of the long-term landscape evolution in the lake basins. The Mongolian lakes we introduce and discuss in this book are distributed in five physiographic regions classified according to their peculiarities of landscapes represented by diverse geology, topography, climate, and hydrological systems.

Eastern Mongolia (see Chap. 6), a scientifically, historically, and economically important region, has been experiencing the East Asian Summer Monsoon (EASM) effect during the Holocene (Rosen et al., 2019), and it contains numerous small lakes of diverse origins, most of which are economically valuable. The geomorphology of lakes in the vast Dornod and Menen plains displays landscape evolution of eastern Mongolia (e.g., tectonic Lakes Buir and Khukh in Chaps. 7 and 8).

In southern Mongolia (see Chap. 9), the long-term landscape evolution of the Govi region is remarkably notable in the Valley of Lakes with the lacustrine geomorphology (e.g., tectonic Lakes Buun Tsagaan and Ulaan in Chaps. 10 and 11). The Govi region stretching over southern and western Mongolia is limited by the Mongolian Altai Mountain Range in the west and northwest, the Khangai Mountain Range in the north and east, and the Govi Altai Mountain Range in the southwest and west. This region also experienced the East Asian Summer Monsoon (EASM) effect during the Marine Oxygen Isotope Stages (MIS) 3 and 5e (Lehmkuhl et al., 2018; Yu et al., 2019).

Western Mongolia (see Chap. 12) is a scientifically, economically, and recreationally important region with the presence of spectacular lakes (e.g., tectonic Lake Uvs in the Depression of Great Lakes in the northernmost corner of the Govi region extending from the southeast to the northwest of Mongolia in Chap. 13 and glacial Lake Khoton in the Mongolian Altai Mountain Range in Chap. 14).

Northern Mongolia (see Chap. 15) is a geotectonically and climatologically particular region because of its link to the Lake Baikal Rift Zone (BRZ). The region contains lakes formed in tectonic rift basins (e.g., Lake Khuvsgul in Chap. 16), remnant lakes of glacial paleolake Darkhad, and ones formed as oxbow lakes (e.g., Lake Khargal in the drainage basins of Eg and Selenge Rivers, major headwaters of the Lake Baikal in Chap. 17).

Central Mongolia (see Chap. 18), a scientifically and historically important region with an ancient capital city called *Khar Khorin*[9] in the Orkhon River valley, hosts the most diverse lakes in the intermontane valley between the Khangai and Khentii Mountain Ranges and river drainage basins in the Khangai Mountain Range (e.g., volcanic Lake Terkhiin Tsagaan in Chap. 19 and oxbow Lake Ugii in Chap. 20). In each region of Mongolia, we can observe a high variation in the diversely originated lakes resulting from the Cenozoic neotectonics and mountain buildings, the Pliocene and Pleistocene volcanic eruptions, and the numerous Pleistocene glaciations (see Chaps. 3 and 4).

Part II also deals with reconstructions of paleoclimate and paleoenvironmental changes in Mongolia inferred from the long-term evolutionary history of geomorphology, sedimentology, and paleogeography of lake basins in each region (see Chap. 21).

Part III provides readers peculiar aspects of scientific, economic, and geotouristic significances of the especially valuable lake basins in Mongolia and their implications on landscapes (see Chaps. 22–24). There are many scientifically valuable lakes that give us important information, as reflected in Chap. 22 (e.g., Lakes Yakhi, Ulaan, Uvs, etc.). The economically most valuable lakes are distributed in the Govi and plain regions, as explained in Chap. 23 (e.g., Lakes Buir, Gurvan Tes, Khyargas, etc.). One of the most attractive touristic landscapes of Mongolia is lakes (see Chap. 24), with examples of Lakes Khar Zurkhnii Khukh, Terkhiin Tsagaan, and Khar Us. Some of these lake basins (e.g., Lakes Khangal, Ulaagchnii Khar, Khuis, etc.) are legally protected (see Chap. 24) with their inclusion in a network of protected drainage basins. This last chapter concisely summarizes geoheritages, geosites, geoparks, and geodiversities in lake basins of Mongolia.

[9] *Khar Khorin* has been often misspelled as *Kara Korum* (or *Khara Khorum*) and *Har Horin* in publications, Wikipedia, etc. *Khar Khorin* is the right English transliteration from Mongolian *Хар Хорин*.

1.4 Importance of the Book

This book provides modern, qualitative, process-oriented approaches and quantitative analytic results-based implications to understand geomorphological, sedimentological, and geochemical evolution of lake basins in Mongolia and past and present climate changes in Mongolia and Eurasia. The book offers descriptions of geology, climate, landform, glacier, permafrost, groundwater, and surface water for each region of Mongolia (see Chaps. 6, 9, 12, 15, and 18). Insights into the interpretation of data obtained from the lake basins in the fields of geomorphology, sedimentology, geochemistry, geochronology, and paleoclimatology are developed from theoretical principles, empirical observations, correlative illustrations, analytic measurements, and conscious hypotheses. Based on the application of a combined compilation of recent Landsat 8 images of the lakes and topographic maps of them in 1970, this book presents enriched results and implications derived from remote sensing together with field measurements and laboratory analyses as evident in Part II. This data compilation owes to research projects by a team at the Laboratory of Geochemistry and Geomorphology (LGG), National University of Mongolia (NUM).

The subject areas presented in this book, especially those in Part II, are developing rapidly, and the quantity and quality as well of the published literature in these areas are growing exponentially with new data revealed every year. The tasks of critically reviewing and summarizing this literature and keeping up to date are challenging because it is a never-ending process. Many contents of this book are not based on our original research. The reference sources that were used or cited are listed at the end of each chapter. We tried to include all related valuable publications up to 2021 regarding the lakes of Mongolia. Nevertheless, we apologize in advance if the essential works of others are missing or misrepresented.

The book is designed for all academics, researchers, and professionals in Earth Science fields of limnology, hydrology, geomorphology, sedimentology, geochemistry, geochronology, paleoclimatology, Quaternary science, and environmental science. This book is the culmination of years of experience in training students in the fields of geomorphology, sedimentology, geochemistry, and paleogeography, and in working with research projects over the last two decades. It intends to provide precise high-resolution multi-proxy data and research-based materials useful for describing and analyzing lake sediments, for interpreting and inferring related landforms and surface processes (e.g., fluvial, aeolian, lacustrine, etc.) in lake basins, and for reconstructing paleoclimate and paleoenvironmental changes in the lake basins of Mongolia.

We hope that the book contents would still be valid for future multi decades for lake studies in Mongolia and Eurasia. It is also hopeful that this book would be reedited with contributions from the next generation of Mongolian geoscientists joining with experts in other disciplines within Earth Science from universities, institutes, research centers, and associated agencies in the future.

References

Academy of Sciences of Mongolia and Academy of Sciences of USSR. (1990). *National Atlas of the Peoples' Republic of Mongolia*. Minsk Publishing, 156 p. [In Russian with English abstract].

Badarch, G., Cunningham, W. D., & Windley, B. F. (2002). A new terrane subdivision for Mongolia: Implications for the Phanerozoic crustal growth of Central Asia. *Journal of Asian Earth Sciences, 21*, 87–110.

Badarch, N., Tsegmid, S., & Tserensodnom, J. (1965). *Landscapes and natural zones of eastern Mongolia*. Press of Mongolian Academy of Sciences, 117 p. [In Mongolian].

Bogoyavlensky, B. A., Batjargal, B., Martinov, V. P., Kozlov, U. P., Borobiev, V. V., & Batsuuri, D. (1989). *Atlas of Lake Khuvsgul: Mongolian People Republic*. State Printing Industry of Geodesy and Cartography, SSSR, 118 p. [In Russian with an abstract in English, German and Mongolian].

Cael, B. B., & Seekell, D. A. (2016). The size-distribution of Earth's lakes. *Scientific Reports, 6*, 29633.

Davaa, G. (2015). *Resource and regime of surface water of Mongolia*. 408 p. [In Mongolian].

Davaa, G., Oyunbaatar, D., & Sugita, M. (2007). *Surface water of Mongolia* (pp. 55–68).

Davaagatan, T., Orkhonselenge, A., Fukushi, K., & Yabe, M. (2015). Sedimentary records in intermontane and prairie lakes in central Mongolia: Preliminary results from Lake Terkhiin Tsagaan and Lake Ugii. *Journal of Geographical Issues, 14*(408), 19–26.

Enkhtaivan, D., Oyungerel, B., Avirmed, E., Renchinmyadag, T., Nyamkhuu, M., Munkhdulam, O., Odbaatar, E., Davaagatan, T., Bayanjargal, B., & Batnyam, T. (2016). *Report on research project: Landscape structure, change, planning and proper zonation (the case study of eastern Mongolia)*. Division of Physical Geography, Institute of Geography and Geoecology, Mongolian Academy of Sciences, 564 p. [In Mongolian].

Gerelsaikhan, D., & Orkhonselenge, A. (2019). Recent lake area changes of Lake Khargal in northern Mongolia. *The Environment, 1*(1), 15–20. [In Mongolian with English abstract].

Gong, D. Y., & Ho, C. H. (2002). The Siberian High and climate change over middle to high latitude Asia. *Theoretical and Applied Climatology, 72*, 1–9.

Hilbig, W. (1995). *The vegetation of Mongolia* (p. 260). SPB Academic Publishing.

Jigj, S. (1975). *Primary feature of Mongolian landforms*. Institute of Geography & Permafrost, Mongolian Academy of Sciences (MAS), 126 p. [In Mongolian].

Lehmkuhl, F., Grunert, J., Hülle, D., Batkhishig, O., & Stauch, G. (2018). Paleolakes in the Gobi region of southern Mongolia. *Quaternary Science Reviews, 179*, 1–23.

Ministry of Environment and Green Development. (2013). *Integrated water management plan of Mongolia*. 339 p.

Murzaev, E. M. (1952). *The description of physical geography of Mongolia* (2nd Ed.). [In Russian].

Orkhonselenge, A., Bulgan, O., Gerelsaikhan, D., Davaagatan, T., & Altansukh, N. (2020). Hydroclimatic fluctuation in Lake Yakhi, eastern Mongolia. *EGU General Assembly 2020. Presentation #EGU2020-686*, Vienna, Austria. https://doi.org/10.5194/egusphere-egu2020-686.

Orkhonselenge, A., Komatsu, G., Davaagatan, T., & Uuganzaya, M. (2019a). Sedimentary records from lakes reconstructing late quaternary paleoclimate changes in Mongolia. *XX INQUA Congress. Presentation #2419*, Dublin, Ireland.

Orkhonselenge, A., Fowel, S., Josh, B., Nancy, G., Camillie, T., & Gerelsaikhan, D. (2019b). *A report from fieldwork measurement and mapping of changes in lake area of Lake Khargal*. Laboratory of Geochemistry and Geomorphology, National University of Mongolia, 15 p. [In Mongolian].

Orkhonselenge, A., Komatsu, G., & Uuganzaya, M. (2018). Climate-driven changes in lake areas for the last half century in the Valley of Lakes, Govi Region, Southern Mongolia. *Natural Science, 10*(7), 263–277.

Panagiotopoulos, F., Shahgedanova, M., Hannachi, A., & Stephenson, D. B. (2005). Observed trends and teleconnections of the Siberian high: A recently declining center of action. *Journal of Climate, 18*, 1411–1422.

Rosen, A. M., Hart, T. C., Farquhar, J., Schneider, J. S., & Yadmaa, T. (2019). Holocene vegetation cycles, land-use, and human adaptations to desertification in the Gobi Desert of Mongolia. *Vegetation History and Archaeobotany, 28*, 295–309.

Sugita, M., Asanuma, J., Tsujimura, M., Mariko, S., Lu, M., Kimura, F., Azzaya, D., & Adyasuren, T. (2007). An overview of the rangelands atmosphere-hydrosphere-biosphere interaction study experiment in northeastern Asia (RAISE). *Journal of Hydrology, 333*, 3–20.

Tsegmid, S. (1969). *Physical geography of Mongolia.* State Press, 759 p. [In Mongolian].

Tserensodnom, J. (1971). *Lakes of Mongolia.* State Publishing, 202 p. [In Mongolian].

Tserensodnom, J. (2000). *A catalog of lakes in Mongolia.* Shuvuun Saaral Publishing, 141 p. [In Mongolian].

Visbeck, M. (2002). The ocean's role in Atlantic climate variability. *Science, 297*, 2223–2224.

Wang, W., Ma, Y., Feng, Z., Narantsetseg, T., Liu, K. B., & Zhai, X. (2011). A prolonged dry mid-Holocene climate revealed by pollen and diatom records from Lake Ugii Nuur in central Mongolia. *Quaternary International, 229*, 74–83.

Yu, K., Lehmkuhl, F., Schlütz, F., Diekmann, B., Mischke, S., Grunert, J., Murad, W., Nottebaum, V., Stauch, G., & Zeeden, C. (2019). Late quaternary environments in the Gobi Desert of Mongolia: Vegetation, hydrological, and palaeoclimate evolution. *Palaeogeography, Palaeoclimatology, Palaeoecology, 514*, 77–91.

Part I
Lake Studies and Types of Lakes in Mongolia

Chapter 2
Lake Studies in Mongolia: An Overview

Abstract In the earliest stage of research, lakes of Mongolia were investigated primarily by Russian scientists. In the nineteenth century, the lakes of Mongolia were studied in the context of physical geography. Physiographic conditions and geneses of lakes were presented with documentation of their geographical locations, and evolutionary history and paleogeography of lake basins were hypothesized based on paleoterraces and paleoshorelines. In the early twentieth century, the lakes of Mongolia began to be studied individually on the genesis, water regime, and morphometric components of lakes in fields of geomorphology and hydrochemistry besides physical geography. Since the late twentieth century, research of lakes has stretched into hydrogeology and hydrobiology. Mongolian scientists, besides the Russian scholars, have conducted lake studies, and they developed fields of limnology and hydrology in Mongolia. During the last half of the twentieth century, lake study of Mongolia extensively covered morphometric features and physical and chemical characteristics. However, in the twenty-first century, study of lakes in Mongolia has been scientifically upgraded with high-resolution multi-proxy data, advanced methods, and new techniques. The research area has rapidly widened and multiplied in the contexts such as lake level fluctuations, spatial and temporal changes of lake area, geomorphological evolution of lake basins, geochemical and sedimentological implications from lake deposits, paleolake reconstructions, and paleoclimate and paleoenvironmental changes. The present study of lakes in Mongolia has expanded beyond limnology and hydrology into geomorphology, sedimentology, geochemistry, geochronology, and paleoclimatology forming interdisciplinary fields. Since the late 1990s and early 2000s, foreign scientists or international research teams have been in the leading positions for the lake study of Mongolia.

Keywords Lake study · Physical geography · Limnology · Hydrology · Climate change · Mongolia

A. Orkhonselenge et al., *Lakes of Mongolia*, Syntheses in Limnogeology,
https://doi.org/10.1007/978-3-030-99120-3_2

2.1 Introduction

Mongolia possesses great inland surface water resources feeding large lakes (e.g., Lakes Baikal,[1] Tari,[2] Dalai,[3] etc.) and rivers (e.g., Yenisei, Amar, Ob Rivers, etc.) within the North Arctic Ocean and Pacific Ocean drainage basins. Among the various water resource components, lake is the largest one (~500 km^3), providing for over 84% (Davaa et al., 2007) or over 75% (Ministry of Environment and Green Development, 2013) of the total freshwater resources (599 km^3) in Mongolia. Although Mongolia has the continental climate ranging from semiarid to arid, it hosts several thousands of large and small lakes.

Lakes in Mongolia developed as early as the Paleozoic Era and were stabilized since the Era Mesozoic (Tsegmid, 1969; Jigj, 1975). However, present lakes formed due to mountain building in the late Neogene and early Quaternary periods (see Chap. 3). The lake formation and evolution in Mongolia are affected by neotectonics, geological structure, climate change, landform feature, river drainage network, and surface processes (Jigj, 1975). These factors play important roles in water temperature, transparency, and mineralization level of lakes (Tserensodnom, 1971). Because lakes are containers, accumulators, and repositories (Davaagatan et al., 2015), lake sediments can be local and regional archives of climate and environmental changes over geological time scales (Orkhonselenge et al., 2015, 2019). Lakes can be considered as an indicator of climate change since they increase or shrink in response to the changing climatological condition (e.g., Brown, 1995; Komatsu et al., 2001; Orkhonselenge et al., 2018a). In particular, lake area is an important indicator for climate change, and its relationship with meteorological factors is critical for understanding the mechanisms that control lake level changes (Kang et al., 2015).

At the beginning of research or in the nineteenth century, foreign travelers and researchers who conducted surveys in Central Asia, particularly the Russians, passed through Mongolia, and they noted geographical positions of natural objects such as lakes, rivers, mountains, sand dunefields, and others. Since then, lakes widely began to attract research efforts. At that time, geographical locations, physiographic conditions, and geneses of lakes were mostly investigated (see Sect. 2.2). In the twentieth century, Mongolian scientists were educated in Russia, and they joined foreign researchers to determine morphometric components of lakes while studying physical geography of the country. However, there was still no study of the hydraulic regime and characteristics of the lakes, which were relevant to the late development of limnology in Mongolia (Tserensodnom, 1971). Since 1962, the Institute of Geography and Permafrost, Mongolian Academy of Sciences (MAS)

---

[1] *Baigali* meaning *nature* has been misspelled as *Baikal* based on the Russian pronunciation. The Baigali is the right English transliteration from Mongolian *Байгаль*.

[2] *Tari* has been misspelled as Torey by the Russians. Tari is the right English transliteration from Mongolian *Тарь*.

[3] *Dalai* meaning *ocean* has been often spelled as *Hulun* in many publications. Dalai is the right English transliteration from Mongolian *Далай*.

started to study hydrophysical and hydrochemical characteristics of lakes in Mongolia, including water temperature, depth, brine resource, and other components (see Sect. 2.3).

Since the late 1990s, international research teams instead of Mongolian scientists have been dominant in lake study of Mongolia (see Sect. 2.4). Lakes of Mongolia are important with various implications because they can provide for the larger Central Asian context and even for the Pan-Eurasian perspective related to its drainage reorganization and lake histories (e.g., Komatsu et al., 2016), a factor that has attracted also foreign scientists to investigate lakes of Mongolia. For instance, among the foreign scientists in the late 1990s, the joint Russian-Mongolian-American expedition in Mongolia investigated lake geomorphology in the Valley of Lakes in southern Mongolia using a combination of satellite imagery and fieldwork (Komatsu et al., 2001). Peck et al. (2001) determined magnetic susceptibility in the Lake Dood sediments within the paleolake Darkhad basin in northern Mongolia during the late Holocene. Moreover, Lehmkuhl and Lang (2001) determined lake level changes in southern Mongolia based on thermoluminescence (TL) and infrared optically stimulated luminescence (IRSL) dating of aeolian and colluvial sediments. Despite the major improvements in our scientific knowledge of lakes in Mongolia with precise results using advanced techniques and methods provided by foreign scholars, there still remains a disadvantage for international scholars (e.g., Wang et al., 2021) regarding incorrect spelling of geographical names and publishing them without any checks by Mongolian scholars. For example, the Govi[4] (see Chap. 9) has been often incorrectly written as *Gobi* all over the world. Lake Khuvsgul[5] (see Chap. 16) has been also commonly misspelled as *Hovsgol* and *Khubsugul* throughout the world over the last century since it was used incorrectly by the Russians. Furthermore, Selenge[6] River has been often incorrectly written as Selenga based on the Russian pronunciation. A wide variation of Mongolian lake names with the Latin spelling are found in the scientific literature, and this can cause confusion in the communication among scholars.

Since the late 1990s and early 2000s, Mongolian physical geographers, hydrologists, and limnologists have joined international research projects on lakes of Mongolia, which have been led by foreign scientists. For instance, physical geographers N. Batnasan and D. Tuvshinjargal participated in research projects led by a German research team at Lake Uvs (see Chap. 13), while geologist P. Khosbayar and geochemist S. Ariunbileg joined an American research team at Lakes Dood, Erkhel, and Telmen and a Chinese research team at Lake Gun. Moreover, geologists and geochemists T. Narantsetseg, Ts. Oyunchimeg, and D. Tumurkhuu joined the

[4] *Govi* has been often spelled as *Gobi* in the international literature. The Govi is the correct English transliteration from Mongolian *Говь*.

[5] *Khuvsgul* has been misspelled as *Hovsgol* and *Khubsugul* in many publications. The Khuvsgul is the right English transliteration from Mongolian *Хөвсгөл*.

[6] *Selenge* has been often incorrectly written as *Selenga* internationally by the Russian pronunciation. Selenge is the correct English transliteration from Mongolian *Сэлэнгэ*.

Japanese-Russian-Mongolian research project in Lake Khuvsgul and paleolake Darkhad basins.

In this chapter we synthesize precise information on representative lake studies conducted in Mongolia from the nineteenth to the twenty-first centuries and their important conclusions (see Sects. 2.2–2.4). Implications from the lakes are integrated in three contexts of lacustrine geomorphology including changes in lake level and area, and geomorphological processes in lake basins (see Sects. 2.4.1–2.4.3), geochemical characteristics of lake water and sediments (see Sect. 2.4.4), and paleoclimate and paleoenvironmental changes recorded in lakes (see Sect. 2.4.5). The findings and results in each context are described following the publication date. Synthesizing these lake studies in Mongolia in systematic and conscious ways aims to provide geoscientific informative archive for helping describe evolution of the lakes under past and present climate changes (see Chap. 21), to give a comprehensive background and motivation to manage proper environmental protection for the lake basins, and to supply valuable data to benefit future advanced studies of lakes in Mongolia (see Chap. 22).

2.2 Lake Studies in the Nineteenth Century

The earliest period of lake study in Mongolia dominantly belongs to the late nineteenth century. At the beginning of the lake study, lakes were not studied independently, but lakes were noted and observed primarily by foreign travelers and researchers, especially Russian scientists, in the context of studying the physical geography of Mongolia (e.g., Murzaev, 1952). Scientists such as N.M. Prjevalsky, G.N. Potanin, M.V. Pevtsov, D.A. Clements, P.K. Kozlov, S.P. Peretolchin, E.D. Michael, V.L. Komarov, V.S. Elpatievsky, G.E.G. Gergemilo, and D. Carruters played important roles in the early limnology of Mongolia (Tserensodnom, 1971).

In 1870–1888, for instance, N.M. Prjevalsky described geographical locations, physiographic conditions, and geneses of lakes in Mongolia during his investigation of physical geography in Central Asia. In 1876–1899, G.N. Potanin provided data regarding area, genesis, water regime, and geographical location of Lakes Dayan, Tal, and Khulam in the Mongolian Altai Mountain Range and Lakes Uureg, Uvs, Khar Us, and Khyargas in the Depression of Great Lakes (Fig. 1.1; see Chap. 12), and he also focused on the brine structure of mineral lakes. He analyzed hydrochemistry of Lakes Uvs, Khyargas, and Sangiin Dalai in the Depression of Great Lakes and concluded that lakes in the Govi region had covered much larger areas than those of the present time.

In 1877, geologist L.A. Yachevsky noted that the water level of Lake Khuvsgul (Fig. 1.1, Table 1.3; see Chap. 16) had stood 40 m lower than the present level, based on observations of sandy levees, oxbow lakes, and peninsulas on its western shore. In 1878–1879, M.V. Pevtsov collected data from lakes between Khangai[7] and Govi Altai Mountain Ranges, and his conclusion stated that present lakes in the Valley of Lakes remained from a paleosea that retreated, based on sedimentary deposits of the lakes because the Govi region previously was a seafloor (see Chap. 3). In addition, he proved that Lake Terkhiin Tsagaan in the Khangai Mountain Range (Fig. 1.1, Table 1.3; see Chap. 19) and small lakes in the Gichgene Mountain Range were volcanic lakes (see Chap. 4). In 1899, G.N. Potanin noted that Lake Buir in eastern Mongolia (Fig. 1.1, Table 1.3; see Chap. 7) had been connected with a neighboring Lake Bayan and they formed a single lake in the past, based on their shorelines.

S.P. Peretolchin initiated detailed lake studies of Mongolia. In 1897, S.P. Peretolchin and E.D. Michael investigated an island Dalain Khui[8] in Lake Khuvsgul (see Chap. 16). In 1897–1902, S.P. Peretolchin determined water discharge, depth, lake deposits, and four terraces in the Lake Khuvsgul basin, and he noted that Mt. Doloon in the northwest of the lake had been an island of a much larger paleolake.

In 1899, P.K. Kozlov provided data on water sources, discharges, and mineralization levels of Lake Khar Us in the Depression of Great Lakes, Lakes Khulam and Tonkhil in the Mongolian Altai Mountain Range (see Chap. 12), and Lakes Shargiin Tsagaan, Biger, Buun Tsagaan, Orog, and Taatsiin Tsagaan in the Valley of Lakes (Fig. 1.1; see Chap. 9). In 1899, A.N. Kaznakov and V.P. Ladigen measured depths of Lakes Khoton (37.7 m) and Khurgan (17.1 m), and they noted that these lakes were glacial lakes (see Chaps. 4 and 14).

2.3 Lake Studies in the Twentieth Century

2.3.1 1900–1950

In the early twentieth century, lakes of Mongolia were extensively studied in the contexts not only of physical geography but also of hydrochemistry and hydrobiology. At this time of the initiation of the development of limnology in Mongolia, Mongolian scientists were emerging. They, while being educated in Russia, began to work alongside the Russian scholars. The progress in Mongolian limnology of that time coincided with the development of limnology in other countries throughout the world during the twentieth century (Meybeck, 1995).

In 1902, V.L. Komarov inferred drops in water level of Lake Khuvsgul, and he also noted that terraces along its beach had been igneous and erosive in origin. In

[7] *Khangai* has been misspelled as *Khangay* (or *Hangay*) and *Hangai* in many publications. Khangai is the right English transliteration from Mongolian *Хангай*.

[8] *Khui* has been misspelled as *Khuu* and/or *Huu* and mistransliterated into English as *Boy* in Wikipedia. Khui, meaning *navel*, is the right English transliteration from Mongolian *Хүй*.

1903, S.P. Peretolchin published a book entitled *Physical Geography of Lake Khuvsgul*. Based on identifications of terraces 3 m, 10 m, 15 m, and 26 m higher than the present-day lake level (Tserensodnom, 1971), Peretolchin claimed that the lake level had decreased gradually due to the glacier retreat of Mt. Munkh[9] Saridag[10] in the north and the intensive erosion of Eg River in the south. In 1903, V.S. Elpatievsky determined the morphometric components of Lake Khuvsgul as 139.0 km long, 45.0 km wide, 1554 m a.s.l. height, and 246.0 m maximum deep (see Chap. 16). In addition, he noticed the thermocline in the lake water based on the optical evidence and water thermal regime. According to his measurement, the water temperature of Lake Khuvsgul was 12°C at the surface and 3.6–4.0°C at the bottom in summer. Furthermore, he noted that the Lake Khuvsgul basin had formed during the uplift of the Bayan (or Bayan Zurkh) Mountain Range in the northwest (see Chap. 15). In 1905–1909, V.V. Sapojnikov investigated the genesis, shoreline, depth, and water regime of Lakes Khoton, Khurgan, Dayan, Tal, Khar, and Onkhot in the Mongolian Altai Mountain Range (see Chap. 12), and he noted that these lakes were glacial lakes in origin (see Chap. 4). In 1910–1911, K. Douglas, a British traveler, indicated the impact of paleoglacier on the genesis, water regime, and morphometric components of Lakes Uvs (see Chap. 13), Uureg, and Achit (Fig. 1.1; see Chap. 12). A geologist, M.P. Price, noted the water level drop of Lake Uureg (Fig. 1.1; see Chap. 12) based on paleolake sediments lying on the terraces 10 m higher than the current water level.

In 1926–1927, V.A. Smirnov investigated hydrochemistry of lakes in Mongolia, and he hypothesized that Lakes Uvs, Khyargas, Airag, Sangiin Dalai, Khar, Durgun, Tsookhor, and Uureg in western Mongolia had been included in a large freshwater lake basin covering an area about 50,000 km^2 in the past and then these lakes had been continuously drying and shrinking since the glacial time to the present (Tsegmid, 1969). In 1927, S.A. Kondratiev investigated the morphology of Lake Telmen basin in the Khangai Mountain Range (Fig. 1.1; see Chap. 18) and surrounding sand dunefields, and he noted that the lake genesis had been related to the drainage network of Ider River, a headwater of Selenge River. He published a book entitled *Lake Telmen and the western Khangai Mountain Range* in 1929 (Tserensodnom, 1971). He continued the lake study on water resource and depth of Lake Khangal in the Khentii[11] Mountain Range (see Chap. 6). In 1928, V.A. Smirnov conducted a survey on evaporation from lakes and mineral waters (or springs) in southern Mongolia, and he concluded that these water resources would continuously be shrinking and that the territory of southern Mongolia would soon shift into a desert environment. In 1929, E.V. Kozlova measured the depth of Lake Orog (Fig. 1.1; see Chap. 9), and she compiled its bathymetric map.

---

[9] *Munkh* has been often misspelled as *Munku* or *Munh* based on the Russians' pronunciation in the international literature. Munkh is the correct English transliteration from Mongolian *Мөнх*.

[10] *Saridag* has been also misspelled as *Sardyk* or *Sardig* by the Russians internationally. Saridag is the correct English transliteration from Mongolian *Сарьдаг*.

[11] *Khentii* has been misspelled as *Khentey* (or *Hentey*) and *Khentei* (or *Hentei*) in many publications. Khentii is the right English transliteration from Mongolian *Хэнтий*.

In 1940–1943, N.D. Bespalov studied genesis, hydrochemical composition, and brine deposition in mineral lakes of Mongolia. In 1940–1944, E.M. Murzaev studied genesis, morphometric feature, water regime, and mineralization level of lakes in Mongolia. He rejected the conclusion that the lakes of Mongolia had shrunk due to aridity, and he claimed that the changes in lake levels had been related to water erosion (Tserensodnom, 1971). Murzaev (1952) summarized that lakes in the Depression of Great Lakes and in the Valley of Lakes had been hydrologically connected among them in the past and that there had been a large drainage basin over a vast region extending from Lake Uvs to Lake Shargiin Tsagaan (Fig. 1.1), based on paleogeographical tracers around the lakes. In other words, these freshwater and saline lakes had been connected through rivers, and their water regimes were unstable (see Chap. 3).

In 1942, the first university of the country, the National University of Mongolia (NUM), was established and initiated national education. In 1942, geologist S.P. Alexeychik investigated the brine resource and hydrochemical composition of saline lakes and the geological structure of lake basins in central Mongolia (see Chap. 18). In 1943, N.A. Marinov, N.E. Nevrozov, and A.E. Pereliman measured brine contents of some lakes in eastern Mongolia (see Chap. 6). In 1943, M.G. Turishiev indicated the brine resource of saline lakes in the Depression of Great Lakes in western Mongolia (see Chap. 12). In 1945, geologist Sh. Tseveg investigated brine resources of saline lakes (see Chap. 23) in eastern and central Mongolia. Since 1948, A. Dashdorj measured depths of lakes in Mongolia and observed their water regime. In 1948–1949, Sh. Tsegmid noted that lakes in the Khentii Mountain Range were glacial lakes in origin (see Chap. 4).

2.3.2 1950–2000

Since the late twentieth century, lakes of Mongolia have been studied primarily by Mongolian researchers besides Russian scientists. In addition to hydromorphology and hydrochemistry of lakes, the research area extended into hydrogeology and hydrobiology. For instance, in 1951–1953, Sh. Tsegmid investigated genesis, paleogeography, present-day changes, and hydrochemistry of lakes in the Mongolian Altai Mountain Range, the Depression of Great Lakes, and the Valley of Lakes. He emphasized that there was no proof regarding the shrinking lakes in the Mongolian Altai Mountain Range and the Depression of Great Lakes in western Mongolia. In addition, he stated that lakes in the Valley of Lakes (see Chap. 9) had stood at 40–45 m higher than the present lake level or at 1380 m a.s.l. in the past, based on paleoshorelines and paleoterraces, and that the lake level had decreased due to the postglacial aridity and neotectonics (Tsegmid, 1955). In 1949–1952 and 1957–1958, hydrologist N.T. Kuznetsov and hydrogeologists P.A. Derevyanko and A.T. Ivanov investigated water regime, discharge, water resource, hydrochemical composition, and water balance of surface water and groundwater including lakes and rivers throughout Mongolia, and they compiled maps of surface water, groundwater, and hydrogeology of Mongolia. In 1956–1958, D. Davaasuren studied the hydrochemical composition of mineral lakes in the Depression of Great Lakes, and he noted that lake salt could be used for food and industry (see Chap. 23). In 1958, A. Avirmed determined the hydrochemical composition of mineral lakes in Mongolia. Since 1958, chemists under the leadership of Sh. Luvsandorj from the Institute of Chemistry, MAS, revealed physical and chemical properties and brine composition of mineral lakes in Mongolia, and they noted that salts from Lakes Gurvan Tes, Dardai, Suuj, and Devteer were largely applicable for commercial usage (Tserensodnom, 1971; Batnasan, 1998). In 1959–1960, A. Dashdorj, A.A. Tomilov, P.F. Bochkarov, and E.P. Nikolaeva conducted a survey on hydrochemistry and hydrobiology in Lake Khuvsgul. Since 1960s physical geographer-limnologist J. Tserensodnom developed an investigation of physiography and hydrology of lakes in Mongolia (Batnasan, 1998).

In 1962, the Institute of Geography and Permafrost, MAS, was established, and it conducted surveys on impact of physiographic factors on water regime, paleogeography, morphometric features, hydrophysical and hydrochemical characteristics, and classification of lakes according to natural zones and importance of lakes. The institute installed hydrological measurement stations at large lakes covering an area over 20.0 km^2 such as Lakes Khuvsgul, Uvs, Airag, Khar Us, Khar, Durgun, Bayan, Sangiin Dalai, Oigon, Telmen, Terkhiin Tsagaan, Ugii, Orog, Buun Tsagaan, Khoton, Khorgon, Dayan, Tolbo, Achit, Uureg, and Buir, and measured depth, water resources, and bathymetric data of the lakes (Tserensodnom, 1971). The Administration Office of Hydrology also installed hydrological and meteorological measurement stations at large Lakes Khuvsgul, Uvs, Airag, Khar Us, and Buir in order to collect data on thermal regime and water regime of the lakes. Badarch et al. (1965) provided morphometric components of large lakes including Lakes Buir, Khukh, Guren, and Khangal in eastern Mongolia (see Chap. 6), and they noted that in eastern Mongolia there were over 300 lakes in total formed in (1) small steppe hollows, (2) wide fluvial valleys, (3) tectonic depressions, (4) thermokarst valleys, and (5) moraine-dammed settings.

In 1967–1968, A. Dulmaa and others from the Institute of Biology, MAS, conducted biological and ecological surveys for flora and fauna in lakes in the Shishkhid (or Darkhad) basin in northern Mongolia and central and eastern Mongolia. In 1967, Ya. Tsend-Ayush investigated the biodiversity of fishes in some lakes of Mongolia. In 1969, Sh. Tsegmid published a book entitled *Physical Geography of Mongolia*, in which lake studies in Mongolia conducted by Russian scientists (e.g., Murzaev, 1952) since the nineteenth century were introduced. In 1971, J. Tserensodnom synthesized results from lakes in Mongolia, and he published a seminal book entitled *Lakes of Mongolia* in which morphometric components (e.g., length, width, area, volume, etc.), genesis, and water regime (e.g., water balance, temperature, transparency, etc.) of Mongolian lakes were provided. During 1971–1986, a Mongolian and Russian joint expedition worked in Lake Khuvsgul region, and they provided characteristics of the lake basin, shoreline feature, water temperature of the lake, hydrochemical composition, geochemistry of its bottom deposits, and biosphere of the lake with their maps in *Atlas of Khuvsgul* by Bogoyavlensky et al. (1989). In 1983,

B. Ariyadagva and G.M. Shpeizer conducted investigations in carbonate lakes of Mongolia (Tserensodnom, 2000). Moreover, in 1987–1992, Mongolian scientists including physical geographers and limnologists J. Tserensodnom, Z. Sanjmyatav, Ts. Sugar, N. Batnasan, O. Tserev, D. Tuvshinjargal, Kh. Tuvshinbayar, L. Sainbayar, and S. Sarantuya from the Institute of Geography and Permafrost, MAS, studied water regime of lakes, hydrothermal balance, and solar insolation in lake basins (Tuvshinjargal, 2001).

With the dissolution of the Soviet Union in 1990 and the democratization of Mongolia, the lake study of Mongolia has been opened for international scholars. For example, international research projects (dominantly German and American) started to study Lakes Uvs and Khuvsgul, and the national and international research teams organized an international conference in Ulaangom, Uvs aimag, in 1999 under the title *Sustainable Development of the Altai-Sayan Ecoregion and Transboundary Nature Conservation Issues*, in which they discussed the impact of global climate change on the Lake Uvs ecosystem and adjacent Altai and Sayan regions and how to protect the ecosystems. At the conference Horn et al. (1999) presented salinity level and hydrochemical composition of the saline Lake Uvs and freshwater Lakes Bayan, Duruu, Shavart, and Baga in western Mongolia, and they noted the major cations Mg^{2+} and Na^{+} in Lake Bayan resulted by biogenic calcite precipitation in the zones of inflowing groundwater. However, the most published data previously obtained by the Soviet geological surveys and research projects in the twentieth century were large but primarily not accessible for international readers (Krivonogov et al., 2012). The international research projects, dominantly German and American, provided and published their results from lakes of Mongolia after the 2000s (see Sect. 2.4) based on their works done in 1990s. For example, the American colleagues (Peck et al., 2001) joined Mongolian geologist P. Khosbayar and geochemist S. Ariunbileg, and they conducted a survey in Lake Dood in the paleolake Darkhad basin in 1998–1999. The joint Russian-Mongolian-American expedition in 1998 focused on paleolake reconstruction in the Valley of Lakes of southern Mongolia (Komatsu et al., 2000, 2001). Among the national researchers beside the foreign scholars, N. Batnasan (1998) summarized the hydrological system, water regime, and evolutionary history of large Govi lakes in the Depression of Great Lakes and Valley of Lakes of Mongolia. Furthermore, Tserensodnom (2000) provided synthesized data on number, area, location, shoreline, width, and length of lakes with an area of greater than 0.1 km^2 in each 21 aimag throughout the country.

2.4 Lake Studies in the Twenty-First Century

2.4.1 Changes in Lake Level

Peck et al. (2002) reconstructed a deep lake since 4.4 cal ka BP based on sedimentological and geomorphological evidence in Lake Telmen basin of central Mongolia. Krivonogov et al. (2003) determined changes in water level of Lake Khuvsgul in northern Mongolia. Schwanghart et al. (2008) found the water level of Lake Ugii in central Mongolia to have been at low elevations during the early Holocene (10.6–7.9 ka BP) and high elevations in the middle Holocene (7.9–4.2 ka BP). Grunert et al. (2009) reconstructed the middle Holocene 9.0 m and 15.0 m deep freshwater pluvial lakes in the Tukhum Els and Khongor Els sand dunefield depressions of the Govi region in southern Mongolia, and the late Pleistocene (39 ka ^{14}C BP) large deep lake and the middle Holocene (5.5 ka BP) shallow lake at the eastern rim of the Mongol Els sand dunefield in western Mongolia. Rudaya et al. (2009) suggested existence of a relatively deep Lake Khoton in western Mongolia prior to 11.5 ka BP, and they reconstructed an increase in the lake level between 11.0 and 8.0 ka BP and shift toward the modern lake level after 8.0 ka BP.

Zhang et al. (2012) reviewed the Holocene hydrology at Lake Gun in central Mongolia, and they found low lake levels between 10.8–7.0, 7.0–5.7, 4.1–3.6, and 3.0–2.5 cal ka BP, intermediate lake levels at the intervening periods, and high lake levels at 2.5–1.6 cal ka BP. Orkhonselenge et al. (2013) showed water level fluctuations of Lake Khuvsgul since the late glacial time, and they found that the lake level had been dependent on two mutually related factors of the continuing incision of the outlet Eg River and the periodic restoration of the Ulkhen Sair fan. Narantsetseg et al. (2013) showed changes in water level, climate, and environments in Lake Dood within the paleolake Darkhad basin during the late glacial and the early to middle Holocene. Oyunbaatar et al. (2017) analyzed water level fluctuation and water balance of Lake Ganga in southeastern Mongolia, and they observed dramatic drops in the lake level and continuous shrinkage of the lake in recent years due to both effects of climate warming and human impacts. Walther et al. (2017) identified the endorheic Lake Khar basin in western Mongolia during the local Last Glacial Maximum (LGM; MIS 2) and that the lake level had remained relatively stable. Yu et al. (2019) considered hydrological variability of Lake Orog in the Govi region of southern Mongolia over the last ~45 ka, and they reconstructed high lake levels during the MIS 3 and early MIS 2 (~35 to ~24 ka) caused by increased precipitation and also a low lake level during the late Pleistocene. Recently, Nottebaum et al. (2021) reconstructed eight paleolake levels in Lake Orog basin based on paleoshoreline identifications with the highest one ~56 m higher than the present providing evidence for a late MIS 5 lake level. Moreover, Klinge et al. (2021) reviewed the late Pleistocene lake level changes in Lakes Khoton, Khurgan, and Ekhen in the Mongolian Altai Mountain Range.

2.4.2 *Changes in Lake Area*

Davaa (2015) estimated changes in area, lake level, and water balance for 185 lakes in Mongolia with an area greater than 3.0 km^2 based on Landsat data acquired in 2010 and 2013, and he noted shrinkage of 295 lakes and disappearance of 50 small lakes by 1999–2002 in comparison to the previous years. Kang et al. (2015) determined changes in lake area and their relations to precipitation for 165 lakes in

Mongolia with an area over 10 km^2 based on the linear regression analysis, and they noted the precedent 2-month precipitation as the best determining factor of change in lake area throughout Mongolia where the regional patterns of precipitation-driven lake area changes varied considerably ($R^2 = 0.028–0.950$), depending on the regional climate regime and hydromorphological characteristics. Sternberg and Paillou (2015) investigated a potential major water resource correlated to large paleolakes in the Govi region of southern Mongolia using advanced remote sensing data, radar images, and optical images, and they indicated that Lake Ulaan having a past lake area of over 19,000 km^2 was a promising site for hydrogeology and water resource.

Kang and Hong (2016) estimated lake areas for the selected 73 Mongolian lakes with an area greater than 6.25 km^2 using the minimum composite Normalized Difference Vegetation Index (NDVI) from Moderate Resolution Imaging Spectroradiometer (MODIS), and they found lake area reductions of slight to moderate rates in semiarid regions and of rapid rates in arid regions. Szumińska (2016) analyzed changes in area of Lakes Buun Tsagaan and Orog in 1974–2013 in the context of climate conditions and permafrost degradation based on satellite images, NOAA climate data, CRU dataset, and principal component analysis (PCA), and she observed that there had been a tendency for decreasing in surface area, intermittent with short episodes of expansion.

Saruulzaya (2017) showed spatial and temporal dynamic changes of thermokarst lakes in permafrost regions of Mongolia during 1962–2007, and she found lakes gaining and losing areas due to the air temperature rise causing permafrost thawing and evaporation. Orkhonselenge et al. (2018a) showed decreases in lake area for six large lakes in the Valley of Lakes of the Govi region in southern Mongolia with variable vulnerability of the lake basins and responses of the Govi landscape to the present climate change. Gerelsaikhan and Orkhonselenge (2019) showed the hydrogeomorphological dynamics of Lake Khargal in the Siberian taiga forest region in northern Mongolia during the past five decades with the gaining area of 1.02 km^2 during 1970–2000 and losing area of 0.35 km^2 during 2000–2018. Furthermore, Orkhonselenge et al. (2020) and Orkhonselenge and Bulgan (2021) showed how rapidly Lake Yakhi in the Dornod plain of eastern Mongolia responded with a decrease of ~62.2 km^2 of the total surface area to the warming since 1987 and drying since 1992 (see Chap. 6).

2.4.3 Geomorphological Processes in Lake Basins

Tuvshinjargal (2001) studied water balance, water resource, and thermal regime of Lake Ugii in central Mongolia. Batima et al. (2004) showed clear trends in regional climate with changes in freezing and thawing dates of the ice coverage observed in lakes of Mongolia. Lee et al. (2011) found that the early to late Holocene sediments from Lake Ulaan in southern Mongolia had been aeolian, whereas the late Pleistocene sediments had deposited by both aeolian and glaciofluvial processes. Fassnacht et al. (2011) noted that Lake Orog in southern Mongolia had been completely dry for many years due to seasonal depletion of water resources derived from Tuin River.

Felauer et al. (2012) noted that aridity in southern Mongolia increased since 4.0 cal ka BP resulted in strengthening of lake desiccation, aeolian processes, and dune remobilization in Lake Bayan Tukhum basin. Stolz et al. (2012) reconstructed fluviolacustrine and aeolian processes for a lake formation at the eastern rim of the Mongol Els, the largest sand dunefield of Mongolia (see Chap. 12), and they revealed that at the Eemian age, a large and deep lake had covered the floodplain formed by converging Shurag and Zavkhan Rivers in western Mongolia. Krivonogov et al. (2012) reconstructed a paleolake history in the Darkhad basin in northern Mongolia since the Pliocene based on geomorphological, sedimentological, and geochronological data. Orkhonselenge et al. (2011, 2013, 2014) found and interpreted three short events of high sedimentation rate in the Borsog Bay of Lake Khuvsgul occurring at 7.4–7.1, 4.8–4.5, and 1.0–0.9 cal ka BP using sedimentological, geochronological, and diatom data.

Davaagatan et al. (2015) and Orkhonselenge and Davaagatan (2016) determined the sedimentary features in Lakes Terkhiin Tsagaan and Ugii in central Mongolia based on physical and chemical properties of the lake sediments. Orkhonselenge et al. (2018b) investigated the sedimentation dynamics in Lake Ulaan during the middle to late Holocene, and they showed a high sedimentation rate of 4.6 cm/ka between 6.0 and 2.7 cal ka BP and a low sedimentation rate of 1.6–1.8 cm/ka after 3.2–2.7 cal ka BP. Matsuyama et al. (2019) identified shoreline features in Lake Olgoi basin and an upstream basin of Lake Buun Tsagaan in the southern Khangai Mountain Range, and they hypothesized that paleo-inflows had likely been linked to a reconstructed paleolake. Agatova et al. (2020) found a sudden drainage formation in Lake Achit in the Depression of Great Lakes in western Mongolia, which had been the last catastrophic flood related to the late Pleistocene ice- and the early Holocene moraine-dammed Lake Ak-Hol in the catchment of the Mogen-Buren River in the Russian Altai Mountain Range.

2.4.4 Geochemistry of Lake Water and Sediments

Peck et al. (2002) found that salinity of Lake Telmen in central Mongolia had been approximately 20 g/L at 7.1–6.3 cal ka BP; however presently it is 4 g/L. Wang et al. (2004) analyzed organic matter, organic $\delta^{13}C$, and magnetic parameters in Lake Gun sediments in central Mongolia, and they found a variation in organic $\delta^{13}C$ linked to climate change and low values of magnetic susceptibility (MS) and Saturation Isothermal Remanent Magnetization (SIRM) due to dissolved fine ferromagnetic particles under cold wetter and possible reducing conditions. Tserenpil et al. (2010) examined physical and chemical properties of some organic matters of peloids from 12 lakes in Mongolia using several analytical techniques (e.g., infrared (IR) and ^{13}C nuclear magnetic resonance (NMR) spectroscopy, gas *chromatography* (GC), MS), and they identified that most lakes, except for Lake Gurvan, contained continental

H_2S sticky peloids with 0.4–3.1% total organic carbon (TOC) and that the biogenic stimulator humic acid varied at 11–56% of total organic matters in the peloids.

Isupov et al. (2011) identified that hypersaline soda closed basin lakes in northwestern Mongolia were highly enriched in uranium-238 (^{238}U, <1 mg/L) based on analyses of major and trace elements in lake water, and they also found that the greatest U resources (~6000 tons) were stored in Lake Khyargas. Linhoff et al. (2011) found that hyper-alkaline soda lakes in eastern Mongolia contain extremely high concentrations of U, Na, Cl, and K, which were correlating to radiogenic Sr in lake water, suggesting that U had been sourced from the local Cretaceous alkaline rhyolites. Zhang et al. (2012) analyzed multi-proxies of $\delta^{13}C_{org}$, TOC, $CaCO_3$, $\delta^{13}C_{carb}$, and $\delta^{18}O_{carb}$ of bulk carbonate in Lake Gun sediments, and they found a rapid rise of $CaCO_3$ at the basal sand-carbonate-rich silt boundary. Narantsetseg et al. (2013) estimated total carbon (TC), total inorganic carbon (TIC), TOC, total nitrogen (TN), total sulfur (TS), $\delta^{13}C$, $\delta^{18}O$, and diatom in Lake Dood sediments in the paleolake Darkhad basin during the late glacial time and the early to middle Holocene. Isupov et al. (2013) found that the U-bearing soda Lake Shaazgai in western Mongolia contained a high concentration of U (~1 mg/L), and they inferred the influence of geological, geomorphological, hydrogeochemical, and climatic factors on the deposition of U in the lake systems.

Gaskova et al. (2015) evaluated highly mineralized closed lakes in western Mongolia using thermodynamic modeling, and they found Lake Shaazgai to be effective of sorption of U and arsenic speciation only at a high concentration of FeOOH sorbent at pH 9.4. Isupov et al. (2015) showed that soda lakes located close to U deposits in western Mongolia contained high concentrations of U (1–3 mg/L), B (<250 mg/L), Br (<1.1 g/L), Li (<600 mg/L), and Sr (<8 mg/L). Szopińska et al. (2016a) presented the hydrochemical characteristics of Lakes Buun Tsagaan and Orog in the Valley of Lakes in southern Mongolia based on the PCA of physical properties including ions, trace elements, pH, electrical conductivity (EC), total dissolved solids (TDS), and TOC, and they found that permafrost thawing could increase the concentration of nitrogen (NH_4^+, NO_3^-), Na^+, Ca^{2+}, Cl^-, and SO_4^{2-}. Szopińska et al. (2016b) summarized the hydrochemistry of river-lake systems (Baidrag River-Lake Buun Tsagaan system and Shargaljuut/Tuin Rivers-Lake Orog system) with TOC, polycyclic aromatic hydrocarbons (PAHs), and other organic compounds (e.g., organic halogen, phthalates, higher alkanes, etc.), and they found that the main factors influencing differences in TOC concentrations in the water bodies were permafrost occurrence. Yu et al. (2017) estimated high-resolution elemental geochemistry of Lake Orog sediments with authigenic productivity (Ca, Cl, $CaCO_3$) in an alkaline environment, allochthonous organic materials (TOC and C/N), and terrigenous input via fluvial inflows and quasi-constant aeolian input in the late Quaternary (Al, Si, K, Ti, and Fe).

Chida et al. (2018) determined the hydrochemistry of Lakes Buun Tsagaan and Orog with high concentrations of dissolved species (e.g., Na^+, SO_4^{2-}, Cl^- excluding Ca^{2+}, and HCO_3^-), and they noted that evaporation of groundwater and river water into Lake Buun Tsagaan had no impacts on the hydrochemical composition of the lake. Kolpakova and Gaskova (2018) found the evaporation and crystallization processes were different for soda, chloride, and sulfate lakes, and they concluded that soda-type lakes (e.g., Lake Shaazgai) maintained unchanged ion composition, while chloride-type (e.g., Lake Davsan) and sulfate-type (e.g., Lake Tonkhil) lakes changed with chemical composition according to the sequence of mineral precipitation. Ariunbileg et al. (2020) provided synthesized data obtained from mineral lakes in eastern (e.g., Lakes Burd, Tsaidam, and Khotont), southern (e.g., Lakes Davsan, Kholbooljin, and Biger), and western (e.g., Lakes Tonkhil, Ikhes, and Oigon) Mongolia in 2009, and they showed macroelements (e.g., Na, K, Mg, Ca) and microelements (e.g., Li, U, Sr, B, Br) in water and sediments of the lakes and also brines (e.g., thenardites, halites, bledites) and minerals (e.g., quartz, feldspars, kaolinites) in the lake sediments. Orkhonselenge and Bulgan (2021) concluded that Lake Yakhi sediments consist of the granite (felsic) and granite basalt (felsic to mafic) source rocks with mostly shale and arkose and that granites, granodiorites, and tonalities dominated in the lake sediments may have been transported from the source rocks enriched in plagioclase and feldspars.

2.4.5 *Lake Studies on Paleoclimate and Paleoenvironmental Changes*

Lacustrine deposits provide valuable insights into local and regional paleoclimate and paleoenvironmental changes (Orkhonselenge et al., 2015, 2019; see Chap. 21). Spatial patterns of the Pleistocene and Holocene paleoclimatic implications learnt from lakes in each region of Mongolia are considered in detail in Chap. 21. Paleoclimatic and paleoenvironmental records from lakes of Mongolia extend to the late Pleistocene.

2.4.5.1 Pleistocene

Grunert et al. (2000) showed in the Lake Uvs basin in western Mongolia (see Chap. 13) a more humid climate with at least one arid phase before the LGM, dry and cold climate during the second half of the LGM, and wet and cold climate during the late glacial of the last Ice Age based on geomorphological evidence and high lake levels. Komatsu et al. (2001) hypothesized that wetter periods corresponding to MIS 3 prior to the LGM corresponding to the reconstructed high lake levels in large stable Lakes Buun Tsagaan, Adgiin Tsagaan, and Orog covering extensive areas in the Valley of Lakes in southern Mongolia. Grunert et al. (2009) found that the period of 17.0–14.0 ka BP coinciding with the aridity in Central Asia at 16.0 ka BP had contributed to the sand accumulation of the Khongor Els sand dunefield in southern Mongolia and that the late Pleistocene glaciation (39.0 ka ^{14}C BP) had generated the aeolian environment of the Mongol Els sand dunefield in western Mongolia. Kashiwaya et al. (2010) reconstructed two episodes during the MIS 11 and 5d for

Lake Khuvsgul basin in northern Mongolia, and they found that the lake sediment had been coarse in grain size at glacial time and fine at interglacial intervals during the MIS 7 to 1.

Felauer et al. (2012) reconstructed the history of climate and environmental conditions in southern Mongolia during the late Quaternary based on the records from Lake Bayan Tukhum, and they revealed nine climatic units during the last 15.0 cal ka BP, representing fluctuations of humid and arid climates. Stolz et al. (2012) hypothesized that the climate in western Mongolia had been quite arid, permitting strong aeolian activity, contributing to the damming of sand dunefields, and forming a large pluvial lake (~2000 km^2) at the eastern rim of the Mongol Els sand dunefield during the Eemian age (OSL dated to 127 ± 10, 119 ± 10, and 114 ± 10 ka BP). Krivonogov et al. (2012) reconstructed the Quaternary glaciation and environmental changes in northern Mongolia based on the history of the paleolake Darkhad (see Chap. 15) and glaciations during the late MIS 5 (Krivonogov et al., 2005), MIS 2 and 3 (Gillespie et al., 2008; Komatsu et al., 2009), and during MIS 4. Narantsetseg et al. (2013) reconstructed that dramatic climate and environmental changes since 14.0 cal ka BP in the paleolake Darkhad basin. Lee et al. (2013) reconstructed high-resolution paleoclimatic changes in the Lake Ulaan basin in southern Mongolia (see Chap. 11) during the last 17,000 years fluctuating with the humid condition during the early and middle Holocene, and the dry condition in the late Holocene. Sun et al. (2013) reconstructed climate changes and the associated vegetation variations in western Mongolia over the past ~22,600 years based on pollen data obtained from Lake Achit (Fig. 1.1).

Walther et al. (2017) reconstructed past and recent environmental changes in western Mongolia based on changes in water level of Lake Khar and dynamics in glaciers, permafrost, and climate at Mt. Tsengel Khairkhan in the Mongolian Altai Mountain Range, and they found the glaciation history from MIS 4/5 to the Little Ice Age. Lehmkuhl et al. (2018) found that the enormous paleolakes in the Govi region maintained by a strong East Asian Summer Monsoon (EASM) reaching southern Mongolia had mostly disappeared during the MIS 6, reappeared during the last interglacial MIS 5e, and lasted probably until the beginning of the last glacial. Yu et al. (2019) reconstructed a relatively humid environment in the Lake Orog basin in southern Mongolia during MIS 3 and the early Holocene and the harsh climatic periods at ~43 ka and ~36 ka, during the global LGM and the Younger Dryas (YD). Mischke et al. (2020) stated that the arid conditions in the Lake Ulaan basin shifted into the humid conditions since 12.3 ka BP causing the rise in lake level and the decrease in salinity. Recently, Nottebaum et al. (2021) reconstructed paleoenvironmental changes since MIS 7 in Lake Orog basin.

2.4.5.2 Holocene

Lehmkuhl and Lang (2001) revealed the evidence for slightly more humid periods around 1.5 ka and 8.5 ka in the early Holocene with a larger extent of the lakes using lacustrine sediments from high shorelines in the Valley of Lakes. Peck et al. (2002)

reconstructed climate change in central Mongolia over the past 7.1 cal ka BP based upon sedimentological and geomorphological evidence recorded in Lake Telmen. Wang et al. (2004) reviewed the Holocene climate changes in central Mongolia, based on records from Lake Gun, noting cold/wet and warm/arid conditions punctuated with a series of abrupt climate shifts linked with those in the North Atlantic Ocean. Prokopenko et al. (2007) reconstructed the Holocene climate change in Lake Khuvsgul basin of northern Mongolia based on diatom and biogenic silica proxies. Schwanghart et al. (2008) reconstructed the Holocene climate change and paleoenvironmental evolution in central Mongolia based upon geochemical and mineralogical properties of Lake Ugii sediments (see Chap. 20).

Wang et al. (2009) reconstructed climate changes in central Mongolia during the past 8.7 cal ka BP with a high-resolution pollen record from Lake Ugii. Moreover, Schwanghart et al. (2009) investigated the Holocene environmental evolution and the middle Holocene Climate Optimum in the Lake Ugii basin. Rudaya et al. (2009) reconstructed the Holocene climate dynamics in western Mongolia based on the palynological and diatom analyses of the sediment cores obtained from Lake Khoton. Shinneman et al. (2010) showed potential changes in the precipitation-evaporation balance with rising temperatures interpreted based on tree-ring records from Lakes Kholboo, Duruu, Zagas, Baga, Takhilt, and Bayan in the Depression of Great Lakes in western Mongolia, and they also demonstrated a negative correlation between temperature and effective moisture (warm-dry and cool-wet conditions) based on the diatom-inferred records at 2.0–0.2 cal ka BP. Wang et al. (2011) presented a high-resolution pollen- and diatom-based bioclimatic reconstruction from Lake Ugii, and they revealed a prolonged dry climate prevailed at 5.8–3.1 ^{14}C BP in central Mongolia.

Zhang et al. (2012) reconstructed the Holocene climate changes in central Mongolia based on ^{14}C dates from Lake Gun, fluctuating with dry conditions during the earliest Holocene, wetter conditions in the early Holocene, significantly drier conditions in the middle Holocene, and wetter conditions in the late Holocene. Tian et al. (2013) reviewed climate changes over the last 1200 years in central Mongolia based on high-resolution multi-proxy data (e.g., pollen, grain size, TOC) from Lake Khuis, and they found the monsoon-westerlies transition area with a greater climate variability using a simulation of the MPI Earth System Model. Kashiwaya et al. (2015) considered the impact of climate change manifested on lacustrine systems in Mongolia during the Holocene, and they highlighted the lacustrine system as a proxy record.

Orkhonselenge et al. (2015, 2019) identified paleoclimate changes recorded in Mongolian lakes noting the periods of warmer/wetter early Holocene, warmer/drier middle Holocene, and cooler/wetter late Holocene drastically varying from region to region of Mongolia. Yu et al. (2017) found that the Holocene aridness in southern Mongolia was dominated by a distinct high productivity alkaline environment relative to the late Pleistocene, and they attributed it to the increased hydrodynamic strength of riverine inflow and/or intensified erosion and weathering of felsic source rocks in the upper catchment of Lake Orog. Orkhonselenge et al. (2018b) reconstructed a climatic shift from humid in the middle Holocene to arid in the late Holocene using the sedimentation dynamics recorded within the Lake Ulaan basin of southern Mongolia.

2.5 Summary

During the nineteenth and early twentieth centuries, geographical location, physiographic conditions, and genesis of lakes in Mongolia were observed and documented in the field of physical geography. Mongolian researchers were educated at foreign, especially Russian, institutions and universities, and they also learned with a course *Limnology* started to be taught at the National University of Mongolia in 1942. So then, Mongolian scientists together with Russian researchers began to study lakes in the country. Since the late twentieth century, lakes of Mongolia were studied individually based on the nation's social and economic interests and scientific orientations in the context of hydrogeology, hydrochemistry, and hydrobiology beside physical geography, hydrology, and limnology. Moreover, permanent hydrological measurement stations were installed at large lakes, and physical geographers and limnologists conducted on bathymetry, shoreline morphometry, and thermal and water regime of lakes. At the same time, hydrochemists and hydrobiologists focused on floral and faunal biodiversities in lakes, hydrochemical composition of lakes, and brine deposit in mineral lakes.

In the beginning of the 1990s, lakes of Mongolia have been widely studied by national and international research teams. For the lake study of Mongolia, foreign scientists have led most research efforts since the late 1990s and early 2000s utilizing advanced methods and techniques and obtaining high-resolution multi-proxy data. In the twenty-first century, lake studies in Mongolia using sophisticated tools, deriving important interpretations and conclusions, have been conducted in the contexts of geomorphology, sedimentology, geochemistry, and paleoclimate and paleoenvironmental changes in lake basins, especially for the geological periods of the late Pleistocene and Holocene.

Present-day studies of lakes in Mongolia are gradually developed to focus more on multidisciplinary and interdisciplinary investigations with related techniques including those in geomorphology (e.g., remote sensing), geochemistry (e.g., magnetic susceptibility), sedimentology (e.g., major and trace elements), palynology (e.g., diatom, pollen), and geochronology (e.g., ^{14}C, OSL). The implications on past and present climate changes from lakes of Mongolia will be discussed in detail for each region in Chap. 21.

References

Agatova, A. R., Nepop, R. K., & Glebova, A. N. (2020). Lake systems in mountain borders of the Great Lakes Basin of Mongolia: From Pleistocene catastrophes to Holocene gradual evolution (case study from southwestern Tuva). *Quaternary International, 538*(10), 53–62.

Ariunbileg, S., Isupov, V. P., Vladimirov, A. G., Kolpakova, M. N., & Kuibida, L. V. (2020). *Chemical evolution and geochemistry of mineralized lakes of Mongolia*. Admon Printing House, 132 p. [In Mongolian].

Badarch, N., Tsegmid, S., & Tserensodnom, J. (1965). *Landscapes and natural zones of eastern Mongolia*. Press of Mongolian Academy of Sciences, 117 p. [In Mongolian].

Batima, P., Batnasan, N., & Bolormaa, B. (2004). Trends in River and Lake Ice in Mongolia. *AIACC Working Paper, 4*, 3–12.

Batnasan, N. (1998). *Hydrological system, water regime, and evolution of large lakes in the Govi region*. Doctorate Thesis in Geography: Hydrology. Institute of Geoecology, Mongolian Academy of Sciences, Ulaanbaatar. 163 p. [In Mongolian].

Bogoyavlensky, B. A., Batjargal, B., Martinov, V. P., Kozlov, U. P., Borobiev, V. V., & Batsuuri, D. (1989). *Atlas of Lake Khuvsgul: Mongolian People Republic*. State Printing Industry of Geodesy and Cartography, SSSR, 118 p. [In Russian with an abstract in English, German and Mongolian].

Brown, A. G. (1995). Vegetation and lake-level change. In K. J. Gregory, L. Starkel, & V. R. Baker (Eds.), *Global continental palaeohydrology* (pp. 131–150). Wiley.

Chida, T., Sekine, Y., Fukushi, K., Matsumiya, H., Solongo, T., Hasebe, N., & Davaadori, J. (2018). Hydrology of subsaline lakes in southern Mongolia: A terrestrial analog study for lacustrine environments and chloride depositions on early mars. In *49th Lunar and Planetary Science Conference (LPI Contrib. No. 2083)*.

Davaa, G. (2015). *Resource and regime of surface water of Mongolia*. 408 p. [In Mongolian].

Davaa, G., Oyunbaatar, D., & Sugita, M. (2007). Surface Water of Mongolia. 55–68.

Davaagatan, T., Orkhonselenge, A., Fukushi, K., & Yabe, M. (2015). Sedimentary records in intermontane and prairie lakes in central Mongolia: Preliminary results from Lake Terkhiin Tsagaan and Lake Ugii. *Journal of Geographical Issues, 14*(408), 19–26.

Fassnacht, S.R., Sukh, T., Fernandez-Gimenez, M., Batbuyan, B. Venable, N.B.H., Laituri, M., & Adiyabadam, G. (2011). Local understanding of hydro-climatic changes in Mongolia. In *IAHS Publication 346: Proceedings of Cold Region Hydrology in a Changing Climate (IUGG)*.

Felauer, T., Schultz, F., Murad, W., Mischke, S., & Lehmkuhl, F. (2012). Late Quaternary climate and landscape evolution in arid Central Asia: A multiproxy study of lake archive Bayan Tohomin Nuur, Gobi desert, southern Mongolia. *Journal of Asian Earth Sciences, 48*, 125–135.

Gaskova, O. L., Isupov, V. P., Vladimirov, A. G., Shvartsev, S. L., & Kolpakova, M. N. (2015). Thermodynamic Modeling of the Behavior of Uranium and Arsenic in Mineralized Shaazgai_ Nuur Lake (Northwest Mongolia). *Doklady Earth Sciences, 465*(1), 1159–1163.

Gerelsaikhan, D., & Orkhonselenge, A. (2019). Recent lake area changes of Lake Khargal in northern Mongolia. *The Environment, 1*(1), 15–20. [In Mongolian with English abstract].

Gillespie, A. R., Burke, R. M., Komatsu, G., & Bayasgalan, A. (2008). Late Pleistocene glaciers in Darhad Basin, northern Mongolia. *Quaternary Research, 69*(2), 169–187.

Grunert, J., Lehmkuhl, F., & Walther, M. (2000). Paleoclimatic evolution of the Uvs Nuur basin and adjacent areas (Western Mongolia). *Quaternary International, 65–66*, 171–192.

Grunert, J., Stolz, C., Hempelmann, N., Hilgers, A., Hülle, D., Lehmkuhl, F., Felauer, T., & Dasch, D. (2009). The evolution of small lake basins in the Gobi desert in Mongolia. *Quaternary Sciences, 29*(4), 678–689.

Horn, W., Paul, M., Davaa, G., Dulmaa, A., & Tseveendorj, N. (1999). Water quality of several lakes and its influencing factors. In *Proceedings of International Conference of Global Change and Uvs Nuur on Sustainable Development of the Altai-Sayan Ecoregion and Transboundary Nature Conservation Issues*. Ulaangom, Uvs aimag, pp. 120–130.

Isupov, V. P., Ariunbileg, S., Razvorotneva, L. I., Lyakhov, N. Z., Shvartsev, S. L., Vladimirov, A. G., Kolpakova, M. N., Shatskaya, S. S., Chupakhina, L. E., Moroz, E. N., & Kuibida, L. V. (2013). Geochemical Model of Uranium Accumulation in Shaazgai-Nuur Lake (Northwestern Mongolia). *Doklady Earth Sciences, 448*(1), 143–148.

Isupov, V. P., Vladimirov, A. G., Lyakhov, N. Z., Shvartsev, S. L., Ariunbileg, S., Kolpakova, M. N., Shatskaya, S. S., Chupakhina, L. E., Kuibida, L. V., & Moroz, E. N. (2011). Uranium in Saline Lakes of Northwestern Mongolia. *Doklady Earth Sciences, 437*(1), 352–356.

Isupov, V., Vladimirov, A., Sodov, A., Kolpakova, M., Shvartsev, S., & Volkova, N. (2015). Hydromineral resources of Saline Lakes of Mongolia and Russian Altai. *Advanced Materials Research, 1085*, 166–170.

Jigj, S. (1975). *Primary feature of Mongolian landforms*. Institute of Geography and Permafrost, Mongolian Academy of Sciences (MAS), 126 p. [In Mongolian].

Kang, S., & Hong, S. Y. (2016). Assessing seasonal and inter-annual variations of lake surface areas in Mongolia during 2000–2011 Using Minimum Composite MODIS NDVI. *PLoS One, 11*(3), e0151395.

Kang, S., Lee, G., Togtock, C., & Jang, K. (2015). Characterizing regional precipitation-driven lake area change in Mongolia. *Journal of Arid Land, 7*(2), 146–158.

Kashiwaya, K., Ochiai, S., Sumino, G., Tsukamoto, T., Szyniszewska, A., Yamamoto, M., Sakaguchi, A., Hasebe, N., Sakai, H., Watanabe, T., & Kawai, T. (2010). Climato-hydrological fluctuations printed in long lacustrine records in Lake Hövsgöl, Mongolia. *Quaternary International, 219*(1–2), 178–187.

Kashiwaya, K., Shen, J., & Kim, J. Y. (2015). *Earth surface processes and environmental changes in East Asia: Records from Lake-Catchment System*. Springer.

Klinge, M., Schlütz, F., Zander, A., Hülle, D., Batkhishig, O., & Lehmkuhl, F. (2021). Late Pleistocene lake level, glaciation and climate change in the Mongolian Altai deduced from sedimentological and palynological archives. *Quaternary Research, 99*, 168–189. https://doi.org/10.1017/qua.2020.67

Kolpakova, M. N., & Gaskova, O. L. (2018). Major ions behaviour during evaporation of different saline type water of Western Mongolian lakes (geochemical modelling). *Hydrology Research, 49*(1), 163–176.

Komatsu, G., Arzhannikov, S. G., Gillespie, A. R., Burke, R. M., Miyamoto, H., & Baker, V. R. (2009). Quaternary paleolake formation and cataclysmic flooding along the upper Yenisei River. *Geomorphology, 104*, 143–164.

Komatsu, G., Baker, V. R., Arzhannikov, S. G., Gallagher, R., Arzhannikova, A. V., Murana, A., & Oguchi, T. (2016). Catastrophic flooding, palaeolakes and late Quaternary drainage reorganization in northern Eurasia. *International Geology Review, 58*, 1693–1722.

Komatsu, G., Brantingham, P. J., Olsen, J. W., & Baker, V. R. (2001). Paleoshoreline geomorphology of Boon Tsagaan Nuur, Tsagaan Nuur and Orog Nuur: The Valley of Lakes, Mongolia. *Geomorphology, 39*, 83–98.

Komatsu, G., Brantingham, P. J., Olsen, J. W., & Hyland, D. C. (2000). A preliminary study of paleoshoreline geomorphology associated with Boon Tsagaan Nuur. In A. P. Derevianko, J. W. Olsen, & D. Tseveendorj (Eds.), *Archaeological studies carried out by the Joint Russian-Mongolian-American expedition in Mongolia in 1997 & 1998* (pp. 380–381). Institute of Archaeology and Ethnography, Siberian Branch of the Russian Academy of Sciences.

Krivonogov, S. K., Bezrukova, E. V., Takahara, H., & Riedel, F. (2003). The Hovsgol Lake in the Late Pleistocene and Holocene: On-land geologic evidences for the change of its level. In K. Kashivaya (Ed.), *Long continental records from Lake Baikal* (pp. 187–206). Springer-Verlag.

Krivonogov, S. K., Sheinkman, V. S., & Mistruykov, A. A. (2005). Stages in the development of the Darhad dammed lake (Northern Mongolia) during the Late Pleistocene and Holocene. *Quaternary International, 136*(1), 83–94.

Krivonogov, S. K., Yi, S., Kashiwaya, K., Kim, J. C., Narantsetseg, T., Oyunchimeg, T., Safonova, I. Y., Kazansky, A. Y., Sitnikova, T., Kim, J. Y., & Hasebe, N. (2012). Solved and unsolved problems of sedimentation, glaciation and paleolakes of the Darhad Basin, Northern Mongolia. *Quaternary Science Reviews, 56*, 142–163.

Lee, M. K., Lee, Y. I., Lim, H. S., Lee, J. I., Choi, J. H., & Yoon, H. I. (2011). Comparison of radiocarbon and OSL dating methods for a Late Quaternary sediment core from Lake Ulaan, Mongolia. *Journal of Paleolimnology, 45*, 127–135.

Lee, M. K., Lee, Y. I., Lim, H. S., Lee, J. I., & Yoon, H. I. (2013). Late Pleistocene–Holocene records from Lake Ulaan, southern Mongolia: Implications for east Asian palaeomonsoonal climate changes. *Journal of Quaternary Science, 28*(4), 370–378.

Lehmkuhl, F., & Lang, A. (2001). Geomorphological investigations and luminescence dating in the southern part of the Khangay and the Valley of the Gobi Lakes (Central Mongolia). *Journal of Quaternary Science, 16*(1), 69–87.

Lehmkuhl, F., Grunert, J., Hülle, D., Batkhishig, O., & Stauch, G. (2018). Paleolakes in the Gobi region of southern Mongolia. *Quaternary Science Reviews, 179*, 1–23.

Linhoff, B. S., Bennett, P. C., Puntsag, T., & Gerel, O. (2011). Geochemical evolution of uraniferous soda lakes in Eastern Mongolia. *Environmental Earth Science, 62*, 171–183. https://doi.org/10.1007/s12665-010-0512-8

Matsuyama, Y., Endo, N., Hasebe, N., & Davaasuren, D. (2019). Paleolake reconstruction and estimation of paleo-inflow in the Olgoi basin, Mongolia, based on GIS and hydraulic analyses. *Island Arc*, 1–3.

Meybeck, M. (1995). Global distribution of lakes. In A. Lerman, D. M. Imboden, & J. R. Gat (Eds.), *Physics and chemistry of lakes* (pp. 1–35). Springer-Verlag.

Ministry of Environment and Green Development. (2013). *Integrated water management plan of Mongolia*. 339 p.

Mischke, S., Lee, M. K., & Lee, Y. I. (2020). Climate history of Southern Mongolia Since 17 ka: The Ostracod, Gastropod and Charophyte Record from Lake Ulaan. *Frontiers in Earth Science, 8*(221), 1–15.

Murzaev, E. M. (1952). *The description of physical geography of Mongolia* (2nd ed.). [In Russian].

Narantsetseg, T., Krivonogov, S. K., Oyunchimeg, T., Uugantsetseg, B., Burr, G. S., Tomurhuu, D., & Dolgorsuren, K. (2013). Late Glacial to Middle Holocene climate and environmental changes as recorded in Lake Dood sediments, Darhad Basin, northern Mongolia. *Quaternary International, 311*, 12–24.

Nottebaum, V., Stauch, G., van der Wal, J. L. N., Zander, A., Schlütz, F., Shumilovskikh, L., Reicherter, K., Batkhishig, O., & Lehmkuhl, F. (2021). Late Quaternary landscape evolution and paleoenvironmental implications from multiple geomorphic dryland systems, Orog Nuur Basin, Mongolia. *Earth Surface Processes and Landforms.*, 1–23. https://doi.org/10.1002/esp.5247

Orkhonselenge, A., & Bulgan, O. (2021). Geochemical studies and lacustrine geomorphology of Lake Yakhi basin in eastern Mongolia. *Géomorphologie: Relief, Processes, Environnement, 27*(3), 231–242. https://doi.org/10.4000/geomorphologie.15873

Orkhonselenge, A., Bulgan, O., Gerelsaikhan, D., Davaagatan, T., & Altansukh, N. (2020). Hydroclimatic fluctuation in Lake Yakhi, eastern Mongolia. *EGU General Assembly 2020. Presentation #EGU2020-686*, Vienna, Austria. : https://doi.org/10.5194/egusphere-egu2020-686.

Orkhonselenge, A., & Davaagatan, T. (2016). Sedimentary Feature in Lake Terkhiin Tsagaan and Lake Ugii, Central Mongolia. *Geophysical Research Abstracts Vol. 18, EGU2016-5666-2, EGU General Assembly 2016*.

Orkhonselenge, A., Komatsu, G., Davaagatan, T., & Uuganzaya, M. (2019). Sedimentary records from lakes reconstructing late quaternary paleoclimate changes in Mongolia. *XX INQUA Congress. Presentation #2419*, Dublin, Ireland.

Orkhonselenge, A., Komatsu, G., & Uuganzaya, M. (2018a). Climate-driven changes in lake areas for the last half century in the Valley of Lakes, Govi Region, Southern Mongolia. *Natural Science, 10*(7), 263–277.

Orkhonselenge, A., Komatsu, G., & Uuganzaya, M. (2018b). Middle to late Holocene sedimentation dynamics and paleoclimatic conditions in the Lake Ulaan basin, southern Mongolia. *Géomorphologie: Relief, Processes, Environnement, 24*(4), 351–363.

Orkhonselenge, A., Krivonogov, S. K., Kashiwaya, K., Nakamura, T., & Davaagatan, T. (2015). Paleoclimate changes in Mongolia: Lake Records. In *Extended Abstracts of International Association of Geomorphologists (IAG) Regional Conference on Gradualism vs Catastrophism in Landscape Evolution*, Barnaul, Russia, p. 175.

Orkhonselenge, A., Krivonogov, S. K., Mino, K., Kashiwaya, K., Safonova, I. Y., Yamamoto, M., Kashima, K., Nakamura, T., & Kim, J. Y. (2013). Holocene sedimentary records from Lake Borsog, eastern shore of Lake Khuvsgul, Mongolia, and their paleoenvironmental implications. *Quaternary International, 290–291*, 95–109.

Orkhonselenge, A., Krivonogov, S. K., Mino, K., Kashiwaya, K., Yamamoto, M., & Nakamura, T. (2014). Holocene landform evolution of lake Khuvsgul basin, Mongolia. *Géomorphologie: Relief, Processus, Environnement, 20*(4), 343–354.

Orkhonselenge, A., Mino, K., Kashiwaya, K., Krivonogov, S. K., Yamamoto, T., Sakaguchi, A., & Nakamura, T. (2011). Holocene hydro-environmental changes in northern Mongolia inferred from lacustrine sediments in Borsog Bay of Lake Khuvsgul. *Journal of Earth Environment, 2*(3), 457–461.

Oyunbaatar, D., Erdenebayar, B., Davaa, G., & Saikhanjargal, D. (2017). Recent changes of water regime and resource of the Ganga Lake and related some socio-economic aspects. *Modern Environmental Science and Engineering, 3*(7), 482–491.

Peck, J. A., Khosbayar, P., Fowell, S. J., Pearce, R. B., Ariunbileg, S., Hansen, B. C. S., & Soninkhishig, N. (2002). Mid to Late Holocene climate change in north central Mongolia as recorded in the sediments of Lake Telmen. *Palaeogeography, Palaeoclimatology, Palaeoecology, 183*, 135–153.

Peck, J. A., Khosbayar, P., King, J. W., Fowell, S. J., Ariunbileg, S., & Soninkhishig, N. (2001). The Latest Holocene sedimentary environmental magnetic record from Lake Dood, Mongolia. *American Geophysical Union, Fall Meeting, Abstract Id. GP41A-0243*.

Prokopenko, A. A., Khursevich, G. K., Bezrukova, E. V., Kuzmin, M. I., Boes, X., Williams, D. F., Fedenya, S. A., Kulagina, N. V., Letunova, P. P., & Abzaeva, A. A. (2007). Paleoenvironmental proxy records from Lake Hovsgol, Mongolia, and a synthesis of Holocene climate change in the Lake Baikal watershed. *Quaternary Research, 68*, 2–17.

Rudaya, N., Tarasov, P., Dorofeyuk, N., Solovieva, N., Kalugin, I., Andreev, A., Daryin, A., Diekmann, B., Riedel, F., Tserendash, N., & Wagner, M. (2009). Holocene environments and climate in the Mongolian Altai reconstructed from the Hoton-Nur pollen and diatom records: A step towards better understanding climate dynamics in Central Asia. *Quaternary Science Reviews, 28*, 540–554.

Saruulzaya, A. (2017). *Thermokarst lake dynamics in the continuous and isolated permafrost zones, Mongolia*. Ph.D. Thesis. Hokkaido University. Sapporo, Japan. 76 p.

Schwanghart, W., Schütt, B., & Walther, M. (2008). Holocene climate evolution of the Ugii Nuur Basin, Mongolia. *Advances in Atmospheric Sciences, 25*(6), 986–998.

Schwanghart, W., Frechen, M., Kuhn, N. J., & Schütt, B. (2009). Holocene environmental changes in the Ugii Nuur basin, Mongolia. *Palaeogeography, Palaeoclimatology, Palaeoecology, 279*, 160–171.

Shinneman, A. L. C., Umbanhowar, C. E., Edlund, M. B., & Soninkhishig, N. (2010). Late-Holocene moisture balance inferred from diatom and lake sediment records in western Mongolia. *The Holocene, 20*(1), 123–138.

Sternberg, T., & Paillou, P. (2015). Mapping potential shallow groundwater in the Gobi Desert using remote sensing: Lake Ulaan Nuur. *Journal of Arid Environments, 118*, 21–27.

Stolz, C., Hülle, D., Hilgers, A., Grunert, J., Lehmkuhl, F., & Dasch, D. (2012). Reconstructing fluvial, lacustrine and aeolian process dynamics in Western Mongolia. *Zeitschrift für Geomorphologie, 56*(3), 267–300.

Sun, A., Feng, Z., Ran, M., & Zhang, C. (2013). Pollen-recorded bioclimatic variations of the last ~22,600 years retrieved from Achit Nuur core in the western Mongolian Plateau. *Quaternary International, 311*, 36–43.

Szopińska, M., Szumińska, D., Polkowska, Ż., Machowiak, K., Lehmann, S., & Chmiel, S. (2016a). The chemistry of river–lake systems in the context of permafrost occurrence (Mongolia, Valley of the Lakes). Part I. Analysis of ion and trace metal concentrations. *Sedimentary Geology, 340*, 74–83.

Szopińska, M., Dymerski, T., Polkowska, Ż., Szumińska, D., & Wolska, L. (2016b). The chemistry of river–lake systems in the context of permafrost occurrence (Mongolia, Valley of the Lakes) Part II. Spatial trends and possible sources of organic composition. *Sedimentary Geology, 340*, 84–95.

Szumińska, D. (2016). Changes in surface area of the Böön tsagaan and Orog lakes (Mongolia, Valley of the lakes, 1974–2013) compared to climate and permafrost changes. *Sedimentary Geology*, 1–37.

Tian, F., Herzschuh, U., Dallmeyer, A., Xu, Q., Mischke, S., & Biskaborn, B. K. (2013). Environmental variability in the monsoonewesterlies transition zone during the last 1200 years: Lake sediment analyses from central Mongolia and supraeregional synthesis. *Quaternary Science Reviews, 73*, 31–47.

Tuvshinjargal, D. (2001). *Thermal regime of Lake Ugii*. Doctorate Thesis in Geography: Hydrology. Institute of Geography, Mongolian Academy of Sciences, Ulaanbaatar. 93 p. [In Mongolian].

Tsegmid, S. (1955). Valley of Lakes in the Mongolian Govi. *Natural Issues, 5*. [In Russian].

Tsegmid, S. (1969). *Physical geography of Mongolia*. State Press, 405 p. [In Mongolian].

Tserenpil, S., Dolmaa, G., & Voronkov, M. G. (2010). Organic matters in healing muds from Mongolia. *Applied Clay Science, 49*, 55–63.

Tserensodnom, J. (1971). *Lakes of Mongolia*. State Publishing, 202 p. [In Mongolian].

Tserensodnom, J. (2000). *A catalog of lakes in Mongolia*. Shuvuun Saaral Publishing, 141 p. [In Mongolian].

Walther, M., Dashtseren, A., Kamp, U., Temujin, K., Meixner, F., Pan, C. G., & Gansukh, Y. (2017). Glaciers, permafrost and lake levels at the Tsengel Khairkhan Massif, Mongolian Altai, During the Late Pleistocene and Holocene. *Geosciences, 7*(73).

Wang, X., Chun, X., Zhou, H., Zhang, Y., Wan, Z., & Dan, D. (2021). Application of standardised growth curves in quartz OSL dating of lacustrine sediments on the Mongolian Plateau. *Quaternary International, 592*, 51–59.

Wang, W., Feng, Z., Lee, X., Zhang, H., Ma, Y., An, C., & Guo, L. (2004). Holocene abrupt climate shifts recorded in Gun Nuur lake core, northern Mongolia. *Chinese Science Bulletin, 49*(5), 520–526.

Wang, W., Ma, Y. Z., Feng, Z. D., Meng, H. W., Sang, Y. L., & Zhai, X. W. (2009). Vegetation and climate changes during the last 8660 cal. a BP in central Mongolia, based on a high-resolution pollen record from Lake Ugii Nuur. *Chinese Science Bulletin, 54*(9), 1579–1589.

Wang, W., Ma, Y., Feng, F., Narantsetseg, T., Liu, K., & Zhai, X. (2011). A prolonged dry mid-Holocene climate revealed by pollen and diatom records from Lake Ugii Nuur in central Mongolia. *Quaternary International, 229*, 74–83.

Yu, K., Lehmkuhl, F., Diekmann, B., Zeeden, C., Nottebaum, V., & Stauch, G. (2017). Geochemical imprints of coupled paleoenvironmental and provenance change in the lacustrine sequence of Orog Nuur, Gobi Desert of Mongolia. *Journal of Paleolimnology, 58*, 511–532.

Yu, K., Lehmkuhl, F., Schlütz, F., Diekmann, B., Mischke, S., Grunert, J., Murad, W., Nottebaum, V., Stauch, G., & Zeeden, C. (2019). Late quaternary environments in the Gobi Desert of Mongolia: Vegetation, hydrological, and palaeoclimate evolution. *Palaeogeography, Palaeoclimatology, Palaeoecology, 514*, 77–91.

Zhang, C., Zhang, W., Feng, F., Mischke, S., Gao, X., Gao, D., & Sun, F. (2012). Holocene hydrological and climatic change on the northern Mongolian Plateau based on multi-proxy records from Lake Gun Nuur. *Palaeogeography, Palaeoclimatology, Palaeoecology, 323–325*, 75–86.

Chapter 3
Formation and Evolution of Lakes in Mongolia

Abstract This chapter considers formation and evolutionary history of lakes in Mongolia spanning over geological time scales. There have been various stages of lake formation in Mongolia. Lakes of Mongolia began to form during the early Paleozoic Era. The lake formation of the late Ordovician and Silurian coincided with the presence of a paleooceanic environment. Many lakes formed during the Mesozoic and Cenozoic Eras due to neotectonics, and they evolved throughout the Mesozoic (particularly during the Jurassic and Cretaceous) and the Cenozoic (particularly during the Neogene and Quaternary). Such lakes were dominantly stable since the Triassic when a terrigenic environment dominated. Lakes of Mongolia multiplied in number during the early Quaternary because high mountains received elevated precipitation contributing to the formation of dense river networks feeding the lakes. Lakes of Mongolia increased during the early Pleistocene, however, decreased in the late Pleistocene due to the glaciation-induced arid climate, neotectonics, and fluvial erosion. Numerous new lakes formed in the Pleistocene and Holocene.

Keywords Formation · Evolution · Change · Paleoocean · Paleosea · Paleolake · Mongolia

3.1 Introduction

Reconstructing the formation and evolution of lakes in Mongolia would bring out further comprehensive understandings in such fields as geomorphology, sedimentology, geochemistry, geochronology, and paleoclimatology of Mongolia and Eurasia. Geophysical, geochemical, and geomorphological processes in lake basins play important roles in lake formation and evolution resulting in differentiation of lake types (see Chap. 4).

Ages of lakes in Mongolia are variable (Tserensodnom, 1971), and there are still uncertainties regarding this point because paleogeography or evolutionary history of lakes in Mongolia has not been studied in detail. Nevertheless, ages of lake have

A. Orkhonselenge et al., *Lakes of Mongolia*, Syntheses in Limnogeology,
https://doi.org/10.1007/978-3-030-99120-3_3

been partly reflected in the geological history of Mongolia and Central Asia (Tsegmid, 1969; Tserensodnom, 1971). Physical geography and genesis of lakes during the late Mesozoic and early Cenozoic were reconstructed in Mongolia, based on paleolake deposits, paleoshorelines, paleoterraces, and connection with landform development by a number of scientists (e.g., E.M. Murzaev, V.M. Sienitsyn, N.T. Kuznetsov, E.A. Efremov, M.M. Elien, E.A. Marinov, R.A. Khasen, Sh. Tsegmid, J. Tserensodnom, S. Jigj). For instance, a paleoshoreline lying 20–30 km away from Lake Uvs (see Chap. 13) was observed by a geologist E.Y. Turshiev, whereas paleolake sediments at the southern shore of Lake Khar Us (see Chap. 12) were found by G.N. Potanin (Tserensodnom, 1971). Paleolake studies including paleogeography and origin of lakes in Mongolia were led by the Russian limnologist D.V. Sevestyanov since 1986, and he published a book entitled *Limnology and Paleolimnology of Mongolia* in 1994 (Batnasan, 1998).

Understanding landform development of Mongolia is a key for interpreting the formation and evolution of lakes in the country. The landform development of Mongolia (Table 3.1) is reconstructed at the first time based on previous results (e.g., Tsegmid, 1969; Jigj, 1975). The first phase of lake formation may have occurred during the late Ordovician and Silurian coinciding with the existence of a paleooceanic environment prevailing throughout the country at that time. The second phase of lake formation occurred in the early Jurassic when large depressions and faults appeared. The third phase of lake formation was found in the late Miocene and coincided with the formation of tectonic rifts and faults (Table 3.1).

A lacustrine landscape had prevailed throughout Mongolia since the late Mesozoic to the early Cenozoic (Murzaev, 1952). There were large lakes, as large as the present-day Aral and Caspian Seas, in the Zuun Bayan and Nemegt basins in eastern and southeastern Govi[1] and in southern Govi Altai Mountain Range. These lakes gradually reduced in area after the Cretaceous when the lakes reached their maximum sizes (Tserensodnom, 1971). However, there has been no study reconstructing in detail these lakes, which archived tracers of the paleoenvironments between the Triassic and Neogene. For example, a large paleolake in the Bor Tsonj basin of southern Govi Altai Mountain Range fed by Bulgan, Bodonch, Uyench, and Barlag Rivers draining the Mongolian Altai Mountain Range was studied by A.T. Ivanov and N.T. Kuznetsov. Two paleolake basins around (1) Lake Tari[2] in southeastern Siberia and (2) Lakes Khukh and Jirmiin Tsagaan in northeastern Mongolia were documented by N.A. Marinov and R.A. Khasyn (Tserensodnom, 1971). However, their ages are still unclear. Instead, lakes formed since the Quaternary have been studied relatively well (see Chaps. 2 and 21).

Relict small lakes, playa basins, and lake sediments left on terraces show that there were many medium- and small-sized lakes beside the larger ones in eastern and southeastern plains and Govi, southern Govi Altai Mountain Range, the Shishkhid

[1] *Govi* has been often spelled as *Gobi* in the international literature. Govi is the correct English transliteration from Mongolian *Говь*.

[2] *Tari* has been misspelled as Torey by the Russians. Tari is the right English transliteration from Mongolian *Тарь*.

Table 3.1 Evolutionary history of landforms and landscapes of Mongolia

Geological time scale			Paleogeography	Geology
Quaternary	Holocene	Late	Deposition of fluvial and lacustrine sediments (~100–250 m thick)	Lacustrine and alluvial deposits
		Early		
	Pleistocene	Late		Dominance of alluvial fan deposits
		Middle	(5) Intensification of river incision (4) Glaciations in the Altai, Khangai, Khuvsgul, and Khentii Mountain Ranges (3) Formation of the third peneplains (2) Formation of intermontane basins (1) Tectonic subsidence	
		Early		

(continued)

Table 3.1 (continued)

Geological time scale			Paleogeography	Geology
Neogene	Pliocene	Late	(2) Rivers followed the faults (1) Reformation of the large basins between the Mongolian Altai, Khangai, and Khentii Mountain Ranges	Alluvial fan deposits (e.g., red and brown clays and gravels) in the tectonic basins in northern Mongolia (e.g., Zaamar basin and Orkhon, Selenge[a], Bayan, Kharaa and Tuul Rivers' valleys in the basin between Khangai and Khentii Mountain Ranges)
		Early	(4) Reformation of lakes (3) Formation of the second peneplains (2) Alpine orogeny along the faults (e.g., intensive in the Altai and Khuvsgul Mountain Ranges, gradual in the Khangai and Khentii Mountain Ranges, and weak in eastern and southern Mongolia) (1) Formation of faults and rifts due to neotectonics	Sedimentary rocks (e.g., 80–100-m-thick gray, greenish gray, and green clays in Orkhon and Selenge Rivers' drainage basins)
	Miocene	Late		
		Early	Erosion of the paleo-peneplains (e.g., the Sayan and Khangai Mountain Ranges in northern Mongolia)	Exposition of basalts
Paleogene	Oligocene			
	Eocene		(4) Rivers' drainage from the Khangai Mountain Range to the Govi region (e.g., the Valley of Lakes) (3) Formation of the first peneplains (especially above 2500 m a.s.l. in the Khangai and Khentii Mountain Ranges) (2) Deposition of sediments in the large basins (1) Intensification of erosion (e.g., in northern Mongolia)	Dominance of terrigenic sedimentary rocks (~180–190 m thick) (e.g., in the northeast of the Mongolian Altai Mountain Range or the Depression of Great Lakes; in western inflow of Yeruu River and the Nalaikh basin in the Khentii Mountain Range)
	Paleocene			
Cretaceous	Late			
	Early		Re-orogeny mountain building	Exposition of intrusive granites and dominance of igneous rocks (e.g., the Khangai Mountain Range in central Mongolia and the Govi Altai Mountain Range in southwestern Mongolia)
Jurassic	Late			
	Middle		(4) Abundance of surface water (3) Dominance of erosive landforms (e.g., in the Khangai Mountain Range) (2) Initial formation of river networks (1) Deposition of sediments in large basins (e.g., in eastern Govi, southern Govi of the Altai, between Mongolian Altai, Khangai, and Khentii Mountain Ranges, etc.)	Presence of sedimentary rocks (e.g., rounded breccia, limestones, sandstones, gritstones in the south of Lake Khar Us and Kharkhiraa Mountain Range; 150–200 m thick in Jargalant and Baatar Khairkhan Mountain Ranges in western Mongolia)
	Early		(4) Formation of lakes (3) Formation of basins (e.g., the Bulnai basin in northern Mongolia, basins between Khangai and Khentii Mountain Ranges in central Mongolia) (2) Formation of faults (1) Re-orogeny mountain building	Dominance of igneous and sedimentary rocks (e.g., granitoids in Tsagaan Olom basin in the Khangai Mountain Range in central Mongolia; intrusive rocks in the Dornod plain in eastern Mongolia)
Triassic	Late		(3) Deposition of sediments (2) Intensification of erosion (1) Dominance of terrigenic plateau	Dominance of sedimentary rocks
	Middle			
	Early			

Permian	Lopingian	Dominance of the plateau	Terrigenic and oceanic (dominantly volcanic rocks) sedimentary rocks (e.g., in central and eastern Mongolia)	Carbonic deposits (sandstones and gritstones) in eastern, southern, and northwestern Mongolia
	Guadalupian	(2) Advance of the paleoocean (1) Hercynian orogeny		Between terrigenic and oceanic deposits in northern Mongolia and shallow oceanic sedimentary rocks in southern Mongolia
	Cisuralian			
Carboniferous	Pennsylvanian	Deposition of sediments (e.g., the Borzon Govi and Galba Govi in eastern Govi and near Mt. Atas Bogd (2695 m a.s.l.) and Mt. Tsagaan Bogd (2480 m a.s.l.) in southern Mongolia)	Dominance of igneous rocks (e.g., in northern Mongolia)	
	Mississippian	(3) Presence of plateau (dominantly the northwestern and eastern Mongolia) (2) Deposition of sediments (e.g., in Orkhon and Selenge Rivers' drainage basins and from Uyench River in western Mongolia to Lake Buir and Khyanganb Mountain Range in eastern Mongolia) (1) Advance of the paleoocean from the northwest to the east of Mongolia (e.g., in the Mongolian Altai Mountain Range, Depression of Great Lakes, Khuvsgul and Bulnai Mountain Ranges, etc.)	Dominance of oceanic deposits (e.g., in the Orkhon River drainage basin, Khangai Mountain Range, and southern Mongolia)	
Devonian	Late	(2) Deposition of sediments (1) Advance of the paleoocean (e.g., in the Khuvsgul and Khan Khukhii Mountain Ranges, Selenge River drainage basin in northern Mongolia, and from southern Khangai Mountain Range to the eastern Govi in southern Mongolia)	Dominance of igneous rocks with molasses deposits (especially sandstones and schists) (2) Volcanic silicate rocks (igneous rocks, siliceous rocks, jasper, etc.) (1) Terrigenic carbonate rocks (limestones, siltstones, argillites, and graywackes)	
	Middle			
	Early			

(continued)

Table 3.1 (continued)

Geological time scale		Paleogeography	Geology
Silurian	Pridoli	(3) Deposition of sediments (2) Advance of paleoocean (e.g., northwestern and southern Mongolia) (1) Caledonian orogeny	Green schists and molasses (e.g., in northern and southern Mongolia)
	Ludlow		
	Wenlock		
	Llandovery		
Ordovician	Late		Sedimentary rocks (e.g., sandstones, siltstones, limestones, schists) in southwestern and southeastern Mongolia
	Middle		
	Early	Salarian uplift	
Cambrian	Furongian		(2) Metamorphic rocks (e.g., limestones) (1) Oceanic deposits with volcanic rocks (e.g., ~3000 m thick near Khodon River in the Shishkhid (or Darkhad) basin)
	Series 3	(2) Subsidence of continent (1) Formation of tectonic rifts	
	Series 2		
	Terreneuvian		
Proterozoic	Neoproterozoic	Orogeny-induced uplifts	

a*Selenge* has been often incorrectly written as *Selenga* internationally by the Russian pronunciation. Selenge is the correct English transliteration from Mongolian *Сэлэнгэ*

b*Khyangan* has been misspelled as *Khingan* (or *Hingan*) and *Hyangan* in many publications. Khyangan is the right English transliteration from Mongolian *Хянган* (or Darkhad) basin, and the Mongolian Altai, Khangai,[3] and Khentii[4] Mountain Ranges during the Quaternary, but most of these lakes shrank or disappeared today. N.T. Kuznetsov concluded that shrinkage, disappearance of lakes, and drops in lake level in Mongolia were not related to the intensive evaporation due to the postglacial arid climate, but to the development of interfluves caused by the uplift and neotectonics in the early Quaternary (Jigj, 1975) and that lakes formed by damming of volcanic lava and moraines disappeared due to fluvial erosion in the late Quaternary.

Further evolution of lakes in Mongolia, especially continuous shrinkage of lakes, is highly debatable to date. The effects of arid climate in Central Asia hypothesized from some lake studies by scientists (e.g., G.E. Grummgrjemailo, N.V. Pavlov, V.A. Smirnov, V.M. Sienetsyn) were rejected by others (e.g., K.K. Markov, E.M. Murzaev, B.B. Polynov, K.M. Krasheninikov, Sh. Tsegmid, Sh. Luvsandorj). For instance, shrinkage of lakes in Mongolia due to arid climate and high elevation above sea level, a conclusion reached by geochemist V.A. Smirnov in 1926, based on drop in lake level and high mineralization of Lakes Uvs and Khyargas, was refused by other scientists (e.g., N.D. Bespalov, S. Lamjav, Sh. Tsegmid, D. Davaasuren, Sh. Luvsandorj).

Paleolakes of Mongolia have not been reconstructed in detail in terms of their initial formations in the early Paleozoic. Recent studies have reconstructed geological histories of lakes, but they cover only the Pliocene and Pleistocene (e.g., Fedotov et al., 2004; Krivonogov et al., 2005). In this chapter, formation and evolution of lakes in Mongolia over long geological time scales are inferred based on the data presented in the literature. The more recent paleoclimate changes contributing to the formation and evolution of lakes are discussed in detail according to each region of Mongolia in Chap. 21.

3.2 Formation of Lakes

In Mongolia, some lakes initially formed during the early Paleozoic Era, and their existences predominantly continued from the Triassic Period (Tsegmid, 1969). Lakes of Mongolia may have initially formed in the late Ordovician and Silurian due to the presence of a paleooceanic environment (Table 3.1). Nevertheless, because the ocean and seawater did not transgress into intercontinental Mongolia during the following 200 Ma since the early Paleozoic, lacustrine and fluvial sediments persistently deposited during that period (Tserensodnom, 1971). Many relict paleolakes formed in intermontane valleys and depressions at the beginning of the Mesozoic. Lakes during the Cretaceous and Neogene multiplied and extended in size much larger than those at the present time.

[3] *Khangai* has been misspelled as *Khangay* (or *Hangay*) and *Hangai* in many publications. Khangai is the right English transliteration from Mongolian *Хангай*.

[4] *Khentii* has been misspelled as *Khentey* (or *Hentey*) and *Khentei* (or *Hentei*) in many publications. Khentii is the right English transliteration from Mongolian *Хэнтий*.

Many lakes of Mongolia originated in the early Paleozoic and Mesozoic Eras, and they continued to exist during the Mesozoic and Cenozoic Eras (Jigj, 1975). For example, lakes in eastern and southern Mongolia (see Chaps. 6 and 9) formed in the late Jurassic and were widely distributed in the early Cretaceous (Tserensodnom, 1971). This is coincident with feeding of lakes in the Valley of Lakes from the southern Khangai Mountain Range (Table 3.1). Lakes also formed in the plains (e.g., Tukhum, Menen, and Dornod plains) and Govi (e.g., Uush, Dolood) of eastern, southeastern, and southern Mongolia prior to the Neogene. In general, a lacustrine environment dominated throughout Mongolia as the first phase in the late Ordovician and Silurian, the second phase in the early Jurassic, and the third phase in the late Miocene (Table 3.1). Many lakes in Mongolia were newly formed in the fourth phase in the late Pliocene and early Pleistocene due to the fact that the Alpine orogeny-induced high mountains building in the early Neogene (Table 3.1) and the mountains received much precipitation contributing to the formation of dense river networks to feed the lakes.

There were many Mongolian lakes that were larger than today, and they existed in the early Quaternary (Murzaev, 1952). For instance, the present-day large lakes in the Depression of Great Lakes of western Mongolia (see Chap. 12) are indeed relict lakes within a great paleolake basin stretched over two large basins of Lakes Uvs and Khyargas (Fig. 1.1). Lake Uvs stood at 1000 m a.s.l. or 257 m higher than the present level at 743 m a.s.l. and covered an area of 16,000 km^2 extending 220 km in length and 110 km in width, while Lake Khyargas stood at 1140 m a.s.l. or 100 m higher than the present level at 1040 m a.s.l. and covered an area of 11,000 km^2 extending 210 km in length and 120 km in width (Murzaev, 1952). Moreover, present-day Khuis Govi, Sharga Govi, and the Zereg valley were the floors of a large paleolake drainage basin (Tsegmid, 1969). Another large paleolake basin reached 1000–1500 m a.s.l. and covered an area of 92,000 km^2 in the Depression of Great Lakes, and this included present-day Lakes Khar Us, Khar, Airag, and Durgun (Fig. 1.1), i.e., the basin was larger in area than the Aral Sea (Tsegmid, 1969). This large paleolake was hypothesized to have existed in the early to middle Pleistocene by E.M. Murzaev and E.V. Devyatkin (Batnasan, 1998). Lake Shargiin Tsagaan in the Sharga Govi (Fig. 1.1), the southernmost corner of the basin, was 300–400 m deep, and this was reconstructed by four to five terraces of the paleolake (Tsegmid, 1969).

The late Neogene and early Quaternary lake sediments and terraces standing at 1450–1500 m a.s.l. around the Baatar Khairkhan Mountain Range, Zereg valley, and Umnu Khairkhan Mountain Range in western Mongolia were described by M.F. Neyburg (Tserensodnom, 1971). Moreover, the late Neogene and early Quaternary lake sediments were also found in the Lake Uvs basin of the Depression of Great Lakes in western Mongolia and in the Lake Shargiin Tsagaan basin of the Valley of Lakes in southern Mongolia by V.P. Maslov, N.S. Zaitsev, E.P. Rachkovsky, and V.A. Kuznetsov (Tserensodnom, 1971). The water level of Lake Buun Tsagaan between Khangai and Govi Altai Mountain Ranges in southern Mongolia (see Chap. 10) stood at 300–340 m higher than the present-day level (Devyatkin, 1981). However, its age is uncertain. Lake Buun Tsagaan was connected with Lake Adgiin Tsagaan, lying 60 km away, and the basin covered the two lakes extended 115 km in length and possibly included Lakes Orog (Fig. 1.1), Taatsiin Tsagaan, and Shar Burd between the two lakes (Tsegmid, 1969; Komatsu et al., 2001). At that time,

Lake Buun Tsagaan stood at 40–45 m higher than the present lake level, while Lake Adgiin Tsagaan stood at 50–60 m higher than the present lake level (Murzaev, 1952; Tsegmid, 1955). However, it is uncertain if Lake Ulaan was included in this basin or was located at the basin's end in the east of the Valley of Lakes, because the basin area in this region was highly deformed and changed due to the postglacial neotectonics (Tserensodnom, 1971). The Neogene deposits and Quaternary terraces in the lake basin imply that Lake Ulaan was ten times larger than the present area, and it stood at 100 m higher than the present lake level (Tsegmid, 1969) lying at 1024 m a.s.l. (Orkhonselenge et al., 2018b; see Chap. 11).

Formation and evolution of larger lakes occurred also around the Khuvsgul[5] Mountain Range in northern Mongolia (see Chap. 15), Khangai and Khentii Mountain Ranges in central Mongolia (see Chap. 18), and the Dornod and Menen plains in eastern Mongolia (see Chap. 6). For example, in eastern Mongolia, Lake Buir (see Chap. 7) connected with neighboring Lake Bayan as a single lake in the past was noted by G.N. Potanin during a survey in Khalkh River, Lake Buir, and Khyangan Mountain Range. Lake Buir was larger than the present, and it decreased in area because Khalkh River draining through the Tamsag Bulag basin and feeding Lake Buir in the southwest during the early Quaternary modified its discharge direction to the northeast of the lake (Tsegmid, 1969). However, there is no indicator to trace of the tectonics-induced channel (e.g., fault line) except for the marshy and peatland areas in the Tamsag Bulag basin and the Menen plain in the southeast of the lake.

A number of lakes formed and evolved (especially disappeared or decreased in area and water level) due to neotectonics (Tsegmid, 1969; Tserensodnom, 2000) during the late Pleistocene postglaciations (Orkhonselenge & Harbor, 2018; Orkhonselenge & Uuganzaya, 2018). Furthermore, landslide-induced, moraine-dammed, and volcanic lava-dammed lakes (see Chap. 4) formed in the high mountains during the late Quaternary, while lakes in the Chuluut, Terkh, and Suman Rivers' valleys of the northern Khangai Mountain Range disappeared (Tsegmid, 1969).

3.3 Evolution of Lakes

Lakes of Mongolia experienced histories of disappearances and reformations due to frequent orogeny, volcanism, glaciation, and climate change over long geological time scales since their initial formation in the late Ordovician and Silurian (Table 3.1). For instance, lakes in southern and western Mongolia (see Chaps. 9 and 12) mostly disappeared during Marine Oxygen Isotope Stage (MIS) 6, and that large paleolakes existed during the last interglacial (MIS 5e) caused by a strong East Asian Summer Monsoon (EASM) that lasted probably until the beginning of the last glacial (Lehmkuhl et al., 2018).

[5] *Khuvsgul* has been misspelled as *Hovsgol* and *Khubsugul* in publications. Khuvsgul is the right English transliteration from Mongolian *Хөвсгөл*.

Looking at a longer time span, it is known that lakes of Mongolia evolved throughout the Mesozoic (in particular, the Jurassic and Cretaceous) and lasted to the Cenozoic (e.g., Paleogene and Neogene). For example, Lakes Khar Us, Tsetseg, Alag, and Tonkhil in the intermontane basin of the Mongolian Altai Mountain Range (see Chap. 12), and Lakes Shargiin Tsagaan, Biger, Taatsiin Tsagaan, and Ulaan in the Valley of Lakes of the Govi (see Chap. 9), are relict lakes of paleolakes that dropped in levels due to changes of river channels.

Lakes of Mongolia shifted to a stable phase during the Neogene, however, they became more widespread in the early Pleistocene and decreased in the late Pleistocene due to glaciation-induced arid climate, neotectonics, and fluvial erosion (Table 3.1). For lakes in eastern, southeastern, and southern Mongolia, lake deposits and fossils show that these lakes decreased in number and size prior to the Miocene. Lakes enlarged and multiplied in northeastern and southeastern Mongolia during the Pliocene. In the early Quaternary, lakes were larger throughout Mongolia and their water levels were more stable than the present day because most lakes were fed by glaciers. The major changes of the lakes were coincident with the postglacial arid climate in the Pleistocene.

Lakes in western Mongolia (see Chap. 12) were in general larger and greater in number than those in eastern Mongolia (see Chap. 6) during the Quaternary. Although lakes in the plains of eastern Mongolia and the Govi region of southern Mongolia (see Chap. 9) were smaller in area, nevertheless they reached 20–30 m deep, or even deeper than 100 m in some cases during the Quaternary (Tsegmid, 1969).

3.3.1 *Tectonics-Induced Evolution*

E.P. Gerasimov and E.M. Lavrenko concluded that neotectonics contributed to terrace formation along lake shorelines, and they highlighted that drop in lake level was not the cause of the formation. In addition, they suggested that when shoreline on one side of the lake (referring to Lakes Khar Us and Khyargas) was uplifted, shoreline on the opposite side was filled with water, so the terraces would be observed at the uplifted side. This phenomenon certainly happened in some lakes, but not for all lakes in Mongolia (Tsegmid, 1969). Moreover, the hypothesis of Lakes Khunguin Khar, Oigon, Telmen, and Sangiin Dalai in the northern Khangai Mountain Range (see Chap. 18) connecting with lakes in the Depression of Great Lakes (see Chap. 12) through Tes River in the past was advocated by Murzaev (1952); however there is no trace of tectonics-induced evolution linked to this connection.

3.3.2 *Climate-Induced Evolution*

Lakes in Mongolia were apparently larger in area and number during the Neogene than those at present. Lakes in Mongolia disappeared, decreased, shrank, salinized, and developed individually since the early Pleistocene due to climate changes (see

Chap. 21), continuous neotectonics, glaciations, river incision, and erosion (Table 3.1). For instance, because of the discharge erosion, many lakes in the Shishkhid (or Darkhad) basin in the Khuvsgul Mountain Range (see Chap. 15); Khag Khongor, Kharaa, and Kherlen Rivers' valleys in the Khentii Mountain Range (see Chap. 6); Chuluut and Khar Us Rivers' valleys in the Khangai Mountain Range (see Chap. 18); and Buyant and Khovd Rivers' valleys in the Mongolian Altai Mountain Range (see Chap. 12) disappeared. However, their dates of disappearance have not been discussed. For the lakes with inflow and outflow such as Lakes Terkhiin Tsagaan and Ugii in the Khangai Mountain Range (see Chaps. 19 and 20); Lakes Tolbo, Khoton (see Chap. 14), Khurgan, Dayan, and Achit in the Mongolian Altai Mountain Range; and Lake Buir in the Dornod plain, their areas, depths, and volumes significantly dropped during the Holocene.

In the Holocene, lakes of Mongolia have been experiencing (1) mineralization/salinization and shrinkage/disappearance shifting to playa lakes due to the aridity-induced evaporation and/or (2) increase in area and level shifting also to freshwater lakes due to the humidity-induced and glaciers and permafrost meltwaters-induced water supplies. These two phenomena are happening under the rising annual and seasonal average air temperatures today (see Part II). For example, large lakes in the Valley of Lakes in southern Mongolia have significantly lost their areas due to the rising temperature over the last 50 years (Orkhonselenge et al., 2018a). Terraces lying at 20–30 m higher than the present-day lake level show that Lake Khuvsgul (see Chap. 16) hosted much greater volumes during the glacial period (Table 3.1). Large lakes in the Mongolian Altai and Khangai Mountain Ranges, the Depression of Great Lakes, and the Valley of Lakes rose their levels and increased areas during the arid climate in the past (Tsegmid, 1969). However, recent studies (e.g., Szumińska, 2016; Orkhonselenge et al., 2018a) are showing the opposite trend of the evolution of lakes during the late Holocene.

3.4 Summary

There have been four major stages of lake formation in Mongolia during the (1) late Ordovician and Silurian, (2) early Jurassic, (3) late Miocene of the Neogene, and (4) early Pleistocene. Lakes of Mongolia initially formed in the early Paleozoic Era, experienced numerous evolution events, and reached the present-day phase since their last evolution occurring during the late Pliocene and early Pleistocene. In general, the lakes have experienced changes in number, area, and water level with expansion/spillover, shrinkage/disappearance, and mineralization/salinization over various geological time scales. Decreases in number and area of lakes and their eventual disappearance accelerated with the emergence of tectonics-induced interfluves and postglacial arid climate. Present large lakes in southern and western Mongolia are relict lakes of even larger paleolakes existing during the late Neogene and early Quaternary. The lakes of the early and middle Pleistocene occupied low elevations and small areas, which were formed due to neotectonics, volcanism, and glaciations when the Depression of Great Lakes and the Valley of Lakes were isolated. Recent studies of lakes in Mongolia (see Chap. 2) document representative lakes of each region (see Part II), and they are placed in the context of reconstruction of paleoclimate changes (see Chap. 21).

References

Batnasan, N. (1998). *Hydrological system, water regime, and evolution of large lakes in the Govi region*. Doctorate Thesis in Geography: Hydrology, Institute of Geoecology, Mongolian Academy of Sciences, Ulaanbaatar. 163 p. [In Mongolian].

Devyatkin, E. V. (1981). *Cenozoic of Inner Asia*. Nauka, 196 p. [in Russian].

Fedotov, A. P., Kazansky, A. Y., Tomurhuu, D., Matasova, G. G., Ziborova, G. A., Zheleznyakova, T. O., Vorobyova, S. S., Phedorin, M. A., Goldberg, E. L., Oyunchimeg, T., Narantsetseg, T., Vologina, E. G., Yuldashev, A. A., Kalugin, I. A., Tomurtogoo, O., & Grachev, M. A. (2004). A 1 My record of paleoclimates from lake Khubsugul, Mongolia. *EOS Transactions, 85*, 387–390.

Jigj, S. (1975). *Primary feature of Mongolian landforms*. Institute of Geography and Permafrost, Mongolian Academy of Sciences (MAS), 126 p. [In Mongolian].

Komatsu, G., Brantingham, P. J., Olsen, J. W., & Baker, V. R. (2001). Paleoshoreline geomorphology of Boon Tsagaan Nuur, Tsagaan Nuur and Orog Nuur: The Valley of Lakes, Mongolia. *Geomorphology, 39*, 83–98.

Krivonogov, S. K., Sheinkman, V. S., & Mistruykov, A. A. (2005). Stages in the development of the Darhad dammed lake (Northern Mongolia) during the Late Pleistocene and Holocene. *Quaternary International, 136*, 83–94.

Lehmkuhl, F., Grunert, J., Hülle, D., Batkhishig, O., & Stauch, G. (2018). Paleolakes in the Gobi region of southern Mongolia. *Quaternary Science Reviews, 179*, 1–23.

Murzaev, E. M. (1952). *The description of physical geography of Mongolia* (2nd Ed.). [In Russian].

Orkhonselenge, A., & Harbor, J. M. (2018). Impacts of modern glacier changes on surface water resources in western and northern Mongolia. *Journal of Water Resource and Protection, 10*(6), 559–576.

Orkhonselenge, A., Komatsu, G., & Uuganzaya, M. (2018a). Climate-driven changes in lake areas for the last half century in the Valley of Lakes, Govi Region, Southern Mongolia. *Natural Science, 10*(7), 263–277.

Orkhonselenge, A., Komatsu, G., & Uuganzaya, M. (2018b). Middle to late Holocene sedimentation dynamics and paleoclimatic conditions in the Lake Ulaan basin, southern Mongolia. *Géomorphologie: Relief, Processes, Environnement, 24*(4), 351–363.

Orkhonselenge, A., & Uuganzaya, M. (2018). Glacial geomorphology of Mt. Asralt Khairkhan and Mt. Baga Khentii Saridag in Khentii Mountain Range, Northeastern Mongolia. *International Journal of Geosciences, 9*(6), 308–319.

Szumińska, D. (2016). Changes in surface area of the Böön tsagaan and Orog lakes (Mongolia, Valley of the lakes, 1974–2013) compared to climate and permafrost changes. *Sedimentary Geology*, 1–37.

Tsegmid, S. (1955). Valley of Lakes in the Mongolian Govi. *Natural Issues, 5*. [In Russian].

Tsegmid, S. (1969). *Physical geography of Mongolia*. State Press, 405 p. [In Mongolian].

Tserensodnom, J. (1971). *Lakes of Mongolia*. State Publishing, 202 p. [In Mongolian].

Tserensodnom, J. (2000). *A catalog of lakes in Mongolia*. Shuvuun Saaral Publishing, 141 p. [In Mongolian].

Chapter 4
Genesis of Lakes in Mongolia

Abstract Lakes of Mongolia are genetically dependent on tectonic setting, volcanism, landscape, and landform evolution. Since the initial formation during the early Paleozoic Era (see Chap. 3), lakes of Mongolia have experienced diverse paths of evolution (e.g., dominant presence in the Triassic, maximum sizes in the Cretaceous, decrease in size since the Paleogene, stabilization of existence in the Neogene reaching the present conditions since the late Pliocene and early Pleistocene). Throughout the Quaternary period, lakes of Mongolia have been dramatically changing in terms of water level, area, and sedimentation dynamics (see Part II). Present lakes of Mongolia are classified to have seven major types of origins, and they can be called tectonic, volcanic, landslide, glacial, karst (and thermokarst), fluvial, and aeolian lakes. Tectonic, volcanic, glacial, and landslide lakes are commonly found in high mountains such as the Mongolian Altai, Govi Altai, Khangai, Khentii, and Khuvsgul Mountain Ranges in western, northern, and central Mongolia. Aeolian, fluvial, and karst lakes are widely distributed in lowlands, valleys, and depressions within the plains and the Govi Desert regions in eastern and southern Mongolia.

Keywords Origin · Genesis · Evolution · Type · Lakes · Mongolia

4.1 Introduction

Lake originates from diverse exogenic and endogenic earth processes over geological time scales. Lakes occur in areas of crustal subsidence such as rift zones, continental sag basins, glacial and periglacial regions, arid and semiarid lowlands, and foreland basins which cannot be inundated by seawater and ocean water. Lakes are defined as water bodies, which can have wave action on the shoreline where wind-induced turbulence plays a major role in mixing the water column (Kuusisto & Hyvärinen, 2000). Lakes cover approximately 1% of the continental area of the

A. Orkhonselenge et al., *Lakes of Mongolia*, Syntheses in Limnogeology,
https://doi.org/10.1007/978-3-030-99120-3_4

Earth (Bridge & Demicco, 2008). Lakes occupy particularly large areas in some places. For instance, many thousands of lakes existed at 50–75% of the Arctic coastal plain in the past, but they cover its 20–40% at the present (Frohn et al., 2005).

There are many types of lakes according to their origin, thermal stratification, oxygen saturation, salinity, seasonal variations in lake volume, level and areas, degree of outflow, and other factors (Kuusisto & Hyvärinen, 2000). Surface processes (e.g., fluvial, aeolian, geomorphological, geochemical, sedimentological, etc.) in lake basins play important roles in differentiating the type and evolution of the lakes. The origin of a lake can be one important classification factor. Hutchinson (1957) defined 11 major types of lakes based on lake's origin as tectonic, volcanic, landslide, meteorite (extraterrestrial impact), glacial, solution, fluvial, aeolian, shoreline, organic, and anthropomorphic lakes, and their 76 subtypes. According to Meybeck (1995) and Bridge and Demicco (2008), lakes can be classified in terms of origin into six types as glacial, tectonic, fluvial, crater, coastal, and other lakes. In the last category *other lakes*, they included karst, wind-deflation, landslide-dammed, and volcanic-material-dammed lakes. Depending on origins, lakes live for long-term and short-term periods (Einsele, 1992). For example, proglacial, ice-contact, graben-structure, and rift zone lakes with high influx of detrital clastics are short-lived, and small and shallow lakes are also shorter-lived due to filling lasting only for a few 1000 years or less (Einsele, 1992).

Lakes of Mongolia are diverse, and each of them has a complex and interesting history (Berkey & Morris, 1927). Some lakes are relicts of the Central Asian inland seas (Tserensodnom, 1971). Some scientists (e.g., E.M. Murzaev, V.M. Sienitsyn, N.T. Kuznetsov, E.A. Efremov, M.M. Elien, E.A. Marinov, R.A. Khasen, Sh. Tsegmid, N. Badarch, etc.) interpreted that geneses of Mongolian lakes were largely attributed to the landform development of the country. For instance, in 1878–1879, M.V. Pevtsov demonstrated that Lake Terkhiin Tsagaan (see Chap. 19) in the Khangai[1] Mountain Range in central Mongolia and small lakes in the Gichgene Mountain Range within the Govi[2] Altai Mountain Range in western Mongolia had volcanic origins (Tsegmid, 1969). In 1899, A.N. Kaznakov and V.P. Ladigen noted that Lakes Khoton and Khurgan (see Chap. 14) in the Mongolian Altai Mountain Range in western Mongolia had glacial origins. In 1902, V.L. Komarov studied the genesis of Lake Khuvsgul[3] (see Chap. 16) in northern Mongolia, and he noted that terraces along its beaches had igneous origins. In 1927, S.A. Kondratiev noted that the Lake Telmen (Fig. 4.1) basin in central Mongolia had an origin related to the network of Ider River (Tserensodnom, 1971), a headwater of Selenge[4] River

[1] *Khangai* has been misspelled as *Khangay* (or *Hangay*) and *Hangai* in many publications. Khangai is the right English transliteration from Mongolian *Хангай*.

[2] *Govi* has been often spelled as *Gobi* in the international literature. Govi is the correct English transliteration from Mongolia *Говь*.

[3] *Khuvsgul* has been misspelled as *Hovsgol* and *Khubsugul* in publications. Khuvsgul is the right English transliteration from Mongolian *Хөвсгөл*.

[4] *Selenge* has been often incorrectly written as *Selenga* internationally by the Russian pronunciation. Selenge is the correct English transliteration from Mongolian *Сэлэнгэ*.

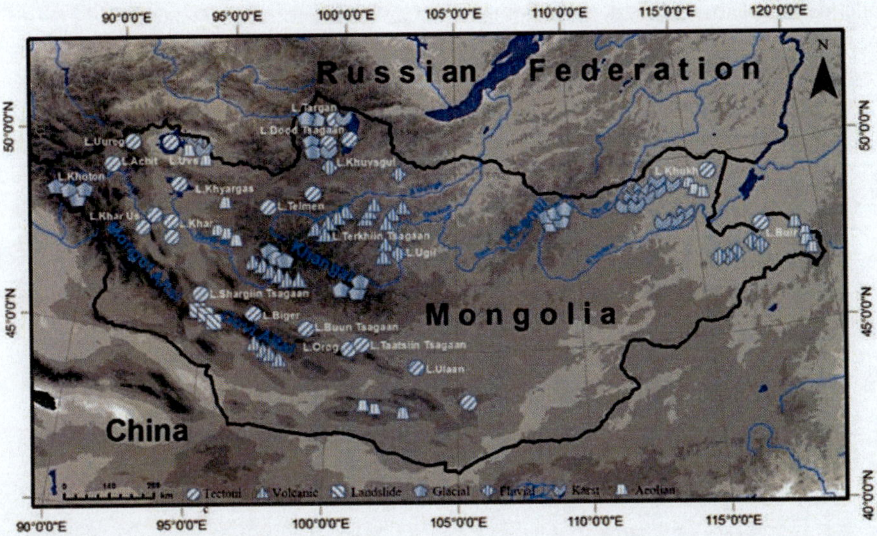

Fig. 4.1 Types of lakes in Mongolia and their spatial distribution

feeding Lake Baikal.[5] However, Lake Telmen formed actually by damming of a paleoterrace tectonically uplifted (Tserensodnom, 2000). In 1948–1949, Sh. Tsegmid noted that lakes in the Khentii[6] Mountain Range in northeastern Mongolia had glacial origins.

The majority of large lakes in Mongolia are fed by rivers, i.e., lake levels directly depend on inflowing river discharges. For instance, inflowing river discharge constitutes from 72% (e.g., for Lake Sangiin Dalai) to 89% (e.g., for Lake Buun Tsagaan in Chap. 10) of water resource of lake systems (Tserensodnom, 1971). Beside the river discharges, rainfall, snow, glaciers, permafrost, and groundwater feed lakes of Mongolia (see Chap. 5). Rainfall constitutes 60–70% of water resource of lakes, and 80–95% of precipitation between April and October and 50–60% of precipitation in July and August contribute to river flooding and lake level rise. Glaciers in high mountains constitute part of surface water resource of Mongolia (Orkhonselenge & Harbor, 2018), i.e., they are one of the major headwaters of lakes. Meltwaters from permafrost and glaciers in spring and summer contribute to a large amount of water resource for the lakes, while groundwater feeds the lakes continuously.

The origins of lake basins in Mongolia are various although many of the present lakes are relict lakes of paleolakes that reached their maximum sizes in the Cretaceous or later (see Chap. 3). There are primarily four modes of lake origin in Mongolia (Berkey & Morris, 1927): (1) lakes occupying fault depressions, (2) lakes

---

[5] *Baigali* meaning *nature* has been misspelled as *Baikal* by the Russian pronunciation globally. Baigali is the right English transliteration from Mongolian *Байгаль*.

[6] *Khentii* has been misspelled as *Khentey* (or *Hentey*) and *Khentei* (or *Hentei*) in many publications. Khentii is the right English transliteration from Mongolian *Хэнтий*.

occupying basins formed by heaping up of loose materials, (3) lakes occupying basins scoured out by eroding agency, and (4) lakes occupying basins with a complex history due partly to erosion, earth movements, and deposition of sediment. However, the most common lakes of Mongolia are shallow playas in deflation hollows, which are dry most of the year (Berkey & Morris, 1927; see Chap. 5). The genesis of lake basins in Mongolia is linked to endogenic and exogenic forces (see Sect. 4.2). The endogenic forces include tectonic and volcanic activities, whereas the exogenic forces include glacial, aeolian, karst, landslide, and fluvial processes. According to Badarch et al. (1965), Tsegmid (1969), and Tserensodnom (1971), lakes of Mongolia can be genetically divided into seven types: tectonic, volcanic, landslide, glacial, karst (and thermokarst), fluvial, and aeolian lakes. In this chapter we aim to describe the genetic description and origins of lakes in Mongolia.

4.2 Genesis of Lakes

4.2.1 Tectonic Lakes

Tectonic lakes are formed by deformation and associated lateral/vertical movements (faulting, tilting, folding, and warping) of Earth's crust (Hutchinson, 1957). There are some large tectonic lakes on Earth occupying rift valleys (e.g., Central African Rift lakes, Lake Baikal), and other tectonic lakes (e.g., Caspian Sea, the Sea of Aral, etc.) occupying basins that have been separated from the sea by tectonic uplift of the seafloor above the sea level (Hutchinson, 1957; Cohen, 2003; Håkanson & Jansson, 1983; Håkanson, 2012). The tectonic action of crustal extension creates a series of parallel grabens and horsts that form elongate basins alternating with mountain ranges (Hutchinson, 1957).

Many large lakes are located in two types of tectonic basins (rift basins and continental sag basins) that have existed for millions of years (Bridge & Demicco, 2008). For example, Lake Baikal in Siberia, the oldest and deepest lake on Earth, occupies a continental rift, whereas Lake Eyre in Australia is located in an intercontinental sag (Bridge & Demicco, 2008). Many lakes in rift basins (e.g., Lake Baikal, Lake Tanganyika, the Dead Sea) are elongated, and their shapes are controlled by faults, whereas other lakes (e.g., Mono Lake, Great Salt Lake) are almost circular and irregularly shaped.

In Mongolia, tectonic lakes occupy depressions originated by tectonic activity (Tsegmid, 1969). Because the Depression of Great Lakes and the Valley of Lakes have tectonic origins (Jigj, 1976), lakes in the depression and valley are generally classified as tectonic lakes (e.g., Lakes Uvs and Buun Tsagaan). It is commonly noted that in Mongolia, lakes that were originally formed due to endogenic forces have been later more influenced by exogenic forces. For instance, small shallow lakes could accumulate over several tens to hundreds of meters thick deposits in paleograben rifts (e.g., relict lakes in the Darkhad basin). According to V.A. Aprodov, tectonic uplifts shape paleograben rifts with hills and terraces; thus lakes are newly formed by blocking or damming effect of these hills and terraces. For example, Lake Telmen

(Fig. 4.1) in central Mongolia and Lakes Tolbo, Tsetseg, and Tonkhil in western Mongolia (see Chap. 12) are included in this type of origin. Lakes formed in tectonic rifts are relatively deep and oval shaped with steep slopes (e.g., Lake Khuvsgul).

Tectonic lakes in Mongolia (Fig. 4.1, Table 4.1) are diverse in their chronologies, and they are as old as the Paleozoic and Mesozoic Eras. However, these lakes have experienced significant evolution due to neotectonics during the late Neogene and early Quaternary (see Chap. 3). There are over ten large tectonic lakes in Mongolia: Lake Buir (see Chap. 7) in eastern Mongolia; Lake Buun Tsagaan (see Chap. 10) in southern Mongolia; Lakes Khyargas, Khar Us, Uureg, Achit (Fig. 4.1; see Chap. 12), and Uvs (see Chap. 13) in western Mongolia; Lake Khuvsgul (see Chap. 16) and Lakes Deed and Dood formed over the site of the paleolake Darkhad of northern Mongolia (see Chap. 15); and Lake Sangiin Dalai in central Mongolia (see Chap. 18).

4.2.2 Volcanic Lakes

Volcanic lakes occupy either large basins (e.g., calderas) formed by volcanism or small local depressions (e.g., craters, maars) filling up with precipitation more rapidly than they empty by evaporation and groundwater discharge (Hutchinson, 1957; Cohen, 2003; Håkanson & Jansson, 1983; Håkanson, 2012). Volcanic lakes are also formed when either rivers or streams are dammed by volcanic lava or lahars (Hutchinson, 1957; Cohen, 2003; Håkanson & Jansson, 1983; Håkanson, 2012).

In Mongolia, some lakes form in (1) volcanic craters and (2) river valleys due to damming of volcanic lava (Tsegmid, 1969; Fig. 4.1, Table 4.1). Volcanic lakes in Mongolia are primarily distributed in the Khangai Mountain Range in central Mongolia (see Chap. 18). Lakes occupying Gichgene, Ul, Shavriin Tsaram, Burkheer, Uran, and Khorgo craters are classified as the volcanic crater lake type.

Table 4.1 Lake distribution in each region of Mongolia according to origin with ranks of elevation and area

#	Regions	Origins	Elevation (m a.s.l.)	Area (km^2)
1	Eastern Mongolia	Fluvial, thermokarst, aeolian, glacial	<1000[1]	0.1–1.0[1], 1.1–5.0[1]
2	Southern Mongolia	Aeolian, volcanic, tectonic	1001–1500[2]	0.1–1.0[2], 1.1–5.0[2], 10.1–20.0[1]
3	Western Mongolia	Tectonic, glacial, aeolian, thermokarst	2001–2500[1], 2501–3000[1], >3001[1]	5.1–10.0[1], 20.1–50.0[1], 50.1–100.0[1], 100.1–500.0[1], >1001.0[1]
4	Northern Mongolia	Tectonic, glacial, fluvial, thermokarst	1501–2000[1]	0.1–1.0[5], 1.1–5.0[5]
5	Central Mongolia	Volcanic, fluvial, tectonic, glacial	1001–1500[1]	0.1–1.0[3], 1.1–5.0[4]

[1–5]Ranks of numbers in the country among the same classes of elevation and area

Some of them are deep and freshwater lakes. It is worth to mention about the 3.7 km circular Tsenkher crater situated in the Govi Altai region, which has been hypothesized to have an impact or volcanic maar origin (Komatsu et al., 2006, 2019; Ormö et al., 2010). The estimated over hundreds of meters thick sediment filled inside the Tsenkher crater would be a great scientific target for investigating the exact origin of the structure and also to extract paleoenvironment achieve spanning a very long time period. There are several tens of lakes formed by volcanic lava damming in the Khanui, Khunui, Orkhon, Chuluut, and Terkh Rivers' valleys of the northern Khangai Mountain Range (Tsegmid, 1969) in central Mongolia. Some of these lakes have outflows and can be deep, and they are freshwater lakes. For instance, Lake Terkhiin Tsagaan (see Chap. 19) in the northern Khangai Mountain Range and Lakes Sharga, Shiree, Bugat, Khaliun, Shanaga, and Shanaa in the southwestern Khangai Mountain Range are included in this lava-dammed volcanic lake type.

4.2.3 Landslide Lakes

Landslide lakes form by blockage of valleys by either mudflows, sediments and rockslides, or screes, and such lakes are common in mountainous regions (Hutchinson, 1957). Landslide lakes are typically short-lived although they may be large and fairly deep (Hutchinson, 1957; Cohen, 2003; Håkanson & Jansson, 1983; Håkanson, 2012).

In Mongolia, lakes forming in intermontane valleys and depressions by landslide or rock failure damming due to slope weathering or tectonic activities are included in this type of origin (Fig. 4.1, Table 4.1). Landslide lakes are widely distributed in the Mongolian and Govi Altai, Khangai, and Khuvsgul Mountain Ranges (Tsegmid, 1969), for example, small lakes (e.g., Lakes Oyut, Nomin, and Binderya) formed by the 50–70-m-high rock damming in Uliastai River and Bituut River valleys due to an earthquake that occurred on December 4, 1957, within the Govi Altai Mountain Range (Tsegmid, 1969). Lakes Oyut, Nomin, and Binderya are small in areas, but they are fed by surface water and groundwater.

4.2.4 Glacial Lakes

Glacial lakes form due to the direct action of glaciers and ice sheets, and they are found in and adjacent areas that were glaciated during the Pleistocene, particularly, since the Last Glacial Maximum (LGM) (Bridge & Demicco, 2008). Most glacial lakes in northern Europe and North America have been either influenced or created by the latest Pleistocene glaciation (Hutchinson, 1957; Håkanson & Jansson, 1983; Roberts, 1998; Cohen, 2003; Håkanson, 2012). For instance, the Laurentide Great Lakes were large and deep, but many others were small waterfilled hollows such as kettle holes (Roberts, 1998).

Different types of glacial lakes are formed in enclosed basins by glacial processes (Hutchinson, 1957; Cohen, 2003; Håkanson & Jansson, 1983; Håkanson, 2012). Generally speaking, glacial lake types are in direct contact with ice, morainic, and outwash sediments, or ones in glacially carved rock basins, depressions, and drift basins (Hutchinson, 1957). Glacial lakes include proglacial lakes, subglacial lakes, finger lakes, epishelf lakes (Hutchinson, 1957; Cohen, 2003; Håkanson & Jansson, 1983; Håkanson, 2012), postglacial lakes, periglacial lakes, and deglacial lakes. Snow avalanche can form lakes in the mountainous hollows, bases of glacial cirques, and intermontane depressions.

In Mongolia, lakes forming in (1) glacial cirques and valleys and (2) river valleys dammed by moraines and outwashes (Tsegmid, 1969) are included in this classification (Fig. 4.1, Table 4.1). In high mountains that experienced glaciations during the Pleistocene, lakes are usually formed by damming of glacial moraines or by plucking and scouring of glaciers. Glacial lakes are mainly distributed today in the Mongolian Altai, Khangai, Khuvsgul, and Khentii Mountain Ranges in western, northern, and central Mongolia (Fig. 1.1). For example, they include Lakes Tolbo, Khoton (see Chap. 14), Khurgan, and Dayan in the Mongolian Altai Mountain Range (see Chap. 12); Lakes Binderya, Khar, and Khukh in the Khangai Mountain Range (see Chap. 18); lakes in the Shishkhid (or Darkhad) basin and Lakes Semsee, Pine, Shar, and Taij in the Khuvsgul Mountain Range (see Chap. 15); and Lakes Khentii, Tsagaan, Khagiin Khar, Khar Zurkhnii Khukh, and Kherkhluur in the Khentii Mountain Range (see Chap. 6). Due to the extensive Pleistocene glaciations in Mongolia (e.g., Tsegmid, 1969; Jigj, 1976; Krivonogov et al., 2005; Gillespie et al., 2008; Komatsu et al., 2009; Rother et al., 2014; Lehmkuhl et al., 2016; Blomdin et al., 2016, 2018; Orkhonselenge, 2016; Orkhonselenge & Harbor, 2018; Orkhonselenge & Uuganzaya, 2018), many glacial lakes in these mountain ranges were formed by the actions of paleo- and modern glaciers and deposition of moraines in glacial cirques and valleys. For example, the enormous paleolake Darkhad in northern Mongolia reached its maximum dimension of approximately 100 km long and 20–40 km wide, and it was formed by ice damming during the Pleistocene.

4.2.5 *Karst (and Thermokarst) Lakes*

In Mongolia, lakes occupying depressions of (1) karst caves and (2) permafrost zones are included in this type of origin (Fig. 4.1; Table 4.1).

4.2.5.1 Karst Lakes

Karst lake is a type of lakes occupying a basin formed by surface dissolution of bedrock (Hutchinson, 1957). The soluble bedrock eroded by precipitation and percolation of water commonly produces cavities, and the cavities frequently collapse to form sinkholes that form karst topography (Hutchinson, 1957). Where groundwater lies near the surface, a sinkhole is filled with water to form a karst lake (Hutchinson, 1957; Håkanson & Jansson, 1983). If such a lake consists of a large area of standing water that occupies an extensive closed depression in limestone, it is also called a karst lake (Neuendorf et al., 2005). Limestone caves often contain pools of standing water, which are known as underground lakes (Hutchinson, 1957).

Karst lakes in Mongolia are small in size and circle shaped with gentle slopes (Tsegmid, 1969). Karst lakes are primarily distributed in valleys and depressions underlain by limestone, dolomites, and gypsum. These lakes are abundant in Mongolia. For example, Lakes Shavart, Chukh, Bus, Mongol Tsagaan, Sum, Yakhi, Bayan Erkhet, and Ukhert in Ulz River and Gal River valleys and Lakes Ikh Tsaidam, Baga Tsaidam, and Tukhum in Kherlen River valley in eastern Mongolia (see Chap. 6), Lakes Eleet and Asgat in Tes River valley in western Mongolia (see Chap. 12), and lakes in Sharga River and Ar Sair River valleys in the Shishkhid (or Darkhad) basin in northern Mongolia (see Chap. 15) are karst lakes.

4.2.5.2 Thermokarst Lakes

Some lakes are inevitable consequences of active layer thawing of permafrost, and such lakes may be considered as a class of thermokarst feature (Anderson & Anderson, 2010). Melting of ice-rich permafrost leads to subsidence of the surface as the excess pore water is expelled because any process promoting thawing of the active layer of permafrost forms a depression, and water accumulates into a lake, and an impermeable bowl forms by the surrounding permafrost. Once the lake becomes deep enough over about 2.0 m, it does not freeze in winter (Anderson & Anderson, 2010).

In Mongolia, the genesis of thermokarst lakes is associated with the distribution of permafrost, i.e., the thermokarst lakes are formed in the active layer of permafrost in high mountain regions. There are 14 lakes formed in thermokarst valleys (Table 4.1). The recent work by Saruulzaya (2017) found thermokarst lakes in seven different places of the Darkhad basin, Mungut River valley, Chuluut River valley, Lake Khongor Ulun basin, Nalaikh basin, Galuut basin, and Erdene in the permanent permafrost region of Mongolia. The work showed spatial and temporal dynamic changes of these thermokarst lakes in terms of number and area between 1962 and 2007 due to the present climate change. For instance, the thermokarst lakes in the Darkhad basin in northern Mongolia increased in number from 296 to 371 and in area from 33.5 to 36.5 km^2, while the thermokarst lakes in Erdene of the Govi region in southern Mongolia decreased in number from 8 to 3 and in area from 0.06 to 0.03 km^2 during the 45 years. This result shows that these two examples of places lying at the northernmost and southernmost permafrost regions directly responded to the rising temperature resulting in ice melting and permafrost thawing and intensive evaporation, respectively.

4.2.6 Fluvial Lakes

Fluvial (or riverine) lakes are formed by running water (Hutchinson, 1957). These lakes include plunge pool lakes, fluviatile dammed lakes, and meander lakes (Hutchinson, 1957; Cohen, 2003; Håkanson & Jansson, 1983; Håkanson, 2012). The fluviatile dammed lakes form where sediments transported from a tributary river block the main river (Schoenherr, 2017). Fluvial lakes lying on the floodplain normally in the vicinity of meandering rivers are called oxbow lakes, and they occupy abandoned river channels (Bridge & Demicco, 2008). They can form in river valleys as a result of meandering (Hutchinson, 1957). The slow-moving river forms a sinuous shape as the outer side of channels is eroded away more rapidly than the inner side. Eventually a horseshoe channel is formed and the river cuts through the narrow neck. This new passage then forms the main channel for the river, and the ends of the channel become tilted up, thus forming a bow-shaped lake (Hutchinson, 1957; Cohen, 2003; Håkanson & Jansson, 1983; Håkanson, 2012). The neck cutoff occurs in single-channel rivers that are sinuous that cut banks of successive bends can meet, leading to cutoff, abandonment of meander loop, and formation of an oxbow lake (Bridge & Demicco, 2008). The oxbow lakes, the most common type of fluvial lake, are crescent-shaped or distinctively curved lakes.

In Mongolia, lakes forming in river valleys are included in this type of origin (Fig. 4.1, Table 4.1). Fluvial lakes primarily form when rivers change their channels, when rivers shrink or disappear, and when rivers are flooded and drained over their floodplains forming oxbow lakes. Fluvial lakes forming due to change in river channels are primarily distributed in large river valleys. For example, lakes of this type are abundant in valleys of Kherlen and Ulz Rivers in eastern Mongolia (see Chap. 6). Moreover, Lake Ugii (see Chap. 20) in central Mongolia and Lakes Erkhel and Khargal (see Chap. 17) in northern Mongolia are oxbow lakes. Lakes forming when rivers disappear are abundant in the Dornod plain and the depression between Matad and Tamsag Bulag basins in eastern Mongolia (Badarch et al., 1965; Tsegmid, 1969; Tserensodnom, 1971). For instance, Lakes Lag, Shar Burd, Bayan, Tsagaan, and Shavart are located in paleoriver valleys between Matad and Tamsag Bulag basins of the Dornod plain (see Chap. 6).

4.2.7 Aeolian Lakes

Aeolian lakes are produced by wind action. They are mainly found in arid environments although some aeolian lakes are relict landforms indicative of arid paleoclimates (Hutchinson, 1957). Aeolian lakes exist in lake basins dammed by wind-blown sands, and they are interdunal lakes that lie between well-oriented sand dunes. In arid to semiarid lowlands, wind-induced deflation leads to morphological depressions (Einsele, 1992), and some aeolian lakes form in those deflation basins originated in previously arid paleoenvironments (Hutchinson, 1957).

In Mongolia, lakes forming by wind erosion in playa depressions are included in this type of origin (Fig. 4.1, Table 4.1). Lakes formed by blockage of river valleys by wind-blown sediments are also included in this type of origin. These aeolian lakes are widely distributed in the Govi and other desertic regions with arid climates in eastern, southern, and western Mongolia. For instance, Lakes Khunt, Khunguin Khar, Mukhar Khungui, Ulaagchnii Khar, Bayan, and Thookhor in Zavkhan, Khungui, Tes, and Nariin Rivers' valleys in western Mongolia (see Chap. 12) are included in the aeolian lakes formed in depressions by depositions of wind-blown sands. Moreover, Lakes Buunii Tsagaan, Tsatsiin Burd, Ulziit, and Dari Gangiin Tsagaan in Khalkh, Numrug, and Mukhar Rivers' valleys in eastern Mongolia (see Chap. 6) are included in the aeolian lakes formed in playa depressions. They are freshwater lakes because of inflows and outflows passing through the sands. Recent studies (Orkhonselenge & Bulgan, 2021; Orkhonselenge et al., in press) show that some lakes in eastern and southern Mongolia are shifting to playa lakes due to the rising temperature and strengthening evaporation.

4.3 Summary

The origin and genesis of lakes in Mongolia are strongly dependent on the tectonic setting, geological structure, landscape type, and landform feature. In total, seven origin types of Mongolian lakes have been identified to date. Tectonic, volcanic, glacial, and landslide lakes are common in high mountains within the Mongolian Altai, Govi Altai, Khangai, Khentii, and Khuvsgul Mountain Ranges in western, northern, and central Mongolia. Aeolian, fluvial, and karst (and thermokarst) lakes are abundant in lowlands, valleys, and depressions within the plains and the Govi Desert regions of eastern and southern Mongolia.

References

Anderson, R. S., & Anderson, S. P. (2010). *Geomorphology: The mechanics and chemistry of landscapes*. Cambridge University Press, 637 p. ISBN: 978-0-521-51978-6.

Badarch, N., Tsegmid, S., & Tserensodnom, J. (1965). *Landscapes and natural zones of eastern Mongolia*. Press of Mongolian Academy of Sciences, 117 p. [In Mongolian].

Berkey, P. C., & Morris, K. F. (1927). *Geology of Mongolia. A reconnaissance report based on the investigations of the years 1922–1923. Natural history of Central Asia* (Vol. 2). The American Museum of Natural History, 475 p.

Blomdin, R., Heyman, J., Stroeven, A. P., Hättestrand, C., Harbor, J. M., Gribenski, N., Jansson, K. N., Petrakov, D. A., Ivanov, Orkhonselenge, A., Rudoy, A. N., & Walther, M. (2016). Glacial geomorphology of the Altai and western Sayan mountains, Central Asia. *Journal of Maps, 12*(1), 123–136.

Blomdin, R., Stroeven, A. P., Harbor, J. M., Gribenski, N., Caffee, M. W., Heyman, J., Rogozhina, I., Ivanov, M. N., Petrakov, D. A., Walther, M., Rudoy, A. N., Zhang, W., Orkhonselenge, A.,

Hättestrand, C., Lifton, N. A., & Jansson, K. N. (2018). Timing and dynamics of glaciation in the Ikh Turgen mountains, Altai Region, High Asia. *Quaternary Geochronology, 47*, 54–71.

Bridge, J., & Demicco, R. (2008). *Earth surface processes, landforms and sediment deposits*. Cambridge University Press, 815 p. ISBN: 978-0-521-85780-2.

Cohen, A. S. (2003). *Paleolimnology: The history and evolution of lake systems*. Oxford University Press.

Einsele, G. (1992). *Sedimentary basins: Evolution, Facies, and Sedimentary Budget*. Springer-Verlag. ISBN: 3-540-54449-6.

Frohn, R. C., Hinkel, K. M., & Eisner, W. R. (2005). Satellite remote sensing classification of thaw lakes and drained thaw lake basins on the North Slope of Alaska. *Remote Sensing of Environment, 97*(1), 116–126.

Gillespie, A. R., Burke, R. M., Komatsu, G., & Bayasgalan, A. (2008). Late Pleistocene glaciers in Darhad Basin, northern Mongolia. *Quaternary Research, 69*, 169–187.

Håkanson, L. (2012). Lakes on earth, different types. In *Encyclopedia of lakes and reservoirs* (pp. 471–472). Springer.

Håkanson, L., & Jansson, J. (1983). *Principles of lake sedimentology*. Springer-Verlag, 320 p.

Hutchinson, G. E. (1957). *A treatise on limnology, v. 1. Geography, physics and chemistry*. Wiley, 1015 p.

Jigj, S. (1976). *A brief description of paleoglaciations and paleoglaciers*. 51 p. [In Mongolian].

Komatsu, G., Arzhannikov, S. G., Gillespie, A. R., Burke, R. M., Miyamoto, H., & Baker, V. R. (2009). Quaternary paleolake formation and cataclysmic flooding along the upper Yenisei River. *Geomorphology, 104*, 143–164.

Komatsu, G., Olsen, J. W., Ormö, J., Di Achille, G., Kring, D. A., & Matsui, T. (2006). The Tsenkher structure in the Gobi-Altai, Mongolia: Geomorphological hints of an impact origin. *Geomorphology, 74*(1–4), 164–180.

Komatsu, G., Ormö, J., Bayaraa, T., Arai, T., Nagao, K., Hidaka, Y., Shirai, N., Ebihara, M., Alwmark, C., Gereltsetseg, L., Tserendug, S., Goto, K., Matsui, T., & Demberel, S. (2019). Further evidence for an impact origin of the Tsenkher structure in the Gobi-Altai, Mongolia: Geology of a 3.7 km crater with a well-preserved ejecta blanket. *Geological Magazine, 156*(1), 1–24.

Krivonogov, S. K., Sheinkman, V. S., & Mistruykov, A. A. (2005). Stages in the development of the Darhad dammed lake (Northern Mongolia) during the Late Pleistocene and Holocene. *Quaternary International, 136*(1), 83–94.

Kuusisto, E., & Hyvärinen, V. (2000). Hydrology of lakes. In P. Heinonen (Ed.), *Hydrological and limnological aspects of lake monitoring* (pp. 4–5). Wiley. ISBN: 978-0-470-51113-8.

Lehmkuhl, F., Klinge, M., Rother, H., & Daniela Hulle, D. (2016). Distribution and timing of Holocene and Pleistocene glacier fluctuations in western Mongolia. *Annals of Glaciology, 57*(71), 1–10.

Meybeck, M. (1995). Global distribution of lakes. In A. Lerman, D. Imboden, & J. Gat (Eds.), *Physics and chemistry of lakes* (2nd ed., pp. 1–35). Springer-Verlag.

Neuendorf, K. K. E., Mehl, J. P., Jr., & Jackson, J. A. (2005). *Glossary of geology, 5th revised and enlarged ed*. Springer. Approx. 900 p. ISBN: 3-540-27951-2.

Orkhonselenge, A. (2016). Glacial Geomorphology of Mt. Munkh Saridag in the Khuvsgul Mountain Range, Northern Mongolia. *Géomorphologie: Relief, Processes, Environnement, 22*(4), 387–396.

Orkhonselenge, A., & Bulgan, O. (2021). Geochemical studies and lacustrine geomorphology of Lake Yakhi basin in eastern Mongolia. *Géomorphologie: Relief, Processes, Environnement, 27*(3), 231–242. https://doi.org/10.4000/geomorphologie.15873

Orkhonselenge, A., & Harbor, J. M. (2018). Impacts of modern glacier changes on surface water resources in western and northern Mongolia. *Journal of Water Resource and Protection, 10*(6), 559–576.

Orkhonselenge, A., & Uuganzaya, M. (2018). Glacial geomorphology of Mt. Asralt Khairkhan and Mt. Baga Khentii Saridag in Khentii Mountain Range, Northeastern Mongolia. *International Journal of Geosciences, 9*(6), 308–319.

Orkhonselenge, A., Uuganzaya, M., & Davaagatan, T. (in press). Late Holocene sedimentation dynamics in the lake Ulaan basin, southern Mongolia. *Environmental Earth Science*.

Ormö, J., Gomez-Ortiz, D., Komatsu, G., Bayaraa, T., & Tserendug, S. (2010). Further evidence for an impact origin of the Tsenkher structure in the Gobi-Altai, Mongolia: Geology of a 3.7 km crater with a well-preserved ejecta blanket. *Meteoritics and Planetary Science, 45*(3), 373–382.

Roberts, N. (1998). *The Holocene: An environmental history* (2nd ed.). Blackwell Publishers, 316 p. ISBN: 0-631-18637-9.

Rother, H., Lehmkuhl, F., Fink, D., & Nottebaum, V. (2014). Surface exposure dating reveals MIS-3 glacial maximum in the Khangai Mountains of Mongolia. *Quaternary Research, 82*, 297–308.

Saruulzaya, A. (2017). *Thermokarst lake dynamics in the continuous and isolated permafrost zones, Mongolia*. Ph.D. Thesis, Hokkaido University, Sapporo, Japan. 76 p.

Schoenherr, A. A. (2017). *A natural history of California: Second Edition*. University of California Press, p. 485. ISBN: 978-0-520-96455-6.

Tsegmid, S. (1969). *Physical geography of Mongolia*. Institute of Geography & Permafrost. Mongolian Academy of Sciences. State Press, 405 pp. [In Mongolian].

Tserensodnom, J. (1971). *Lakes of Mongolia*. State Publishing, 202 p. [In Mongolian].

Tserensodnom, J. (2000). *A catalog of lakes in Mongolia*. Shuvuun Saaral Publishing, 141 p. [In Mongolian].

Chapter 5
Classification of Lakes in Mongolia

Abstract Lakes of Mongolia are classified into six types depending on altitude, area, stability, depth, salinity, and presence of outlet. 34.1% of the total lakes in Mongolia are mountainous lakes, while 65.9% are prairie lakes (see Chap. 1). Mountainous, high-altitude, and large lakes are abundantly distributed in western and northern Mongolia, whereas prairie, low-altitude, and small lakes occur in eastern, southern, and central Mongolia. 74.6% of the lakes in Mongolia are located at the elevations below 1500 m a.s.l., and 99.5% of the lakes in Mongolia occupy an area <100.0 km^2, and they are dominantly distributed in eastern, southern, and central Mongolia. 56.3% of the lakes are ephemeral lakes, and these lakes occur in eastern, southern, and central Mongolia. 96.9% of the lakes in Mongolia are shallow lakes (<10.0 m depth) dominantly distributed in eastern and southern Mongolia, whereas the deep lakes are located in western and northern Mongolia. 52.8% of the lakes are saline lakes abundantly distributed in eastern, southern, and central Mongolia, whereas freshwater lakes are widely distributed in western, southern, and northern Mongolia. Most lakes in Mongolia (70.3%) are closed lakes, and they commonly occur in eastern and southern Mongolia.

Keywords Lakes · Types · Classes · Groups · Category · Mongolia

5.1 Introduction

In lake classification, origins of formation and types of process should be accounted (Einsele, 1992). Lakes were firstly classified by F.A. Forel in 1901 based on thermal stratification of lake water meaning that how the lake supports floral and faunal life with oxygen. In 1912, Walther classified desert lakes based on their water supply into three groups: (1) rain-fed lakes, (2) spring lakes, and (3) terminal lakes (Berkey & Morris, 1927). In 1957, Hutchinson classified lakes comprehensively based on their origins into 11 major types and 76 subtypes and presented their morphometric characteristics and distributions (Cohen, 2003; Håkanson & Jansson, 1983; Håkanson, 2012). Lakes are commonly classified in terms of (1) salinity (fresh,

A. Orkhonselenge et al., *Lakes of Mongolia*, Syntheses in Limnogeology,
https://doi.org/10.1007/978-3-030-99120-3_5

brackish, and saline), (2) whether to have outlet (hydrologically open and closed), and (3) origin (glacial, tectonic, fluvial, crater, coastal, solution, wind-deflation, landslide-dammed, and volcanic-material-dammed) (Bridge & Demicco, 2008).

Large and deep lakes, especially those forming by crustal subsidence and low-terrigenous influx, are longer-lived for considerable geological time scales. The pronounced, repeated vertical facies in sediment within this type of lakes illustrate how sensitive lacustrine sediments are to climate changes (Einsele, 1992). In small lakes infilling by sediment progressively reduces water depth, while encroachment of plants from the shore reduces lake area (Roberts, 1998). The regionality of lake area change in Mongolia agrees with the evidence from lake level data indicating that the distinct lake area change was primarily related to the precipitation change as observed in many regions of Mongolia, and the regionality also suggests the importance of regional climate regime and hydromorphological characteristics affecting the relationships between lake area and precipitation change (Kang et al., 2015).

In Mongolia, water resource, water level and hydrochemistry of lakes are highly dependent on river discharges controlled by mountain orientation, elevation, and latitude. For instance, rivers draining from the mountains above 1000 m a.s.l. in the Mongolian Altai, Khangai,[1] Khentii,[2] and Khuvsgul[3] Mountain Ranges in western, northern, and central Mongolia primarily feed lakes. However, rivers draining from the mountains at above 2000–2500 m a.s.l. in the Govi Altai Mountain Range cannot reach nor feed lakes in the Valley of Lakes because of the geographical location of the range positioned at low latitudes. Large lakes form in high mountainous regions in western and northern Mongolia where dominantly rivers feed the lakes. The mountains above 1000 m a.s.l. receive high precipitations of 250–500 mm/year (see Chaps. 12 and 15–18), and evaporation from the lakes is lower than those in lowland regions below 1000 m a.s.l. The lowland regions receive low precipitations that are less than 150 mm/year (see Chaps. 9–11 and 13). Lakes in northern Mongolia are fed by 52.1% of the total discharge in Mongolia, belonging to the North Arctic Ocean drainage basin, whereas lakes in southern, western, and central Mongolia are fed by 32.0% of the total discharge in Mongolia, and they are included in the Central Asian internal drainage basin. Lakes in eastern Mongolia (see Chap. 6) are fed by 15.9% of the total discharge in Mongolia, and they belong to the Pacific Ocean drainage basin.

Lakes of Mongolia have been classified according to landscape type, area, and altitude in previous studies (Tsegmid, 1969; Tserensodnom, 1971, 2000). For instance, Mongolian lakes are primarily divided into two groups: mountainous and prairie in terms of landscape type. Among the total lakes in Mongolia, mountainous lakes cover their 34.1%, while prairie lakes are 65.9% (Tserensodnom, 1971; see

---

[1] *Khangai* has been misspelled as *Khangay* (or *Hangay*) and *Hangai* in many publications. Khangai is the right English transliteration from Mongolian *Хангай*.

[2] *Khentii* has been misspelled as *Khentey* (or *Hentey*) and *Khentei* (or *Hentei*) in many publications. Khentii is the right English transliteration from Mongolian *Хэнтий*.

[3] *Khuvsgul* has been misspelled as *Hovsgol* and *Khubsugul* in publications. Khuvsgul is the right English transliteration from Mongolian *Хөвсгөл*.

Chap. 1). However, lakes of Mongolia have not been classified based on thermal stratification (i.e., hypolimnion, metalimnion, epilimnion, meromictic, and holomictic lakes) nor nutrition level of the lakes (i.e., oligotrophic, mesotrophic, eutrophic, and hypertrophic lakes). Moreover, to date there is not any classification of lakes in Mongolia in terms of stability, depth, salinity, and outlet. This is due to a lack of detailed studies of the lakes, with the exception of general descriptions about some lakes (e.g., Lake Uvs as the largest and a saline lake (see Chap. 13); Lake Khuvsgul as the deepest and a freshwater lake (see Chap. 16); etc.). Lakes of Mongolia are diverse in terms of their landscape feature (Tserensodnom, 1971), origin (see Chap. 4), hydrothermal balance and mineralization (see Sect. 5.6), and water interconnections (see Sect. 5.7).

In this chapter, in addition to the previous classification of lakes based on altitude (see Sect. 5.2) and area (see Sect. 5.3), a new classification of lakes in Mongolia is considered in contexts of stability (see Sect. 5.4), depth (see Sect. 5.5), salinity (see Sect. 5.6), and presence of outlet (see Sect. 5.7). Here, the previous classification of lakes is renewed statistically, and the groups of lakes in each theme are classified according to each region of Mongolia (Fig. 1.1).

5.2 Altitude

Mongolia is a highly elevated country with its elevations distributing between 552 m a.s.l. and 4374 m a.s.l. (Academy of Sciences of Mongolia and Academy of Sciences of USSR, 1990) or 4653 m a.s.l. (Tsegmid, 1969; Jigj, 1975) with an average of 1850 m a.s.l. Some lakes are situated above 3000 m a.s.l.; however many lakes are located below 1000 m a.s.l. In Mongolia, 74.6% of the total lakes are lying below 1500 m a.s.l., and they are distributed in eastern, southern, and central Mongolia, while 25.4% of the total lakes are lying above 1500 m a.s.l. and they are distributed in western and northern Mongolia (Table 5.1). Here, we classify a group of lakes based on altitude according to each region of Mongolia (Fig. 1.1).

5.2.1 High-Altitude Lakes

Lakes lying at 1500–2000 m a.s.l. constitute 8.8% of the total lakes in Mongolia, and they are dominantly distributed in northern, southern, and western Mongolia (Table 5.1). Lakes elevated in the range of 2000–2500 m a.s.l. constitute also 8.8% of lakes in Mongolia, and they mainly occur in western and northern Mongolia (Table 5.1). Lakes at the range of 2500–3000 m a.s.l. constitute 6.7% of lakes in Mongolia, and they are primarily situated in western Mongolia (Table 5.1). There are 33 lakes elevated above 3000 m a.s.l., constituting 1.1% of lakes in the country, and they occur only in western Mongolia (Table 5.1).

Table 5.1 Mongolian lakes classified according to altitude for each region

#	Altitude (m a.s.l.)	Eastern Mongolia		Southern Mongolia		Western Mongolia		Northern Mongolia		Central Mongolia		Total number of lakes	Percentage
		Number	Area (km^2)	Number	Area (km^2)	Number	Area (km^2)	Number	Area (km^2)	Number	Area (km^2)		
1	<1000	648	1308.5	159	149.2	15	3390.7	30	27.9	4	4.1	856	28.0
2	1000–1500	237	140.4	481	959.0	116	5032.8	29	41.7	563	2054.0	1426	46.6
	<1500	885	–	640	–	131	–	59	–	567	–	2282	
3	1500–2000	–	–	79	69.8	79	529.8	84	3158.3	29	29.2	271	8.8
4	2000–2500	–	–	27	43.1	135	436.2	89	68.4	19	85.8	270	8.8
5	2500–3000	–	–	10	24.9	189	152.7	4	1.1	2	7.7	205	6.7
6	>3000	–	–	–	–	33	7.0	–	–	–	–	33	1.1
	>1500	–	–	153	–	727	–	97	–	23	–	1221	
	Total number	885	–	756	–	567	–	236	–	617	–	3061	100
	Percentage	28.9	–	24.7	–	18.5	–	7.7	–	20.2	–	100	
	Total area	–	1448.9	–	1246.0	–	9549.2	–	3297.4	–	461.8	16,003.3	
	Percentage	–	9.1	–	7.8	–	59.6	–	20.6	–	2.9	100	

5.2.2 *Low-Altitude Lakes*

There are 856 lakes situated below 1000 m a.s.l., and they constitute 28.0% of the total lakes in Mongolia (Table 5.1). 75.7% of these lakes (or 648 lakes) are located in eastern Mongolia (Table 5.1). There are 1426 lakes lying at 1000–1500 m a.s.l. and they constitute 46.6% of lakes in the country (Table 5.1). Many of these lakes dominantly occur in central and southern Mongolia. Overall, there are 2282 lakes situated below 1500 m a.s.l., and most of them occur in the eastern, southern, and central Mongolia (Table 5.1).

5.3 Area

Lakes in the world range in size from less than a few hundred square meters up to very large ones such as Lake Superior (82,410 km^2) in North America and Lake Baikal[4] (31,500 km^2) in Russia (Bridge & Demicco, 2008). The patterns of the lake area changes have not been similar across the country (Kang et al., 2015), i.e., the temporal variation of lake area has been significant in each region of Mongolia. The lake area in the Govi[5] region of southern Mongolia and in the Dornod plain in eastern Mongolia is highly sensitive to air temperature and precipitation (Orkhonselenge et al., 2018a; Orkhonselenge & Bulgan, 2021). Because Mongolia has experienced a semiarid to arid climate during the recent period (see Chap. 21), lakes are not exceptionally large. Nevertheless, there are many large (Fig. 1.1) and small lakes.

There are approximately 3061 lakes each covering an area over 0.1 km^2, and they amount to a total area of 15,640 km^2 (Tsegmid, 1969; Table 5.2) or 16,003.3 km^2

Table 5.2 Mongolian lakes classified according to area. Modified from Tsegmid (1969)

#	Lakes	Area (km^2)	Total area of lakes (km^2)	Total area of lakes (%)	Total number of lakes (%)
1	Very small	0.1–1.0	1021.0	6.5	85.0
2	Smaller	1.1–10.0	1182.0	7.6	13.1
3	Small	10.1–50.0	583.0	3.7	1.0
4	Medium	50.1–100.0	723.0	4.6	0.3
5	Large	100.1–1000.0	3041.0	19.5	0.4
6	Very large	>1001.0	9090.0	58.1	0.2
	Total		15,640.0	100.0	100.0

[4] *Baigali* meaning *nature* has been misspelled as *Baikal* based on Russian pronunciation. The Baigali is the right English transliteration from Mongolian *Байгаль*.

[5] *Govi* has been often spelled as *Gobi* in the international literature. Govi is the correct English transliteration from Mongolian *Говь*.

(Tserensodnom, 2000; Tables 5.1 and 5.3). The total area of the lakes covers over 1% of the territory of Mongolia. 85.0% of lakes in Mongolia are very small lakes occupying an area of 0.1–1.0 km^2, and 13.1% of lakes in Mongolia are smaller lakes occupying an area of 1.1–10.0 km^2. 77.6% of the total area of lakes is occupied by large lakes in the range of 100.1–1000.0 km^2 and very large lakes over 1001.0 km^2 (Table 5.2). The lakes in the categories of small (10.1–50.0 km^2) and medium (50.1–100.0 km^2) constitute only low percentages in total number and total area (Table 5.2).

5.3.1 *Large-Sized Lakes*

Only 0.5% of the total lakes in Mongolia cover an area over 100.0 km^2, and the large-sized lakes are dominantly distributed in western Mongolia (Table 5.3). For instance, Lakes Achit, Uureg, Buun Tsagaan, and Telmen each occupy an area over 100.0 km^2. Two lakes, Lakes Buir and Khar, occupying an area over 500.0 km^2 are located in eastern and western Mongolia (Table 5.3; see Chaps. 6 and 12). In addition, four lakes occupying an area over 1000.0 km^2 are distributed in western and northern Mongolia (Table 5.3; see Chaps. 12 and 15). The Lakes Khar Us, Khyargas, Uvs, and Khuvsgul with an area over 1000.0 km^2 are considered as very large lakes not only in Mongolia but also in Central Asia (Tsegmid, 1969).

5.3.2 *Small-Sized Lakes*

In Mongolia, 98.1% (Table 5.2) or 97.9% (Table 5.3) of the total lakes occupy an area less than 10.0 km^2, and these 2997 lakes are primarily distributed in eastern, southern, and central Mongolia (Table 5.3). 99.5% of lakes cover an area less than 100.0 km^2 each, and these 3047 lakes are dominantly distributed in eastern, southern, and central Mongolia (Table 5.3). Small-sized lakes have been rapidly changing in number in the Govi, plain, thermokarst, and permafrost regions due to the abrupt temperature rise during the past five decades (see Chap. 4). For instance, thermokarst lakes in the Khongor Ulun basin of the eastern Mongolian Altai permanent permafrost region increased in number from 10 to 13 and in area from 1.1 to 1.24 km^2, while the thermokarst lakes in the Galuut basin of the southern Khangai intermittent permafrost region decreased in number from 108 to 64 and in area from 4.09 km^2 to 3.67 km^2 between 1962 and 2007 (Saruulzaya, 2017).

Table 5.3 Mongolian lakes classified according to area for each region

#	Area (km^2)	Eastern Mongolia		Southern Mongolia		Western Mongolia		Northern Mongolia		Central Mongolia		Total number of lakes	Total area of lakes
		Number	Area (km^2)	Number	Area (km^2)	Number	Area (km^2)	Number	Area (km^2)	Number	Area (km^2)		
1	0.1–1.0	707	248.6	658	228.9	452	153.0	185	75.0	544	187.6	2546	893.1
2	1.1–5.0	155	266.9	68	156.5	67	146.9	36	72.6	63	128.0	389	770.9
3	5.1–10.0	14	100.8	16	116.9	20	133.5	5	37.6	7	48.6	62	437.4
	<10.0	876	–	742	–	539	–	226	–	614	–	2997	
4	10.1–20.0	5	75.8	11	318.6	7	98.6	5	76.0	1	10.9	29	579.9
5	20.1–50.0	2	44.5	1	33.1	4	111.4	2	472	1	25.9	10	262.1
6	50.1–100.0	1	97.3	–	–	8	544.8	1	64.0	1	61.0	11	767.1
	<100.0	884	–	754	–	558	–	234	–	617	–	3047	
7	100.1–500.0	–	–	2	392.0	5	1177.0	1	165.0	–	–	8	1734
8	500.1–1000.0	1	615.0	–	–	1	575.0	–	–	–	–	2	1190
9	>1001.0	–	–	–	–	3	6609.0	1	2760.0	–	–	4	9369
	>100.0	1	–	2	–	8	–	2	–	–	–	14	
	Total number	885	–	756	–	567	–	236	–	617	–	3061	
	Percentage	28.9	–	24.7	–	18.5	–	7.7	–	20.2	–	100	
	Total area	–	1448.9	–	1246.0	–	9549.2	–	3297.4	–	461.8	–	16,003.3
	Percentage	–	9.1	–	7.8	–	59.6	–	20.6	–	2.9	–	100

5.4 Stability

Lakes of Mongolia are classified based on stability as perennial and ephemeral lakes according to each region of Mongolia (Table 5.4). In this classification, 1142 lakes occupying an area greater than 0.5 km^2 among the total lakes of Mongolia are included. Table 5.4 shows relatively deep and large lakes listed as perennial, whereas a general classification for shrinking, shallow, and small lakes is listed as ephemeral.

5.4.1 *Perennial Lakes*

A perennial lake has water in its basin throughout the year, and it is not subject to extreme fluctuations of its water level (e.g., Neuendorf et al., 2005; Poehls & Smith, 2009). In Mongolia, 43.7% of the total lakes are perennial lakes, and they dominantly occur in the mountainous regions of western and northern Mongolia even though such lakes exist also in eastern and southern Mongolia (Table 5.4).

5.4.2 *Ephemeral Lakes*

A lake is known as intermittent if it fills with water and dries up or disappears seasonally (Poehls & Smith, 2009). Large ephemeral lakes may be as deep as 5 m during floods (Bridge & Demicco, 2008). Since 1995 the drying-up phenomenon of lakes in the Govi region has been reported (Ministry of Environment and Green Development, 2013). The intermittent or ephemeral lakes have been increasing recently in the Govi region of southern Mongolia due to the temperature rise since 1995 and the precipitation fluctuation since 1975 (Orkhonselenge et al., 2018a). This phenomenon is observed for lakes also in the plains of eastern Mongolia (Orkhonselenge et al., 2019).

Dry or playa lakes can be included in this type of lake (Komatsu et al., 2007). Playa lake is a typically shallow, intermittent lake that occupies a playa basin in wet seasons or wet years but subsequently drying up in an arid or semiarid condition

Table 5.4 Mongolian lakes classified according to stability for each region

#	Regions	Perennial lakes		Ephemeral lakes	
		Number	Percentage	Number	Percentage
1	Eastern Mongolia	163	14.3	183	16.0
2	Southern Mongolia	91	8.0	177	15.5
3	Western Mongolia	116	10.2	97	8.5
4	Northern Mongolia	51	4.5	70	6.1
5	Central Mongolia	77	6.7	117	10.2
	Total	498	43.7	644	56.3

(e.g., Last & Smol, 2001; Neuendorf et al., 2005). In eastern and southern Mongolia, playa lakes are multiplying in recent years due to the climate change during the last 50 years (Orkhonselenge & Bulgan, 2021; Orkhonselenge et al., in press). Table 5.4 shows that 56.3% of the total lakes are ephemeral lakes. The ephemeral lakes are dominantly distributed in eastern, southern, and central Mongolia.

5.5 Depth

Lake depths are variable from a few tens to a few hundreds and even over 1000 meters. For instance, the deep Lake Baikal is 1640 m in depth, but Lake Chad is only ~8 m deep (Bridge & Demicco, 2008). Lake depths are often greater during glacial times but lower during interglacial times (Bridge & Demicco, 2008). For example, the Great Lakes region of the North America contained significantly larger lakes (e.g., glacial Lakes Agassiz and Missoula) during the Pleistocene glacial periods (Bridge & Demicco, 2008). The 8.2 ka climate event resulted the catastrophic draining of glacial Lakes Agassiz and Ojibway through the Hudson Strait into the Labrador Sea (Anderson & Anderson, 2010). Moreover, a paleolake Darkhad was a deep lake with a more than a hundred meters depth in the Darkhad basin of northern Mongolia during the Pleistocene glacial periods (see Chap. 15).

Changes in lake depth are highly controlled by input and output (see Sect. 5.7). Most lakes in Mongolia have been dropping their levels since the 2000s (Ministry of Environment and Green Development, 2013). This is particularly the case for the lakes in the Depression of Great Lakes and the Valley of Lakes in the Govi region because the ratio of evaporation to the inflow has been increasing (Ministry of Environment and Green Development, 2013).

Here, lakes of Mongolia are classified, based on their depths as deep and shallow lakes according to each region of Mongolia (Table 5.5). In this classification, 1142 lakes occupying an area greater than 0.5 km^2 among the total lakes of Mongolia are included. In Table 5.5, there is a classification for lakes with a depth over 10.0 m listed as deep, but lakes with a depth less than 10.0 m are listed as shallow.

Table 5.5 Mongolian lakes classified according to depth for each region

		Deep lakes		Shallow lakes	
#	Regions	Number	Percentage	Number	Percentage
1	Eastern Mongolia	1	0.1	346	30.3
2	Southern Mongolia	6	0.5	261	22.9
3	Western Mongolia	15	1.3	198	17.3
4	Northern Mongolia	10	0.9	109	9.5
5	Central Mongolia	3	0.3	193	16.9
	Total	35	3.1	1107	96.9

5.5.1 *Deep Lakes*

Lakes formed in tectonic grabens or rifts (see Chap. 4) are deep. For instance, Lake Khuvsgul (see Chap. 16) residing in the Lake Baikal tectonic rift basin is the deepest in Mongolia (262.4 m). Glacial lakes can be hundreds of meters deep, and the Great Lakes and Finger Lakes in North America are the examples (Bridge & Demicco, 2008). The paleolake Darkhad is also an example. Table 5.5 shows that 3.1% of the total lakes are deep lakes. The deep lakes are dominantly distributed in western and northern Mongolia (Table 5.5).

5.5.2 *Shallow Lakes*

Shallow lakes vary in their depths seasonally depending on precipitation. Table 5.5 shows that 96.9% of the total lakes are shallow lakes. The shallow lakes are dominantly distributed in eastern and southern Mongolia (Table 5.5). For example, Lake Ulaan in the Valley of Lakes of southern Mongolia (see Chap. 11) rapidly covers a large area when Ongi River reaches it during the high precipitation, and it often shifts into a playa environment during the dry season with the annual average precipitation of less than 100 mm/yr. because of its shallowness (Orkhonselenge et al., 2018b, in press).

5.6 Salinity

Lakes can be hydrochemically classified based on their water regarding its chemical composition (salinity) (Bridge & Demicco, 2008). Here, lakes of Mongolia are classified, based on hydrochemical composition and mineralization level of lake water, as freshwater and saline (highly mineralized, brackish, and salt) lakes according to each region of Mongolia (Table 5.6). In this classification, 1142 lakes occupying an area greater than 0.5 km^2 are included.

Table 5.6 Mongolian lakes classified according to salinity for each region

#	Regions	Freshwater lakes		Brackish and saline lakes	
		Number	Percentage	Number	Percentage
1	Eastern Mongolia	94	8.2	252	22.1
2	Southern Mongolia	131	11.4	138	12.2
3	Western Mongolia	148	12.9	65	5.7
4	Northern Mongolia	101	8.8	18	1.6
5	Central Mongolia	68	5.9	127	11.2
	Total	542	47.2	600	52.8

5.6.1 Freshwater Lakes

Lakes tend to remain fresh if they have sufficient inflow and outflow. Large freshwater lakes in temperate and tropical climates are fed by groundwater and streams through inlets, whereas small freshwater lakes may be entirely fed by groundwater (Bridge & Demicco, 2008). Freshwater lakes in rift basins often have hydrothermal input of calcium-chloride-type brines (Bridge & Demicco, 2008).

The total freshwater resources in Mongolia are estimated to be over 500 km^3, and its approximately 75% (Ministry of Environment and Green Development, 2013) or 84% (Davaa et al., 2007) is stored in lakes. Large freshwater lakes in the country are Lakes Khuvsgul, Buir, Khar, and Khar Us. Lake Khuvsgul alone stores nearly 66.4% (Ministry of Environment and Green Development, 2013) or 74% (Davaa et al., 2007) of the total freshwater in the country (see Chap. 16). Table 5.6 shows that 47.2% of the total lakes are freshwater lakes. The freshwater lakes are dominantly distributed in western, southern, and northern Mongolia (Table 5.6).

5.6.2 Saline Lake

Saline lake, also known as brine lake, is an inland body of water situated in arid and/or semiarid regions, having no outlet to the sea and containing a high concentration of dissolved salts, principally NaCl (Last & Smol, 2001; Neuendorf et al., 2005). Alkali lake, also known as salt flat, is a lake on the other scale of large and deep saline lakes (Last & Smol, 2001; Neuendorf et al., 2005). Shallow saline lakes occupy lowlands in arid regions and groundwater discharge zones. Such lakes are typically classifiable as either playas or playa lakes because they are periodically flooded by either rain and then dry up during drier intervals, leaving accumulations of brines and evaporitic minerals (e.g., Last & Smol, 2001; Neuendorf et al., 2005; Komatsu et al., 2007; Orkhonselenge & Bulgan, 2021).

Saline lakes commonly have more than 5000 ppm total dissolved solids (TDS), and brines in saline lakes may reach 400,000 ppm TDS (Bridge & Demicco, 2008). Many lakes in tectonic basins are saline lakes although some of the lakes have freshwater to brackish water (Bridge & Demicco, 2008). The brackish lakes have no outlet and concentrations of TDS increase over time due to evaporation of the water.

The total volume of brackish water lakes in Mongolia is about 90 km^3 (Ministry of Environment and Green Development, 2013). The largest saline lakes are Lakes Uvs and Khyargas. Table 5.6 shows that 52.8% of the total lakes are saline lakes (see Chap. 23). The saline lakes are widely distributed in eastern, southern, and central Mongolia (Table 5.6). Many of highly salinized and mineralized lakes in eastern, southern, and western Mongolia have hydrochemically changed in recent years (Ariunbileg et al., 2020; see Chap. 23).

5.7 Presence of Outlet

Significant inflow sources are precipitation onto the lake, runoff carried by rivers, streams and creeks, groundwater channels, snow coverage, and glacial and permafrost meltwaters. Outflow modes include evaporation from the lake, outlet and groundwater flows, and other extractions of lake water (e.g., human activities). Lakes that have outlet of water are called hydrologically open, whereas those that do not have such drainage are called hydrologically closed (Bridge & Demicco, 2008).

Here, lakes of Mongolia are classified, based on the presence and absence of outlet of water, as open and closed lakes according to each region of Mongolia (Table 5.7). In the classification, 1142 lakes occupying an area greater than 0.5 km^2 among the total lakes of Mongolia are included. Table 5.7 shows a classification of lakes with outflow listed as an open, and lakes without outflow listed as a closed.

5.7.1 *Open Lakes*

For open lakes, their water level is controlled by the level of outlet, and lakes with outlets can maintain a nearly constant water level for long periods of time (Bridge & Demicco, 2008). For example, the outlet Eg River in northern Mongolia is one of the mutually related factors causing the level of Lake Khuvsgul to be controlled with its continuous incision (Orkhonselenge et al., 2013). Outlet is also one indicator for freshwater lakes. Table 5.7 shows that 29.7% of the total lakes are open lakes. The open lakes occur dominantly in western and eastern Mongolia.

5.7.2 *Closed Lakes*

For closed lakes, water level fluctuates due to variations in water supply and evaporation (Bridge & Demicco, 2008). In Mongolia, 70.3% of the total lakes are closed lakes (Table 5.7). The closed lakes have diverse origins (see Chap. 4), and they are found dominantly in southern and eastern Mongolia (Table 5.7). The closed lakes

Table 5.7 Mongolian lakes classified according to the presence of outlet for each region

		Open lakes		Closed lakes	
#	Regions	Number	Percentage	Number	Percentage
1	Eastern Mongolia	94	8.2	254	22.2
2	Southern Mongolia	59	5.2	326	28.6
3	Western Mongolia	117	10.3	96	8.4
4	Northern Mongolia	45	3.9	74	6.5
5	Central Mongolia	24	2.1	53	4.6
	Total	339	29.7	803	70.3

are highly salinized and mineralized (see Sect. 5.7). For example, the closed Lake Telmen in central Mongolia hosts high concentration of brines with the quantity of 6.8 g/L (Tserensodnom, 2000) or 4 g/L at present (Peck et al., 2002). Moreover, closed small Lakes Bor Ovoo (or Borvoi) and Burd in the Valley of Lakes in Govi region are enriched with Na, K, Mg, Ca, and Cl ions, and Li, S, and U deposits (Ariunbileg et al., 2020; see Chap. 23).

5.8 Summary

Lakes of Mongolia are classified into six types based on altitude, area, stability, depth, salinity, and presence of outlet according to each region of Mongolia. 34.1% of the total lakes in Mongolia are mountainous lakes, while 65.9% are prairie lakes. Mountainous, high-altitude, and large lakes occur in western and northern Mongolia, whereas prairie, low-altitude, and small lakes are located in eastern, southern, and central Mongolia. 74.6% of the lakes in Mongolia lie below the elevation of 1500 m a.s.l., and they exist in eastern, southern, and central Mongolia. 97.9% of the lakes in Mongolia occupy an area of up to 10.0 km^2, and such lakes are dominantly distributed in eastern, southern, and central Mongolia. 56.3% of the lakes occupying an area over 0.5 km^2 are ephemeral lakes existing in eastern, southern, and central Mongolia. 96.9% of the lakes are shallow lakes dominantly distributed in eastern and southern Mongolia, while the deep lakes are found in western and northern Mongolia. 52.8% of the lakes are saline lakes abundantly distributed in eastern, southern, and central Mongolia, whereas the freshwater lakes are dominantly distributed in western, southern, and northern Mongolia. 70.3% of the lakes are closed lakes mainly occupying the southern and eastern Mongolia, while the open lakes exist commonly in western and eastern Mongolia.

References

Academy of Sciences of Mongolia and Academy of Sciences of USSR. (1990). *National Atlas of the Mongolian People's Republic*. [In Russian].

Anderson, R. S., & Anderson, S. P. (2010). *Geomorphology: The mechanics and chemistry of landscapes*. Cambridge University Press, 637 p. ISBN: 978-0-521-51978-6.

Ariunbileg, S., Isupov, V. P., Vladimirov, A. G., Kolpakova, M. N., & Kuibida, L. V. (2020). *Chemical evolution and geochemistry of mineralized lakes of Mongolia*. Admon Printing House, 132 p. [In Mongolian].

Berkey, P. C., & Morris, K. F. (1927). *Geology of Mongolia. A reconnaissance report based on the investigations of the years 1922–1923. Natural History of Central Asia* (Vol. 2). The American Museum of Natural History. 475 p.

Bridge, J., & Demicco, R. (2008). *Earth surface processes, landforms and sediment deposits*. Cambridge University Press, 815 p. ISBN: 978-0-521-85780-2.

Cohen, A. S. (2003). *Paleolimnology: The history and evolution of lake systems*. Oxford University Press.

Davaa, G., Oyunbaatar, D., & Sugita, M. (2007). *Surface water of Mongolia*. pp. 55–68.

Einsele, G. (1992). *Sedimentary basins: Evolution, facies, and sedimentary budget*. Springer-Verlag. ISBN: 3-540-54449-6.

Håkanson, L., & Jansson, J. (1983). *Principles of lake sedimentology*. Springer-Verlag, 320 p.

Håkanson, L. (2012). Lakes on earth, different types. In *Encyclopedia of lakes and reservoirs* (pp. 471–472). Springer.

Jigj, S. (1975). *Primary feature of Mongolian landforms*. Institute of Geography & Permafrost. Mongolian Academy of Sciences (MAS), 126 p. [In Mongolian].

Kang, S., Lee, G., Togtock, C., & Jang, K. (2015). Characterizing regional precipitation-driven lake area change in Mongolia. *Journal of Arid Land, 7*(2), 146–158.

Komatsu, G., Ori, G. G., Marinangeli, L., & Moersch, J. E. (2007). Playa environments on earth: Possible analogs for mars. In M. G. Chapman (Ed.), *The geology of mars: Evidence from earth-based analogs* (pp. 322–348). Cambridge University Press. ISBN: 9780521832922.

Last, W. M., & Smol, J. P. (2001). *Tracking environmental change using lake sediments. Volume 2: Physical and geochemical methods*. Kluwer Academic Publishers, 504 p.

Ministry of Environment and Green Development. (2013). *Integrated Water Management Plan of Mongolia*. 339 p.

Neuendorf, K. K. E., Mehl, J. P., Jr., & Jackson, J. A. (2005). *Glossary of geology* (5th revised and enlarged ed.). Berlin: Springer. Approx. 900 p. ISBN: 3-540-27951-2.

Orkhonselenge, A., & Bulgan, O. (2021). Geochemical studies and lacustrine geomorphology of Lake Yakhi basin in eastern Mongolia. *Géomorphologie: Relief, Processes, Environnement, 27*(3), 231–242. https://doi.org/10.4000/geomorphologie.15873

Orkhonselenge, A., Davaanyam, S., Altanbagana, M., & Bolorchuluun, C. (2019). *Long-term changes in ecosystems in natural zones of Mongolia: A report of Research Project funded by Science and Technology Foundation (ShUT_A/349_005/2013)*. Laboratory of Geochemistry and Geomorphology, National University of Mongolia. NUM Publishing, 384 p. [In Mongolian].

Orkhonselenge, A., Komatsu, G., & Uuganzaya, M. (2018a). Climate-driven changes in lake areas for the last half century in the Valley of Lakes, Govi Region, Southern Mongolia. *Natural Science, 10*(7), 263–277.

Orkhonselenge, A., Komatsu, G., & Uuganzaya, M. (2018b). Middle to late Holocene sedimentation dynamics and paleoclimatic conditions in the Lake Ulaan basin, southern Mongolia. *Géomorphologie: Relief, Processus, Environnement, 24*(4), 351–363.

Orkhonselenge, A., Krivonogov, S. K., Mino, K., Kashiwaya, K., Safonova, I. Y., Yamamoto, M., Kashima, K., Nakamura, T., & Kim, J. Y. (2013). Holocene sedimentary records from Lake Borsog, eastern shore of Lake Khuvsgul, Mongolia, and their paleoenvironmental implications. *Quaternary International, 290–291*, 95–109.

Orkhonselenge, A., Uuganzaya, M., & Davaagatan, T. (in press). Late Holocene sedimentation dynamics in the Lake Ulaan basin, southern Mongolia. *Environmental Earth Science*.

Peck, J. A., Khosbayar, P., Fowell, S. J., Pearce, R. B., Ariunbileg, S., Hansen, B. C. S., & Soninkhishig, N. (2002). Mid to Late Holocene climate change in north central Mongolia as recorded in the sediments of Lake Telmen. *Palaeogeography, Palaeoclimatology, Palaeoecology, 183*, 135–153.

Poehls, D. J., & Smith, G. J. (2009). *Encyclopedia dictionary of hydrogeology* (1st ed.). 358 p.

Roberts, N. (1998). *The Holocene: An environmental history* (2nd ed.). Blackwell Publishers Ltd, 316 p. ISBN: 0-631-18637-9.

Saruulzaya, A. (2017). *Thermokarst lake dynamics in the continuous and isolated permafrost zones, Mongolia*. Ph.D. Thesis, Hokkaido University, Sapporo, Japan. 76 p.

Tsegmid, S. (1969). *Physical Geography of Mongolia*. State Press, 405 p. [In Mongolian].

Tserensodnom, J. (1971). *Lakes of Mongolia*. State Publishing, 202 p. [In Mongolian].

Tserensodnom, J. (2000). *A catalog of lakes in Mongolia*. Shuvuun Saaral Publishing, 141 p. [In Mongolian].

Part II
Landscape Evolution of Large Lakes in Mongolia and Their Paleoclimate Records

Chapter 6
Landscape, Lake Distribution, and Evolution in Eastern Mongolia

Abstract Eastern Mongolia is characterized by a unique landscape of the endless vast Dornod plain encompassed by the Khentii Mountain Range in the northwest and the Khyangan Mountain Range in the southeast, with a large lake called Lake Buir and many small lakes, which is in stark contrast to other regions of Mongolia. Lakes, rivers, and streams in the region play important role in the surface water resource in the Pacific Ocean drainage basin. In eastern Mongolia the greatest number of lakes in the country (28.9%) are distributed, and these lakes are dominantly fluvial, thermokarst, aeolian, and glacial lakes. Lakes residing in the valleys and depressions in this region under the arid climate have primarily fluvial and thermokarst origins with exceptions of a few tectonic and moraine-dammed lakes. The lakes in the east, south, and west of this region are shallow and saline, and they are strongly dependent on precipitation. However, lakes in the north of the region are deep and freshwater, and they are fed by precipitation, runoff, permafrost meltwater, and groundwater. Many lakes in this region are shifting to playa environments in recent years due to the rising temperature and strengthening evaporation. Eastern Mongolia was under the strong influence of the East Asian Summer Monsoon during the Pleistocene and early Holocene, but it has been under the influence of the East Asian Winter Monsoon since the middle Holocene. This shift in the atmospheric circulation is followed by the arid climate in the late Holocene and the present warming is causing drought. These recent conditions are distinctly reflected in the evolution of landscape and surface water, particularly for lakes of eastern Mongolia today.

Keywords Dornod · Menen · Khentii · Khyangan · Kherlen · Buir · Khukh · Eastern Mongolia

A. Orkhonselenge et al., *Lakes of Mongolia*, Syntheses in Limnogeology,
https://doi.org/10.1007/978-3-030-99120-3_6

6.1 Introduction

Eastern Mongolia covers an area of 286,209 km^2 (or 18.2%) of the total territory of Mongolia (Badarch et al., 1965; Enkhtaivan et al., 2016). In terms of the physiographical classification of Mongolia, in addition to the Dornod plain, the Khentii[1] subzone is included within this region (Tsegmid, 1969). In a new classification, this region is included in the Daguur Mongol steppe zone in the north and the intermittent desert region of Central Asia in the south. Eastern Mongolia presents spectacular steppe landscapes and extraordinary landforms within the country because this region contains a complex system of mountains, plains, and depressions consisting of nine different landscape types from high mountains to desert steppe. The landscape in the region (see Sect. 6.2) consists of the eastern Khentii Mountain Range (Fig. 6.1) in the northwest, Khyangan[2] Mountain Range in the southeast, the Dornod plain (Fig. 6.2) in the center, and lowlands and depressions throughout the region. This region hosts one of the Central Asian interfluves, the Khentii Mountain Range, which feeds large rivers such as Kherlen, Onon, Balj, and Ulz Rivers, headwaters of Amar River in eastern Siberia (Jigj, 1975), and large lakes such as Lakes Tari[3] and Dalai[4] in the southeastern Siberia.

Eastern Mongolia (Fig. 1.1) has been less investigated scientifically compared to other regions of Mongolia. Physiographically, eastern Mongolia consists of three parts: (1) the Khentii Mountain Range (Fig. 6.1) and its branch mountains, (2) the Dornod plain (Fig. 6.2), and (3) the Govi[5], and they were studied by German scientists (e.g., H. Richter, H. Bartel, G. Haase) in 1959 (see Sect. 6.2). Then 19 landscape types were determined in eastern Mongolia by physical geographer Sh. Tsegmid in 1962–1963 (Tsegmid, 1969). Badarch et al. (1965) described the entire landscape and natural zones throughout eastern Mongolia. In the late 1980s, a joint Mongolian and Russian geological expedition noted that the present-day landscape in eastern Mongolia had formed due to the intensifying aridification and retreating of the forest line caused by the East Asian Summer Monsoon (EASM) at the transition time from the late Pleistocene to the early Holocene, and earlier lake levels in eastern Mongolia had dropped during the Pleistocene glaciation. Moreover, in 1994–1995 physical geographers from the Institute of Geography studied the landscape and geomorphology of eastern Mongolia, and they compiled a map on these subjects. In 2014–2016 physical geographers from the Institute of Geography and

[1] *Khentii* has has been misspelled as *Khentey* (or *Hentey*) and *Khentei* (or *Hentei*) in many publications. Khentii is the right English transliteration from Mongolian *Хэнтий*.

[2] *Khyangan* has has been misspelled as *Khingan* (or *Hingan*) and *Hyangan* in many publications. Khyangan is the right English transliteration from Mongolian *Хянган*.

[3] *Tari* has been misspelled as Torey by the Russians. Tari is the right English transliteration from Mongolia *Тарь*.

[4] *Dalai* meaning *ocean* has been often spelled as *Hulun* in many publications. Dalai is the right English transliteration from Mongolian *Далай*.

[5] *Govi* has been often spelled as *Gobi* in the international literature. Govi is the correct English transliteration from Mongolian *Говь*.

Fig. 6.1 (**a**) Mt. Dunkhen (2121 m a.s.l.) in the Khentii Mountain Range viewing toward the east. (**b**) A permafrost valley between Mt. Dunkhen (2121 m a.s.l.) and Mt. Ikh Must (1917 m a.s.l.) in the Khentii Mountain Range viewing toward the north. Photos by A. Orkhonselenge on August 20, 2008

Geoecology (Enkhtaivan et al., 2016) assessed the landscape structure and its dynamic changes in eastern Mongolia over the last 25 years. Moreover, Orkhonselenge et al. (2019) evaluated landform features, present surface water resources, recent climate changes, and valuable geoheritages that are to be classified as protected areas in eastern Mongolia.

In eastern Mongolia lakes (see Sect. 6.3), as much as rivers and streams play important role in the surface water resource in the Pacific Ocean drainage basin.

Fig. 6.2 The Dornod plain in eastern Mongolia viewing toward the (**a**) southeast and (**b**) south. Photos by A. Orkhonselenge on July 28, 2014

However, lakes in eastern Mongolia have not been investigated in detail, despite the fact that landform and landscape were studied relatively well by Badarch et al. (1965), Tsegmid (1969), Jigj (1975), and Enkhtaivan et al. (2016). Badarch et al. (1965) noted that there were more than 300 lakes formed in small depressions and fluvial valleys within the Dornod plain, and they also provided morphometric values of large lakes (e.g., Lakes Buir, Khukh, Guren, Khangal, etc.). Luvsandorj and Tumur (1969) conducted a hydrochemical investigation in lakes of eastern Mongolia. Tserensodnom (1971, 2000) described that there were hundreds of lakes in the Dornod plain and they primarily occupied an area less than 10.0 km^2 distributing along the paleo- and modern-day river valleys and small depressions between hills. However, only Lake Buir (see Sect. 6.3; Chap. 7) covers an area over 100.0 km^2 in eastern Mongolia where it constitutes 60% of the total surface water resource in the region. There are 885 lakes in total covering an area over 0.1 km^2 each, and they constitute 28.9% of the total lakes covering the same area range in the whole country (see Sect. 6.3 and Chaps. 1 and 5).

In eastern Mongolia some lakes were studied in terms of hydrochemistry for seeking balneological properties (Ariyadagva et al., 1980). Most lakes in the Dornod plain are saturated with saline solutions, and brines are crystallized on their bottoms if there is no outflow (Tserensodnom, 1971). Linhoff et al. (2011) found extremely high concentration of uranium (U) in shallow, groundwater-fed hyper-alkaline soda lakes in eastern Mongolia, and they concluded that the high concentrations of U, Na, Cl, and K were correlated to radiogenic strontium (Sr) in the lake water suggesting that the U had been sourced from the local Cretaceous alkaline rhyolites. Recently, Ariunbileg et al. (2020) showed the hydrochemical and geochemical results for 19 small lakes covering an area of 1–2 km^2 and a depth of 0.2–1.0 m in a tectonic depression within the Dornod plain. Most lakes are fed by rainfall and groundwater, and some lakes often shrink due to climate changes with examples including Lakes Yakhi, Khukh, Toson, and Burd (Orkhonselenge et al., 2019; see Sect. 6.4). Davaa (2015) showed that the water temperature of Lake Buir had decreased by −0.14 °C/year before 1989 and risen by 0.18 °C/year since the 1990s. Kang et al. (2015) noted a generally decreasing pattern of lake area in the Choibalsan region, where the lake area in 2001 was only one half of 1994. Moreover, Oyunbaatar et al. (2017) analyzed morphometric values, water level, and water balance of Lake Ganga in southeastern Mongolia, and they observed a dramatic drop in water level, recent decrease in lake area, and continuous shrinkage of the lake due to both effects of climate warming and human impacts (see Sect. 6.4). A recent study noted that Lake Yakhi shows a decreasing trend in an area with a loss of ~62.2 km^2 or 63.9% of the total area during the past five decades (Orkhonselenge & Bulgan, 2021; see Sect. 6.4).

Eastern Mongolia had been a major center of human activities in Eurasia since the twelfth century and of unique cultural routes in the world where man, over the eight centuries, interacted with mountains, hills, plains, river courses, lakes, swamps, forests, etc. These morphologies have been integrated into the natural landscape forming a unique and harmonious entity in the Khentii Mountain Range (Fig. 6.1). This region includes two World Heritage sites (UNESCO, 2020), namely, (1) Mt. Burkhan Khaldun and the surrounding sacred landscape as a cultural site and (2) the Dauria landscape as a natural site (see Chap. 24). The Mt. Burkhan Khaldun, registered as the World Heritage in 2015, is located at the border between the Siberian coniferous taiga forest in the central Khentii Mountain Range and the vast Central Asian steppe. The Dauria, registered as the World Heritage in 2017, is an outstanding example of the steppe region, extending from the taiga to the desert.

In this chapter, landscape features, lakes, lacustrine environments, and their evolution in eastern Mongolia are discussed. Lakes representing eastern Mongolia are discussed in detail in the following chapters (see Chaps. 7 and 8).

6.2 Landscape

6.2.1 Geology

In eastern Mongolia the Jurassic and Cretaceous sedimentary rocks and the Neogene deposits are dominant, and the Neogene and Quaternary basalts are widely distributed around the Dari Ganga region in the southern Dornod plain (Tsegmid, 1969). In the region the tectonic-denudation landforms (see Sect. 6.2.3) are underlain by the Paleozoic and Mesozoic rocks, whereas the denudation and tectonic-depositional landforms are underlain by the Mesozoic and Cenozoic rocks. The volcanic landform is underlain by the Pliocene and Pleistocene igneous rocks. The erosional plains formed during the late Paleozoic and early Mesozoic occur in the northeast of eastern Mongolia, while the depositional plains formed during the Cretaceous and Neogene occupy in the south of the region (Enkhtaivan et al., 2016).

The Khentii Mountain Range (Fig. 6.1) is predominantly composed of the early Paleozoic and Mesozoic granites, and it is characterized by dome-shaped peaks with steep gorges and cliffs, and gentle valleys formed during the Pleistocene glaciations (Murzaev, 1952; Tsegmid, 1969). In the southern Khentii Mountain Range, the Gorkhi formation made of feldspathic arenite sandstones derived from continental margin to continental island arc settings together with plutonic source rocks enriched in saussuritized plagioclase is linked to the subduction of the paleo-Asian oceanic plate within the Central Asian Orogenic Belt (CAOB) during the middle Devonian to early Carboniferous (Hara et al., 2013). However, in northeastern Mongolia, the Ereendavaa metamorphic terrane in the Ereen Mountain Range in the east of the Khentii Mountain Range formed mainly during the early Paleozoic (495–464 Ma) and late Paleozoic to early Mesozoic (295–172 Ma) with the Precambrian rocks as the protoliths of these rocks (Miao et al., 2017). In the Khyangan Mountain Range, the Paleozoic rocks are dominant (Badarch et al., 1965).

6.2.2 Climate

The climate in eastern Mongolia is characterized mainly by the landscape types of forest steppe to desert steppe with mountains, hills, plains, valleys, and depressions. The climate of the Khentii Mountain Range is characterized by large annual temperature variations, and the average precipitation is 200–300 mm/year in the Khyangan Mountain Range (Academy of Sciences of Mongolia and Academy of Sciences of USSR, 1990). Davi et al. (2013) highlighted that eastern Mongolia had been strongly drying due to the global warming, causing a long drought since the late 1900s. A significant trend in the large-scale drying (1951–2005) is found across eastern Mongolia (Li et al., 2009). Eastern Mongolia is intensely warming and drying in summer, and this region will face a major desertification due to the continuously strengthening evaporation (Enkhtaivan et al., 2016).

In the region, the annual average air temperature ranges from –0.9°C to 1.5°C (Jambaajamts, 1989). The average air temperature ranges are −15 °C to −25 °C in January and 20–25 °C in July, and the average precipitation is 200–300 mm/year (Tserensodnom, 1971). The precipitation is 188–391 mm/year, and its 93–96% falls in summer, while 4–7% falls in winter (Altantsetseg & Munkhjargal, 2006). Another source states that the average annual precipitation is 250–350 mm and evaporation is 600–1000 mm (Ministry of Environment and Green Development, 2013). The Dornod plain is characterized by the average wind velocity of 4.1 m/s, average humidity of 1300 GPa/day, soil surface temperature of 1–3 °C/year, and evaporation of 934 mm/year (Davaa, 2015). Generally speaking, the values of 54% evaporation and 46% humidity are indicated in the plain. The easternmost corner of Mongolia, including Khalkh River and Numrug River valleys and the Tamsag Bulag basin, is characterized by the average wind velocity of 3.9 m/s, average humidity of 1293 GPa/day, soil surface temperature of 1.8–5.0 °C/year, and evaporation of 1050 mm/year. In this area, the values of 51–61% evaporation and 39–49% humidity have been reported.

The precipitation ranges in eastern Mongolia (Enkhtaivan et al., 2016) show that the precipitation increased by 32.7% in winter and decreased by 25.3% in summer during the period of last three decades (Table 6.1). Over the last 30 years, the snow cover was greater in 2004 and 2007–2010 and lower in 1982, 1984, 1985, 1988, 1997, 1998, and 2011 than the annual average.

Table 6.1 Air temperature, precipitation, air pressure, and wind velocity in eastern Mongolia

Sectors	Landscape	Average air temperature (°C)				Average precipitation (mm/year)	Air pressure (GPa/day)	Wind velocity (m/s)
		Annual	January	July	Maximum			
North, northwest	Mountain	–2.0–1.1	–35.4 to –43.3	18.8		>400 350–400	850	0.8–2.3 to 2.8–3.4
	Plain	–0.1–1.8	–45.2 to –46.3		32.6–36.9	250–300	870	3.2–3.9 to 4.2–5.2
East, northeast	Steppe	–1.7–0.5 0.39–2.2	–23.4 to –30.3		35.1–47.9	320–420 243–297	849–900 920–945	5.0–10.0 to 12.0–40.0
South, southwest	Steppe	0.5–2.6	–17.0 to –23.0	19.0–23.0	38.1–39.7	< 250	825–890	3.3–3.7 to 10.0–15.0
	Desert steppe		–38.0 to –43.0		40.0	108.7–280.6	891–933	5.1–5.2 to 20.0–40.0

6.2.3 Landform

The landform in eastern Mongolia consists of the Dornod plain in the center, the eastern Khentii Mountain Range in the northwest, and the Khyangan Mountain Range in the southeast, and Kherlen, Onon, Ulz, and Khalkh Rivers' drainage basins (Table 1.2). The region's elevation is 1200 m a.s.l. on average ranging from 560 m a.s.l. of Lake Khukh (see Chap. 8) to 2800 m a.s.l. of Mt. Asralt Khairkhan in the Khentii Mountain Range (Badarch et al., 1965). The landform in the region is divided into five types: (1) tectonic-denudation, (2) denudation, (3) tectonic-depositional, (4) erosional, and (5) volcanic (Badarch et al., 1965).

The Dornod plain (Fig. 6.2), dominant in the region, has attracted relatively limited attention by researchers. The Dornod plain shifted to a continental regime during the Mesozoic Era, and the present plain was established during the Pliocene and Pleistocene by tectonics (Badarch et al., 1965; see Chap. 3). In this plain, hills are elevated up to 1800 m a.s.l. (Jigj, 1975).

In the northeastern Dornod plain, a depression of Lake Khukh (560 m a.s.l.), the lowest place in Mongolia, is located (see Chap. 8). Low mountains, hills, and depressions are distributed in a wide valley extended from the north to the south in the plain. The highest mountain in the valley is Mt. Vangiin Tsagaan elevated at 1099 m a.s.l. (Jigj, 1975). There are numerous depressions in the wide plain between lowlands and hills. For instance, Menen plain, Tamsag plain, Matad plain, Targan plain, Baruun Agit Bor plain, Tsagaan Daravgain Khooloi, Shuvuut Khooloi, Khaltar Govi, Galba Govi, Khorkhoit Enger, Kharganat, Khoshuu Khooloi, Duulga Valley, Tsagaan Chuluut Valley, and Chuluut Ganga Valley are included in those depressions. Some of the depressions may have been related to paleo-depressions judging from their narrow, shallow, and elongated shapes without any terraces and knick points. They are bordered with low mountains and hills such as Mt. Khaalt (933 m a.s.l.), Mt. Shankhchin Ovoo (917 m a.s.l.), Mt. Khavtsal (940 m a.s.l.), Mt. Kharaat Undur (898 m a.s.l.), Mt. Tsagaan (1052 m a.s.l.), Mt. Khutagtiin Undur (1059 m a.s.l.), Mt. Baruun Suul (943 m a.s.l.), and Mt. Builst Ukhaa (934 m a.s.l.).

In the southwestern Dornod plain, there are mountains higher than those in the east. For instance, there are Mt. Adaatsag (1804 m a.s.l.), Baga Gazriin Chuluu (1768 m a.s.l.), Ikh Gazriin Chuluu (1706 m a.s.l.), Ikh Nart (1280 m a.s.l.), Mt. Choir (1696 m a.s.l.), Mt. Darkhan (1716 m a.s.l.), Mt. Bayan Khutag (1580 m a.s.l.), Mt. Munkh Khaan (1607 m a.s.l.), Mt. Gal Shar (1597 m a.s.l.), Ikh Chuluut (1247 m a.s.l.), Mt. Matad (1051 m a.s.l.), Mt. Tsagaan (1101 m a.s.l.), Khar Undur (1132 m a.s.l.), and Mt. Khukh (1046 m a.s.l.).

The Khentii Mountain Range (Fig. 6.1), the second major landform in the region, extends along the southwest-northeast direction from its most southwestern tip, Mt. Bogd, to the Russian border in the northeast of Mongolia, with an average elevation of 1600 m a.s.l. (Orkhonselenge et al., 2021). The Khentii Mountain Range was glaciated twice to three times (Jigj, 1976) during the Pleistocene (Tsegmid, 1969), similar to the Altai, Khangai, and Khuvsgul Mountain Ranges in other regions. The glaciations were centered on Mt. Asralt Khairkhan, Mt. Baga Khentii Saridag[6], Mt.

[6] *Saridag* has been often misspelled as *Sardyk* or Sardig by the Russian pronunciation in the international literature. Saridag is the correct English transliteration from Mongolian *Сарьдаг*.

Fig. 6.3 (**a**) Terelj River flowing between Mt. Bayan Bulagiin Zoo (2115 m a.s.l.) and Mt. Ikh Must (1917 m a.s.l.) in the Khentii Mountain Range viewing toward the southwest. Photo by A. Orkhonselenge on August 20, 2008. After Orkhonselenge et al. (2021). (**b**) Kherlen River flowing below the bridge in the Chinggis city, Khentii aimag, viewing toward the southwest. Photo by A. Orkhonselenge on July 12, 2020. The arrows indicate the flow directions of the rivers

Khiid Saridag, Mt. Kherkhluur Saridag, and headwaters of Minj, Zakhar, Iluur, Terelj (Fig. 6.3a), and Kherlen (Fig. 6.3b) Rivers, covered by 50–100-m-thick glaciers (Jigj, 1976). However, unlike the mountain ranges of other regions, the Khentii Mountain Range does not host modern glaciers (Orkhonselenge & Uuganzaya, 2018; see Sect. 6.2.4). The eastern Khentii Mountain Range is included in this region (Fig. 1.1), but the western Khentii Mountain Range is included in central Mongolia (see Chap. 18).

The Khyangan Mountain Range, the third major landform in the region, elevated at 1000–1500 m a.s.l., extends from the northwest to the southeast in the easternmost corner of Mongolia. Badarch et al. (1965) noted that there were independent mountains and ranges such as Gutai and Ereen Mountain Ranges and Mt. Ulzii Saikhan where the Paleozoic metamorphic schist, sandstones, and granites are abundant. There are low mountains (e.g., Mt. Sogoot, Mt. Zaraa (1351 m a.s.l.), Mt. Bayankheer (1492 m a.s.l.), and Mt. Khan Chandmani (1384 m a.s.l.)), and the low mountains have gradual slopes and rounded peaks. The mountains are dissected by water erosion.

6.2.4 Glaciers and Permafrost

Glaciers are one of the surface water resources in Mongolia and one of the major headwaters of lakes. In eastern Mongolia, traces of the late Pleistocene glaciers are found in the Khentii Mountain Range. For instance, in the Khentii Mountain Range, lateral and terminal moraines, hummocky terrains, glacial cirques (Badarch et al., 1965; Jigj, 1976), U-shaped valley, hanging valley, fluvioglacial terraces (Tsegmid, 1969), and glacial lakes (see Sect. 6.3) are abundant (Tserensodnom, 1971; Orkhonselenge & Uuganzaya, 2018), and their formations are probably linked to the maximum glaciation during Marine Oxygen Isotope Stage (MIS) 2 (Khandsuren et al., 2019). Although the total paleoglacier area in the Khentii Mountain Range remains uncertain to date, Jigj (1976) calculated the total area of the paleoglaciers as 1760 km^2, based on aerial photos, topographical maps, and field data. Moreover, Orkhonselenge and Uuganzaya (2018) estimated the maximum glaciation extents to be 10.0 km^2 on Mt. Asralt Khairkhan and 0.7 km^2 on Mt. Baga Khentii Saridag in the Khentii Mountain Range where only two types of glacial landforms including glacial cirques and terminal moraines were observed based on mapping over 30 m resolution ASTER DEMs and Landsat 8 satellite imagery. The spatial distribution of the glacial landforms indicates that the Khentii Mountain Range experienced intermontane valley glaciations with small ice caps centered on the high mountains. The glacial cirques mark numerous glacial erosional landforms, while the terminal moraines indicate glacial advances and depositional landforms at the valley outlets (Orkhonselenge & Uuganzaya, 2018).

Permafrost is a near-surface water resource. In eastern Mongolia perennial permafrost (Fig. 6.1b) is indicated in lake depressions with deposits of silts and sands that are lying at 1000–1300 m a.s.l.; however, ice wedge is rare in the permafrost (Enkhtaivan et al., 2016). The perennial permafrost is also revealed in the eastern, southeastern, and southern mountains of the Khentii Mountain Range and at 1300 m a.s.l. near Lake Kholboo on Mt. Dari Ganga. Most regions in eastern Mongolia contain seasonal permafrost (Tumurbaatar, 2004; Table 6.2). Impacts of the less snow cover, the sparse vegetation cover, and the annual air temperature above 0 °C throughout the year due to the long-lasting solar insolation in eastern Mongolia are felt in the formation of the seasonal permafrost. In addition to the

Table 6.2 Seasonal freezing and thawing depths in eastern Mongolia

#	Regions	Materials	Depth (m)	Regions	Materials	Depth (m)
1	Steppe	Clays	2.1–2.9	Northern (e.g., Ereen Tsav)	Clays and silts	2.7
		Sands	3.0–3.5		Sands	2.9
2	Eastern	Clays	2.5–2.6		Gravels	3.8
		Sands	2.7–2.8	Central (e.g., Sergelen)	Clays and silts	2.6
3	Southern	Clays	2.2–2.3		Sands	2.8
		Sands	2.7–2.8		Gravels	3.6
4	Western	Clays	2.6–2.7	Southern (e.g., Erdene Tsagaan)	Clays and silts	2.4
		Sands	2.9–3.2		Sands	2.6–2.8
5	Northwestern	Clays	3.2–3.3		Gravels	3.4
		Sands	3.5–3.6		–	–

Table 6.3 Cold and warm permafrost in river drainage basins of eastern Mongolia

Places	Drainage basins	Area of the basins (km^2)	Permafrost in the basins (%)		AMGT (°C)	ALT (m)
			Cold permafrost (AMGT <-2 °C)	Warm permafrost (−2 °C < AMGT < 0 °C)		
Eastern Khentii	Kherlen River	106,487	1.6	16.1	−0.9	1.3–3.5
	Onon River	27,870	2.8	64.6	−1.2	2.6–3.6
Dornod	Ulz River	37,462	0.0	0.0	–	0
	Khalkh River	23,443	0.0	0.0	−0.5	0

landform, the physiographic and stratigraphical conditions affect seasonal thawing and freezing of permafrost (Table 6.2). The seasonal permafrost continues to be present from the beginning of November to the end of March in eastern Mongolia (Tumurbaatar, 2004). In the plains, the upper limit of freezing drops and the thawing rate rises from the north to the south. An occasional permafrost phenomenon occurs in the Khyangan Mountain Range (Jambaljav et al., 2016). The permafrost in large river drainage basins of the region is defined as cold and warm depending on annual mean ground temperature (AMGT) by Zorigt et al. (2020), and it is shown with its active layer thickness (ALT) in Table 6.3.

6.2.5 *Groundwater*

In eastern Mongolia, the groundwater reservoir in the late Cretaceous lithology constitutes a major resource. There is abundant artesian groundwater related to the Cretaceous and Neogene deposits in extensive depressions within the Dornod plain

Table 6.4 Water resource of the lake and river drainage basins and the plain in eastern Mongolia

#	Drainage basins	Area (km²)	Type	Surface water resource (million m³/year)			Groundwater resource (million m³/year)	
				Total resource	Environmental	Possible use	Potential exploitable	Exploitable resource
1	Onon River	28,241	SW	1480	1221	259.0	344	0.6
2	Ulz River	37,961	SW	130	107	22.7	320	26.4
3	Kherlen River	107,906	SW	567	507	59.5	721	43.9
4	Lake Buir-Khalkh River	23,756	SW	1023	920	102.3	198	1.1
5	Menen plain	54,082	GW	0	0	0	168	0.1
	Total	251,946		3200	2755	443.5	1751	72.1

SW surface water; *GW* groundwater

(Tserensodnom, 1971). The groundwater resides in the Paleozoic metamorphic rocks, intrusive and effusive rocks, and the Mesozoic and Cenozoic sedimentary rocks consisting of sandstones and gritstones distributed in the intermontane valleys and river valleys (Tsegmid, 1969). The groundwater in the Paleozoic and Mesozoic rocks has the resource quantity of about 0.003–0.03 L/s. The primary sources of groundwater are precipitation and surface water. The total volume of groundwater in eastern Mongolia is 2.81 km³; however only its 0.82 km³ is available for use (Enkhtaivan et al., 2016). Eastern Mongolia hosts five basins (Table 6.4) among the total drainage basins of 29 existing in Mongolia (Ministry of Environment and Green Development, 2013), and these basins are characterized by the presence of groundwater.

An important issue in these basins is the decrease of groundwater resources as a result of over usage for mining, livestock, and irrigation of crops (the chemical composition of the groundwater making the water unsuitable for drinking). In the north of the region, the groundwater resources are annually renewed, mainly through infiltration of surface water, while in the south of the region, shallow groundwater resources are annually renewed but deep ones are not renewed and considered as fossil.

The Ulz River drainage basin (Fig. 6.4b; Tables 6.3 and 6.4) represents the northern hydrogeological system, while the Menen plain drainage basin (Table 6.4) borders the southern system of eastern Mongolia (Enkhtaivan et al., 2016). In the Ulz River drainage basin, the groundwater resource is dominantly accumulated in alluviums along the river valleys, and it consists of two types of hydrogeological strata: (1) the Quaternary alluviums with a high water capacity range of 0.14–0.17 L/s and (2) the Paleozoic igneous and sedimentary rocks with a low water capacity range of 0.001–0.02 L/s. In the Menen plain drainage basin, the groundwater hosted in the Neogene clays and silts is dominant. The groundwater with a water capacity range of 0.003–1.0 L/s is indicated at 4.2 m depth, but 15 L/s is found at 6.6 m depth in the well.

Fig. 6.4 (**a**) Onon River viewing toward the south and the hydrological measurement point on its western beach in the Binder soum, Khentii aimag. Photo by A. Orkhonselenge on July 28, 2014. (**b**) Ulz River viewing toward the southeast near Mt. Kharaat (1312 m a.s.l.). Photo by A. Orkhonselenge on July 30, 2014. The arrows indicate the flow directions of the rivers

6.2.6 Surface Water

The surface water area constitutes 11.5% of the total area in eastern Mongolia (Enkhtaivan et al., 2016). The Khentii Mountain Range provides large river watersheds between the North Arctic and Pacific Ocean drainage basins, i.e., the large rivers in the range feed other large rivers (e.g., Amar and Shilka Rivers) and lakes (e.g., Lakes Tari and Dalai) in the southeastern Siberia. The mountains in the range feed significantly large rivers flowing to the north, northeast, east, and southeast in this region, and south, southwest, west, and northwest in central Mongolia (see

Table 6.5 The numbers of surface water resources of eastern Mongolia during the last three decades

#	Type	Total	Disappeared
1	River	1062	117
2	Stream	1529	423
3	Spring	49	–
4	Lake	1829	332

Table 6.6 Morphometric components of large rivers in eastern Mongolia

#	Rivers	Area (km^2)	Length (km)	Average elevation of basin (m a.s.l.)	River network density (km/km^2)
1	Kherlen	107,040	1213	1071	0.10
2	Onon	27,998	570	1090	0.46
3	Ulz	37,391	496	720	0.32
4	Khalkh	7440	264	720	0.34

Chap. 18). In the region, there are abundant small rivers, streams, springs, lakes (Table 6.5), and swamps representing peatlands and wetlands in Daguur Mongol. The total volume of surface water in eastern Mongolia (Table 6.5) is estimated to be 8.34 km^3; however only its 0.55 km^3 is available for use (Enkhtaivan et al., 2016).

Dense river drainage networks occur in the Khentii and Khyangan Mountain Ranges, but they are rare in the plains (especially in the southern Matad and Tamsag plains and hills around the Mt. Dari Ganga). There are large rivers (Tables 6.3, 6.4, and 6.6) such as Kherlen (Fig. 6.3b), Onon (Fig. 6.4a), Ulz (Fig. 6.4b), Terelj (Fig. 6.3a), Tuul, Munj, Balj, Gal (Figs. 6.5 and 6.6b), Mukhar, Teel, Adag, Eg, Barkh, Khurkh, Tsenkher, Murun, and Shuus in the eastern Khentii Mountain Range, and Khalkh and Numrug Rivers in the Khyangan Mountain Range. In the Ulz River drainage basin, the drainage network of surface water is dense and/or developed well. There were dense river drainage networks in the plains in the past, but such rivers often shrink due to the arid climate prevailing since the postglaciations in the Quaternary Period (Tsegmid, 1969). For instance, Onon River's water level (Fig. 6.4a) dropped from 275 cm in 2003 to 194 cm in 2007, whereas in the river the ice thickness rose from 1.55 m in 2009 to 1.94 m in 2014 (Orkhonselenge et al., 2019). Kherlen, Onon, and Khalkh Rivers (Tables 6.3, 6.4, and 6.6) are fed by groundwater for about 15–25% of their inventories (Tserensodnom, 1971).

Lakes (see Sect. 6.3) are widely distributed and have diverse origins (see Chap. 4); however they have small areas (see Chap. 5). Lakes Tukhum, Ikh Delger, Baga Delger, Bayan Erkhet, Temeet, and Dungai Tsagaan are located in the Yakhi National Park where Gal River feeding Lake Yakhi (Figs. 6.5 and 6.6a) is the largest water resource. Small rivers and streams (e.g., Nariin, Kharaat, Buduun Ukhaa, Nogoon Khoshuu) are in the Toson Khulstai National Park, and they often shrink during dry periods (Table 6.5). There are many small springs and lakes in the Ugtam National Park.

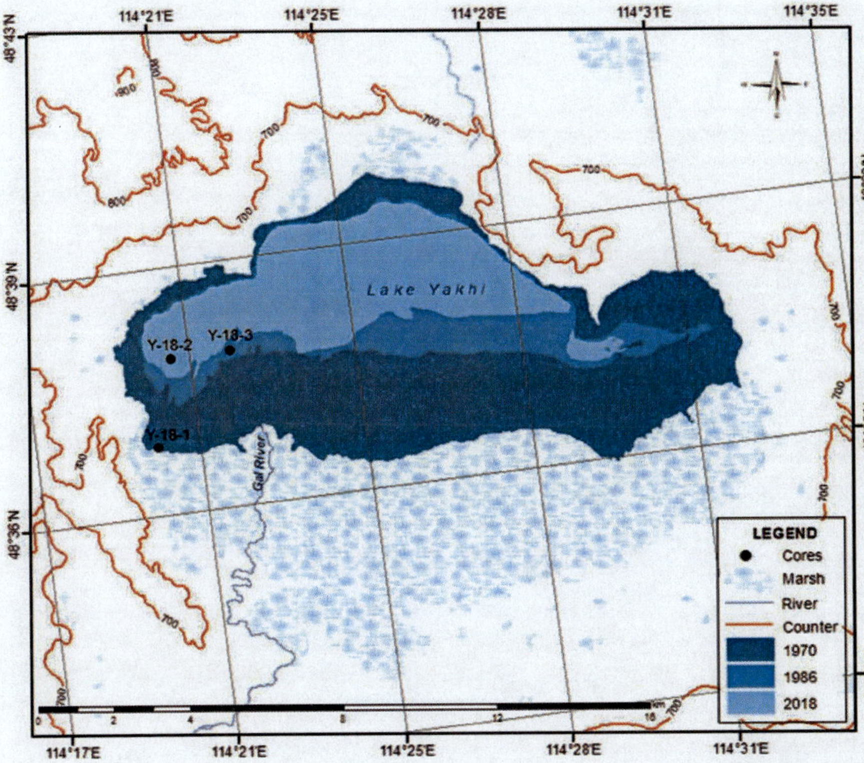

Fig. 6.5 Spatial distributions of areas of Lake Yakhi in 1970, 1986, and 2018 with locations of sedimentary cores (Y18-1 to Y18-3) for geochemical study (Fig. 23.3). After Orkhonselenge and Bulgan (2021)

6.3 Lakes

Lakes in eastern Mongolia exist primarily in depressions (see Chap. 4) without outlet, and they are dominantly fed by precipitation and groundwater (see Chap. 5). There are estimated about 1497 lakes (Enkhtaivan et al., 2016), over 300 lakes (Badarch et al., 1965), 346 lakes covering an area greater than 0.5 km^2 (Tserensodnom, 2000), or 885 lakes covering an area greater than 0.1 km^2 (see Chaps. 1 and 5) including the very large Lake Buir (see Chap. 7); large lakes such as Lakes Khukh (see Chap. 8), Yakhi (see Sect. 6.4), Guren, Ikh Tsaidam, Bayan Erkhet, Tsats, Khunt, Gert, and Ganga (see Sect. 6.4); and many small lakes such as Lakes Khangal (see Chap. 24), Khar Zurkhnii Khukh (see Chap. 24), Khagiin Khar (see Chap. 24), Khurkhree, Galuut, Angirt, Ikh Davst, Duruu, Chukh, Khaichiin Yagaan, Bus, Galuutai Khukh Nuden, Toson, and Burd (see Sect. 6.4). Lakes in the Khalkh and Numrug Rivers' drainage basins, Menen and Tamsag Bulag plains, and Dari Ganga region were connected by large and small rivers and streams in the past

Fig. 6.6 (**a**) Exposed floor of Lake Yakhi viewing toward the east. Photo by A. Orkhonselenge on April 15, 2018. Modified from Orkhonselenge and Bulgan (2021). (**b**) Gal River viewing toward the southwest. Photo by O. Bulgan on September 2, 2019

(Enkhtaivan et al., 2016). The small lakes such as Lakes Baruun Bayan, Dund Bayan, Baruun Davst, Zuun Davst, Burd, Shar Burd, Gurvan, Tsaidam, and Khotont between the Lake Yakhi and Kherlen River in the Dornod plain situated at the elevation of 732–762 m a.s.l. are relict closed saline lakes of a large paleolake formed in the late Cretaceous and Neogene tectonic depression, and these lakes are enriched in Ca, Na, U, Br, and B because these lakes are located in the U ore basin deposited in volcanic rocks (Ariunbileg et al., 2020).

Lakes in eastern Mongolia are classified in terms of their origins (Badarch et al., 1965): (1) lakes in small depressions, (2) lakes in fluvial valleys, (3) tectonic lakes, (4) thermokarst lakes, and (5) moraine-dammed lakes. Lakes formed in river valleys and small depressions, which are underlain by the Jurassic and Cretaceous sedimentary rocks and the Neogene deposits, amount to about 90% of the total lakes in the region, and they are distributed in Kherlen, Onon, Ulz, and Khalkh River valleys and paleoriver valleys in the south of the region. Lakes Lag, Shar Burd, Bayan, Tsagaan, and Shavart occur in paleoriver valleys between Matad and Tamsag basins in the Dornod plain. These lakes are very shallow and unstable with fluctuating water levels (see Chap. 5) because they are filled with water during the summer precipitations, but they shrink in spring and autumn. Tectonic, thermokarst, and moraine-dammed lakes occur in high mountains in the northwest and southeast of the region (see Chap. 4), and they are deep and freshwater. For instance, Lakes Khangal, Guren, and Buir reside in tectonic depressions, whereas Lakes Khentii, Khagiin Khar, Khar Zurkhnii Khukh, Kherkhluur, Mungun, Khurgan, and Tsagaan were formed by moraine damming (Jigj, 1975) or are glacial lakes (see Chap. 4). The Lake Khagiin Khar was dammed by moraines at the global Last Glacial Maximum (LGM; 21–19 ka), and the lake reached its maximum paleoshoreline level at 1840 m a.s.l. (Oh et al., 2019). Lake Shivkhii in the Eg River valley formed in a thermokarst depression, whereas Lakes Tuuz, Khar Nuden, Galuut, and Bus in the Ulz River valley and other lakes (e.g., Lakes Mongol Tsagaan, and Shavart) formed in karst caverns. Thermokarst lakes such as Lakes Shavart, Chukh, Bus, Mongol Tsagaan, Sum, Yakhi, Bayan Erkhet, and Ukhert are distributed in the Ulz River and Gal River valleys (see Chap. 4). Lakes Ikh Tsaidam, Baga Tsaidam, and Tukhum are distributed in the Kherlen River valley. Fluvial lakes are abundant in the Kherlen River and Ulz River valleys. There are aeolian lakes such as Lakes Buunii Tsagaan, Tsatsiin Burd, Ulziit, and Dari Gangiin Tsagaan in the Khalkh, Numrug, and Mukhar River valleys (see Chap. 4).

Lakes in eastern Mongolia were divided into three groups by T.V. Nikolaeva in 1985 as follows: (1) lakes in the Lake Khukh basin, (2) lakes in arid depressions, and (3) lakes related to modern-day river networks (Enkhtaivan et al., 2016). The lakes in the Lake Khukh basin occur typically in a closed depression without outlet, and Lakes Khukh, Jirmiin Tsagaan, and Tari are included in this type of lakes. The lakes in arid depressions occupy a long (more than 100 km) and narrow (7–15 km) depression with steep slopes surrounded by hills, and Lakes Suman, Sav, and Shavart are included in this type of lakes. The lakes related to the modern-day river networks exist with a close connection of Ulz and Duch Rivers and their tributaries, where Lakes Duruu, Galuut, and Bust are included.

Lakes in eastern Mongolia (Fig. 1.1) are the highest in number but generally small in area (see Chap. 1). In this region, there are 648 lakes located below 1000 m a.s.l. (see Chap. 4), and the lakes constitute 75.7% of the total 856 lakes with the same elevation range in Mongolia (see Chap. 5). 97.4% of the total 885 lakes in the region (Table 1.2) cover an area less than 5.0 km^2 (Table 4.1), and they constitute 29.4% of the total lakes covering the same area range in the whole country (see Chap. 5). Although many small lakes are fed by groundwater, lakes in the

Table 6.7 Morphometric components of lakes in the southern and southwestern parts of eastern Mongolia

#	Lakes	Elevation (m a.s.l.)	Area (km^2)	Length (km)	Depth (m)	Volume (km^3)
1	Duut	1224	0.5	1.1	0.7	0.3
2	Ganga	1394	2.2	2.1	1.5	7.2
3	Sumt	1260	2.7	2.2	0.7	1.9
4	Erdene	1260	2.0	2.1	0.4	8.0
5	Kholboo	1320	0.07	0.5	0.3	0.02
6	Zuun Kholboo	1308	1.44	1.8	0.7	1.0
7	Tsagaan	1309	1.68	2.1	0.7	1.1
8	Khoshmogt	1313	0.18	0.7	0.4	0.07
9	Ulaan	1299	0.99	1.1	0.4	0.4
10	Zegst	1230	0.25	1.0	0.7	0.18

south of the region are fed only by precipitation. Table 6.7 shows lakes in the south and southwest of the region where some small lakes in the valleys (e.g., Khaltar Govi, Khoshuu Khooloi, etc.) are saline (see Chap. 5).

6.4 Lake Evolution

In the last decades, lakes in eastern Mongolia have been facing challenges regarding shrinkage and disappearance caused by the global warming. Some lakes often shrink and shift to marshy and playa environments during the drying time. A total of 332 out of 1829 large and small lakes have shrunk during the periods of the last three decades (Table 6.5). For example, Orkhonselenge et al. (2019) noted that most of the 18 lakes in the Binder soum of Khentii aimag have shrunk since 2003 and only Lake Onon Tsagaan has remained at the same size today. The lakes in eastern Mongolia cover an area of approximately 1000 km^2, and about 60% of this area (or 610 km^2) belongs to a single large lake, Lake Buir (see Chap. 7). Other lakes are relatively small (see Chap. 5), and their water regimes are highly dependent on annual and seasonal precipitations (Badarch et al., 1965; Orkhonselenge et al., 2019). The two largest, Lakes Buir and Khukh, are discussed in detail in Chaps. 7 and 8, whereas in this section other small lakes are briefly described as examples.

6.4.1 Lake Yakhi

Lake Yakhi is located at the elevation of 670 m a.s.l. in the Dornod plain (Fig. 1.1). There was a small marsh in the Yakhi Govi in the 1950s, and the marsh developed to become a lake covering an area of 97.2 km^2 by 1964 due to the increase of water

Table 6.8 Temporal changes in area of Lake Yakhi. After Orkhonselenge and Bulgan (2021)

#	Year	Area (km²)	Authors
1	1964	97.2	Tserensodnom (2000)
2	1970	79.72	Orkhonselenge et al. (2020)
3	1971	91.2	Tserensodnom (1971)
4	1986	53.76	Orkhonselenge et al. (2020)
5	2010	0.0	Davaa (2015)
6	2013	33.4	Davaa (2015)
7	2018	35.03	Orkhonselenge et al. (2020)

level since the 1960s (Table 6.8). The lake expands for 20.4 km long and 11.9 km wide at the maximum (Tserensodnom, 2000). Lake Yakhi is formed in a dry depression extended over 100 km in length and 7–15 km in width (Tsegmid, 1969), and it is surrounded by low mountains in the northwest, north, and east, and a wide plain in the south (Fig. 6.5). The lake (Fig. 6.6a) is fed by Gal River in the southwest (Figs. 6.5 and 6.6b) and streams Sumiin Bulag and Modot Bulag. The lake depth is 1.2 m on average and 2.5 m at the maximum (Davaa, 2015). The water level is unstable, and it is highly dependent on precipitation (Orkhonselenge et al., 2020).

Lake Yakhi covers an area of 35.03 km² (Table 6.8). The lake lost an area of 43.44 km² during 1964–1986 except for 1971, and 18.73 km² during 1986–2018, i.e., it has lost an area by ~62.2 km² or 63.9% of the total area during the last five decades (Table 6.8 and Fig. 6.5; Orkhonselenge & Bulgan, 2021). The hydrological fluctuations of Lake Yakhi show that the lake has shifted to a playa phase recently (Fig. 6.6a). The lake evolution may have been strongly contributed by the local climate trend with the rising annual average air temperature since 1987 and the falling annual precipitation since 1992 (Orkhonselenge & Bulgan, 2021). The linear regression analysis shows that the lake area between 1986 and 2018 was moderately related to temperature ($R^2 = 0.731$) and precipitation ($R^2 = 0.6872$). The correlation analysis shows that the lake area during 1986–2018 had strong negative correlations with the temperature ($r = -0.855$) and precipitation ($r = -0.829$).

6.4.2 *Lake Ganga*

Lake Ganga is located at the elevation of 1394 m a.s.l. in the south of the region, and it extends for 2.1 km long and is 1.5 m deep, covering an area of 2.2 km² (Table 6.7). Lake Ganga, a freshwater lake, was originally formed by damming of windblown sands. The lake is fed by about 21 streams with a total water resource of 29 L/s per year. Among the streams, the largest Orgikh stream (6.3–12.0 L/s) and Dagshin stream (14.8–19.0 L/s) are included in the Central Asian internal drainage basin. The lake level decreases due to intense sand accumulation in the lake, water pollution by livestock, and human impact.

Lake Ganga's water level started to fall in 2008, and the lake shrank significantly in 2016 (Enkhtaivan et al., 2016). The year of 2016 had the hottest summer in the Lake Ganga basin exceeding 30 °C with rainless days from the late July to the early September since the 1940s (Oyunbaatar et al., 2017). Because of such extreme weather conditions, at the margin of the lake, the depth of 40–50 cm decreased to 18–25 cm, while the depth of 1.5 m in the center dropped to 0.5 m between 2007 and 2011 when the lake area decreased by 0.53 km^2 or 25% (Oyunbaatar et al., 2017). Lake Ganga's water level dropped by 50–60 cm in recent 3–4 years when about 31,000 livestock gathered in the lake basin seriously affecting the water quality and morphometry of the lake. In recent years shrinkage of the lake is strongly influenced by the livestock watering in addition to the aridity.

6.4.3 Lake Toson

Lake Toson (Fig. 6.7a) is located at the elevation of 1232 m a.s.l. in the west of the region, and it is connected with Lake Burd (Fig. 6.7b) by Avarga and Khug Rivers. The lake mineralization value is 9.52 g/L (Tserensodnom, 2000), and the lake water (pH = 9.2) is hydrochemically characterized by HCO_3^-, Ca^{2+}, Na^++K^+, Cl^-, and SO_4^{2-} ions (Orkhonselenge et al., 2019; Table 6.9). The Lake Toson was often polluted due to the human activity for using leech, and the lake water level had fallen due to the decreased precipitation and the increased temperature since 2000 (Orkhonselenge et al., 2019). The Normalized Difference Water Index (NDWI) shows that the lake area decreased in 1998, 2006, 2009, and 2010 (Gerelsaikhan & Orkhonselenge, 2019) due to (1) the strengthened aridification related to evaporation and (2) the over usage of dark clay in the lake for therapy (Orkhonselenge et al., 2019). Because the Avarga and Khug Rivers also shrank due to the human activity, at present local nomads use only groundwater.

6.4.4 Lake Burd

Lake Burd is located at the elevation of 1230 m a.s.l. in the west of the region, and it lies at 300 m away to the northeast of Lake Toson. The lake (Fig. 6.7b) extends over 1.0 km long, 0.8 km wide, and 5.3 km along the shoreline length. Lake Burd is circle shaped, and it covers an area of 0.9 km^2. An outlet of the lake is Burd River, a headwater of Avarga River. The lake water (pH = 7.9) temperature is 18–21 °C in July, and clay and sands are found in its bottom (Tserensodnom, 2000). In recent years the hydrochemistry of the lake has changed due to the human activity and climate change, and HCO_3^-, Cl^-, Ca^{2+}, and SO_4^{2-} are dominant in the lake water (Orkhonselenge et al., 2019; Table 6.9). The spatial patterns in 1998, 2006, 2009, and 2010 estimated by NDWI show that the lake area dramatically decreased between 1998 and 2010 and that the abrupt drop in the lake area is related to the dryness since 2000 recorded at the Delgerkhaan meteorological station near the lake (Gerelsaikhan & Orkhonselenge, 2019).

Fig. 6.7 (**a**) Lake Toson viewing toward the southeast. (**b**) Northern beach of Lake Burd viewing toward the east. Photos by A. Orkhonselenge on August 1, 2014

Table 6.9 Hydrochemical composition of Lakes Toson and Burd. After Orkhonselenge et al. (2019)

	Lake Toson		Lake Burd			Lake Toson		Lake Burd	
Anion	mg/dm^3	%	mg/dm^3	%	Cation	mg/dm^3	%	mg/dm^3	%
Cl^-	1.562	28.22	56.8	35.3	Na^++K^+	4.01	98.3	10.3	9.9
SO_4^{2-}	1.08	14.43	35	16.1	Ca^{2+}	3.53	1.3	48.1	53.0
NO_2^-	–	–	–	–	Mg^{2+}	0.73	0.4	19.5	35.3
NO_3^-	0.2	–	0.2	–	NH_4^+	0.30	–	1.5	1.8
CO_3^{2-}	0.825	17.64	–	–	Fe^{2+}	–	–	–	–
HCO_3^-	3.776	39.70	134.2	48.6	Fe^{3+}	–	–	–	–
Total	7.4	100.0	226.2	100.0	Total	8.57	100.0	79.4	100.0

6.5 Summary

Eastern Mongolia is characterized by the wide Dornod plain throughout the region, which is surrounded by the Khentii and Khyangan Mountain Ranges in the northwest and southeast. This region hosts a large lake and many small lakes formed in tectonic valleys and depressions, fluvial valleys, and thermokarst caverns and by moraine damming. In eastern Mongolia, the highest number of lakes in the country (28.9%) occupy small areas of the total lake area (9.0%), and fluvial, thermokarst, aeolian, and glacial lakes are predominant. Lakes in the south and west of the region are shallow and saline, and they are highly dependent on precipitation, whereas lakes in the north and east of the region are deep and freshwater fed by precipitation, runoff, and groundwater. In recent years, many lakes in eastern Mongolia are mineralizing and shifting into marshy and playa environments. For instance, the water levels of small Lakes Toson and Burd have been falling, their areas are also decreasing, and they are hydrochemically salinizing. Such changes of surface water, especially in lakes, have intensified since the 2000s due to the rapid climate change of the rising air temperature and strengthening evaporation. The climate in eastern Mongolia is strengthening its tendency to shift from semiarid into arid during the late Holocene. The drying trend is hypothesized to continue in the next 100 years. The climate intensely modifies the landscape and lacustrine environments of eastern Mongolia.

References

Academy of Sciences of Mongolia and Academy of Sciences of USSR. (1990). *National Atlas of the People's Republic of Mongolia.* 156 p. [In Russian].

Altantsetseg, J., & Munkhjargal, E. (2006). Regime and change of air temperature in eastern regions of Mongolia. In *Proceedings of climate change in eastern regions of Mongolia, Ulaanbaatar*, pp. 25–32.

Ariunbileg, S., Isupov, V. P., Vladimirov, A. G., Kolpakova, M. N., & Kuibida, L. V. (2020). *Chemical evolution and geochemistry of mineralized lakes of Mongolia.* Admon Printing House, 132 p. [In Mongolian].

Ariyadagva, B., Dorjsuren, P., Shpeizer, G. M., & Pisarsky, B. N. (1980). *Balneological condition in hydrochemistry for some lakes in eastern Mongolia.* Scientific Journal of Institute of Chemistry, [In Mongolian].

Badarch, N., Tsegmid, S., & Tserensodnom, J. (1965). *Landscapes and natural zones of eastern Mongolia.* Press of Mongolian Academy of Sciences, 117 p. [In Mongolian].

Davaa, G. (2015). *Resource and regime of surface water of Mongolia.* 408 p. [In Mongolian].

Davi, N. K., Pederson, N., Leland, C., Nachin, B., Suran, B., & Jacoby, G. C. (2013). Is eastern Mongolia drying? A long-term perspective of a multidecadal trend. *Water Resources Research, 49*, 151–158.

Enkhtaivan, D., Oyungerel, B., Avirmed, E., Renchinmyadag, T., Nyamkhuu, M., Munkhdulam, O., Odbaatar, E., Davaagatan, T., Bayanjargal, B., & Batnyam, T. (2016). *Report on research project: Landscape structure, change, planning and proper zonation (the case study of eastern Mongolia).* Division of Physical Geography, Institute of Geography and Geoecology, Mongolian Academy of Sciences, 564 p. [In Mongolian].

Gerelsaikhan, D., & Orkhonselenge, A. (2019). Water pollution and area changes of Lakes Toson and Burd. In *Proceedings of The Environment-2019*, Ulaanbaatar, p. 48. [In Mongolian].

Hara, H., Kurihara, T., Tsukada, K., Kon, Y., Uchino, T., Suzuki, T., Takeuchi, M., Nakane, Y., Nuramkhaan, M., & Chuluun, M. (2013). Provenance and origins of a Late Paleozoic accretionary complex within the Khangai–Khentei belt in the Central Asian Orogenic Belt, central Mongolia. *Journal of Asian Earth Sciences, 75*, 141–157.

Jambaajamts, C. (1989). *Climate of Mongolia*. [In Mongolian].

Jambaljav, Y., Gansukh, Y., Temuujin, K., Tsogt-Erdene, G., Undrakhtsetseg, T., Saruulzaya, A., Amarbayasgalan, Y., Dashtseren, A., & Narangerel, S. (2016). *Report of new map of permafrost distribution in Mongolia: Scale 1:1000000*. Institute of Geography and Geoecology, Mongolian Academy of Sciences, 90 p.

Jigj, S. (1975). *Primary feature of Mongolian landforms*. Institute of Geography and Permafrost. Mongolian Academy of Sciences (MAS), 126 p. [In Mongolian].

Jigj, S. (1976). *A brief description of paleoglaciations and paleoglaciers*. 51 p. [In Mongolian].

Kang, S., Lee, G., Togtock, C., & Jang, K. (2015). Characterizing regional precipitation-driven lake area change in Mongolia. *Journal of Arid Land, 7*(2), 146–158.

Khandsuren, P., Seong, Y. B., Oh, J. S., Rhee, H. H., Sandag, K., & Yu, B. Y. (2019). Late quaternary glacial history of Khentey Mountains, Central Mongolia. *Boreas, 48*(3), 779–799.

Li, J., Cook, E. R., D'Arrigo, R., Chen, F., & Gou, X. (2009). Moisture variability across China and Mongolia: 1951–2005. *Climate Dynamics, 32*, 1173–1186.

Linhoff, B. S., Bennett, P. C., Puntsag, T., & Gerel, O. (2011). Geochemical evolution of uraniferous soda lakes in Eastern Mongolia. *Environmental Earth Science, 62*, 171–183.

Luvsandorj, S., & Tumur, S. (1969). Hydrochemical characteristics of lakes in the Dornod plain in eastern Mongolia. *Journal of Mongolian Academy of Sciences*.

Miao, L., Zhang, F., Baatar, M., Zhu, M., & Anaad, C. (2017). SHRIMP zircon U-Pb ages and tectonic implications of igneous events in the Ereendavaa metamorphic terrane in NE Mongolia. *Journal of Asian Earth Sciences, 144*, 243–260.

Ministry of Environment and Green Development. (2013). *Integrated water management plan of Mongolia*. 339 p.

Murzaev, E.M. (1952). *The description of physical geography of Mongolia* (2nd Ed.). [In Russian].

Oh, J. S., Seoung, Y. B., Hong, S. C., & Yu, B. Y. (2019). Paleo-shoreline changes in moraine dammed Lake Khagiin Khar, Khentey Mountains, Central Mongolia. *Journal of Mountain Science, 16*(6), 1215–1230.

Orkhonselenge, A., & Bulgan, O. (2021). Geochemical studies and lacustrine geomorphology of Lake Yakhi basin in eastern Mongolia. *Géomorphologie: Relief, Processes, Environnement, 27*(3), 231–242. https://doi.org/10.4000/geomorphologie.15873

Orkhonselenge A., Bulgan O., Gerelsaikhan D., Davaagatan T., & Altansukh N. (2020). Hydroclimatic fluctuation in Lake Yakhi, eastern Mongolia. *EGU General Assembly. EGU2020-686 presentation*.

Orkhonselenge, A., Davaanyam, S., Altanbagana, M., & Bolorchuluun, C. (2019). *Long-term changes in ecosystems in natural zones of Mongolia: A report of Research Project funded by Science and Technology Foundation (ShUT_A/349_005/2013)*. Laboratory of Geochemistry and Geomorphology, National University of Mongolia. NUM Publishing, 384 p. [In Mongolian].

Orkhonselenge, A., & Uuganzaya, M. (2018). Glacial Geomorphology of Mt. Asralt Khairkhan and Mt. Baga Khentii Saridag in Khentii Mountain Range, Northeastern Mongolia. *International Journal of Geosciences, 9*(6), 308–319.

Orkhonselenge, A., Uuganzaya, M., Davaagatan, T., & Enkhbayar, G. (2021). Late Holocene Peatland Evolution in Terelj and Tuul Rivers Drainage Basins in the Khentii Mountain Range of Northeastern Mongolia. *Water, 13*, 562. https://doi.org/10.3390/w13040562

Oyunbaatar, D., Erdenebayar, B., Davaa, G., & Saikhanjargal, D. (2017). Recent changes of water regime and resource of the Ganga Lake and related some socio-economic aspects. *Modern Environmental Science and Engineering, 3*(7), 482–491.

Tsegmid, S. (1969). *Physical geography of Mongolia*. State Press, 405 p. [In Mongolian].

Tserensodnom, J. (1971). *Lakes of Mongolia*. State Publishing, 202 p. [In Mongolian].

Tserensodnom, J. (2000). *A catalog of lakes in Mongolia*. Shuvuun Saaral Publishing, 141 p. [In Mongolian].

Tumurbaatar, D. (2004). *Seasonal and Permanent Permafrost of Mongolia*. Mongolian Academy of Sciences, 147 p. [In Mongolian].

UNESCO. (2020). *World Heritage List*. Retrieved from https://whc.unesco.org/en/list/&order=country#alphaM.

Zorigt, M., Myagmar, K., Orkhonselenge, A., van Beek, E., Kwadijk, J., Tsogtbayar, J., Yamkhin, J., & Dechinlkhundev, D. (2020). Modelling permafrost distribution over the river basins of Mongolia using remote sensing and analytical approaches. *Environmental Earth Sciences, 79*(308), 1–11.

Chapter 7
Lake Buir

Abstract Lake Buir represents a major surface water resource in eastern Mongolia. The lake plays an important role in the hydrological system of the Pacific Ocean drainage basin. Lake Buir gained an area by 10.43 km^2 during the last 50 years between 1965 and 2015. This increase in area may have been related to the recent slight rise of the annual precipitation with peaks in 1968–1974, 1987–1998, and 2002–2018 as observed in the lake basin. The regression analysis shows that the lake area during 1965–2015 had no clear linear relationships with the annual average precipitation ($R^2 = 0.0267$) and annual average air temperature ($R^2 = 0.0223$). The correlation analysis shows that the area of Lake Buir between 1965 and 2015 had a weak positive correlation with precipitation ($r = 0.163$) and a weak negative correlation with temperature ($r = -0.149$). The climate trend shows that Lake Buir basin has become gradually warmer during the last six decades and more humid during the last three decades compared with the previous three decades. The local climate trend implies that the impact of the East Asian Summer Monsoon may still reach the Lake Buir basin today. Geomorphological, geochronological, and paleoclimatological investigations in the Lake Buir basin will help to reconstruct the Quaternary, especially Holocene climate change in eastern Mongolia.

Keywords Lake Buir · Khalkh River · Dornod plain · Climate change · Eastern Mongolia

7.1 Introduction

Lake Buir, a tectonic lake, is the largest lake in eastern Mongolia, and it is located at the border between Mongolia and China (Fig. 1.1). Early studies observed and noted physiography and morpho-dynamic changes of Lake Buir. Physiography of Lake Buir was initially defined by G.N. Potanin (1899), V.A. Kormazov (1928), and E.M. Murzaev (1949); however data regarding depth, bathymetry, and deposits were not provided (Tsegmid, 1969; see Chap. 2). Russian geographer G.N. Potanin

A. Orkhonselenge et al., *Lakes of Mongolia*, Syntheses in Limnogeology,
https://doi.org/10.1007/978-3-030-99120-3_7

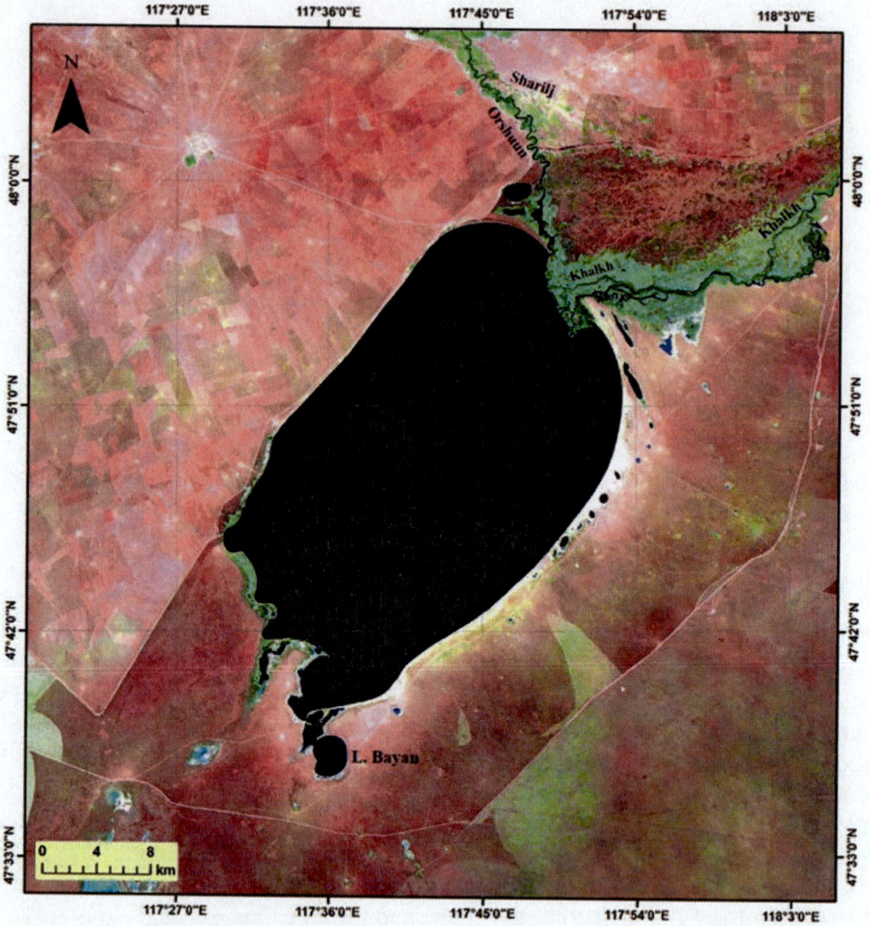

Fig. 7.1 Landsat 8 image of Lake Buir on September 14, 2015 (path of 124 and row of 27). Modified from Orkhonselenge and Uuganzaya (unpublished)

conducted survey on Khalkh River, Lake Buir, and Khyangan[1] Mountain Range in eastern Mongolia, and he noted based on their shorelines that Lake Buir had been connected with the neighboring Lake Bayan in the south (Fig. 7.1) forming a single lake in the past (Tserensodnom, 1971).

Later on, morpho-dynamic changes of Lake Buir were studied by Badarch et al. (1965), Tsegmid (1969), Tserensodnom (1971, 2000), and Enkhtaivan et al. (2016). According to Jigj (1975), sediments began to be deposited in Lake Buir in the early Carboniferous (see Chap. 3). In 1962, J. Tserensodnom provided bathymetric data of Lake Buir for the first time (Davaa, 2015). Since then, in 1985 Ts. Sugar and O. Tserev and in 1989 N. Dashdeleg and T. Baatar provided morphometric values

[1] *Khyangan* has been misspelled as *Khingan* (or *Hingan*) and *Hyangan* in many publications. Khyangan is the right English transliteration from Mongolian *Хянган*.

regarding area, depth, volume, shoreline, and mineralization (Davaa, 2015). Lake Buir is classified as a plain lake (see Chap. 5) based on its morphometric values (e.g., shoreline, length, width) and the plain landscape (see Chap. 6). The Lake Buir level has been measured since 1966 but for its water balance since 1974 (Davaa, 2015). Nonetheless, comprehensive studies in Lake Buir have not been conducted in terms of geomorphology, geochemistry, sedimentology, and paleoclimatology.

Lake Buir is included in the Pacific Ocean drainage basin, and it plays an important role in the surface water of eastern Mongolia. It contains over 60% of the total surface water resources (about 1000 km^2) in this region (see Sect. 7.6). Sedimentary records from Lake Buir can reconstruct a regional pattern on paleoclimate and paleoenvironmental changes in eastern Mongolia and southeastern Siberia over various geological time scales. Understanding and reconstructing past climate and environmental changes in eastern Mongolia and Eurasia are important for the prediction of impacts of further climate change scenarios. Because there are still insufficient data for inferring the paleoclimate changes in eastern Mongolia (see Chap. 21), geomorphological and paleogeographical implications gained from this region will be added to the paleoclimatic data in Mongolia and climatically sensitive adjacent regions in Eurasia.

In this chapter, previously collected data on physiography and hydrology of Lake Buir are synthesized, and changes in the lake area over the last five decades are discussed in detail.

7.2 Physiographic Condition

Lake Buir is located at the elevation of 581 m a.s.l. in the easternmost corner of Mongolia (Fig. 1.1), and it is surrounded by the Del Khoshuu plain (665 m a.s.l.) in the east and southeast, marshy area (633 m a.s.l.) with numerous small saline lakes in the south and southwest, and the Menen plain (606 m a.s.l.) and the Dornod plain (647 m a.s.l.) in the west (Orshikh et al., 2014). Morphometric values of Lake Buir (e.g., depth, area, etc.) were previously provided by Badarch et al. (1965), Tsegmid (1969), and Tserensodnom (1971, 2000), as shown in Table 1.3. Lake Buir covers an area of 625.4 km^2 (Orkhonselenge and Uuganzaya, unpublished; see Sect. 7.6), extending over 40.0 km in the maximum length, about 21.0 km in the maximum width, and 118.2 km along the shoreline length (Table 1.3).

In comparison to the large area of Lake Buir, its depth is shallow and water volume is low (Tsegmid, 1969; Table 7.1). Its average depth is 6.0 m, and the maximum depth is 10.0 m (Table 7.1). Others report that the water depth is 7–8 m on average, and it reaches 10.1 m in the deepest part at the center of the lake (Badarch et al., 1965; Tserensodnom, 1971). The maximum depth of 10.2 m was measured by Ts. Sugar and O. Tserev in 1985 and of 10.4 m by N. Dashdeleg and T. Baatar in 1989 (Davaa, 2015). Recent studies show that the water depth is 10.0 m on average and 15.0 m at the maximum in the center of the lake (Enkhtaivan et al., 2016).

The lake bottom sediment consists of dark muds, clays, and sands. Fine sands are widely distributed on the beach and shallow parts of the lake, but clays and muds with odors of H_2S occur in other parts of the lake (Tsegmid, 1969; Tserensodnom, 1971).

Table 7.1 Bathymetric data with corresponding area and volume of Lake Buir

#	Water depth (m)	Area (km^2)	Volume (km^3)
1	0.0	614.7	3.7
2	2.0	549.0	2.6
3	4.0	426.0	1.6
4	6.0	319.6	0.8
5	8.0	239.4	0.3
6	10.0	45.4	0.02

7.3 Geological and Geomorphological Settings

Lake Buir is underlain by the Neogene and Quaternary deposits with marine molasses in the Cenozoic platform structure (Academy of Sciences of Mongolia and Academy of Sciences of USSR, 1990). The lake covers geological formations of a continental platform consisting of the Cenozoic red and gray terrigenic sands, gravels, and pebbles (Academy of Sciences of Mongolia and Academy of Sciences of USSR, 1990). Lake Buir is surrounded by the Paleogene and Neogene system along the transection from the northeast to the southwest, and middle and late Quaternary deposits in the west and south (Academy of Sciences of Mongolia and Academy of Sciences of USSR, 1990).

Lake Buir is formed in a tectonic graben (Badarch et al., 1965) of the Dornod plain (Fig. 1.1). Lake Buir is included in the plain lake category (see Chap. 5) due to its morphology and landscape (see Chap. 6). The lake floor is relatively flat in the center. The beach slope is gentle, and shorelines are smooth and flat (Tserensodnom, 1971). Some beaches and inlets are isolated by sandy bays (Fig. 7.2). High terraces are developed on the northern, northeastern, and eastern beaches. Transgressions of Lake Buir formed terraces in the northeast, and the highest terrace is 6–10 m above the lake level, whereas the lowest is 2–3 m (Tserensodnom, 1971). The highest terrace is elevated at 593 m a.s.l., and it is covered by dense plants, whereas the lowest one is elevated at 585 m a.s.l. and it is bumpy (Enkhtaivan et al., 2016). Low hills between them are elevated at 588 m a.s.l., and they are 3–8 m high and covered by bushy plants (Enkhtaivan et al., 2016).

Lake Buir is oval shaped with its major axis elongated in the direction of northeast to southwest (Fig. 7.1). A wide delta is formed in the downstream of Khalkh River on the northeastern beach (Fig. 7.1). There is a 3–5 m high sandy dam (Fig. 7.2) formed on the northeastern, eastern, and southeastern beaches during the past cycles of transgressions and regressions, and the dam is transected by small lakes, some of which are connected with Lake Buir (Tsegmid, 1969; Tserensodnom, 1971). There are small Lakes Khujirt and Olgoi on the northern beach and many small lakes (e.g., Lakes Nariin, Khar, Zuun Zakh, Dund Zakh, Baruun Zakh, Takhir, and Khukh Us) formed on the eastern beach (Fig. 7.1). These lakes are isolated from Lake Buir by 6–10-m-high sandy hills (Enkhtaivan et al., 2016, Fig. 7.2). These lakes on the southeastern beach are circle shaped, and they extend from the northeast to the southwest along the beach of the main lake (Fig. 7.1). In the northeast,

Fig. 7.2 View of sand dunes on the northeastern shore of Lake Buir viewing toward the (**a**) southeast and (**b**) southwest. Photos by T. Davaagatan on June 24, 2016

east, southeast, and south of the lake, aeolian and fluvial erosion are active and form small sand dunes (Fig. 7.2), and the sand dunes range from 0.9 to 1.1 km in width (Enkhtaivan et al., 2016).

7.4 Climate Condition

The local climate in the Lake Buir basin is recorded at the Khalkh Gol meteorological station, positioned at about 60 km east of the lake. The annual average air temperature was −0.1 °C, and the annual average precipitation was 301.5 mm between

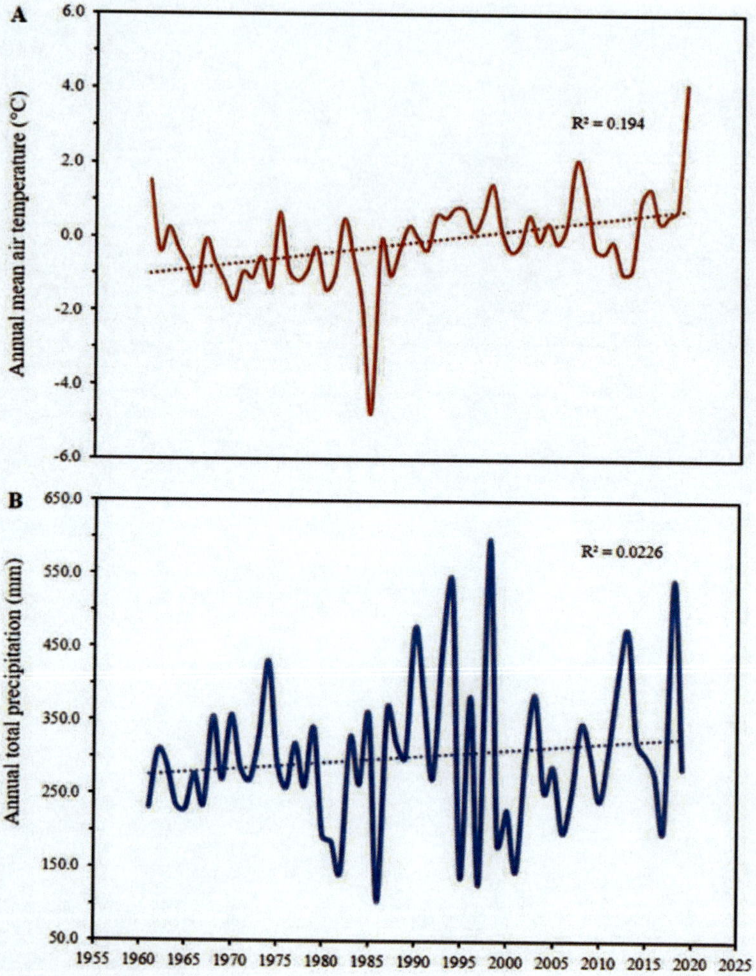

Fig. 7.3 Annual average (**a**) air temperature and (**b**) precipitation recorded at the Khalkh Gol meteorological station during 1961–2019

1961 and 2019 (Fig. 7.3). The warmest year was recorded with the annual average air temperature of 2.1 °C in 2007 (Fig. 7.3a). Other warmer years were recorded with the annual average air temperatures of 1.4 °C in 1998, 1.3 °C in 2008, 1.0 °C in 2014, and 1.3 °C in 2015. The coldest year was recorded with the annual average air temperature of −4.9 °C in 1985 (Fig. 7.3a). The air temperature variation shows a warming trend from 1989 to 2019 with intermittent or continuous rates.

The driest year was in 1986 when the annual average precipitation was recorded at 105.1 mm (Fig. 7.3b). Other drier years occurred in 1964–1966 with the annual average precipitation of 229.5–277.2 mm, 1980–1982 with 145.4–195.1 mm, and 1999–2001 with 146.3–229.3 mm. Three humid periods were found in 1968–1974,

1987–1998, and 2002–2018 (Fig. 7.3b). Their peaks were recorded at 431.1 mm in 1974, 477.5 mm in 1990, 536.7 mm in 1994, 598.1 mm in 1998, 473.3 mm in 2013, and 542.5 mm in 2018. The local climate data show a combination of climate trends in arid and cool and/or humid and warm within the Lake Buir basin during the last 60 years (Fig. 7.3).

7.5 Hydrological Condition

The major water resource of Lake Buir is river inflow. Lake Buir is primarily fed by the Khalkh and Tatuur Rivers in the northeast (Figs. 7.1 and 7.4) draining from an area of 20,200 km^2 (Murzaev, 1952). Lake Buir feeds only Orshuun River (Figs. 7.1 and 7.4), one of the inflows for Lake Dalai,[2] in the northeastern China. In the past, Lake Buir also fed the Kholboo stream in the southwest. This stream feeds Lake Bayan (Fig. 7.1); however, the stream activity often diminishes, and it drains only during the years of high precipitation (Tsegmid, 1969). Today the stream disappeared (Figs. 7.1 and 7.4). The water volume of the lake is 3.7 km^3 (Tsegmid, 1969; Table 7.1) or 8.7 km^3 (Davaa, 2015) on average.

The water level of Lake Buir varies depending on annual rainfall and discharge of the inflowing rivers, and it reaches its maximum in August (Tsegmid, 1969). The water level dropped by 1.0–1.5 m since Sharilj River (Figs. 7.1 and 7.4), an inflow of Khalkh River, changed its direction toward the Orshuun River (Murzaev, 1952). Lake Buir is fed by the river inflow at 56.6%, groundwater at 33.5%, and precipitation at 9.9%. The lake loses its water due to river outflow at 62.4% and evaporation at 37.6% (Davaa, 2015).

Table 7.2 shows that water temperature of Lake Buir ranges with depth (Tserensodnom, 1971). The annual maximum temperature of the lake water decreased by −0.14 °C/year in 1966–1989 and increased by 0.18 °C/year in 1990–2008 (Davaa, 2015). The temperature trend of the lake water coincided with the variation in the annual mean air temperature during 1961–2019 (see Sect. 7.4). Lake Buir is frozen, and it is covered by a 1.1–1.5-m-thick ice between November and April (Table 7.2).

Lake Buir is a freshwater lake. The hydrochemical composition of Lake Buir depends on inflows. The mineralization value of the lake is 302.1 mg/L, consisting of HCO_3^- at 208.7 mg/L, Na^++K^+ at 54.1 mg/L, Ca^{+2} at 29.6 mg/L, and other ions at 9.7 mg/L (Tsegmid, 1969). In 1950s, the value of mineralization was about 240.0 mg/L in the downstream of Khalkh River (Tserensodnom, 1971). According to the Institute of Livestock, in 1965, the mineralization value ranged between 298.7–365.9 mg/L in the west and 200.5–294.1 mg/L in the east of the lake (Tsegmid, 1969; Tserensodnom, 1971). Table 7.3 shows the ion composition of the lake water, which was determined by Tsend (1966).

---

[2] *Dalai* has been often spelled as *Hulun* in many publications. Dalai is the right English transliteration from Mongolian *Далай* meaning *ocean*.

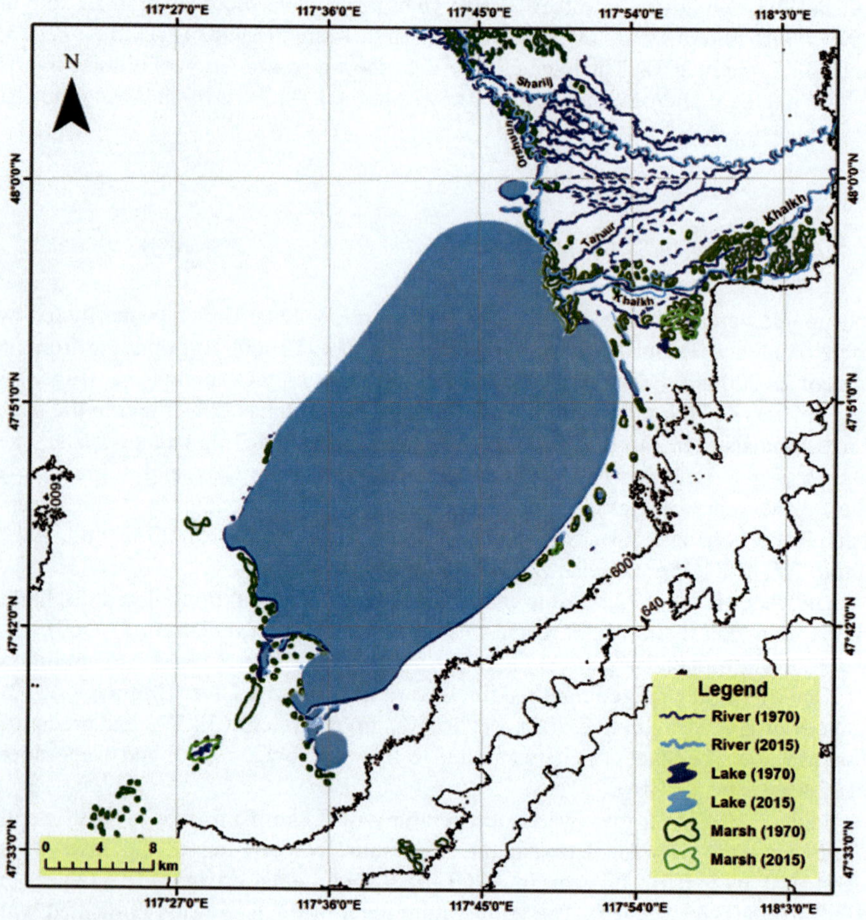

Fig. 7.4 Spatial distributions of areas and marshes of Lake Buir in 1970 and 2015. Modified from Orkhonselenge and Uuganzaya (unpublished)

Table 7.2 Water temperature and ice thickness of Lake Buir

#	Position		June–August (°C)	January (°C)	Ice thickness (m)
1	Surface	Center	25.0–28.0	0.5–1.0	1.1–1.2
		Margin	–	–	1.5
2	5–6 m depth		–	2.0	–
3	Bottom		18.0–20.0	2.0–5.0	–

Table 7.3 Hydrochemical composition of Lake Buir

Ions	HCO_3^-	CO_3^{-2}	SO_4^{-2}	Cl^-	Ca^{+2}	Mg^{+2}	Na^++K^+
mg/L	184.4	1.1	17.1	22. 1	28.7	12.5	36.4

7.6 Changes in Area

Lake Buir has been experiencing a relative rise of its area since 1965 with multiple repeated fluctuations (Table 7.4). The changes of the Lake Buir area during the past 50 years show two trends in times of gaining area in 1965–1970 and 1979–2013, and times of losing area in 1970–1979 and 2013–2015 (Table 7.4). The expansion and reduction in area for Lake Buir over the last five decades may have been related to the local climate trend of the gradually increasing temperature (Fig. 7.3a) and the slightly rising precipitation (Fig. 7.3b).

The increase in area by 11.16 km^2 in 1965–1970 (Table 7.4) shows that Lake Buir responded to the increase of precipitation from 231.5 mm/year to 357.6 mm/year in those years (Fig. 7.3b) when the low temperatures of −0.1 °C to −1.7 °C were recorded (Fig. 7.3a). At that time, the Lake Buir basin may have been humid with low evaporation. The decrease in area by 24.69 km^2 in 1970–1979 coincided with low precipitations, with an exception in 1974, and low temperatures, with an exception in 1975. By 1998, the lake enlarged its area by 15.53 km^2 compared with that in 1979 (Table 7.4). The increase coincided with low precipitations in 1980–1986 and high precipitations in 1987–1998, except for the drops in 1995 and 1997 (Fig. 7.3b). The period matched with low temperatures in 1979–1991 and high temperatures in 1992–1998 (Fig. 7.3a). The last period of the precipitation may have contributed to the rise of the lake area.

By 2010, the lake expanded its area by 3.6 km^2 compared with 1998, and by 2013 it also gained an area by 11.6 km^2 in comparison to 2010 (Table 7.4). Overall, by 2015 the lake enlarged in area over the previous years, except for those in 1970 and 2013. The long-term decrease of area by 0.73 km^2 in 1970–2015 (Fig. 7.4) and the short-term decrease of area by 6.77 km^2 in 2013–2015 were found (Table 7.4). The last decrease is in agreement with the rise of temperature from −0.9 to 1.3 °C (Fig. 7.3a) and abrupt decrease in precipitation in those years (Fig. 7.3b).

Overall, Lake Buir gained its area ending in 2015 by 10.43 km^2 since 1965 and by 23.96 km^2 (~24.0 km^2) since 1979 (Table 7.4). The gains coincided with the remarkable rise of the water volume of Lake Buir by ~5.0 km^3 over the last 30 years (see Sect. 7.5). The noticeable increase of the lake area may be consistent with the

Table 7.4 Temporal changes in area of Lake Buir

#	Year	Area (km^2)	Authors
1	1965	610.0	Badarch et al. (1965)
2	1970	626.16	Orkhonselenge and Uuganzaya (Unpublished)
3	1971	615.0	Tserensodnom (1971)
4	1979	601.47	Davaa (2015)
5	1998	617.0	Davaa (2015)
6	2010	620.6	Davaa (2015)
7	2013	632.2	Davaa (2015)
8	2015	625.43	Orkhonselenge and Uuganzaya (Unpublished)

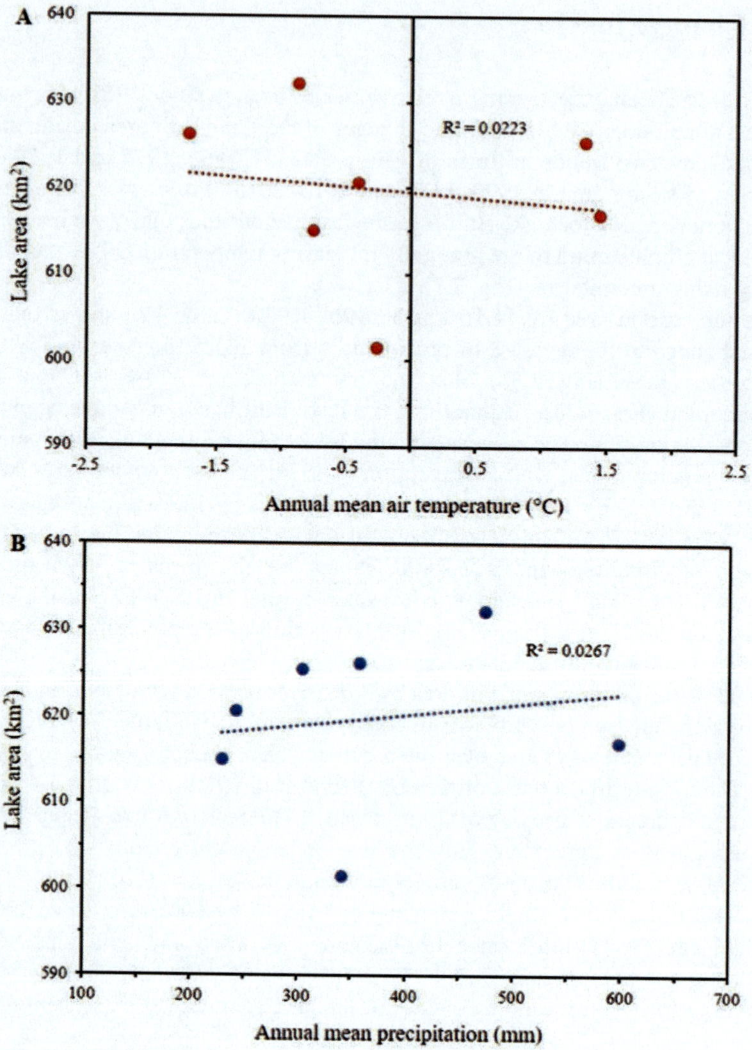

Fig. 7.5 Relationships between area of Lake Buir and annual average (**a**) air temperature and (**b**) precipitation during 1965–2015

slight rise in precipitation during the last three decades, although the air temperature has been gradually rising during the past six decades.

The linear regression analysis shows that the area of Lake Buir during 1965–2015 did not exhibit clear relationships with the annual average air temperature ($R^2 = 0.0223$ in Fig. 7.5a) nor precipitation ($R^2 = 0.0267$ in Fig. 7.5b). The correlation analysis shows that the lake area between 1965 and 2015 had a weak positive correlation with precipitation ($r = 0.163$) and a weak negative correlation with temperature ($r = -0.149$). For the lake area, the positive correlation with precipitation and the negative correlation with temperature were also observed in Lake Khukh in the same region although the lake lost its area during the last 35 years (see Chap. 8).

The enlarging Lake Buir differs from the lakes in the Valley of Lakes in southern Mongolia (see Chaps. 9–11), where the lakes have been sensitive to the fluctuating precipitation since 1975 and the rising temperature since 1995 (Orkhonselenge et al., 2018). However, expanded lakes over the past five decades, showing the same trend in the correlation with precipitation and temperature, are also observed for lakes in the mountainous regions of western (see Chap. 14), northern (see Chap. 16), and central (see Chap. 19) Mongolia, where the lakes have strong correlation with feeding of glaciers and permafrost meltwaters under the rising temperature conditions.

7.7 Summary

Lake Buir experienced significant fluctuations of its area extension during the last 50 years. The overall gain in area by 10.43 km^2 between 1965 and 2015 is noted. The linear regression analysis shows that the lake area between 1965 and 2015 had no clear relationship with the annual average precipitation (R^2 = 0.0267) and air temperature (R^2 = 0.0223). Nevertheless, the lake area during 1965–2015 had a weak positive correlation with precipitation ($r = 0.163$) and a weak negative correlation with temperature ($r = -0.149$). The Lake Buir basin was relatively arid with the annual precipitation less than 477.5 mm/year until 1994, and cool with the annual average air temperature lower than 0.0 °C until 1992. The lake basin has recently become more humid (up to 598.1 mm/year) and warmer (up to 2.1 °C) than in the previous 30 years. The local climate trend shows that the impact on the lake basin of the cool and arid climate was replaced by that of the warm and humid climate since the beginning of the 1990s. Lake Buir represents an important archive site to reconstruct the atmospheric circulation shift in the past as it is experiencing both the East Asian Summer and Winter Monsoons. In the Lake Buir basin, further geomorphological, geochronological, and paleoclimatological studies are needed.

References

Academy of Sciences of Mongolia and Academy of Sciences of USSR. (1990). *National Atlas of the Mongolian People's Republic*. 144 p. [In Russian].

Badarch, N., Tsegmid, S., & Tserensodnom, J. (1965). *Landscapes and natural zones of eastern Mongolia*. Press of Mongolian Academy of Sciences, 117 p. [In Mongolian].

Davaa, G. (2015). *Resource and regime of surface water of Mongolia*. 408 p. [In Mongolian].

Enkhtaivan, D., Oyungerel, B., Avirmed, E., Renchinmyadag, T., Nyamkhuu, M., Munkhdulam, O., Odbaatar, E., Davaagatan, T., Bayanjargal, B., & Batnyam, T. (2016). *Landscape structure, change, planning and proper zonation (the case study in Eastern Mongolia): Report of Research Project funded by Science and Technology Foundation*. Institute of Geography and Geoecology, Mongolian Academy of Sciences, 564 p. [In Mongolian].

Jigj, S. (1975). *Primary feature of Mongolian landforms*. Institute of Geography and Permafrost. Mongolian Academy of Sciences (MAS), 126 p. [In Mongolian].
Murzaev, E. M. (1952). *The description of physical geography of Mongolia* (2nd ed.) [In Russian].
Orkhonselenge, A., Komatsu, G., & Uuganzaya, M. (2018). Climate-driven changes in lake areas for the last half century in the Valley of Lakes, Govi region, southern Mongolia. *Natural Science, 10*(7), 263–277.
Orshikh, N., Enkhbayar, G., Garig, G., Enkhbat, R., & Chantuu, B. (2014). *Atlas of Mongolia. 1:800,000*. Monsudar Press, 117 p. [In Mongolian].
Tsegmid, S. (1969). *Physical geography of Mongolia*. State Press, 405 p. [In Mongolian].
Tsend, N. (1966). *Chemical composition of lakes in eastern Mongolia* (p. 15). Institute of Livestock. [In Mongolian].
Tserensodnom, J. (1971). *Lakes of Mongolia*. State Publishing, 192 p. [In Mongolian].
Tserensodnom, J. (2000). *A catalog of lakes in Mongolia*. Shuvuun Saaral Publishing, 141 p. [In Mongolian].

Chapter 8
Lake Khukh

Abstract Lake Khukh, lying at the lowest elevation of Mongolia, is one of geotectonically, geomorphologically, and climatologically important sites in eastern Mongolia for inferring the regional landform evolution and reconstructing the impact of the East Asian Winter Monsoon on the Quaternary climate change over geological time scales. Lake Khukh, the second largest lake in eastern Mongolia, provides valuable surface water resources in the Dornod plain. Lake Khukh lost an area of 23.8 km^2 over the past 45 years since 1970, and the area decrease may have been related to the regional and local climate changes. The local climate trend shows that the Lake Khukh basin has gradually warmed as indicated by the rise of the annual average air temperature and dried as shown by the drop of the annual average precipitation during the last 35 years. The linear regression analysis shows that the lake area during 1989–2015 was moderately related to temperature ($R^2 = 0.6705$), but it had no clear relationship with precipitation ($R^2 = 0.0453$). The correlation analysis shows that the lake area between 1989 and 2015 had a strong negative correlation with temperature ($r = -0.818$) and a weak positive correlation with precipitation ($r = 0.212$). Paleohydrological and paleoclimatological implications from Lake Khukh in eastern Mongolia will lead to the reconstruction of regional climate changes in the past and at the present, and the prediction of their fluctuations in the future.

Keywords Lake Khukh · Ulz River · Dornod plain · Climate change · Eastern Mongolia

8.1 Introduction

Lake Khukh, a tectonic lake, is located in the northeastern corner of Mongolia (Fig. 1.1), and it is remarkably noted as the lowest place in Mongolia, lying at 560 m a.s.l. Lake Khukh is included in the Pacific Ocean drainage basin, and it plays an important role in the surface water resources in the Dornod plain of eastern Mongolia. However, Lake Khukh has not attracted much attention of scientists including

A. Orkhonselenge et al., *Lakes of Mongolia*, Syntheses in Limnogeology,
https://doi.org/10.1007/978-3-030-99120-3_8

geomorphologists and paleogeographers. To date, some researchers (e.g., Badarch et al., 1965; Tsegmid, 1969; Tserensodnom, 1971; Davaa, 2015; Enkhtaivan et al., 2016; Orkhonselenge et al., 2019) provided only the general description on morphometric values of the lake (physiography, level, area, volume, hydrochemistry, and mineralization) without detailed investigations on geology, geomorphology, sedimentology, geochemistry, paleohydrology, and paleoclimatology.

Lakes in the Lake Khukh basin were classified as an independent group of lakes of eastern Mongolia by T.V. Nikolaeva in 1985 (Enkhtaivan et al., 2016). For instance, Lake Khukh and other lakes (e.g., Lakes Duruu, Tsagaan, Shavart, Jirmiin Tsagaan, Khar, and Khyart on the Mongolian side and Lake Tari,[1] on the Russian side) are all included in the group of lakes distributing in the Lake Khukh basin, which occupies a closed depression without outlet (see Chap. 6). Geomorphological and hydrological dynamics of the lakes lead to the landform evolution in the regional flat plains and the local depression.

Sedimentary records from Lake Khukh can reconstruct the regional surface processes and landform evolution over various geological time scales and give regional paleogeographical implications because of its unique geographical location within a tectonic depression between the Siberian and Sino-Korean Cratons (see Chap. 1). Lake Khukh also represents the long- and short-term changes in surface water and paleohydrological networks of eastern Mongolia because it records regional and local hydrological fluctuations of the region. Moreover, Lake Khukh deposits can shed light on the climate change in eastern Mongolia and Eurasia in the past, especially during the Quaternary period because the lake basin stood under the impacts of the East Asian Summer Monsoon (EASM) in the past, and of the East Asian Winter Monsoon (EAWM) more recently.

In this chapter, previously presented data on physiography and hydrology on Lake Khukh are synthesized, and changes in lake area over the last nearly eight decades are discussed in detail.

8.2 Physiographic Condition

Lake Khukh (Fig. 8.1) is located at the elevation of 552 m a.s.l. (Badarch et al., 1965; Tserensodnom, 1971) or 560 m a.s.l. (Tserensodnom, 2000; Orshikh et al., 2014; Orkhonselenge et al., 2019) in a depression within the Dornod plain (Fig. 1.1). Lake Khukh is the second largest lake in the Dornod plain, and it plays an important role in the surface water resources in eastern Mongolia. The major water resource of Lake Khukh is rainfall and runoff. Lake Khukh (Fig. 8.2) is fed by Shuuduu River in the northeast, but the lake does not feed any outflow rivers

[1] *Tari* has been misspelled as Torey by the Russians. The Tari is the right English transliteration from Mongolian *Тарь*.

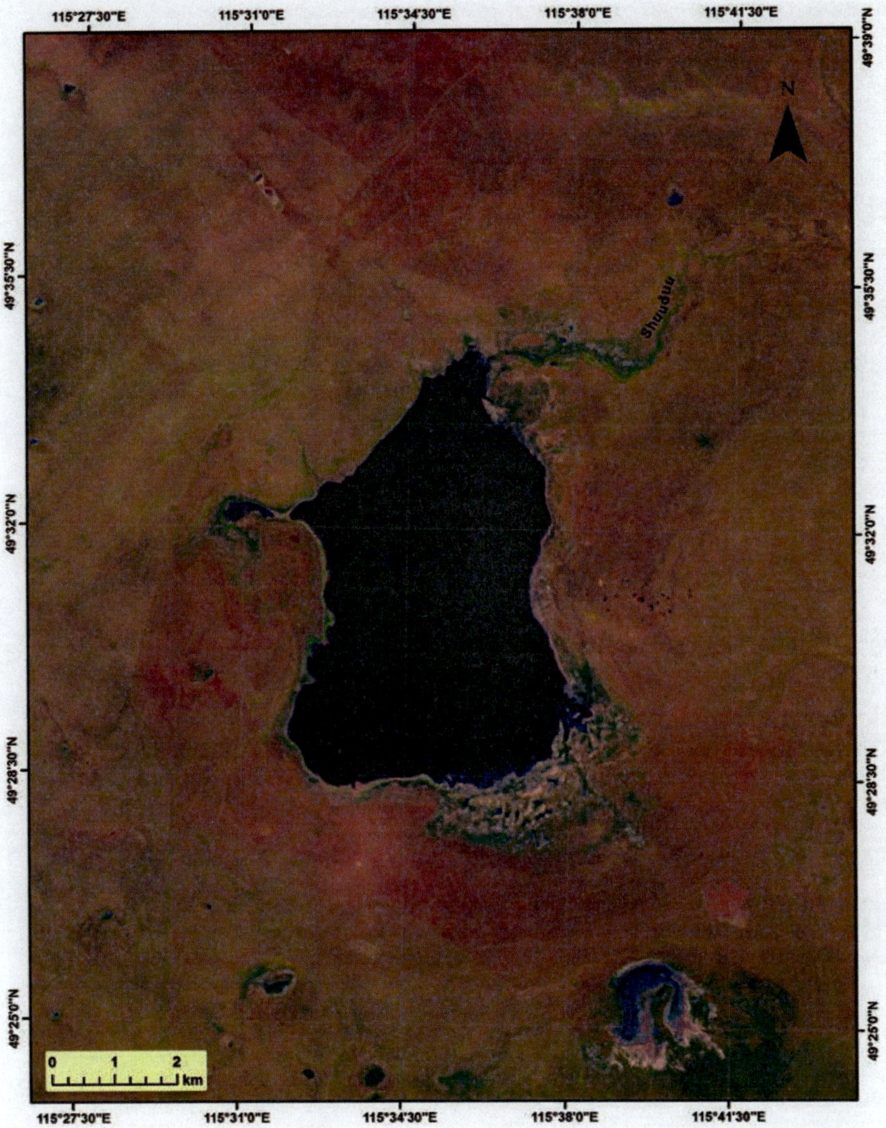

Fig. 8.1 Landsat 8 image of Lake Khukh on August 27, 2015 (path of 126 and row of 26). Modified from Orkhonselenge and Uuganzaya (unpublished)

(Fig. 8.1). The headwater of the Shuuduu River is Teel River, a distributary of Ulz River (see Chap. 6; Sect. 8.5). The lake is called *Khukh* meaning *blue*.

Lake Khukh extends over 10.5 km in length and 8.2 km in width, and it is 5–6 m in depth (Badarch et al., 1965). However, some reported that the morphometric values of the lake reached 14.5 km in length, 10.5 km in width, 14.0 m in depth, and

Fig. 8.2 (**a**) Lake Khukh viewing toward the west and (**b**) a marshy or swampy part at the exposed margin on the western beach of Lake Khukh. Photos by A. Orkhonselenge on July 30, 2014

52.6 km along the shoreline length (Tserensodnom, 1971) as shown in Table 1.3. Another source reported that the lake was 5.0 km wide on average (8.0 km maximum) and 3.2 m deep on average (8.0 m maximum) (Davaa, 2015). The lake covers an area of 51.6 km^2 (Tserensodnom, 2000) or 56.7 km^2 (Orkhonselenge and Uuganzaya, unpublished; see Sect. 8.6). The evaporation from the lake water surface is very high because the duration of solar insolation is long and the wind velocity in the lake basin is high. Lake Khukh sediments consist of sands and gravels at margins and muds in the center of the lake (Tserensodnom, 1971).

8.3 Geological and Geomorphological Settings

Lake Khukh occurs at the border between the late Paleozoic plain geosynclinals in the north and the Rhipean plain geosynclinals in the south (Academy of Sciences of Mongolia and Academy of Sciences of USSR, 1990). Lake Khukh resides over the geological formation of a continental platform consisting of the Cenozoic red and gray terrigenic sands, gravels, and pebbles, and it is surrounded by the Paleozoic and Mesozoic effusive sandstones and siltstones with marine molasses in the south, the Proterozoic and Paleozoic siliceous shale and green shale in the southwest, and the Devonian terrigenic tuff graywacke in the west, northwest, and north (Academy of Sciences of Mongolia and Academy of Sciences of USSR, 1990). Lake Khukh is underlain by the middle and late Quaternary rocks, and the lake basin is surrounded by the late Jurassic and early Cretaceous sedimentary rocks in the southeast and south, the early and middle Proterozoic rocks in south and southwest, and the Devonian rocks in the west, northwest, north, and northeast (Academy of Sciences of Mongolia and Academy of Sciences of USSR, 1990).

Lake Khukh (Fig. 8.2a) is a plain lake based on its morphometric values (e.g., area, depth, shoreline shape, etc.) and landscape (see Chap. 6). The lake floor is uniformly concave (Badarch et al., 1965). Lake Khukh shorelines are uniform without any levees and bays because of its location in a depression within the Dornod plain (Enkhtaivan et al., 2016). There is no island nor peninsula in the lake basin. However, a small, abandoned lake was newly formed in the west of the lake (Fig. 8.1), and bumpy marshes were observed at the exposed margin on the western beach (Fig. 8.2b).

Lake Khukh is surrounded by low mountains including Mt. Ovoot Tolgoi (726 m a.s.l.) in the northeast, Mt. Khar Khoshuu (628 m a.s.l.) in the east, Mt. Zuun Khyart (688 m a.s.l.) and Mt. Dund Khyart (733 m a.s.l.) in the southeast, Mt. Ulaan Tolgoi (665 m a.s.l.) in the southwest, Mt. Khar Tolgoi (703 m a.s.l.) in the west, Mt. Duruu (743 m a.s.l.) and Mt. Suul Undur Tolgoi (707 m a.s.l.) in the northwest, and Mt. Durvuljin Tolgoi (684 m a.s.l.) and Mt. Tsogt (702 m a.s.l.) in the north (Orshikh et al., 2014).

8.4 Climate Condition

The local climate data recorded at the Ereen Tsav meteorological station, located about 35.2 km north of Lake Khukh, show fluctuations of the annual average air temperature and precipitation within the lake basin (Fig. 8.3). The warmest years were recorded with the annual average air temperatures of 2.1 °C in 1997 and 1.8 °C in 2019 (Fig. 8.3a). Other warm years were recorded in 1989–1995, 1998, 1999, 2008, and 2014–2018. The cold years were recorded during 2000–2013 with the annual average air temperature under 0 °C with the lowest peak of −2.5 °C in 2006 (Fig. 8.3a).

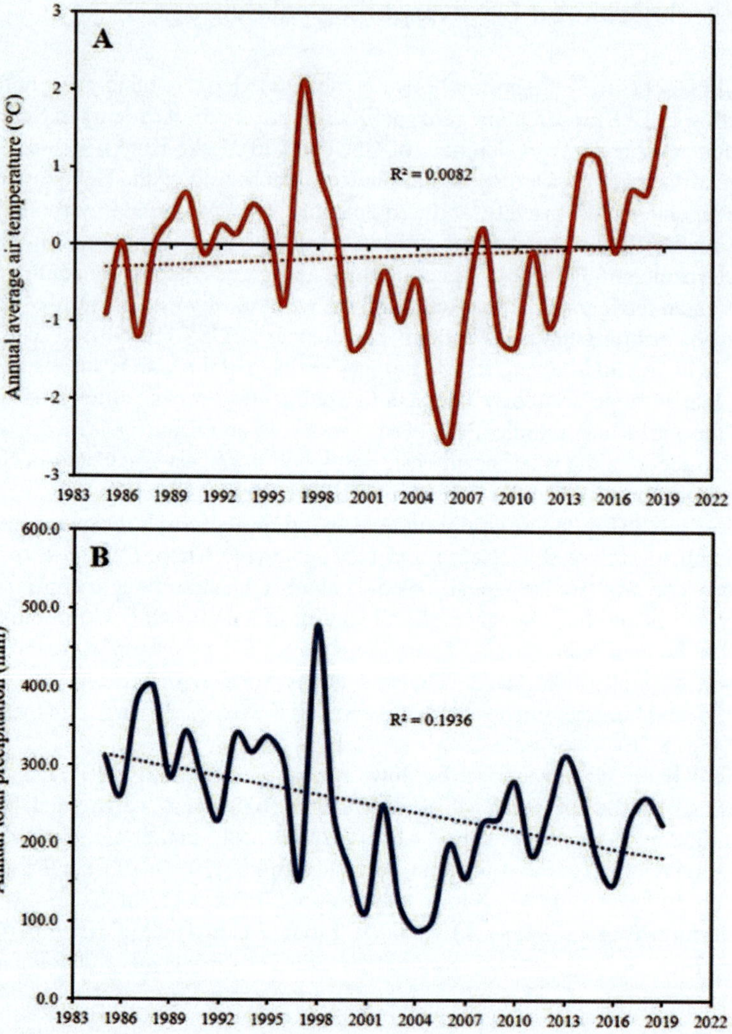

Fig. 8.3 Annual average (**a**) air temperature and (**b**) precipitation recorded at the Ereen Tsav meteorological station during the last 35 years

The annual average precipitation is gradually dropping year by year (Fig. 8.3b). The annual average precipitation was recorded to be 241.7 mm/year on average during the last 35 years except for 1998 (Fig. 8.3b). The most humid year was recorded in 1998 with the precipitation of 480.3 mm/year. Arid years were recorded between 2000 and 2007 with the annual precipitation less than 200 mm/year (Fig. 8.3b).

8.5 Hydrological Condition

The water level of Lake Khukh is strongly dependent on precipitation. The lake level stood at 532 m a.s.l. in 1942 when the lake area was 21.5 km^2 (see Sect. 8.6). The lake level rose to be at 566.7 m a.s.l. in 1964 when the lake covered an area of 94.9 km^2 (see Sect. 8.6). At that time, Ulz River (see Chap. 6) became flooded and Teel River drained from the Ulz River as a distributary. The Teel River contributed to the increases in the lake level and area, and then the over-discharged water of Lake Khukh reached the railway[2] dam in the west (Tserensodnom, 1971).

Table 8.1 shows bathymetric data of Lake Khukh with the area and volume measured by Tserensodnom (1971). The depth varies mainly from 10 to 12 m throughout the lake, except for that in central part of the lake. The lake volume is 0.7 km^3 (Tserensodnom, 1971) or 0.26 km^3 (Davaa, 2015). The water temperature reaches 20 °C during the warmest period in summer (Tserensodnom, 1971) ranging between 20.0 °C and 22.5 °C in July, and it reaches 5.7 °C at the bottom in March (Tserensodnom, 2000). Water temperature inversion occurs, and the amplitude of the maximum and minimum temperatures reaches 4 °C (Tserensodnom, 2000). Lake Khukh is frozen from the end of October to April, and the ice reaches 1.4–1.5 m thick at the beaches and 1.1–1.2 m thick in the center of the lake (Tserensodnom, 1971).

The value of mineralization of Lake Khukh is 1.2–1.4 g/L on average (Table 8.2), and it is higher in the southwestern part than those in other parts of the lake (Tsend, 1966). The ion concentration of the lake water is relatively variable, and HCO_3 and Na ions are dominant (Table 8.2). The mineralization value depends on the water depth, and it increased by 0.73 g/L between 1964 and 1989 (Table 8.3).

Table 8.1 Bathymetric data with corresponding area and volume of Lake Khukh

#	Depth (m)	Area (km^2)	Volume (m^3)
1	2.0	81.3	544
2	4.0	67.8	395
3	6.0	60.2	267
4	8.0	47.6	159
5	10.0	40.9	70
6	12.0	27.5	2

Table 8.2 Hydrochemical composition of Lake Khukh

Ions (g/L)	HCO_3^-	CO_3^{-2}	Cl^-	Ca^{+2}	Mg^{+2}	Na^+
1.2–1.4	0.47–0.65	0.06–0.09	0.19–0.21	0.02–0.03	0.03–0.04	0.31–0.36

[2] The railway runs from the Choibalsan city in eastern Mongolia to the Nerchinsky Zavod city in Russia near the border between Russia and China. It passes through the harbors Ereen Tsav in Mongolia and Solov'evsk in Russia at the border between Mongolia and Russia.

Table 8.3 Temporal changes in water level, area, and mineralization value of Lake Khukh

#	Year	Water level (m a.s.l.)	Area (km^2)	Mineralization (g/L)	Authors
1	1942	532	21.5	–	Tserensodnom (2000)
2	1964	566.7	94.9	0.87	Tserensodnom (1971, 2000)
3	1966	–	–	1.2–1.4	Tsend (1966)
4	1967	565	75.4	1.36	Tserensodnom (2000)
5	1970	–	80.53	–	Orkhonselenge and Uuganzaya, unpublished
6	1989	560	51.6	1.6	Tserensodnom (2000)
7	2010	–	69.4	–	Davaa (2015)
8	2013	–	60.6	–	Davaa (2015)
9	2014	560	–	–	Orkhonselenge et al. (2019)
10	2015	–	56.75	–	Orkhonselenge and Uuganzaya, unpublished

8.6 Changes in Area

Lake Khukh has experienced significant changes in area during the last nearly eight decades (Table 8.3). For instance, Lake Khukh abruptly expanded its area by 73.4 km^2 in 1964 with a pulse of incoming water mass due to a major flooding event at that time. The lake area fell for 19.5 km^2 between 1964 and 1967, and it rose by 5.13 km^2 between 1967 and 1970 (Table 8.3). Then, the lake area dropped again by 28.9 km^2 between 1970 and 1989, and it rose by 17.8 km^2 between 1989 and 2010 (Table 8.3). The lake experienced repeated fluctuations of its area with increases and decreases during each 20 years between 1942 and 2010. The lake area is still dropping since 2010, and the drop reached 12.65 km^2 between 2010 and 2015 (Table 8.3).

Generally speaking, Lake Khukh increased its area for 35.25 km^2 over the last 73 years, but the lake gained an area of 59.03 km^2 during the first 28 years, and it lost an area of 23.78 km^2 during the last 45 years (Table 8.3, Fig. 8.4). The lake area rise during the first period may be affected by the abrupt flooding mentioned above (see Sect. 8.5). However, the drop of the area during the last period, especially since 1989, may have been related to the continuous fall in the annual precipitation (Fig. 8.3b) with the peak period of the falling during 2000–2007. The rapid drop in area since 2010 may have been significantly related to the abrupt climb of the annual average air temperature in 2014–2018 (Fig. 8.3a) even though there was a slight climb also of precipitation. The drop during the last period coincided with the lake volume decreasing by 0.44 km^3 during the last 44 years with a rate of 10 m^3/year (see Sect. 8.5). The recent lake area drop was clearly observed during the field work (Orkhonselenge et al., 2019) with the rapid shift of the lake margin into marshy areas (Fig. 8.2b) and terraced shorelines. Figure 8.4 shows that the lake lost larger areas on southeastern, southern, and western shores than those on other shores. Due to the rapid shrinkage of the lake, a new small, abandoned lake was formed in the west of the lake by 2015 (Fig. 8.4).

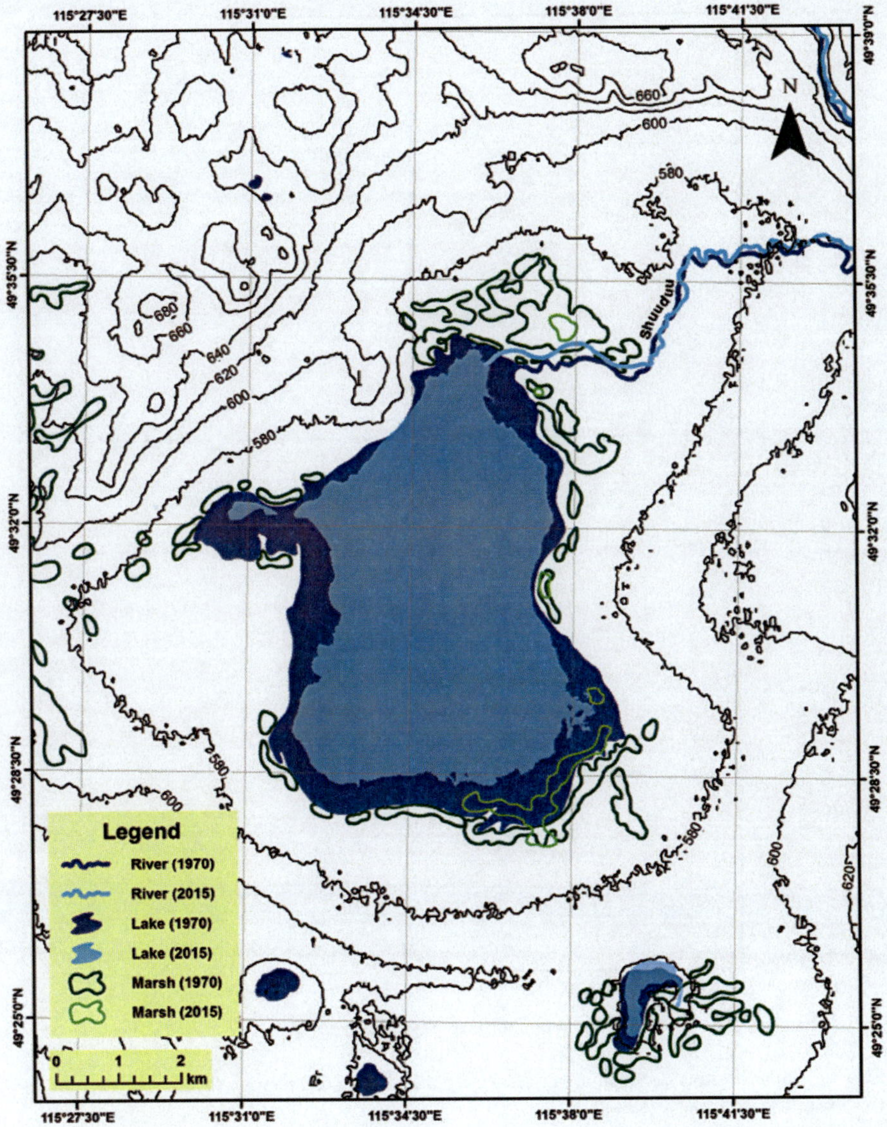

Fig. 8.4 Spatial distributions of areas and marshes of Lake Khukh in 1970 and 2015. Modified from Orkhonselenge and Uuganzaya (unpublished)

The linear regression analysis (Fig. 8.5) shows that the area of Lake Khukh between 1989 and 2015 was moderately related to the annual average air temperature (R^2 = 0.6705) but it had no clear relationship with the annual precipitation (R^2 = 0.0453). The correlation analysis shows that the lake area during 1989–2015 had a strong negative correlation with temperature ($r = -0.818$) and a weak positive

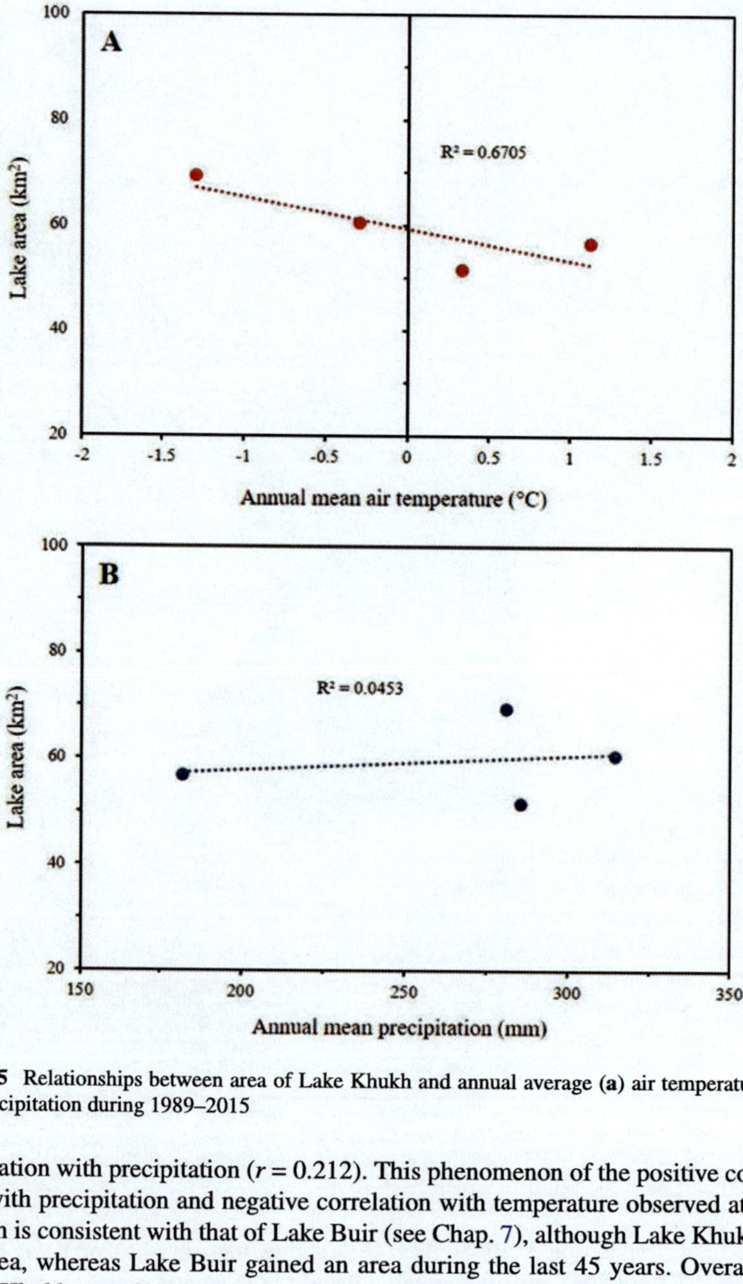

Fig. 8.5 Relationships between area of Lake Khukh and annual average (**a**) air temperature and (**b**) precipitation during 1989–2015

correlation with precipitation ($r = 0.212$). This phenomenon of the positive correlation with precipitation and negative correlation with temperature observed at Lake Khukh is consistent with that of Lake Buir (see Chap. 7), although Lake Khukh lost the area, whereas Lake Buir gained an area during the last 45 years. Overall, the Lake Khukh area is dependent on both temperature and precipitation, and it is strongly negatively correlated with temperature, similar to Govi[3] lakes in southern Mongolia (see Chaps. 9–11).

8.7 Summary

Lake Khukh in eastern Mongolia is very intriguing geotectonically, geomorphologically, and climatologically. The lake records hydrological fluctuations of eastern Mongolia during the Quaternary. Lake Khukh showed an increase in area during the first 28 years of the last eight decades linked to abrupt floodings and a decrease in area during the last 45 years due to the local climate change. In the lake basin, the annual average air temperature has been gradually rising, but the annual precipitation has been falling during the last 35 years. The linear regression analysis shows that the area of Lake Khukh during 1989–2015 was moderately related to temperature ($R^2 = 0.6705$) but not clearly related to precipitation ($R^2 = 0.0453$). Nevertheless, the correlation analysis shows that the lake area between 1989 and 2015 had a strong negative correlation with temperature ($r = -0.818$) and a weak positive correlation with precipitation ($r = 0.212$). Geomorphological, geochemical, and geochronological records from Lake Khukh can provide information useful for assessing the impact of the East Asian Winter Monsoon on the climatic and hydrological systems in eastern Mongolia and for reconstructing paleoclimate changes in Mongolia and Eurasia.

References

Academy of Sciences of Mongolia and Academy of Sciences of USSR. (1990). *National Atlas of the Mongolian People's Republic*. 144 p. [In Russian].

Badarch, N., Tsegmid, S., & Tserensodnom, J. (1965). *Landscapes and natural zones of eastern Mongolia*. Press of Mongolian Academy of Sciences, 117 p. [In Mongolian].

Davaa, G. (2015). *Resource and regime of surface water of Mongolia*. 408 p. [In Mongolian].

Enkhtaivan, D., Oyungerel, B., Avirmed, E., Renchinmyadag, T., Nyamkhuu, M., Munkhdulam, O., Odbaatar, E., Davaagatan, T., Bayanjargal, B., & Batnyam, T. (2016). *Landscape Structure, Change, Planning and Proper Zonation (the case study in eastern Mongolia): Report of Research Project funded by Science and Technology Foundation*. Institute of Geography and Geoecology, Mongolian Academy of Sciences, 564 p. [In Mongolian].

Orkhonselenge, A., Davaanyam, S., Altanbagana, M., & Bolorchuluun, C. (2019). *Long-term changes in ecosystems in natural zones of Mongolia: A report of Research Project funded by Science and Technology Foundation (ShUT_A/349_005/2013)*. Laboratory of Geochemistry and Geomorphology, National University of Mongolia. NUM Publishing, 384 p. [In Mongolian].

Orshikh, N., Enkhbayar, G., Garig, G., Enkhbat, R., & Chantuu, B. (2014). *Atlas of Mongolia. 1:800,000*. Monsudar Press, 117 p. [In Mongolian].

[3] *Govi* has been often spelled as *Gobi* in the international literature. Govi is the correct English transliteration from Mongolian *Говь*.

Tsegmid, S. (1969). *Physical geography of Mongolia.* State Press, 405 p. [In Mongolian].
Tsend, N. (1966). Chemical composition of lakes in eastern Mongolia. *Institute of Livestock, 15.* [In Mongolian].
Tserensodnom, J. (1971). *Lakes of Mongolia.* State Publishing, 192 p. [In Mongolian].
Tserensodnom, J. (2000). *A catalog of lakes in Mongolia.* Shuvuun Saaral Publishing, 141 p. [In Mongolian].

Chapter 9
Landscape, Lake Distribution, and Evolution in Southern Mongolia

Abstract Landscape in southern Mongolia, specifically the Govi shows an outstanding variety as the result of the region's long-term geotectonic evolution and climate changes over geological time scales. In southern Mongolia, the number of lakes is second in the country (24.7%), and they are mainly aeolian, volcanic, and tectonic lakes. Lakes in the Govi are important and useful archives for reconstructing paleoclimate and paleoenvironmental changes in southern Mongolia because they preserve information on hydrological and geomorphological processes. The recent significant changes of lake areas in southern Mongolia indicate how rapid the Govi landscape has been responding and how sensitive the lakes are to the present climatic warming. The warming-induced aridity strengthened in the Govi region during the last century has been controlling the landscape evolution and also disappearance and reappearance of the lakes. Sedimentological and geochronological data obtained from the Govi contribute to our understanding of the paleogeographical histories in Mongolia and Central Asia.

Keywords Govi desert · Govi Altai · Valley of Lakes · Paleolakes · Playa lakes · Southern Mongolia

9.1 Introduction

Southern Mongolia hosts a great variety of landscapes making it an outstanding region from geomorphological and paleogeographical points of view. Its geological history since the Mesozoic Era has given a fundamental imprint on the geomorphology of the region, forming the large Govi Desert[1] and the high mountains

[1] *Govi* has been often spelled as *Gobi* in the international literature. *Govi* is the correct English transliteration from Mongolian *Говь*.

A. Orkhonselenge et al., *Lakes of Mongolia*, Syntheses in Limnogeology,
https://doi.org/10.1007/978-3-030-99120-3_9

of the Govi Altai Mountain Range that experienced glaciation during the Pleistocene.

In southern Mongolia, the Jurassic and Cretaceous sedimentary rocks are dominant (Tsegmid, 1969) although the continuity of the geological story is interrupted with many hiatus (Berkey & Morris, 1927). Aeolian, loess-like sediments are distributed as mantles of silt covering the sandy and gravelly Govi region in which paleolakes represent valuable archives about the environmental changes during the late Quaternary (Lehmkuhl, 2014).

Southern Mongolia (Fig. 1.1) is climatologically important because the intensive variations of the Mongolian High Pressure System (MHPS) have imposed a strong control on the regional climate of Central Asia during the Quaternary (Owen et al., 1997). The climate in the Govi region (see Sect. 9.2) was cool and temperate in the late Precambrian, cool and arid in the Paleozoic, warm and moderately humid in the Jurassic, semiarid in the Cretaceous and Paleogene, humid in the Neogene, and arid in the Quaternary (Berkey & Morris, 1927). Lately, southern Mongolia was influenced by the East Asian Summer Monsoon (EASM) between 9.0 and 4.0 ka BP (Lee et al., 2013).

Physiographically speaking, southern Mongolia consists of the Valley of Lakes in the north, Govi Altai Mountain Range in the west (Fig. 9.1), and the Govi Desert for the rest (see Sect. 9.2). The Govi Desert occupies a broad, shallow, basin-like depression in southern Mongolia. The term *Govi*, meaning an open plain with three sides, was described by Berkey and Morris in their book entitled *The peneplains of Mongolia* published in 1924 (Jigj, 1975). Tectonic faults in the Valley of Lakes, active at rates similar to those in the Govi Altai and Khangai[2] Mountain Ranges, have played a major role in the drainage network evolution since the middle Pleistocene (van der Wal et al., 2020). Enormous paleolakes in the Govi region were

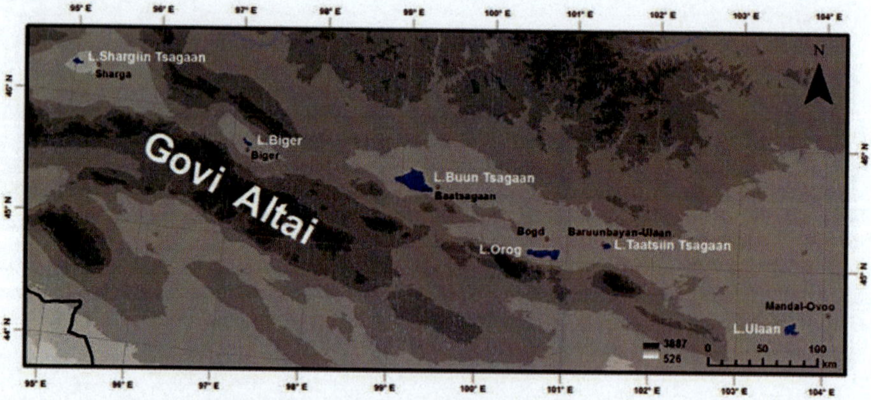

Fig. 9.1 Large lakes of the Valley of Lakes in the Govi region with locations of the meteorological stations near the lakes. After Orkhonselenge et al. (2018a)

[2] *Khangai* has been misspelled as *Khangay* (or *Hangay*) and *Hangai* in many publications. Khangai is the right English transliteration from Mongolian *Хангай*.

maintained by a strong EASM reaching southern Mongolia, and these paleolakes mostly disappeared during the Marine Oxygen Isotope Stage (MIS) 6, and it reappeared during the last interglacial MIS 5e and lasted probably until the beginning of the last glacial (Lehmkuhl et al., 2018).

Lakes in southern Mongolia reflect conditions of paleoenvironments in the Govi region. There were early Cretaceous continental saline lakes between the Govi Altai Mountain Range in the west and Mt. Dari Ganga in the east (Devyatkin, 1981). Because climate change is the major driving force resulting in the environmental changes in lake basins (e.g., Andrén et al., 2015; Orkhonselenge et al., 2015; Komatsu et al., 2016), interpreting sequences of lake sediments is important for understanding long-term climate changes and landscape evolution of the lake basins (Davaagatan et al., 2015; Orkhonselenge et al., 2019). Furthermore, history of expansion and shrinkage of a lake provides a proxy for inferring short-term climate and environmental changes (see Sect. 9.3). Therefore, investigating lake landscape evolution and revealing changes of lakes' areas in southern Mongolia are important for understanding the past and present climate and environmental changes in Mongolia and Central Asia (Orkhonselenge et al., 2018a).

Lakes in southern Mongolia have attracted much attention in terms of Quaternary geomorphology, sedimentology, geochemistry, and climatology. For instance, the late Pleistocene and Holocene paleoclimates in the region (see Chap. 21) were reconstructed to be dry at the late glacial, humid at 11.0–4.0 ka cal BP, and arid since 4.0 ka cal BP (Hülle et al., 2010; Felauer et al., 2012).

The recent climate change in the Govi region is significantly impacting surface water, especially lakes. Many rivers have become intermittent, and Lakes Orog, Taatsiin Tsagaan, Adgiin Tsagaan, Khaya, and Ulaan shrank during the dry period lasting for 16 years between 1996 and 2012 (Institute of Meteorology, Hydrology and Environment, 2013). The lake disappearance and reappearance and temporal changes of lake areas in the Govi region showed large temporal variations, and the lake area changes showed high correlations with precipitation changes (Kang et al., 2015). For instance, the record in Dalanzadgad of the Govi region showed the most dramatic decrease of lake areas for some lakes (Fig. 1 in Kang et al., 2015) between 1991 and 2000, and over those 10 years, the lake areas decreased over 80%, from 47.1 to 5.9 km^2 (Kang et al., 2015). The changes in area during the past 44–45 years for six large lakes in the Valley of Lakes in the Govi (Fig. 9.1) correlate significantly with the rising temperature since 1995 and the fluctuating precipitation since 1975 (Orkhonselenge et al., 2018a).

In recent years, Lakes Buun Tsagaan (e.g., Batnasan, 1998; Komatsu et al., 2001; Lehmkuhl & Lang, 2001; see Chap. 10), Orog (e.g., Komatsu et al., 2001; Szumińska, 2016; Yu et al., 2017), and Ulaan (e.g., Lee et al., 2013; Holguin & Sternberg, 2016; Orkhonselenge et al., 2018b, in press; see Chap. 11) have been discussed in detail. For instance, Krylov et al. (2011) indicated Lake Orog as hypertrophic during high stand, oligotrophic during low stand, and mesotrophic when the tributary inflows dried up based on inter-annual differences in the zooplankton structure. In this chapter, landscape features and lacustrine environments in the Govi region of southern Mongolia are introduced. Paleoclimate changes recorded from lakes in this region are discussed in detail in Chap. 21.

9.2 Landscape

9.2.1 Geology

Southern Mongolia is underlain by the Precambrian basement block, Silurian-Devonian oceanic, Proterozoic-Devonian turbidite basin coinciding with the Main Mongolian Lineament (Fig. 1.1) and the Proterozoic-Carboniferous lake zone from the south to the north (Guy et al., 2014). The Govi region lies on the Mesozoic and Cenozoic sedimentary basins underlain by the Devonian and Carboniferous mobile belt with the Ordovician and Silurian mobile belt in the north and the middle to late Paleozoic volcano-plutonic associations with the Paleozoic and Mesozoic granitoids in the south (Kröner & Rojas-Agramonte, 2014).

The Hercynian juvenile crust, which began forming in the Silurian-Devonian (~420 Ma) under the conditions of ensimatic island arcs, back-arc basins, and oceanic plateau and islands, governed the structure of the major part of the southern Mongolian segment of Hercynides in the Central Asian foldbelt and structures of the Hercynian paleo-Asian ocean including terranes of the southern Govi microcontinent, which were transformed into a continental block in the late Devonian–early Carboniferous owing to accretion, folding, and metamorphism at the boundaries of heterogeneous structural zones (Yarmolyuk et al., 2007).

The tectonostratigraphic terrane map of Mongolia (Badarch et al., 2002) shows the Govi Altai back-arc/forearc basin, Mandal Ovoo island arc, and Gurvan Saikhan island arc in the northern and central Govi; the Nukhet Davaa back-arc/forearc basin, Khutag Uul cratonal block, Sulin Kheer accretionary wedge, and Duulgant back-arc/forearc basin in the eastern Govi; the Zuulun alluvial basin, Atas Bogd back-arc/forearc basin, Khashaat island arc, and Tsagaan Uul cratonal block in the southern Govi (Fig. 9.2); and the Tseel metamorphic rocks, Baitag island arc, and Baraan back-arc/forearc basin in the western Govi.

The eastern Govi is underlain by the late Jurassic–early Cretaceous extensional province of prerift, synrift, and postrift of Mesozoic stratigraphic mega-sequences formed prior to 155 Ma and bound by unconformities upward from basal alluvial and fluvial conglomerate to fluvial and lacustrine sandstone and mudstone, where the subsidence outstripped sediment supply and fresh to saline lacustrine environments expanding southward with time (Graham et al., 2001). The distinct phases of deformation were recorded in the Paleozoic basement rocks and Mesozoic-Cenozoic basin fill of the eastern Govi basin (Heumann et al., 2018).

The central Govi is dominantly composed of the Devonian-Carboniferous volcanic and sedimentary rocks with the Mesozoic-Cenozoic sediments (Zhu et al., 2016).

The western Govi consists of the Neoproterozoic ophiolites of the Khan Taishir formation in the north, orthogneisses and mica schists with eclogites of the Alag Khad metamorphic complex and marbles of the Maikhan Tsakhir formation in the center, and the Zamt Mountain Range formation in the south (Javkhlan et al., 2013) with ca. 470–290 Ma metaigneous and meta-volcaniclastic rocks in Gurvan Saikhan, Zuulun, Nemegt, Bayanleg-Khatuu, and Tseel Mountain Ranges (Helo et al., 2006; Fig. 9.2).

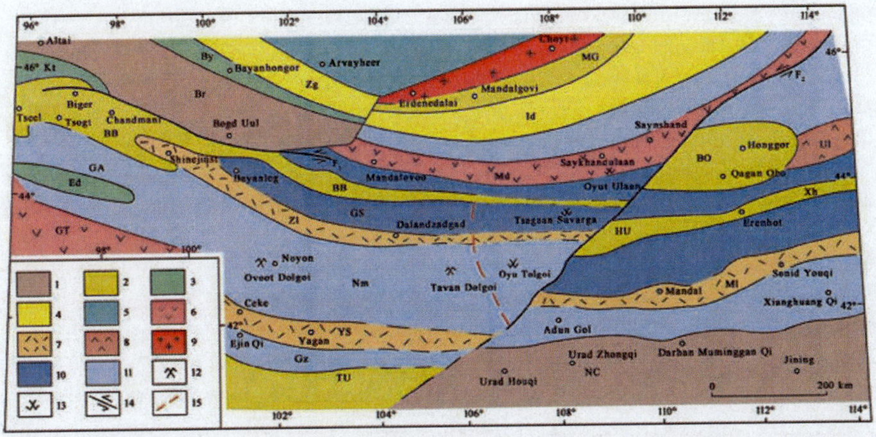

Fig. 9.2 Trench-arc-basin systems of the southern belt of Central Asian Orogenic Belt (CAOB) bordering Mongolia and China. (1) Cratons (Br, Baidrag; NC, North China); (2) Continental blocks (BB, Baga Bogd; MG, Middle Govi (or Dundgovi); BO, Barga Ovoo; HU, Khutag Uul; Xh, Xilinhot; TU, Tsagaan Uul); (3) Ophiolite/late Paleozoic spreading centers (By, Bayan Khongor; Kt, Khan Taishir; Ed, Edren); (4) Accretionary wedges (passive continental margin) (Zg, Zag; Id, Idermeg); (5) Passive continental margins of the Mongolia-Okhotsk ocean; (6) Carboniferous magmatic arcs (Md, Mandal Ovoo; GT, Govi-Tienshan); (7) (Silurian-)Devonian arcs (Zl, Zuulun; Ml, Mandal; YS, Yagaan-Suoguozhuo); (8) Early–middle Permian arcs (Ul, Ulaankhot); (9) Middle Govi (or Dundgovi) Permian-Triassic volcano-plutonic belt; (10) Forearc basins (GS, Gurvan Saikhan); (11) Back-arc basins (Nm, Nemegt; GA, Govi Altai; Gz, Guaizihu); (12) Coal mine; (13) Porphyry Cu-Au-Mo ore deposit; (14) Shear fractures (F1, Govi Altai fault; F2, Zuun Bayan fault); (15) Present boundary between the south Govi basin and the allochthonous magmatic arc. Modified from Wu (2014). Some notes about correct spelling of names from Mongolian Cyrillic into English: Arvaikheer not Arvayheer, Choir not Choyr, Khongor not Honggor, Tolgoi not Dolgoi, Tsagaan Ovoo not Qagan Obo, Ereenkhot not Erenhot, Sainshand not Saynshand, and Saikhandulaan not Saykhandulaan

The northern Govi is composed of lithostratigraphic formations of three basaltic layers erupted in the Valley of Lakes in the early (31.5 Ma) and late (27.0–28.0 Ma) Oligocene and the middle Miocene (13.0–12.5 Ma) (Hock et al., 1999).

9.2.2 *Climate*

In the Govi region, the average air temperature is −15 to −25 °C in January and 20–25 °C in July, and the average precipitation is less than 100 mm/year (Tserensodnom, 1971). The annual average precipitation is 50–150 mm and the open water evaporation is 1000–1300 mm (Ministry of Environment and Green Development, 2013). The wind velocity is 20–25 m/s on average, but it reached the maximum of 26–40 m/s in 2011 (Amarjargal, 2016).

The observed warming trend shows that the annual average air temperature rose by 1.6–1.7 °C in the Govi Desert from 1940 to 2001, and the intense drought spells in recent years are most likely due to the increased temperature and decreased precipitation (Batima et al., 2005). A drought record in 1970–2006 across the Govi region showed cyclical fluctuations with broadly wetter conditions during the 1970s and 1990s, a notably drier period in the 1980s, and alternating wet-dry episodes in the 2000s, with a particularly dry year in 2006 (Sternberg et al., 2011).

The Govi region, where the annual average precipitation is below 200 mm, is characterized by the relatively high inter-annual variability of precipitation (CV = 35.5%) and by the decreasing temporal precipitation variability with the annual average precipitation ($r = -0.70$) (Kang et al., 2015). The climate in the Valley of Lakes of the northern Govi (Fig. 9.3) shows trends of temperature rising since 1995 and precipitation fluctuations since 1975 (Orkhonselenge et al., 2018a). The rise in the annual average air temperature has strengthened by 1.7–2.5 °C, and the Govi region is expected to extend 600 km toward the north by the end of this century due to the increases of precipitation by 20–50 mm, evaporation by 200–300 mm, and temperature by 5–6 °C in Mongolia (Dulamsuren, 2016).

9.2.3 Landform

The Govi region was mountainous and rugged with rivers flowing in all directions to the surrounding seas (Berkey & Morris, 1927). The region was uplifted, warped, and faulted, and the streams turned inland, and deposits accumulated on bottoms of the inland basins. Since the early Cretaceous, the region acquired a continental condition (see Chap. 3). After the long-term erosion, the ancient continent was worn down to the base level, and the traces of this old peneplain are some of the dissected remnants of the older rock floor (Berkey & Morris, 1927). There are gorges, gulches, escarpments, and uplands, where strata are exposed, in some places throughout considerable thickness even down into the ancient rock floor.

The landscape of this region consists of the southern Khangai Mountain Range, Govi Altai Mountain Range, Valley of Lakes, Govi, and other desertic regions (Table 1.2). The Govi landscapes vary from high mountains in the west to desert with oasis in the east and south. Here, landforms are the results of warping and uplifting, and an example is the Govi Altai Mountain Range (Fig. 9.1) that is a complex uplifted fault block (Berkey & Morris, 1927). The Govi Altai extends for over 1000 km from the Gichgene Mountain Range in the northwest to Mt. Khurkh in the southeast (Murzaev, 1952; Tsegmid, 1969; Jigj, 1975). This range is characterized by gorges, canyons, valleys, and alluvial fans (Jigj, 1975), and Mt. Terguun Bogd (3957 m a.s.l.), peak of the range, has steep gorges with sharp tops. Steep gorges with streams (e.g., in the Zuun Saikhan Mountain Range in Fig. 9.4, the southeastern tip of the Govi Altai) are typically found in the Govi Altai. There are high mountains including Mt. Baitag Bogd, Mt. Ikh Khavtaga, and Mt. Takhiin Shar elevated at 2300–3100 m a.s.l. in the southwestern Govi, and Mt. Atas Bogd, Mt. Chinggis, Mt. Shar Khuls, and Mt. Tsagaan Bogd elevated at 1700–2600 m a.s.l. in the eastern Govi. The eastern Govi is characterized by plains and depressions at the elevations of 900–1600 m a.s.l.

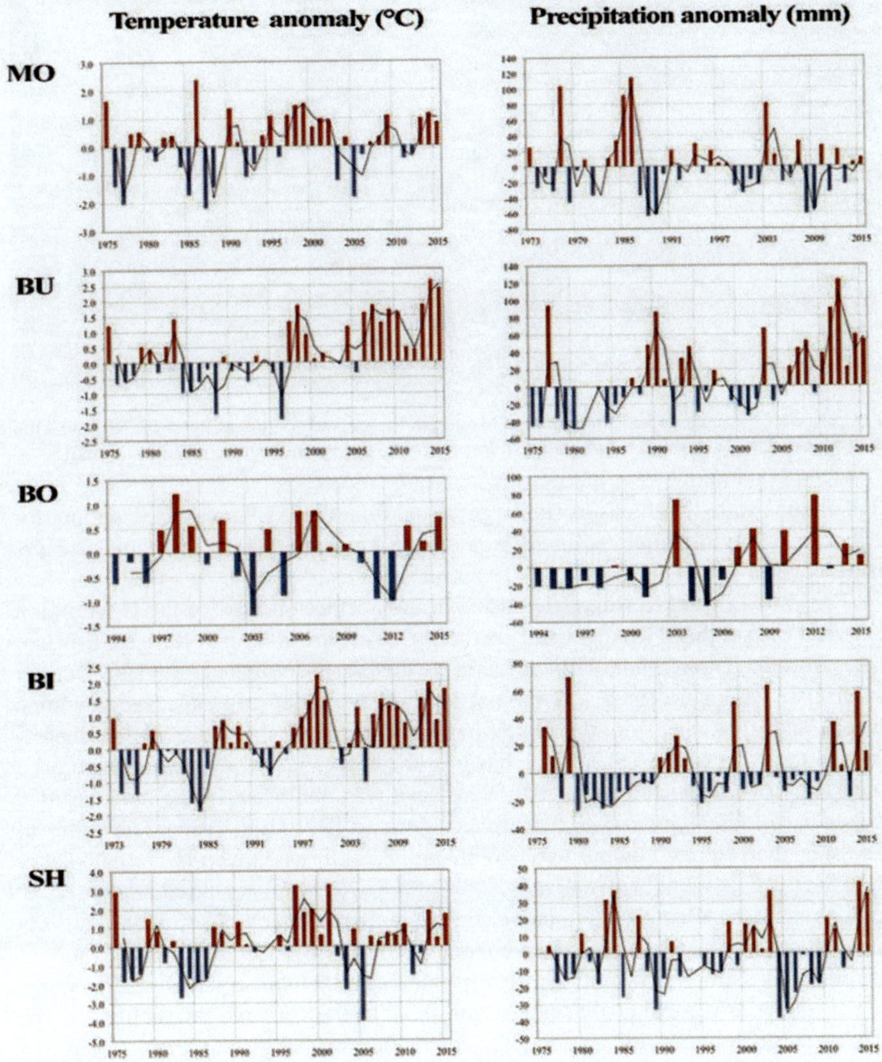

Fig. 9.3 Standard deviation of annual total air temperature and precipitation compared to annual average air temperature and precipitation recorded at the MO (Mandal Ovoo), BU (Baruunbayan-Ulaan), BO (Bogd), BI (Biger), and SH (Sharga) meteorological stations in the Valley of Lakes in the Govi region (Fig. 9.1). After Orkhonselenge et al. (2018a)

The Govi Desert occupies throughout the region. Govi is numbered for about 500 (e.g., Khuis, Nariin Khukh, Borzon, Galba, etc.) including 33 large Govi. For instance, there are Govi environments found in the eastern (e.g., Dolood, Uush, Ulziit, Deliin Gashuun, Dulaan, etc.), southern (e.g., Zagiin Us, Oni, etc.), western (e.g., Khonin Us, Nomin, Mongon Gar, Zakhui, Zarman, etc.), and northern (e.g., Tevsh) parts of the region. These Govi environments are abundant with sand

Fig. 9.4 Steep gorges in the Zuun Saikhan Mountain Range, the southeastern tip of the Govi Altai Mountain Range, viewing toward the west. Photos by A. Orkhonselenge on October 25, 2015

dunefields spotted with oasis. The largest sand dunefield is Khongor, extending for 185 km in length and covering an area of 925 km^2, and the second largest one is Bird, covering an area of 493 km^2 (Tsegmid, 1969).

The Valley of Lakes (Fig. 9.1) occurs at the northern border of the Govi region between the southern Khangai and northern Govi Altai Mountain Ranges. It is isolated from the Depression of Great Lakes by the presence of the Darvi and Khasagt Khairkhan Mountain Ranges in the northwest. The Valley of Lakes has a tectonic origin, and it extends from the Lake Shargiin Tsagaan basin of the Sharga depression in the west to the Lake Ulaan basin in the east (Fig. 9.1). Geomorphology of the Valley of Lakes studied by T.V. Nikolaeva was published in the book entitled *Geomorphological features of Central Mongolia* in 1971 (Jigj, 1975). The Valley of Lakes is characterized by numerous remnant lakes of once much larger paleolakes (see Sect. 9.3).

The southern Khangai Mountain Range separates two tectonic-structural units at the foot of the Khangai where the Neogene tectonics produced for the initial mountain massif (Klimek, 1980). Other less prominent hills mark areas of gentler warping and smaller vertical displacement, and some of them are older fault blocks. The region stretches out with a succession of plains, marking the location of minor enclosed basins where down-warped areas have caught the sediments washed down from the uplifted segments (Berkey & Morris, 1927).

9.2.4 *Glaciers and Permafrost*

Pleistocene glaciations in Mongolia (Tsegmid, 1969; Jigj, 1976) were imprinted on the Govi Altai Mountain Range, and they are correlating with the late Quaternary glaciations recorded in the Mongolian Altai Mountain Range. There is no trace of modern glaciers in the Govi Altai, and the modern glacial cirques observed in the

Table 9.1 Seasonal freezing and thawing depths in the Govi region of southern Mongolia

Regions	Govi		Eastern Govi		Southern Govi		Western Govi		Northern Govi	
Types	Clays	Sands	Clays	Sands	Clays	Sands	Clays	Sands	Clays	Sands
Depth (m)	1.7–2.0	2.5–3.0	2.2–2.4	2.4–2.6	1.7–2.1	2.2–2.4	2.5–2.6	2.8–3.0	2.5–2.7	2.7–2.9

Ikh Bogd Mountain Range elevated at 3957 m a.s.l., a branch of the Govi Altai, have not been proven (Tsegmid, 1969). The glaciations of the Khangai Mountain Range during the MIS 3 (e.g., Rother et al., 2014; see Chap. 18) have been reflected in the hydrological system in the Valley of Lakes (e.g., Lehmkuhl et al., 2018).

Perennial permafrost occurs only at the top of Mt. Terguun Bogd (Tsegmid, 1969) in the Ikh Bogd Mountain Range. There was a phase of permafrost development under the annual average air temperature of below approximately −6 °C in the Govi region during the latter part of the Last Glacial Maximum (LGM, after ~22.0–15.0 ka BP), which resulted in cryoturbated sediments and ice-wedge casts, and the permafrost degradation occurred during the late Pleistocene (13.0–10.0 ka BP) and disappeared during the early Holocene (Owen et al., 1998).

The seasonal permafrost is distributed at 1.7–2.0 m depth in clays and 2.5–3.0 m in sands of the Govi region (Tumurbaatar, 2004; Table 9.1). The permafrost is maintained up to the end of March (Tumurbaatar, 2004), but no evidence for the perennial permafrost in southern Mongolia has been found (Jambaljav et al., 2016). The intermittent permafrost, in which thermokarst lakes decreased in number from eight to three and in area from 0.06 km^2 to 0.03 km^2 between 1962 and 2007, was found in Erdene in front of the Gichgene Mountain Range (Saruulzaya, 2017). The permafrost in river drainage basins and in the Govi regions of southern Mongolia is defined as cold and warm depending on annual mean ground temperature (AMGT) by Zorigt et al. (2020), and it is shown with the active layer thickness (ALT) in Table 9.2. We note that in June of 2019, permafrost was observed below 70 cm depth within a playa Lake Sain Usnii Ulaan in the Govi region (Fig. 9.5). A small vegetated pingo (Fig. 9.5a) produced fluid muds with meltwater derived from the permafrost thawed by tire and feet (Fig. 9.5b).

9.2.5 Groundwater

In the Govi region, groundwater resides in the Paleozoic metamorphic rocks, intrusive and effusive rocks, and the Mesozoic and Cenozoic sedimentary rocks consisting of sands, gravels, sandstones, and shales in intermontane valleys and river valleys (Tsegmid, 1969). Artesian groundwater in the Cretaceous and Neogene deposits occurs in the Valley of Lakes (Tserensodnom, 1971).

The Govi region hosts intrusive groundwater and the Quaternary lacustrine-alluvial groundwater (Tsegmid, 1969). The intrusive groundwater is found at 14.8–152.0 m depth, whereas the lacustrine-alluvial groundwater is found at 1.5–3.0 m depth and it is fed by precipitation and soil water (Tsegmid, 1969).

Table 9.2 Cold and warm permafrost in the river drainage basins and Govi regions of southern Mongolia

Places	Basins	Area of the basins (km^2)	Permafrost in the basins (%)		AMGT (°C)	ALT (m)
			Cold permafrost (AMGT < -2 °C)	Warm permafrost (−2 °C < AMGT <0 °C)		
Southern Khangai	Baidrag River	35,153	30.1	10.1	−0.5	0.9–3.2
	Ongi River	39,202	0.0	0.0	0.0	0
	Taats River	25,092	0.2	4.9	−0.6	0.8–3.3
	Tuin River	15,529	17.6	6.2	−0.9	0.9–3.3
Govi	Southwestern Govi	218,250	0.0	0.0	–	0
	Galba-Uush-Dolood Govi	14,0416	0.0	0.0	–	0
	Khuis Govi	42,460	2.6	10.1	−0.9	0.9–3.2
	Northern Govi (or Dundgovi)	178,182	0.0	0.0	–	0

Fig. 9.5 (**a**) Occasional permafrost (dotted line) excavated in the playa Lake Sain Usnii Ulaan viewing toward the west and (**b**) its fluid muds (dotted line). Photos by A. Orkhonselenge on June 4, 2019

Table 9.3 Water resources of the lake and river drainage basins and the Govi regions in southern Mongolia

#	Drainage basins	Area (km^2)	Type	Surface water resource (million m^3/year)			Groundwater resource (million m^3/year)	
				Total resource	Environmental	Possible use	Potential exploitable	Exploitable resource
1	Northern Govi (or Dundgovi)	180,555	GW	–	–	–	433	46.7
2	Galba-Uush-Dolood Govi	142,287	GW	–	–	–	352	59.0
3	Ongi River	39,724	SW/GW	26	25	1.0	294	5.8
4	Southwestern Govi	221,156	SW/GW	–	–	–	337	65.5
5	Taats River	25,425	SW/GW	22	21	0.9	61	0.5
6	Lake Orog-Tuin River	15,735	SW/GW	66	63	2.6	33	5.9
7	Lake Buun Tsagaan-Baidrag River	35,622	SW/GW	303	280	22.7	174	2.9
	Total	479,949		417	389	27.2	1251	139.6

SW surface water; *GW* groundwater

The groundwater decreases depending on changes of surface water resources. For instance, the groundwater level dropped by 0.5 m in the Ekhiin Gol oasis during 1997–2009 (Davaa, 2010). The groundwater level decreased by 0.36 m in the Ekhiin Gol oasis and by 2.3 m in Arvaikheer in 1997–2012; however the level stabilized in the Ekhiin Gol oasis and increased in Arvaikheer in 2013 (Institute of Meteorology, Hydrology and Environment, 2013). The decreases of the groundwater level and lake area were strengthened due to a severe drought in the eastern Govi region since the late 1990s (Kang et al., 2015).

Table 9.3 shows groundwater resources in the region where shallow groundwater resources are annually recovered, but deep groundwater resources are considered as fossil (Ministry of Environment and Green Development, 2013). In the drainage basins, groundwater resources are decreasing as the results of mining operations, livestock watering, and irrigation of crops during the last two to three decades.

9.2.6 *Surface Water*

Large Baidrag, Tuin, Taats, and Ongi Rivers (Tables 9.3 and 9.4), large and small lakes (see Sect. 9.3), and intermittent streams constitute the surface water resource of the Govi region (Table 9.3). The large rivers drain the southern Khangai Mountain

Table 9.4 Morphometric components of large rivers in the Govi region

#	Rivers	Length (km)	Drainage area (km²)	Width (m)	Depth (m)	Discharge rate (m/s)	Discharge (m³/s)	Tributary rivers	Outlets
1	Baidrag	295[a] 310[c]	28,277	30–60	0.3–1.5	1.0–1.5	14.8[a] 142[b] 970[c]	Zag, Ulziit	L. Buun Tsagaan
2	Tuin	243	9410	25–80 150–200[d]	0.5–1.5	–	5.14[a] 218[b] 334[c]	Shargaljuut	L. Orog
3	Taats	200	9190	10–50	–	0.8–1.2	1.1[a] 15.5[b] 38.9[c]	Small rivers, streams	L. Taatsiin Tsagaan
4	Ongi	435	52,920	15–30	0.4–1.7	1.2–1.8	4.3[a]	Small rivers, streams	L. Ulaan

[a]Annual average
[b]Maximum average
[c]Maximum
[d]Flooding

Range, and they feed lakes in the Valley of Lakes (Table 9.4, Fig. 9.1). However, their discharges are decreasing at the present time (Davaa, 2015). For instance, Ongi River has frequently shrunk, and it has shifted to an intermittent discharge to feed Lake Ulaan (Orkhonselenge et al., in press). In the Govi rivers, their water temperature is expected to rise 2.5–2.7 °C in 2020, 3.2–3.5 °C in 2050, and 3.9–4.2 °C in 2080 (Davaa, 2010).

9.3 Govi Lakes

It is remarkable to note that there exist abundant large lakes in the Govi regions even though these regions are characterized with low precipitation, sparse river network, and high evaporation. This is related to the following factors: first, the Govi regions are low-elevated basins collecting water through the valley and depressions between the Altai and the Khangai Mountain Ranges, and second, the large lakes in the Govi region are relicts of larger paleolakes (see Chap. 3). The earlier larger lakes in the eastern Govi (or Dornogovi) and northern Govi (or Dundgovi) regions became divided into smaller lakes today (see Sect. 9.4).

The Govi lakes in southern Mongolia (Fig. 1.1) are large in number but small in area (Table 1.2; see Chap. 5). Volcanic lakes such as Lakes Sharga, Shiree, Bugat, Khaliun, Shanaga, and Shanaa are distributed in the southwestern Khangai Mountain Range (see Chap. 4). Tectonic lakes are distributed in the Valley of Lakes underlain by the Jurassic and Cretaceous sedimentary rocks and the Neogene deposits (see Chap. 3). There are many aeolian and intermittent playa lakes in the Govi region (see Chap. 4). For example, Lake Tukhum is an aeolian lake.

9.3.1 Paleolakes

There were many stages in the development of lakes within the Govi region (Berkey & Morris, 1927). Rivers draining the southern Khangai Mountain Range were enlarged during the early Jurassic, and they fed lakes in the Valley of Lakes intensely since the Paleogene period (Jigj, 1975; see Chap. 3).

The middle Holocene 9.0-m and 15.0-m-deep freshwater pluvial lakes in Tukhum Els and Khongor Els sand dunefield depressions of the Govi region were reconstructed by Grunert et al. (2009). Lake-level fluctuations at about 8.5 ka BP and a slightly humid period around 1.5 ka BP in southern Mongolia were reconstructed (Lehmkuhl & Lang, 2001). Large paleolakes existed in the Govi region during the last interglacial (MIS 5e), and most of the lakes were relatively small or even disappeared during the considerably weaker EASM during the MIS 3. Then the lakes lasted probably until the beginning of the LGM, and they had the highest lake levels in the middle Holocene (Lehmkuhl et al., 2018) when high sedimentation rate of 4.6 cm/ka was estimated in the Lake Ulaan basin (Orkhonselenge et al., 2018b; see Chap. 11).

The desiccation of lakes in southern Mongolia strengthened since 4.0 cal ka BP (Felauer et al., 2012). In the late Holocene, lake levels were generally low, and the sedimentation rate was estimated as 1.8 cm/ka in the Lake Ulaan basin (Orkhonselenge et al., 2018b). Lakes in the Valley of Lakes have rapidly changed their dimensions due to the shallowness, large graben flat floors and gentle beaches (Batnasan, 1998), and the rising temperature (Orkhonselenge et al., 2018a, in press).

9.3.2 Modern Lakes

There are lakes occupying an area greater than 100 km^2 (Table 9.5) and many smaller lakes in the Govi region today (see Chap. 5). At the present time, lakes are unstable depending on annual and seasonal precipitations. The water level of lakes (e.g., Lakes Orog, Taatsiin Tsagaan, Adgiin Tsagaan, and Ulaan) dropped by 1–2 m due to the decrease of precipitation in the Govi region since the 1980s (Batnasan, 1998). Lake areas in the hot-and-dry Govi region may vary with the complex interplay of hydrological balance among precipitation, evaporation, inflow, and outflow of runoff and groundwater (Kang et al., 2015). Many small lakes are fed by groundwater, and they are saline (see Chap. 5).

Many lakes in the Govi region are shallow playas today (e.g., Ariunbileg et al., 2020; Orkhonselenge et al., in press; Fig. 9.6) in deflation hollows, which are dry most of the year, and they are at least partially surrounded by sand dunefields (Berkey & Morris, 1927). The remnant lakes of the larger paleolakes in the Valley of Lakes are facing shrinkage due to the present global warming (Orkhonselenge et al., 2018a). These lakes normally develop with the landscape as the main controlling factor (Batnasan, 1998). In the last 50 years, many lakes shrank or disappeared

Table 9.5 Morphometric components of lakes in the Govi region

#	Lakes	Elevation (m a.s.l.)	Area (km^2)	Depth (m)		Volume (m^3)
				Average	Maximum	
1	Khukh	2649	17.2	15.5	43.0	206.0
2	Adgiin Tsagaan	1285	11.5	0.8	1.5	9.0
3	Tsagaan	1284	9.1	0.2	0.7	2.0
4	Khun	953	16.7	0.3	1.1	5.0
5	Zagiin Shal	1032	33.1	0.3	1.0	3.0
6	Buun Tsagaan	1312	241.12	9.9	16.0	638.5
7	Orog	1217	49.65	3.0	6.3	420.0
8	Alag	1036	6.8	0.7	1.6	4.0
9	Tonkhil	2062	6.2	1.0	1.6	5.0
10	Khulma	2234	8.5	0.7	2.0	6.0
11	Biger	1287	7.4	0.9	2.0	1.0
12	Buur	1149	10.6	0.6	1.5	6.0
13	Khutag	2108	8.1	0.3	0.8	2.0
14	Ikhes	1604	19.3	0.6	1.7	11.0
15	Shargiin Tsagaan	2964	15.7	0.5	1.3	8.0
16	Tamcha	2446	6.6	0.6	1.5	0.4
17	Sangiin Dalai	1355	15.5	1.6	3.8	25.0
18	Agiin Ulaan	943	8.8	0.4	1.0	4.0
19	Nuden	990	11.6	0.5	1.2	5.0
20	Ulaan Usnii Govi	950	10.4	0.5	1.1	5.0
21	Shokhoi	1040	7.5	0.4	0.9	3.0
22	Sulinkheer	1007	15.1	0.5	1.2	8.0
23	Khutag	1010	5.5	0.3	0.8	2.0
24	Tenger	1077	5.3	0.4	0.9	2.0
25	Bayan Khoshuu	1200	5.1	0.6	1.1	3.0
26	Khavtsgain Ulaan	1040	5.0	0.3	0.9	2.0
27	Jargalan	1043	5.3	0.6	1.3	3.0
28	Gashuun	1030	7.6	0.4	1.2	0.3
29	Zuun Gyalgar	1010	8.0	0.7	1.2	6.0
30	Ulaan	1322	5.6	0.7	1.2	4.0
31	Shanagan	1100	5.5	0.4	1.1	2.0
32	Taatsiin Tsagaan	1234	9.9	1.0	2.3	11.0
33	Ulaan	1320	6.2	0.7	1.2	4.0
34	Ulaan	1520	6.9	0.6	1.1	4.0
35	Ulaan	1024	44.16	0.9	1.6	158.0

also due to the strong human usage of water resources (Lehmkuhl et al., 2018; Orkhonselenge et al., in press).

Among the total 756 lakes of the region (Table 1.2), there are 481 lakes elevated at the range of 1001–1500 m a.s.l. (Table 4.1), and these lakes constitute 33.7% of the total 1426 lakes with the same elevation range in Mongolia (see Chap. 5). Out of the 756 lakes, 96.0% of them covers an area less than 5.0 km^2, while 1.5% of

Fig. 9.6 Exposed floors of Lakes (**a**) Orog, (**b**) Sain Bulangiin Ulaan viewing toward the south, (**c**) Doloodiin Ulaan viewing toward the west, and (**d**) Dersen Bulangiin Ulaan viewing toward the east in eastern Govi. Photo by S. Ariunbileg on August 6, 2009, for (**a**) and by A. Orkhonselenge on June 3–4, 2019, for (**b–d**)

them cover an area of 10.1–20.0 km^2 (Table 4.1). The lakes constitute 24.8% and 37.9% of the total lakes covering the same areas in each class within the whole country (see Chap. 5).

In southern Mongolia, there are 393 lakes covering an area greater than 0.5 km^2 each, or 756 lakes cover an area greater than 0.1 km^2 each and 1246 km^2 in total (see Chap. 5). Table 9.5 shows lakes covering an area larger than 5.0 km^2, and two large lakes among them are discussed in detail (see Chaps. 10 and 11). The three largest lakes occupied collectively an area of 888.6 km^2 or 71.3% of the combined area of all the lakes in the region, but they have shifted into an area of only 334.93 km^2 or 26.9% (Table 9.5). Today, it is shown that lakes in the Govi region are very sensitive to the recent rising annual average air temperature trends (Orkhonselenge et al., 2018a), and they are shifting into playa environments (Orkhonselenge et al., in press; see Sect. 9.4).

9.4 Lake Evolution

In the last 50 years, most lakes in the region decreased their areas (Orkhonselenge et al., 2018a) or completely disappeared (Figs. 9.5 and 9.6) due to the strong human usage of water resources for mining operation and livestock in addition to the influence of the strengthening temperature rise and dryness (Orkhonselenge et al., in press; see Sect. 9.2.2).

Lakes in the Govi region have frequently disappeared and reappeared (see Chaps. 10 and 11). For instance, Lakes Taatsiin Tsagaan (see Sect. 9.4.1) and Orog (see Sect. 9.4.2) disappeared in 1980–1984 (Batnasan, 1999). Revealing lake area changes is a key for inferring the history of local climate changes within the lake basins, and it also contributes to understanding the impact of climate change on the wider-scale hydrological regime within the region. For instance, most of the large Govi lakes (over an area of 6 km^2) existed in 1991 had disappeared by 2000, all of the lakes had disappeared by 2002, but nine small lakes were detected again in 2006 (Kang et al., 2015). Moreover, a decreasing trend in surface area, intermittent with short episodes of expansion, has been observed in Lakes Buun Tsagaan (see Chap. 10) and Orog (see Sect. 9.4.2) during the last 40 years (Orkhonselenge and Uuganzaya, unpublished), and the rapid decrease in the surface areas of those lakes since 1995 is related to the increase of the annual average air temperature observed at the Bayankhongor meteorological station (Szumińska, 2016).

A significant hydrology-induced regional decreasing trend of the lake surface area was observed for large lakes in the Govi region between 2000 and 2011 (Kang & Hong, 2016). In the eastern Govi (e.g., DOR with 4.8 °C and 126 mm), the changes in lake area are sensitive to precipitation for the precedent 2 months ($R^2 = 0.90$) and current year ($R^2 = 0.54$). It is that (1) the small and shallow morphology of lakes could result in high sensitivity and (2) the effect of the severe drought conditions found in this region since the late 1990s draws substantial vulnerability of lakes (Kang et al., 2015). In the southern Govi (e.g., DAL with 4.8 °C and 116 mm), the changes in lake area were not accountable with both monthly ($R^2 = 0.26$) and annual ($R^2 = 0.028$) precipitations, but they were highly correlated with temperature (Kang et al., 2015). In the northern Govi (e.g., ARV with 1.5 °C and 227 mm), the changes in lake area were explained well by both monthly ($R^2 = 0.95$) and annual ($R^2 = 0.64$) precipitations. The results suggest a determinant role of precedent precipitation on changes in lake area in the eastern and northern Govi. It has been noted that the changes in lake area in the eastern and central Govi, where the lake disappearance and reappearance showed more variation, are highly related to precipitation (Kang et al., 2015).

All the large lakes in the Valley of Lakes (Fig. 9.1) decreased in their areas between 1970 and 2015, and this was linked with the continuous rise in the temperature (Fig. 9.3), not with the gradual increase of precipitation with fluctuations (Orkhonselenge et al., 2018a). For example, three largest Lakes Buun Tsagaan, Ulaan (see Chaps. 10 and 11), and Orog lost large areas of >10.0 km^2 during the last 44–45 years, whereas three smaller Lakes Taatsiin Tsagaan, Shargiin Tsagaan, and Biger (see Sects. 9.4.1–9.4.4) decreased their areas of <10.0 km^2 (Orkhonselenge et al., 2018a; Table 9.6; Fig. 9.7). These changes in area for the lakes indicate how directly and significantly the lakes have been responding to the temperature rise since 1995 and the precipitation fluctuations since 1975 (Figs. 9.3 and 9.7). The recent shrinkage of the lakes has contributed to the high mineralization, a shift of

Table 9.6 Temporal changes in areas of large lakes in the Valley of Lakes. After Orkhonselenge et al. (2018a)

#	Lakes	Areas (km^2)			Details
		1970	2014	2015	
1	Ulaan	62.38	44.16		Chapter 11
2	Taatsiin Tsagaan	18.66		10.67	Section 9.4.1
3	Orog	145.21		49.65	Section 9.4.2
4	Buun Tsagaan	324.29	241.12		Chapter 10
5	Biger	8.95	8.45		Section 9.4.3
6	Shargiin Tsagaan	16.14		11.12	Section 9.4.4

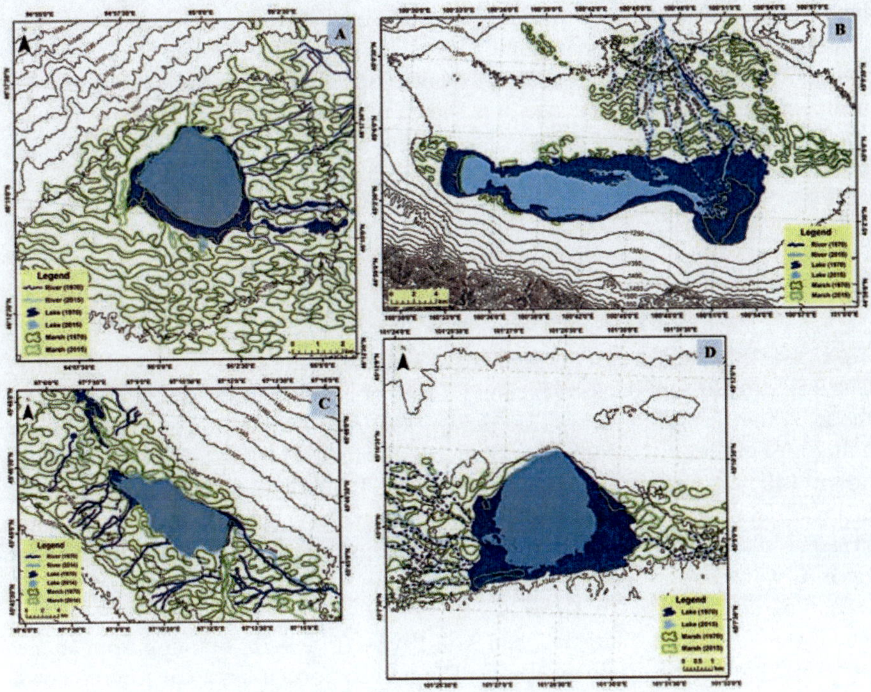

Fig. 9.7 Spatial distributions of areas and marshes of Lakes (**a**) Taatsiin Tsagaan, (**b**) Orog, (**c**) Biger, and (**d**) Shargiin Tsagaan during the last 44–45 years. Modified from Orkhonselenge et al. (2018a)

the cation ratio from $Na^{+}+K^{+} > Mg^{+2} > Ca^{+2}$ to $Mg^{2+} > Ca^{2+} > Na^{+}+K^{+}$ and increase in concentrations of S-As and Sr (Ariunbileg et al., 2017). For instance, Lake Biger covered by a thick layer of soda was observed in 2009, and waters of Lakes Biger, Kholbooljin, and Davst have been highly mineralized, whereas saline Lakes Burd, Zeges, and Bor Ovoo contain high concentrations of uranium (U) and lithium (Li) (Ariunbileg et al., 2020).

9.4.1 Lake Taatsiin Tsagaan

Lake Taatsiin Tsagaan is located at the elevation of 1234 m a.s.l. in the Valley of Lakes between the Khangai and Govi Altai Mountain Ranges (Fig. 9.1). The lake extends 4.1 km in length, 2 km on average or 3.5 km in the maximum width, and 11.0 km along the shoreline. The lake depth is 2.0 m on average and 2.3 m in the maximum (Tserensodnom, 2000; Table 9.5). The lake is fed by rainfall, overflow, and groundwater. During the heavy rainfall, the Taats River draining the Khangai Mountain Range flows into the lake.

Lake Taatsiin Tsagaan occupies an area of 9.9 km^2 (Table 9.5). The lake area decreased by 7.9 km^2 between 1970 and 2015 (Table 9.6, Fig. 9.7a). The decrease in the lake area may have been connected to the intensification of rising air temperature since 1997 although precipitation increased in 1987–1997 and since 2006 (BU in Fig. 9.3) as observed at the Baruunbayan-Ulaan meteorological station (Fig. 9.1) positioning 1.7 km from the lake. It is shown that the lake area is highly dependent on air temperature, and not on fluctuation of precipitation, i.e., the lake is very sensitive to the present warming (Orkhonselenge et al., 2018a).

9.4.2 Lake Orog

Lake Orog is located at the elevation of 1217 m a.s.l. in the Valley of Lakes (Fig. 9.1). The lake extends for 30.0 km (Tsegmid, 1969) or 31.8 km (Tserensodnom, 1971) in the maximum length, 8–9 km (Tsegmid, 1969) or 7.7 km in the maximum width, and 75.0 km along the shoreline (Tserensodnom, 1971). Lake Orog occupies an area of 140 km^2 (Tsegmid, 1969; Tserensodnom, 1971) or 49.65 km^2 (Orkhonselenge et al., 2018a). The lake depth is 3.0 m on average and 6.3 m in the maximum (Tserensodnom, 2000; Table 9.5). The lake level strongly varies due to evaporation from its water surface and runoff changing from year to year. Lake Orog, a saline lake, is currently fed by Tui River (see Sect. 9.2.6) and its distributary rivers such as Zuun, Dund, Baruun, Tsutgalan, and Khii Rivers (Fig. 9.7b) draining from an area of 10,500 km^2 (Tserensodnom, 1971). The inflows form a large delta in the downstream of Tui River, and the lake has no outlet (Fig. 9.7b). The mineralization value of Lake Orog is 5.5 g/L (Batnasan, 1998) or 1.75–5.01 g/L (Tserensodnom, 2000). The cation ratio of $Ca^{2+} > Na^{+}+K^{+} > Mg^{2+}$ is dominant in the lake water (Ariunbileg et al., 2017), and their concentrations have risen in the recent years (Ariunbileg et al., 2020).

Lake Orog fluctuated its area significantly during the last eight decades (Table 9.7). For instance, the lake expanded from 1952 to 1970 due to the phenomena linked to an earthquake occurred in 1957 in the Govi Altai Mountain Range (Tsegmid, 1969). The lake decreased for 95.56 km^2 or 65.8% during 1970–2015 when the southeastern, eastern, northeastern, northern, and western shores of the lake were dominantly lost (Orkhonselenge et al., 2018a; Fig. 9.7b). In other words, the lake began to decrease its area from 1971, and it shrank in the 1980s and waterlogged in the 1990s. The lake increased by 23.87 km^2 from 1969/1971 to 1992 and reshrank in the 2000s (Table 9.7). The lake floor was exposed in 2009 (Fig. 9.6a) when the annual precipitation was 45.8 mm (BO in Fig. 9.3) as observed in the Bogd meteorological station (Fig. 9.1). The lake expanded by 78.8 km^2 between 2009 and 2013 and reduced by 29.15 km^2 between 2013 and 2015 (Table 9.7). Because of the limited climate data available since 1994, the lake area in recent years between 2004 and 2015 show weak linear relationships with the annual average precipitation ($R^2 = 0.2183$) and air temperature ($R^2 = 0.191$). The correlation analysis shows that the lake area during 2004–2015 had weak positive correlations with precipitation (r = 0.467) and temperature ($r = 0.437$).

Lake Orog responded more directly and significantly to the temperature rise since 1997 probably resulting in intensive evaporation. Lake Orog consists primarily of two depressions, the elongated deeper and larger western one and the smaller eastern one (Tserensodnom, 1971). However, the lake may be isolated into three parts, a small circle-shaped one in the west, a large one in the center, and a small playa in the east (Fig. 9.7b), in the next few years. In other words, the abrupt decrease of lake area during the last 45 years with the shrinkages during the early to the end of the 1980s and the early to the end of the 2000s shows that Lake Orog may be shifting toward a playa lake condition in the next few years. The geomorphological and hydrological evidence from Lake Orog show the significant response of the lake to the climate change in this century.

Table 9.7 Temporal changes in area of Lake Orog

#	Year	Area (km^2)	Authors
1	1951	155.0	Bespalov (1951)
2	1952	130.0	Murzaev (1952)
3	1969	140.0	Tsegmid (1969)
4	1970	145.21	Orkhonselenge et al. (2018a)
5	1970	146.0	Szumińska (2016)
6	1971	140.0	Tserensodnom (1971)
7	1980–1984	0.0	Batnasan (1999)
8	1986–1989	0.0	Batnasan (1998)
9	1988–1989	0.0	Szumińska (2016)
10	1992	163.87	Batnasan (1998)
11	2004	0.0	Szumińska (2016)
12	2009	0.0	Ariunbileg et al. (2020)
13	2010	29.0	Davaa (2015)
14	2013	78.8	Davaa (2015)
15	2013	74.0	Szumińska (2016)
16	2015	49.65	Orkhonselenge et al. (2018a, 2018b)

9.4.3 Lake Biger

Lake Biger is located at the elevation of 1287 m a.s.l. in an intermontane basin surrounded by high mountains between the Khangai and Govi Altai Mountain Ranges (Fig. 9.1). The lake extends for 5.7 km in length, 1.2 km on average or 2.3 km in the maximum width, and 13.8 km along the shoreline. The lake depth is 0.9 m on average and 2.0 m in the maximum (Tserensodnom, 2000; Table 9.5). The saline lake is fed by rainfall and groundwater. Numerous streams flow into the lake. Shorelines are situated on the lowlands and are abundantly covered by marshes (Tserensodnom, 1971; Fig. 9.7c).

Lake Biger occupies an area of 7.4 km^2 (Table 9.5). The lake area reduced by 0.5 km^2 between 1970 and 2014 (Table 9.6, Fig. 9.7c). The small decrease of lake area shows that Lake Biger is similar to other lakes in the Valley of Lakes being affected by the rising temperature since 1987 and decreasing precipitation since 1980 (BI in Fig. 9.3) as observed in the Biger meteorological station (Fig. 9.1) located 3.3 km from the lake. The data shows that decrease of lake area may have started since 1980 and strengthened since 1996. It is implied that although the gradual increase in precipitation appeared since 1980, the lake area sensitively responds to the rising temperature (Orkhonselenge et al., 2018a).

9.4.4 Lake Shargiin Tsagaan

Lake Shargiin Tsagaan is located at the elevation of 964 m a.s.l. in a wide depression of Sharga Govi in an intermontane basin surrounded by high mountains, i.e., it is situated in the northwesternmost tip of the Valley of Lakes between the Khangai and Govi Altai Mountain Ranges (Fig. 9.1). The lake extends for 5.0 km in length, 3.0 km on average or 4.8 km in the maximum width, and 16.2 km along the shoreline (Tserensodnom, 1971). The lake depth is 0.5 m on average and 1.3 m in the maximum (Tserensodnom, 2000; Table 9.5). The lake is fed by rainfall and groundwater, and the Bayan, Sharga, and Khoid Rivers flow into the lake. The lake level is unstable as the lake level highly oscillates during the heavy rainfall. Shores are swampy and marshy due to its position on lowland (Fig. 9.7d).

Lake Taatsiin Tsagaan occupies an area of 15.7 km^2 (Table 9.5). However, the lake area decreased by 5.0 km^2 between 1970 and 2015 (Table 9.6, Fig. 9.7d). The decrease in lake area indicates a relationship with the rising temperature since 1987 and the abruptly dropping precipitation since 2004 (SH in Fig. 9.3) observed at the Sharga meteorological station (Fig. 9.1) situated 18.2 km from the lake. In other words, the decrease of lake area may have begun since 1976 when precipitation fluctuated with frequencies of ~5 years and intensified since 1987 (Orkhonselenge et al., 2018a).

9.5 Summary

Southern Mongolia is a spectacular region characterized by the Govi landscape. The warming and drying climate in the region has strengthened during the last century. The water regime has rapidly responded to the warming and dryness, and sensitivity of the hydrological system has been reflected in dropping of the groundwater level, disappearance and retreat of permafrost, shrinkage of surface water, and decrease of lake levels and areas. In southern Mongolia, lakes are the second highest number in the country (24.7%), and they occupy small areas of the total lake area (7.8%). Aeolian, volcanic, and tectonic lakes are abundant. During the last half century, lakes in the Govi region have abruptly lost their areas due to the rising temperature since 1995 and fluctuating precipitation since 1975. Patterns of the warming-induced reduction in lake areas and shifting to playa and saline lakes are observed for many lakes in southern Mongolia. Geomorphological, sedimentological, geochemical, and geochronological information from the Govi region could provide a more complete understanding of the landscape evolution in lake basins of southern Mongolia and expand our knowledge of the paleoclimatic history of Mongolia and Central Asia (see Chap. 21).

References

Amarjargal, D. (2016). *A relationship between weather condition and dust mobility in the Govi and steppe regions of Mongolia*. Ph.D. Thesis, National University of Mongolia, Ulaanbaatar. [In Mongolian].

Andrén, E., Klimaschewski, A., Self, A. E., Amour, N. S., Andreev, A. A., Bennett, K. D., Conley, D. J., Edwards, T. W. D., Solovieva, N., & Hammarlund, D. (2015). Holocene climate and environmental change in north-eastern Kamchatka (Russian Far East), inferred from a multi-proxy study of lake sediments. *Global and Planetary Change, 134*, 41–54.

Ariunbileg, S., Isupov, V. P., Vladimirov, A. G., Kolpakova, M. N., & Kuibida, L. V. (2020). *Chemical evolution and geochemistry of mineralized lakes of Mongolia*. Ulaanbaatar, 132 p. [In Mongolian].

Ariunbileg, S., Isupov, V. P., Vladimirov, A. G., Orkhonselenge, A., & Shatskaya, S. S. (2017). Impact of climate change on hydrogeochemical characteristics of lakes in the Valley of Lakes, southern Mongolia. *Goldschmidt2017: Presentation #2169*. Paris.

Badarch, G., Cunningham, W. D., & Windley, B. F. (2002). A new terrane subdivision for Mongolia: Implications for the Phanerozoic crustal growth of Central Asia. *Journal of Asian Earth Sciences, 21*, 87–110.

Batima, P., Natsagdorj, L., Gomboluudev, P., & Erdenetsetseg, B. (2005). Observed climate change in Mongolia. *AIACC Working Paper*, 12 p. Ulaanbaatar.

Batnasan, N. (1998). *Hydrological system, water regime, and evolution of large lakes in the Govi region*. Doctorate Thesis in Geography: Hydrology. Institute of Geoecology, Mongolian Academy of Sciences, Ulaanbaatar. 163 p. [In Mongolian].

Batnasan, N. (1999). Morphology and long-term water level fluctuation of Lake Uvs. In *Proceedings of International Conference of Global Change and Uvs Nuur on Sustainable Development of the Altai-Sayan Ecoregion and Transboundary Nature Conservation Issues, Ulaangom, Uvs aimag*, pp. 94–108. [In Mongolian with English and Russian abstract].

Berkey, P. C., & Morris, K. F. (1927). *Geology of Mongolia. A reconnaissance report based on the investigations of the years 1922–1923. Natural History of Central Asia* (Vol. 2). The American Museum of Natural History, 475 p.

Bespalov, N. D. (1951). Soil of Mongolia. AH. Vol. 41. Moscow. [In Russian].

Davaa, G. (2010). Climate change impacts on water resources in Mongolia. In *Proceedings of Consultative Meeting on Integration of Climate Change Adaptation into Sustainable Development in Mongolia.* Institute for Global Environmental Strategies (IGES), pp. 30–36.

Davaa, G. (2015). *Resource and regime of surface water of Mongolia.* 408 p. [In Mongolian].

Davaagatan, T., Orkhonselenge, A., Fukushi, K., & Yabe, M. (2015). Sedimentary records in intermontane and prairie lakes in central Mongolia: Preliminary results from Lake Terkhiin Tsagaan and Lake Ugii. *Journal of Geographical Issues, 14*(408), 19–26.

Devyatkin, E. V. (1981). *The Cenozoic of Inner Asia: Stratigraphy, geochronology, correlation.* 196 p. Transaction of Joint Soviet–Mongolian Geological Expedition, 27, Nauka, Moscow. [In Russian].

Dulamsuren, D. (2016). *Extreme air temperature changes and future of natural zone in Mongolia.* Ph.D. Thesis, National University of Mongolia, Ulaanbaatar. [In Mongolian].

Felauer, T., Schultz, F., Murad, W., Mischke, S., & Lehmkuhl, F. (2012). Late Quaternary climate and landscape evolution in arid Central Asia: A multiproxy study of lake archive Bayan Tohomin Nuur, Gobi desert, southern Mongolia. *Journal of Asian Earth Sciences, 48*, 125–135.

Graham, S. A., Hendrix, M. S., Johnson, C. L., Badamgarav, D., Badarch, G., Amory, J., Porter, M., Barsbold, R., Webb, L. E., & Hacker, B. R. (2001). Sedimentary record and tectonic implications of Mesozoic rifting in southeast Mongolia. *Geological Society of America Bulletin, 113*(12), 1560–1579.

Grunert, J., Stolz, C., Hempelmann, N., Hilgers, A., Hülle, D., Lehmkuhl, F., Felauer, T., & Dasch, D. (2009). The evolution of small lake basins in the Gobi desert in Mongolia. *Quaternary Sciences, 29*(4), 678–689.

Guy, A., Schulmann, K., Clauer, N., Hasalová, P., Seltmann, R., Armstrong, R., Lexa, O., & Benedicto, A. (2014). Late Paleozoic–Mesozoic tectonic evolution of the Trans-Altai and South Gobi Zones in southern Mongolia based on structural and geochronological data. *Gondwana Research, 25*, 309–337.

Helo, C., Hegner, E., Kröner, A., Badarch, G., Tomurtogoo, O., Windley, B. F., & Dulski, P. (2006). Geochemical signature of Paleozoic accretionary complexes of the Central Asian Orogenic Belt in South Mongolia: Constraints on arc environments and crustal growth. *Chemical Geology, 227*, 236–257.

Heumann, M. J., Johnson, C. L., & Webb, L. E. (2018). Plate interior polyphase fault systems and sedimentary basin evolution: A case study of the East Gobi Basin and East Gobi Fault Zone, southeastern Mongolia. *Journal of Asian Earth Sciences, 151*, 343–358.

Hock, V., Daxner-Hock, G., Schmid, H. P., Badamgarav, D., Frank, W., Furtmuller, G., Montag, O., Barsbold, R., Khand, Y., & Sodov, J. (1999). Oligocene-Miocene sediments, fossils and basalts from the Valley of Lakes (Central Mongolia)—An integrated study. *Mitt. Osterr. Geol. Ges., 90*, 83–125.

Holguin, L. R., & Sternberg, T. (2016). A GIS based approach to Holocene hydrology and social connectivity in the Gobi Desert, Mongolia. *Archaeological Research in Asia, 39*, 1–9.

Hülle, D., Hilgers, A., Radtke, U., Stolz, C., Hempelmann, N., Grunert, J., Felauer, T., & Lehmkuhl, F. (2010). OSL dating of sediments from the Gobi Desert, Southern Mongolia. *Quaternary Geochronology, 5*, 107–113.

Institute of Meteorology, Hydrology and Environment. (2013). *Water and weather of Mongolia.* 27 p. [In Mongolian].

Jambaljav, Y., Gansukh, Y., Temuujin, K., Tsogt-Erdene, G., Undrakhtsetseg, T., Saruulzaya, A., Amarbayasgalan, Y., Dashtseren, A., & Narangerel, S. (2016). *Report of new map of permafrost distribution in Mongolia: Scale 1:1000000.* Institute of Geography and Geoecology, Mongolian Academy of Sciences, 90 p.

Javkhlan, O., Takasu, A., Bat-Ulzii, D., & Kabir, M. F. (2013). Metamorphic pressure-temperature evolution of garnet-chloritoid schists from the Lake Zone, SW Mongolia. *Journal of Mineralogical and Petrological Sciences, 108*, 255–266.

Jigj, S. (1975). *Primary feature of Mongolian landforms*. Institute of Geography & Permafrost. Mongolian Academy of Sciences (MAS), 126 p. [In Mongolian].

Jigj, S. (1976). *A brief description of paleoglaciations and paleoglaciers*. 51 p. [In Mongolian].

Kang, S., & Hong, S. Y. (2016). Assessing Seasonal and Inter-Annual Variations of Lake Surface Areas in Mongolia during 2000–2011 Using Minimum Composite MODIS NDVI. *PLoS One, 11*(3), e0151395.

Kang, S., Lee, G., Togtokh, C., & Jang, K. (2015). Characterizing regional precipitation-driven lake area change in Mongolia. *Journal of Arid Land, 7*(2), 146–158.

Klimek, K. (1980). Major physic-geographical features of the southern slope of the Khangai Mountains. In K. Klimek & L. Starkel (Eds.), *Vertical zonality in the southern Khangai Mountains (Mongolia)* (pp. 9–13).

Komatsu, G., Brantingham, P. J., Olsen, J. W., & Baker, V. R. (2001). Paleoshoreline geomorphology of Boon Tsagaan Nuur, Tsagaan Nuur and Orog Nuur: The Valley of Lakes, Mongolia. *Geomorphology, 39*, 83–98.

Komatsu, G., Baker, V. R., Arzhannikov, S., Gallagher, R., Arzhannikova, A. V., Murana, A., & Oguchi, T. (2016). Catastrophic flooding, palaeolakes and late Quaternary drainage reorganization in northern Eurasia. *International Geology Review, 58*, 1693–1722.

Kröner, A., & Rojas-Agramonte, Y. (2014). The Altaids as seen by Eduard Suess, and Present Thinking on the Late Mesoproterozoic to Palaeozoic Evolution of Central Asia. *Austrian Journal of Earth Sciences, 107*(1), 156–168.

Krylov, A. V., Mendsaikhan, B., & Gantstsooz, B. (2011). Zooplankton of the drying lake Orog (Mongolia). *Inland Water Biology, 4*(2), 179–181.

Lee, M. K., Lee, Y. I., Lim, H. S., Lee, J. I., & Yoon, H. I. (2013). Late Pleistocene–Holocene records from Lake Ulaan, southern Mongolia: Implications for east Asian palaeomonsoonal climate changes. *Journal of Quaternary Science, 28*(4), 370–378.

Lehmkuhl, F. (2014). Aeolian dust in mountain areas of Tibet and Mongolia. *ProScience, 1*, 44–50.

Lehmkuhl, F., & Lang, A. (2001). Geomorphological investigations and luminescence dating in the southern part of the Khangay and the Valley of the Gobi Lakes (Central Mongolia). *Journal of Quaternary Science, 16*(1), 69–87.

Lehmkuhl, F., Grunert, J., Hülle, D., Batkhishig, O., & Stauch, G. (2018). Paleolakes in the Gobi region of southern Mongolia. *Quaternary Science Reviews, 179*, 1–23.

Ministry of Environment and Green Development. (2013). *Integrated water management plan of Mongolia.* 339 p.

Murzaev, E. M. (1952). *The description of physical geography of Mongolia* (2nd ed.). [In Russian].

Orkhonselenge, A., Komatsu, G., Davaagatan, T., & Uuganzaya, M. (2019). Sedimentary records from lakes reconstructing late quaternary paleoclimate changes in Mongolia. In *XX INQUA Congress. Presentation #2419*, Dublin, Ireland.

Orkhonselenge, A., Komatsu, G., & Uuganzaya, M. (2018a). Climate-driven changes in lake areas for the last half century in the Valley of Lakes, Govi region, southern Mongolia. *Natural Science, 10*(7), 263–277.

Orkhonselenge, A., Komatsu, G., & Uuganzaya, M. (2018b). Middle to late Holocene sedimentation dynamics and paleoclimatic conditions in the Lake Ulaan basin, southern Mongolia. *Géomorphologie: Relief, Processus, Environnement, 24*(4), 351–363.

Orkhonselenge, A., Krivonogov, S. K., Kashiwaya, K., Nakamura, T., & Davaagatan, T. (2015). Paleoclimate changes in Mongolia: Lake records. In *Extended Abstracts of International Association of Geomorphologists (IAG) Regional Conference on Gradualism vs Catastrophism in Landscape Evolution*, Barnaul, Russia, p. 175.

Orkhonselenge, A., Uuganzaya, M., & Davaagatan, T. (in press). Late Holocene sedimentation dynamics in the Lake Ulaan basin, southern Mongolia. *Environmental Earth Science*.

Owen, L. A., Windley, B. F., Cunningham, W. D., Badamgarav, J., & Dorjnamjaa, D. (1997). Quaternary alluvial fans in the Gobi of southern Mongolia: Evidence for neotectonics and climate change. *Journal of Quaternary Science, 12*(3), 239–252.

Owen, L. A., Richards, B., Rhodes, E. J., Cunningham, W. D., Windley, B. F., Badamgarav, J., & Dorjnamjaa, D. (1998). Relic permafrost structures in the Gobi of Mongolia: Age and significance. *Journal of Quaternary Science, 13*(6), 539–547.

Rother, H., Lehmkuhl, F., Fink, D., & Nottebaum, V. (2014). Surface exposure dating reveals MIS-3 glacial maximum in the Khangai Mountains of Mongolia. *Quaternary Research, 82*, 297–308.

Saruulzaya, A. (2017). *Thermokarst lake dynamics in the continuous and isolated permafrost zones, Mongolia.* Ph.D. Thesis, Hokkaido University, Sapporo, Japan. 76 p.

Sternberg, T., Thomas, D., & Middleton, N. (2011). Drought dynamics on the Mongolian steppe, 1970–2006. *International Journal of Climatology, 31*, 1823–1830.

Szumińska, D. (2016). Changes in surface area of the Böön tsagaan and Orog lakes (Mongolia, Valley of the lakes, 1974–2013) compared to climate and permafrost changes. *Sedimentary Geology*, 1–37.

Tsegmid, S. (1969). *Physical geography of Mongolia.* State Press, 405 p. [In Mongolian].

Tserensodnom, J. (1971). *Lakes of Mongolia.* State Publishing, 202 p. In Mongolian].

Tserensodnom, J. (2000). *A catalog of lakes in Mongolia.* Shuvuun Saaral Publishing, 141 p. [In Mongolian].

Tumurbaatar, D. (2004). *Seasonal and permanent permafrost of Mongolia.* Mongolian Academy of Sciences, 147 p. [In Mongolian].

van der Wal, J. L. N., Nottebaum, V., Stauch, G., Lehmkuhl, F., & Reicherter, K. (2020). Geomorphological evidence of active faulting in low seismicity regions—Examples from the Valley of Lakes, southern Mongolia. *EGU2020–949 presentation.*

Wu, G. Y. (2014). Tectonopalaeogeography and palaeotectonics. Palinspastic reconstruction and geological evolution of Permian residual marine basins bordering China and Mongolia. *Journal of Palaeogeography, 3*(2), 219–232.

Yarmolyuk, V. V., Kovach, V. P., Kovalenko, V. I., Terent'eva, L. B., Kozakov, I. K., Kotov, A. B., & Eenjin, G. (2007). Isotopic composition of the Hercynian Crust of Southern Mongolia: Substantiation of the Hercynian Juvenile Crust-Forming Event. *Doklady Earth Sciences, 417*(8), 1178–1182.

Yu, K., Lehmkuhl, F., Diekmann, B., Zeeden, C., Nottebaum, V., & Stauch, G. (2017). Geochemical imprints of coupled paleoenvironmental and provenance change in the lacustrine sequence of Orog Nuur, Gobi Desert of Mongolia. *Journal of Paleolimnology, 58*, 511–532.

Zhu, M., Miao, L., Baatar, M., Zhang, F., Anaad, C., Yang, S., & Li, X. (2016). Late Paleozoic magmatic record of Middle Gobi area, South Mongolia and its implications for tectonic evolution: Evidences from zircon U–Pb dating and geochemistry. *Journal of Asian Earth Sciences, 115*, 507–519.

Zorigt, M., Myagmar, K., Orkhonselenge, A., van Beek, E., Kwadijk, J., Tsogtbayar, J., Yamkhin, J., & Dechinlkhundev, D. (2020). Modelling permafrost distribution over the river basins of Mongolia using remote sensing and analytical approaches. *Environmental Earth Sciences, 79*(308), 1–11.

Chapter 10
Lake Buun Tsagaan

Abstract Lake Buun Tsagaan is the largest lake in the Govi region within southern Mongolia, and this lake constitutes a major portion of the region's surface water resources. Lake Buun Tsagaan plays an essential role in the hydrological system of the Central Asian internal drainage basin through Baidrag River. The lake has experienced fluctuations of its area coverage, water level, and hydrochemical condition during the last half century, and it continuously lost its area for 83.17 km^2 or 25.6% from 1970 to 2014. The significant modifications of the lake are controlled by local and regional climate changes. The linear regression analysis shows that the lake area between 1992 and 2014 was strongly related to the annual average air temperature ($R^2 = 0.7847$) but weakly related to the annual precipitation ($R^2 = 0.3612$). The correlation analysis shows that the lake area during 1992–2014 had a strong negative correlation with air temperature ($r = -0.885$) and a moderately negative correlation with precipitation ($r = -0.601$). The sensitivities of Lake Buun Tsagaan to the climate changes imply that geomorphological, geochemical, and geochronological data from this lake can provide us much needed knowledge for the better understanding of its paleohydrological histories and the lake basin evolution over geological time scales.

Keywords Lake Buun Tsagaan · Baidrag River · Valley of Lakes · Govi region · Climate change · Southern Mongolia

10.1 Introduction

Lake Buun Tsagaan is one of the major lakes in the Valley of Lakes, which are located at the northern border of the Govi[1] region within southern Mongolia (Fig. 1.1). Lake Buun Tsagaan is a valuable lake in terms of its hydrological system, especially for the surface water resource of southern Mongolia.

[1] *Govi* has been often spelled as *Gobi* in the international literature. *Govi* is the correct English transliteration from Mongolian *Говь*.

A. Orkhonselenge et al., *Lakes of Mongolia*, Syntheses in Limnogeology,
https://doi.org/10.1007/978-3-030-99120-3_10

Lake Buun Tsagaan has been investigated in numerous studies. The general physical geography of Lake Buun Tsagaan was described by Tsegmid (1969). More information on morphometric values (e.g., area, length, width, depth, etc.) of the lake (see Chap. 9) was provided by Tserensodnom (1971, 2000). Bathymetric data from Lake Buun Tsagaan were provided by N. Batnasan, Ts. Sugar, and other researchers (Batnasan, 1998). Lake Buun Tsagaan was classified as a plain lake (Tserensodnom, 1971; see Chap. 1).

Previous studies observed paleoshorelines, paleoterraces, and floral and faunal fossils deriving paleogeographical implications and hypotheses in the lake basin and adjacent areas (see Chap. 2). For instance, Lake Buun Tsagaan had a hydrological connection with Lake Adgiin Tsagaan to the east, and the whole basin included the lakes extending 115 km in length (Murzaev, 1952; Tsegmid, 1969). At that time of the extensive lake, the water level of Lake Buun Tsagaan was 300–340 m (Devyatkin, 1981) and 40–45 m (Batnasan, 1998) higher than that at the present. The lake level of Lake Buun Tsagaan rose for about 15–20 m in the late glacial and Holocene (Lehmkuhl & Lang, 2001). Moreover, Komatsu et al. (2001) identified a variety of paleoshoreline features of Lake Adgiin Tsagaan, such as relict beach ridges, wave-cut terraces, spits, and fan deltas.

In recent years, Lake Buun Tsagaan has attracted much scientific attention in terms of evolution of the lacustrine basin and reconstruction of paleoclimate and modern climate changes in southern Mongolia. For instance, changes in the Lake Buun Tsagaan surface area in 1974–2013 were analyzed by Szumińska (2016) in the context of climate conditions and permafrost degradation. She noted a tendency for decreasing in surface area, intermittent with short episodes of resupply of water in the lake and significant impacts on the lake water supply by annual precipitation fluctuation, increase in air temperature, and snow cover thickness. The changes in area of Lake Buun Tsagaan during the last 45 years were estimated, and a decreasing trend of the lake area responding to the present climatic warming was shown by Orkhonselenge et al. (2018).

Geochemical and hydrometeorological processes were shown to be impacting on the hydrochemistry in Lake Buun Tsagaan (Szopińska et al., 2016a), and increases in the concentrations of NH_4^+, NO_3^-, Na^++Ca^{2+}, and Cl^- + SO_4^{2-} driven by permafrost thawing were found. The hydrochemistry of the Baidrag River-Lake Buun Tsagaan system was summarized by Szopińska et al. (2016b), and the impact of permafrost occurrence on TOC concentrations in the lake water was found. Moreover, the hydrochemistry of Lake Buun Tsagaan was investigated by Chida et al. (2018), and supplies of groundwater and river water into the lake were suggested to be the controlling factors.

Geochemical and geochronological records from Lake Buun Tsagaan are important for revealing the temporal pattern of paleoclimate changes in southern Mongolia. The paleoclimate implications from the lake are represented in detail in Chap. 21. In this chapter, recent research results from Lake Buun Tsagaan are synthesized, and changes in area of the lake during the last 45 years are discussed in detail.

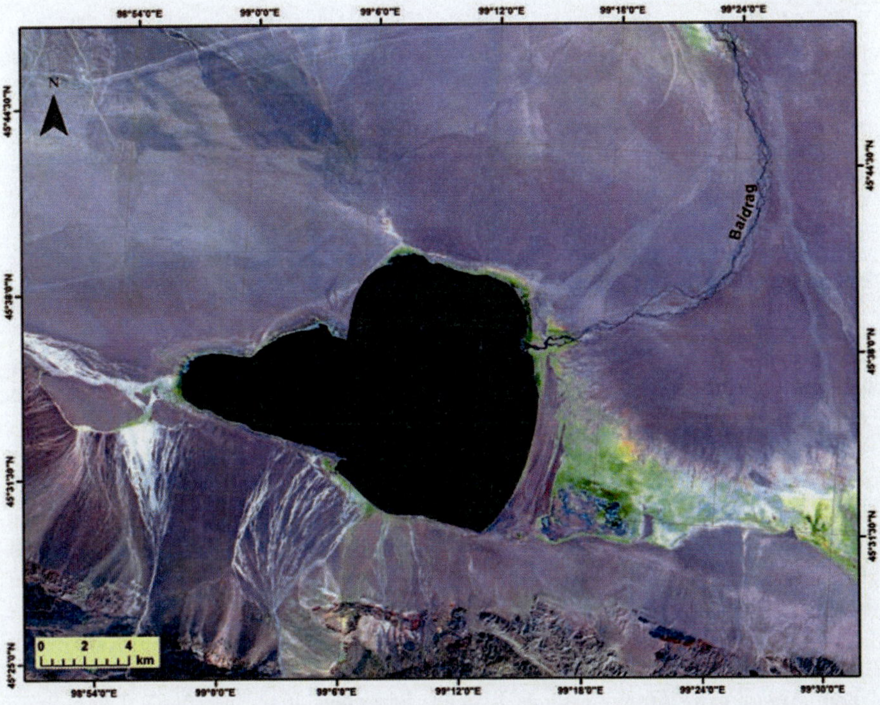

Fig. 10.1 Landsat 8 image of Lake Buun Tsagaan on August 30, 2014 (path of 136 and row of 28). Modified from Orkhonselenge et al. (2018)

10.2 Physiographic Condition

Lake Buun Tsagaan is located at the elevation level of 1336 m a.s.l. (Tsegmid, 1969) or 1312 m a.s.l. (Tserensodnom, 1971) in the center of the Valley of Lakes between Khangai[2] and Govi Altai Mountain Ranges in southern Mongolia (Figs. 1.1 and 9.1).

Lake Buun Tsagaan (Fig. 10.1) extends for 24.0 km (Tsegmid, 1969) or 23.5 km (Tserensodnom, 1971) in the maximum length, 10.7 km in width on average (Tsegmid, 1969) or 19.0 km in the maximum width (Tserensodnom, 1971), and 81.0 km along the shoreline length (Table 1.3). The lake occupies an area of 240.0 km^2 (Tsegmid, 1969), 252.1 km^2 (Tserensodnom, 1971), or 241.12 km^2 (Orkhonselenge et al., 2018; Table 9.5; see Sect. 10.6).

The water depth is about 10–15 m (Tsegmid, 1969), 9.9 m on average reaching the maximum 16.0 m in the center (Table 9.5), and 12.0–14.0 m in other parts of the lake (Tserensodnom, 1971). Its water volume is 0.638 km^3 (Tables 1.3 and 9.5) and the percentage volumes according to the depths are shown in Table 10.1.

[2] *Khangai* has been misspelled as *Khangay* (or *Hangay*) and *Hangai* in many publications. Khangai is the right English transliteration from Mongolian *Хангай*.

Table 10.1 Bathymetric data with corresponding area and volume of Lake Buun Tsagaan

#	Depth (m)	Area (%)	Volume (%)
1	3.0	84.0	70.0
2	6.0	70.0	46.0
3	9.0	52.0	27.0
4	12.0	37.0	13.0
5	15.0	23.0	3.0

10.3 Geological and Geomorphological Settings

Lake Buun Tsagaan, a tectonic lake, resides on the Neogene and Quaternary molassic deposits, and it is filled with the late Quaternary and modern-day deposits (Academy of Sciences of Mongolia and Academy of Sciences of USSR, 1990). The lake basin is tectonostratigraphically underlain by an island arc terrane (Badarch et al., 2002). Regarding geological formation, it covers a continental platform consisting of the Cenozoic red and gray terrigenic sands, gravels, and pebbles, and it borders with the Mesozoic granodiorites and granites and Devonian terrigenic tuff graywacke in the southeast (Academy of Sciences of Mongolia and Academy of Sciences of USSR, 1990).

Lake Buun Tsagaan (Fig. 10.2) is surrounded by the Khoid Tal plain in the north; the Ereen Khargant sand dunefield in the northeast and east; the Khar, Dund, and Ulaan Argalant Mountain Ranges that are branches of the Govi Altai Mountain Range, in the southeast, south, and southwest; and the Ulaan Shaliin Khooloi sand dunefield in the west and northwest. There are about 5 beach ridges observable along the northeastern shoreline, while about 15 beach ridges are preserved along the southeastern shoreline (Tsegmid, 1969; Tserensodnom, 1971). The terraces on the northern and eastern shores of Lake Buun Tsagaan imply higher paleolake shorelines.

The paleolake Buun Tsagaan was likely connected to the paleolake Adgiin Tsagaan during the late Quaternary period (Komatsu et al., 2001). At the present time, the Lake Adgiin Tsagaan is lying as a small ephemeral lake 60 km to the east at less than 1300 m a.s.l. Between Lakes Buun Tsagaan and Adgiin Tsagaan, there are many small freshwater and saline lakes (Tsegmid, 1969).

10.4 Climate Condition

The climate in the Lake Buun Tsagaan basin is led by the regional trend in the Valley of Lakes of the Govi region (see Chap. 9). In the lake basin, the precipitation decreased by 25.0 mm/year compared to the inter-annual average rainfall of 70.0 mm/year (Batnasan & Tserensodnom, 1987).

The local climate data recorded at the Bayan Sair meteorological station in Baatsagaan soum (Fig. 9.1), located at 13.6 km east of the lake, show the annual

Fig. 10.2 Lake Buun Tsagaan viewing toward the (**a**) south and (**b**) west. Photos by S. Ariunbileg on August 12, 2009

average air temperature of 3.44 °C and the annual average precipitation of 86.2 mm/year between 1985 and 2019 (Fig. 10.3). However, the climate data in 1988–1990 and 1997–1999 were intermittent (Fig. 10.3). The annual average air temperature has been continuously rising except for 2002 (Fig. 10.3a), but the annual average precipitation has often fluctuated during the last three decades (Fig. 10.3b).

10.5 Hydrological Condition

The major water sources of Lake Buun Tsagaan are rainfall, runoff, and groundwater draining from an area of 33,500 km^2 (Tsegmid, 1969). In addition, meltwaters from the permafrost and modern glaciers in the southern Khangai Mountain Range (see Chap. 9) are the additional sources. The lake is primarily fed by Baidrag River (Figs. 10.1 and 10.4) draining the southern Khangai Mountain Range in the north

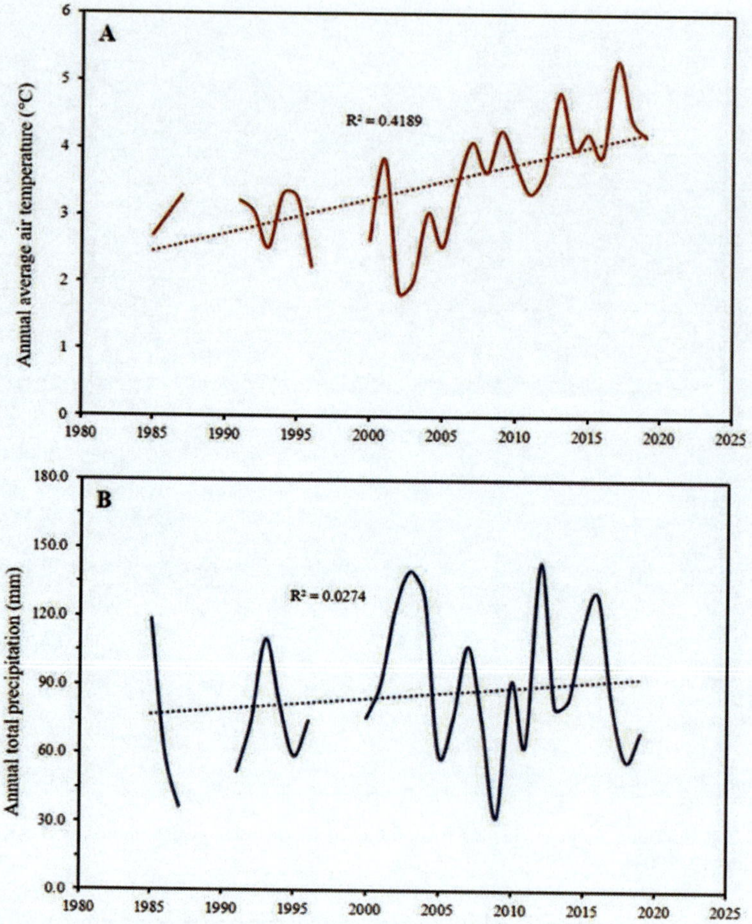

Fig. 10.3 Annual average **(a)** air temperature and **(b)** precipitation recorded at the Bayan Sair meteorological station between 1985 and 2019

and northeast and by the streams draining the Govi Altai Mountain Range in the south and southwest. The Baidrag River enters the lake with a prominent deltaic body, and the lake has no outlet (Figs. 10.1 and 10.4).

The water level of Lake Buun Tsagaan is highly dependent on the Baidrag River and precipitation. In an average year, the lake level reaches the highest during the maximum precipitation in July and August (Batnasan, 1998). The lake water level dropped by approximately 2.0 m (Batnasan, 1998) or 6.0 m from 1998 to 2009 (Davaa, 2010). The increased level of Lake Buun Tsagaan in the last decade contributed to the inundation of islands and small isolated lakes in the lake basin, and Lake Buun Tsagaan extended toward the Baatsagaan soum in the east where a small lake of 1–2 m in depth and covering an area about 10.0 km² formed (Tserensodnom,

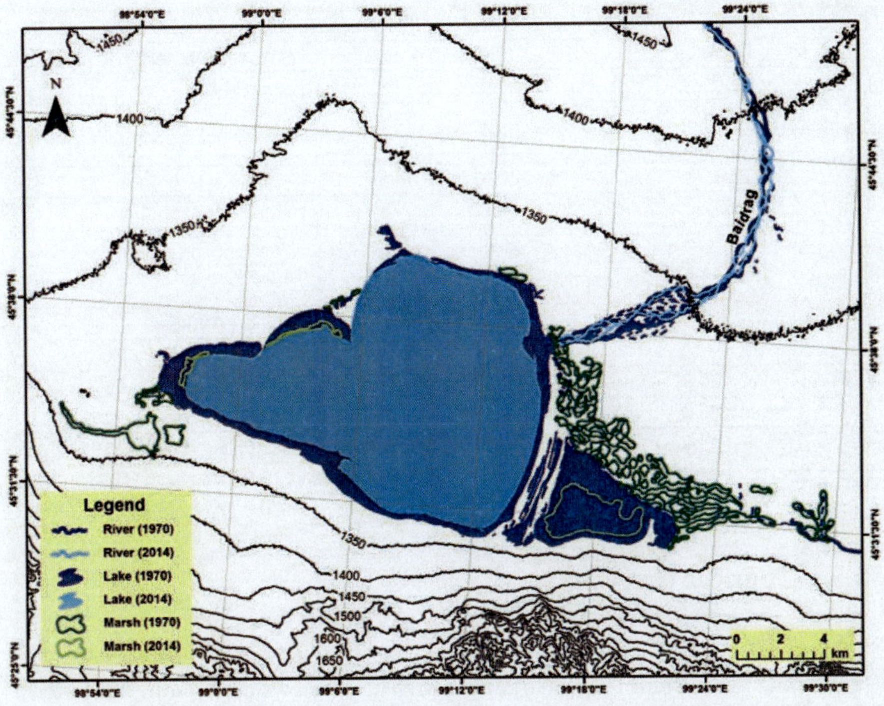

Fig. 10.4 Spatial distributions of areas and marshes of Lake Buun Tsagaan in 1970 and 2014. Modified from Orkhonselenge et al. (2018)

1971). However the increased lake level in 1959 caused residents to leave the Baatsagaan soum (Fig. 9.1), an administrative unit (Batnasan, 1998). The water level dropping of Lake Buun Tsagaan in 1986–1989 was caused by human activity, and it waterlogged in 1994–1996 (Batnasan, 1998). For instance, the installation of the Khersen draining system in the Baidrag River valley caused a decline of running water flowing into Lake Buun Tsagaan by 16.4% (Batnasan, 1998). The present-day lake level is estimated to be slightly higher than 1300 m a.s.l. (e.g., Komatsu et al., 2001).

The Baidrag River and evaporation from the lake water surface play important roles in the water regime of Lake Buun Tsagaan. The evaporation is 920–930 mm/year; however it has been increased by 40 mm/year during the period of the last decade (Batnasan & Tserensodnom, 1987). The lake water temperature is shown in Table 10.2. The lake is frozen between November (Tsegmid, 1969) or December (Tserensodnom, 1971) and April, and the ice thickness is usually less than 1.2 m during the period (Tsegmid, 1969). The mineralization value of the lake is shown in Table 10.3, and the value increased in 1930–1940 and 1980, and decreased in 1960 (Batnasan, 1998). The cation ratio of $Ca^{2+} > Na^{+}+K^{+} > Mg^{2+}$ is dominant in the lake water (Ariunbileg et al., 2017), and their concentrations are increased in recent years (Ariunbileg et al., 2020; Table 10.3).

Table 10.2 Water temperature of Lake Buun Tsagaan

#	Location	May (°C)	July and August (°C)
1	Center	12.0	22.0–24.0
2	Shore	13.0	–
	Water depth (m)		
3	12.0	10.0–11.0	–
4	15.0	9.2	–

Table 10.3 Hydrochemical composition of Lake Buun Tsagaan

#	Ions (mg/L)	Na^{+}+K^{+}	Ca^{2+}	Mg^{2+}	CO_3^{2-}	HCO_3^{-}	SO_4^{2-}	Cl^{-}	Authors
1	573.8	–	–	–	–	–	222.0	140.0	Bespalov (1951)
2	304.1	38.0	36.0	8.5	–	140.3	60.0	21.3	Tsend (1965)
3	399.0 (west)	95.0	3.0	25.0	5.0	45.0	142.0	86.0	Luvsandorj (1968)
4	386.0 (east)	88.0	5.0	25.0	6.0	43.0	181.0	87.0	Luvsandorj (1968)
5	531.0	158.0	4.0	17.0	–	67.0	168.0	117.0	Batnasan (1998)
6	–	1425.8	120	192.4	–	–	–	600.0	Ariunbileg et al. (2020)

10.6 Changes in Area

The area changes of Lake Buun Tsagaan are shown between 1969 and 2014 (Table 10.4). Lake Buun Tsagaan's surface area has decreased with some short-term fluctuations since 1970 (Table 10.4). Such fluctuations during the past 45 years may imply impact of climate changes.

The detailed information on the fluctuations in area of Lake Buun Tsagaan between 1969 and 1985 (Table 10.4) is not accessible for assessing their correlation with the local climate data between 1985 and 2019 (Fig. 10.3). Nevertheless, it is shown that the area increased by 23.87 km² from 1969 to 1992 (Table 10.4). An increase is also found for the period of 1991–1995 (Szumińska, 2016). The abrupt increase in area by 84.29 km² between 1969 and 1970 (Table 10.4) coincided with the observations noted by Tsegmid (1969) and Tserensodnom (1971). That may have been correlated to the flooding event of the year (see Sect. 10.5).

The area of the lake continuously decreased since 1992 (Table 10.4) and since 1996 (Szumińska, 2016). The decrease by 16.67 km² between 1992 and 2010 may have been caused by the low annual average precipitation less than 100.0 mm in 1986–1987, 1991–1992, 1994–1996, 2000–2001, 2005–2006, and 2008–2010 (Fig. 10.3b), and it coincided with the high annual average air temperatures of 3.0–3.3 °C in 1991–1995 and of 3.8–4.2 °C in 2006–2010 (Fig. 10.3a). The area reduced by 83.17 km² or 25.6% from 1970 to 2014 (Orkhonselenge et al., 2018; Table 10.4; Fig. 10.4), but Szumińska (2016) showed that the decrease had been 14.0% since 1970.

The linear regression analysis (Fig. 10.5) shows that the lake area between 1992 and 2014 was strongly related to the annual average air temperature ($R^2 = 0.7847$),

Table 10.4 Temporal changes in area of Lake Buun Tsagaan

#	Year	Area (km²)	Authors
1	1969	240.0	Tsegmid (1969)
2	1970	324.29	Orkhonselenge et al. (2018)
3	1970	288.0	Szumińska (2016)
4	1971	252.1	Tserensodnom (1971)
5	1992	263.87	Batnasan (1998)
6	2010	247.2	Davaa (2015)
7	2013	240.2	Davaa (2015)
8	2014	241.12	Orkhonselenge et al. (2018)

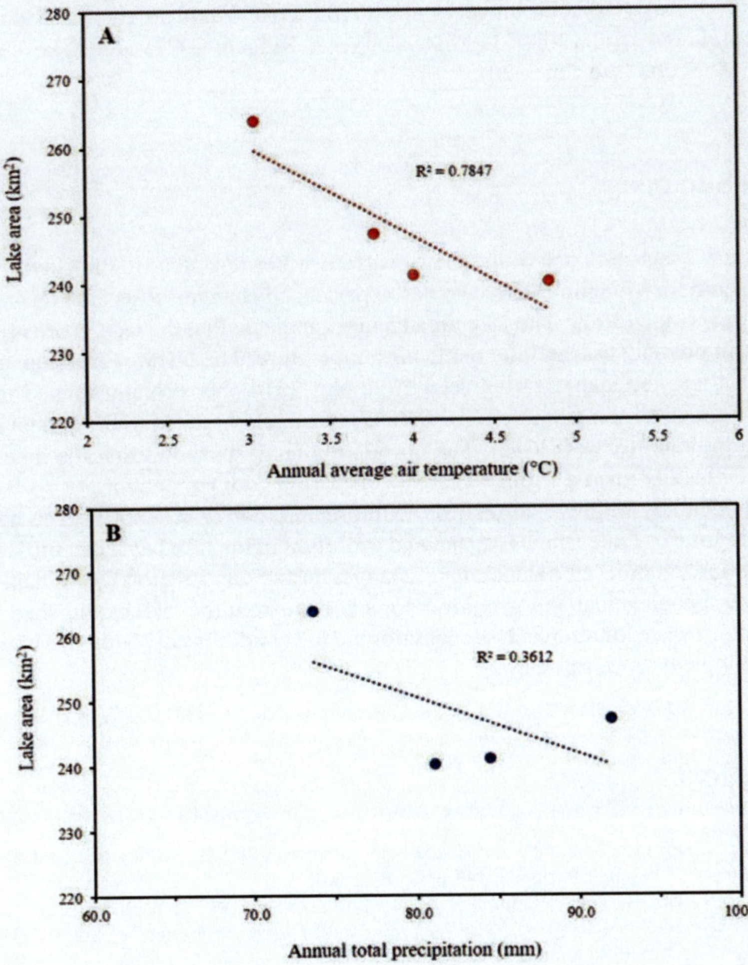

Fig. 10.5 Relationships between area of Lake Buun Tsagaan and annual average (**a**) air temperature and (**b**) precipitation during 1992–2014

but weakly related to the annual average precipitation ($R^2 = 0.3612$). The correlation analysis shows that the lake area during 1992–2014 had a strong negative correlation with temperature ($r = -0.885$) and a moderate negative correlation with precipitation ($r = -0.601$).

The decrease in lake area shows that Lake Buun Tsagaan responded significantly to the precipitation drop and the temperature rise. This trend coincides with the low levels of many other lakes in the Govi region (Orkhonselenge et al., 2018) due to the intensified aridity since the 1990s (Batnasan, 1998). The relationships between climate factors and lake areas indicate controlling factors of the precipitation to be 38.9% and of the temperature to be 28.7% (Szumińska, 2016). The reduction in area of Lake Buun Tsagaan following the falling precipitation and the rising air temperature is in agreement with the trend of Lake Orog (see Chap. 9) and other lakes in the Valley of Lakes (Orkhonselenge et al., 2018), Lake Yakhi in eastern Mongolia (Orkhonselenge et al., 2020; Orkhonselenge & Bulgan, 2021) and Lake Ugii in central Mongolia (see Chap. 20).

10.7 Summary

Lake Buun Tsagaan is one of the major surface water resources in the Govi region within southern Mongolia. The lake has experienced changes of its area during the last 45 years since 1969. The lake area changes indicate that the lake is sensitive to the falling precipitation and the rising air temperature. The linear regression analysis shows that the lake area between 1992 and 2014 was strongly related to the annual average air temperature ($R^2 = 0.7847$) but weakly related to the annual average precipitation ($R^2 = 0.3612$). The correlation analysis shows that the lake area during 1992–2014 had a strong negative correlation with temperature ($r = -0.885$) and a moderately negative correlation with precipitation ($r = -0.601$). The hydrological regime of Lake Buun Tsagaan and evolution in the lake basin can provide us valuable information on paleoclimate changes in the Govi region (see Chap. 21). Moreover, geochemical and geochronological data from the lake would shed light on the landscape evolution and paleoenvironmental conditions of southern Mongolia over geological time scales.

References

Academy of Sciences of Mongolia and Academy of Sciences of USSR. (1990). National Atlas of the Mongolian People's Republic. 144 p. [In Russian].

Ariunbileg, S., Isupov, V. P., Vladimirov, A. G., Kolpakova, M. N., & Kuibida, L. V. (2020). *Chemical evolution and geochemistry of mineralized lakes of Mongolia*. Admon Printing House, 132 p. [In Mongolian].

Ariunbileg, S., Isupov, V. P., Vladimirov, A. G., Orkhonselenge, A., & Shatskaya, S. S. (2017). Impact of Climate Change on Hydrogeochemical characteristics of lakes in the Valley of Lakes, southern Mongolia. *Goldschmidt2017: Presentation #2169*, Paris.

Badarch, G., Cunningham, W. D., & Windley, B. F. (2002). A new terrane subdivision for Mongolia: Implications for the Phanerozoic crustal growth of Central Asia. *Journal of Asian Earth Sciences, 21*, 87–110.

Batnasan, N. (1998). *Hydrological system, water regime, and evolution of large lakes in the Govi region*. Doctorate Thesis in Geography: Hydrology, Institute of Geoecology, Mongolian Academy of Sciences, Ulaanbaatar. 163 p. [In Mongolian].

Batnasan, N., & Tserensodnom, J. (1987). Aspects in water regime of Lake Buun Tsagaan. *Mongolian Geographical Issues, 27*, 18–24.

Bespalov, N. D. (1951). *Soil of Mongolia*. (AH. vol. 41). [In Russian].

Chida, T., Sekine, Y., Fukushi, K., Matsumiya, H., Solongo, T., Hasebe, N., & Davaadori, J. (2018). Hydrology of subsaline lakes in Southern Mongolia: A terrestrial analog study for lacustrine environments and chloride depositions on early mars. In *49th Lunar and Planetary Science Conference (LPI Contrib. No. 2083)*.

Davaa, G. (2010). Climate change impacts on water resources in Mongolia. In *Proceedings of Consultative Meeting on Integration of Climate Change Adaptation into Sustainable Development in Mongolia. Institute for Global Environmental Strategies (IGES)*, pp. 30–36.

Davaa, G. (2015). *Resource and regime of surface water of Mongolia*. 408 p. [In Mongolian].

Devyatkin, E. V. (1981). *The Cenozoic of Inner Asia: Stratigraphy, geochronology, correlation*. 196 p. Transaction of Joint Soviet–Mongolian Geological Expedition, 27, Nauka, Moscow. [In Russian].

Komatsu, G., Brantingham, P. J., Olsen, J. W., & Baker, V. R. (2001). Paleoshoreline geomorphology of Boon Tsagaan Nuur, Tsagaan Nuur and Orog Nuur: The Valley of Lakes, Mongolia. *Geomorphology, 39*, 83–98.

Lehmkuhl, F., & Lang, A. (2001). Geomorphological investigations and luminescence dating in the southern part of the Khangay and the Valley of the Gobi Lakes (Central Mongolia). *Journal of Quaternary Science, 16*(1), 69–87.

Luvsandorj, S. (1968). *Mineral lakes of Mongolia*. Mongolian Academy of Sciences. [In Mongolian].

Murzaev, E. M. (1952). *The description of physical geography of Mongolia* (2nd ed.). [In Russian].

Orkhonselenge, A., & Bulgan, O. (2021). Geochemical studies and lacustrine geomorphology of Lake Yakhi basin in eastern Mongolia. *Géomorphologie: Relief, Processes, Environnement, 27*(3), 231–242. https://doi.org/10.4000/geomorphologie.15873

Orkhonselenge, A., Bulgan, O., Gerelsaikhan, D., Davaagatan, T., & Altansukh, N. (2020). Hydroclimatic fluctuation in Lake Yakhi, eastern Mongolia. In *EGU General Assembly 2020. EGU2020–686 presentation*.

Orkhonselenge, A., Komatsu, G., & Uuganzaya, M. (2018). Climate-driven changes in lake areas for the last half century in the Valley of Lakes, Govi Region, Southern Mongolia. *Natural Science, 10*(7), 263–277.

Szopińska, M., Szumińska, D., Polkowska, Ż., Machowiak, K., Lehmann, S., & Chmiel, S. (2016a). The chemistry of river–lake systems in the context of permafrost occurrence (Mongolia, Valley of the Lakes). Part I. Analysis of ion and trace metal concentrations. *Sedimentary Geology, 340*, 74–83.

Szopińska, M., Dymerski, T., Polkowska, Ż., Szumińska, D., & Wolska, L. (2016b). The chemistry of river–lake systems in the context of permafrost occurrence (Mongolia, Valley of the Lakes) Part II. Spatial trends and possible sources of organic composition. *Sedimentary Geology, 340*, 84–95.

Szumińska, D. (2016). Changes in surface area of the Böön tsagaan and Orog lakes (Mongolia, Valley of the lakes, 1974–2013) compared to climate and permafrost changes. *Sedimentary Geology, 340*, 62–73.

Tsegmid, S. (1969). *Physical geography of Mongolia*. State Press, 405 p. [In Mongolian].

Tsend, N. (1965). *Water chemical composition of lakes in the Govi region*. Institute of Livestock.

Tserensodnom, J. (1971). *Lakes of Mongolia*. State Publishing, 192 p. [In Mongolian].

Tserensodnom, J. (2000). *A catalog of lakes in Mongolia*. Shuvuun Saaral Publishing, 141 p. [In Mongolian].

Chapter 11
Lake Ulaan

Abstract Lake Ulaan is a terminal lake at the eastern end of the Valley of Lakes in the northernmost Govi region, and the lake plays a valuable role in surface water resource of southern Mongolia. Lake Ulaan basin has been strengthened with the warming and drying since the mid-1970s. Lake Ulaan experienced changes of its area and sedimentation dynamics during the Holocene. The lake lost an area of 18.2 km^2 during the last four to five decades, and it disappeared between the mid-1980s and the mid-2010s. In short, Lake Ulaan has been a playa lake during the last half century. The playa lake condition coincides with regional climate change and local warming since 1975 and drying since 1974. The linear regression analysis shows that the lake area between 1986 and 2014 had no clear relationships with the annual average air temperature ($R^2 = 0.0833$) and precipitation ($R^2 = 0.0063$). The correlation analysis shows that the lake area during 1986–2014 had a weak positive correlation with air temperature ($r = 0.288$) but no clear correlation with precipitation ($r = 0.07$). The geochemical analysis shows the dominance of alkali and alkaline earth metals in Lake Ulaan sediments and the higher degree of chemical weathering in the lake margin than the lake center. The chemical maturity of the Lake Ulaan sediments shows the shift from semiarid to arid climate conditions in the lake basin. Lake Ulaan contributes to the understanding of the response of the local hydrological system to climate change, and it helps in revealing the landscape evolution of the Valley of Lakes in southern Mongolia over geological time scales.

Keywords Lake Ulaan · Ongi River · Valley of Lakes · Govi region · Climate change · Southern Mongolia

A. Orkhonselenge et al., *Lakes of Mongolia*, Syntheses in Limnogeology,
https://doi.org/10.1007/978-3-030-99120-3_11

11.1 Introduction

Lake Ulaan, the third largest lake in the Govi[1] region, plays an important role in surface water resources of southern Mongolia and the hydrological system of the Central Asian internal drainage basin. Lake Ulaan is situated at the easternmost tip of the Valley of Lakes between the Khangai[2] and Govi Altai Mountain Ranges (Fig. 1.1). As one of complex regional archives (see Chap. 9), Lake Ulaan in the Govi region provides an opportunity to reconstruct geomorphological and geochemical evolutions and paleoclimate changes in southern Mongolia during the geological time scale.

Lake Ulaan has attracted much scientific attention during the last decade. To date, numerous studies have been conducted in terms of limnology, geomorphology, sedimentology, geochronology, and paleoclimatology. For instance, in the Lake Ulaan basin, it was found that reddish sediments and sandstones overlay rhyolite, andesite, or basalt flows (Berkey & Morris, 1927). Water level of Lake Ulaan dropped by 1–2 m when the precipitation in the Govi region decreased in 1952–1953 and 1986–1989 (Batnasan, 1998). The radiocarbon and optically stimulated luminescence methods were applied for studying sediments from Lake Ulaan (Lee et al., 2011). Furthermore, Lee et al. (2013) found that the lake sediments were transported in the Lake Ulaan basin by local westerly winds during the 16.7–15.0 ka BP and also the last 11.2 ka BP, and by fluvial processes between 15.0 and 11.2 ka BP. The Lake Ulaan basin was influenced by the East Asian Summer Monsoon (EASM) between 11.3 and 3.0 ka BP (Lee et al., 2013).

A series of geomorphological evidence for the presence of a paleolake, paleochannels, and paleoshorelines were identified by Sternberg and Paillou (2015). A paleo-hydrological system in the Lake Ulaan basin was hypothesized by Holguin and Sternberg (2016). Orkhonselenge et al. (2018a) noted that the lake was shrinking due to the present global warming. A paleo-Lake Ulaan occupying an area of approximately 500 km^2 in the middle Holocene was reconstructed by Lehmkuhl et al. (2018). The sedimentation dynamics and paleoclimate changes in Lake Ulaan basin in the middle to late Holocene basin were investigated by Orkhonselenge et al. (2018b, in press). Moreover, the paleoclimate changes in the Lake Ulaan basin during the late Pleistocene and Holocene causing the lake level and salinity fluctuations were reconstructed by Mischke et al. (2020).

In this chapter, previous research results from the Lake Ulaan basin are synthesized, and changes in area of the lake during the last over six decades, the response of the lake to climate change, geochemical review, and evolution of the lake are discussed in detail. The paleoclimate implications from Lake Ulaan are represented in detail in Chap. 21.

---

[1] *Govi* has been often spelled as *Gobi* in the international literature. *Govi* is the correct English transliteration from Mongolian *Говь*.

[2] *Khangai* has been misspelled as *Khangay* (or *Hangay*) and *Hangai* in many publications. Khangai is the right English transliteration from Mongolian *Хангай*.

11.2 Physiographic Condition

Lake Ulaan is located at the elevation of 1024 m a.s.l. in the eastern end of the Valley of Lakes at the northern border of the Govi region (Figs. 1.1 and 9.1). Morphometric components of the lake (e.g., altitude, area, width, length, depth, and shoreline) are shown in Table 1.3. The lake extends for 45.0 km in the maximum length, 29.0 km and 40.0 km wide on average and the maximum, respectively (Tsegmid, 1969), and 87.0 km along the shoreline length (Tserensodnom, 1971). During high rainfall the lake's depth reaches 0.9 m on average and up to 1.6 m (Tserensodnom, 2000; Table 9.5).

Lake Ulaan (Fig. 11.1) covers an area of 175.0 km^2 (Murzaev, 1952) and 62.38 km^2 or 44.16 km^2 (Orkhonselenge et al., 2018a; see Sect. 11.6), but at present, the lake floor is generally exposed and covered by reeds and sedges (Fig. 11.2a). The inflowing rivers (see Sect. 11.5) transport alluvial sands and gravels into the lake (Tsegmid, 1969). The lake sediments consist of yellowish red to brown fine silts at the surface and dark reddish brown clays at the subsurface layers (Orkhonselenge et al., 2018b). Because diverse sediments filling in the lake basin radiate red ray from its water surface, the lake is called *Ulaan* (Tsegmid, 1969) meaning *red* (Orkhonselenge et al., 2018a, 2018b).

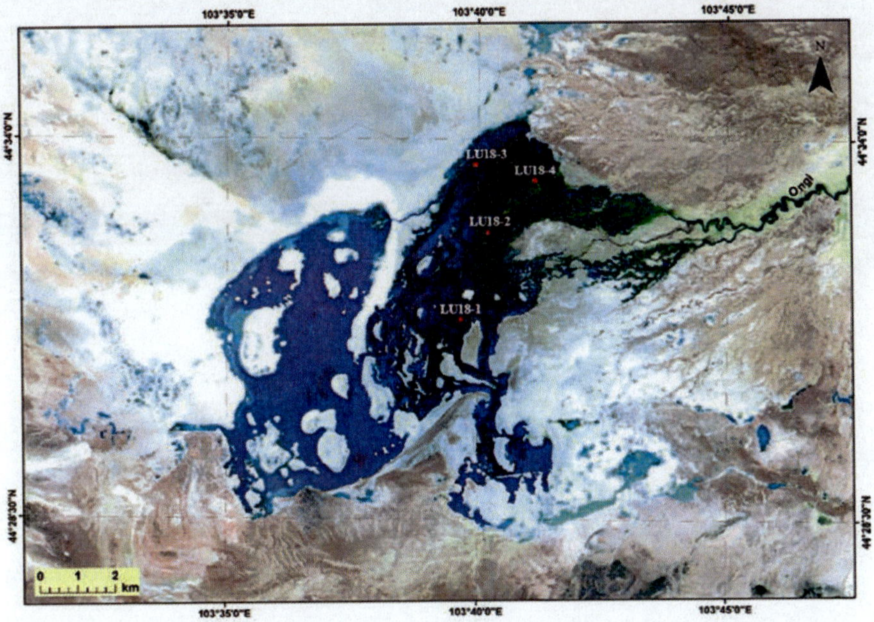

Fig. 11.1 Landsat 8 image of Lake Ulaan on June 6, 2014 (path of 133 and row of 29) with locations of sedimentary cores (LU18-1 to LU18-4) for geochemical study (see Sect. 11.7). Modified from Orkhonselenge et al. (2018b)

Fig. 11.2 (**a**) The exposed vegetated lake floor of Lake Ulaan viewing toward the south. (**b**) The dried out channel of Ongi River viewing toward the northwest. The arrow indicates the river flow direction. Photos by A. Orkhonselenge on June 18, 2018

11.3 Geological and Geomorphological Settings

Lake Ulaan resides on the Neogene and Quaternary molasses deposits, and it is geologically filled on its bottom with the middle to late Quaternary deposits (Academy of Sciences of Mongolian and Academy of Sciences of USSR, 1990). The geology of the Lake Ulaan basin is tectonostratigraphically included in the Mandal Ovoo island arc terrane consisting mainly of the middle to late Paleozoic oceanic ophiolites, tholeiitic to calc-alkaline volcanic and volcanoclastic rocks overlain by the nonmarine sedimentary rocks (Badarch & Tumurtogoo, 2001; Badarch et al., 2002).

Lake Ulaan covers a continental platform consisting of the Cenozoic red and gray terrigenic sands, gravels, and pebbles, and it borders with the Cenozoic limestone basalts and the Devonian terrigenic tuff greywacke in the southeast (Academy of Sciences of Mongolian and Academy of Sciences of USSR, 1990). Hills to the south and west of Lake Ulaan are composed mainly of the Silurian limestone and dolomite and the Cretaceous basaltic and clastic sedimentary rocks, while to the north and east, the bedrock is dominated by the Archaean and Paleoproterozoic metamorphic complexes and the Paleozoic and Mesozoic sedimentary rocks (Academy of Sciences of Mongolian and Academy of Sciences of USSR, 1990).

Lake Ulaan is classified as a tectonic lake (see Chap. 5). The modern topography of the Lake Ulaan basin formed in a late Mesozoic rift, filled with terrigenic red deposits of the upper Cretaceous Baruun Bayan formation (Narantsetseg et al., 2011). The late Mesozoic rift is a graben-synclinal structure surrounded by the Paleozoic linear uplifts extended along the latitude (Narantsetseg et al., 2011). Because the lake basin is the lowest one in the Valley of Lakes (Fig. 9.1), thick alluvial and lacustrine deposits were accumulated in the past (Tsegmid, 1969). Much of the Lake Ulaan sediments deposited during the Neogene when its water level was almost 100 m higher than the present, and the lake area was several ten times larger than today (Tserensodnom, 1971). The lake is hypothesized as a center of deposition for a whole lake basin in the Valley of Lakes during the Quaternary time (Tserensodnom, 2000). The higher sedimentation rate of 4.6 cm/ka at 6.0–2.7 cal ka BP and the lower sedimentation rate of 1.8 cm/ka after 3.2–2.7 cal ka BP were recorded in the Lake Ulaan basin (Orkhonselenge et al., 2018b).

The Lake Ulaan basin is marshy, and it is located between relatively low hills at 1050–1110 m a.s.l. (Fig. 9.1). The lake is encompassed by lowlands, hills, and mountains elevated at 1050–1600 m a.s.l., for example, Mt. Bor Khairkhan (1233 m a.s.l.) in the northwest and Mt. Ikh Khongor (1202 m a.s.l.) and Mt. Maikhan Tolgoi (1232 m a.s.l.) in the south (Orkhonselenge et al., 2018b). The marshy areas with scattered reeds and sedges (Fig. 11.2a) are widely distributed in the lake basin. The lake's beaches are flat plain and swampy, and there are sand dunefields in the northwest and north (Tsegmid, 1969). During torrential rains, flash floods transport sands and gravels into the lake by the Ongi River (Fig. 11.2b) and some other streams (Orkhonselenge et al., 2018a).

11.4 Climate Condition

The climate in the Lake Ulaan basin is led by the regional trend in the Valley of Lakes of the Govi region (see Chap. 9). The climate of the Lake Ulaan basin is extremely arid and/or desert-like today (Orkhonselenge et al., 2018a, 2018b). The westerly and southwesterly winds are dominant in the Govi region including the Lake Ulaan basin. The local climate data recorded at the Mandal Ovoo meteorological station (MO in Fig. 9.3) located 23.1 km northeast of the lake (Fig. 9.1) span between 1973 and 2015.

The local climate record during 1973–2015 (Fig. 11.3) shows the annual average air temperature of 5.4 °C between 1975 and 2015 and the annual average precipitation of 69.1 mm between 1973 and 2015. Most of the total annual precipitation occurs between June and August (Orkhonselenge et al., 2018a, 2018b). The warmest year was recorded in 1986 when the annual average air temperature reached

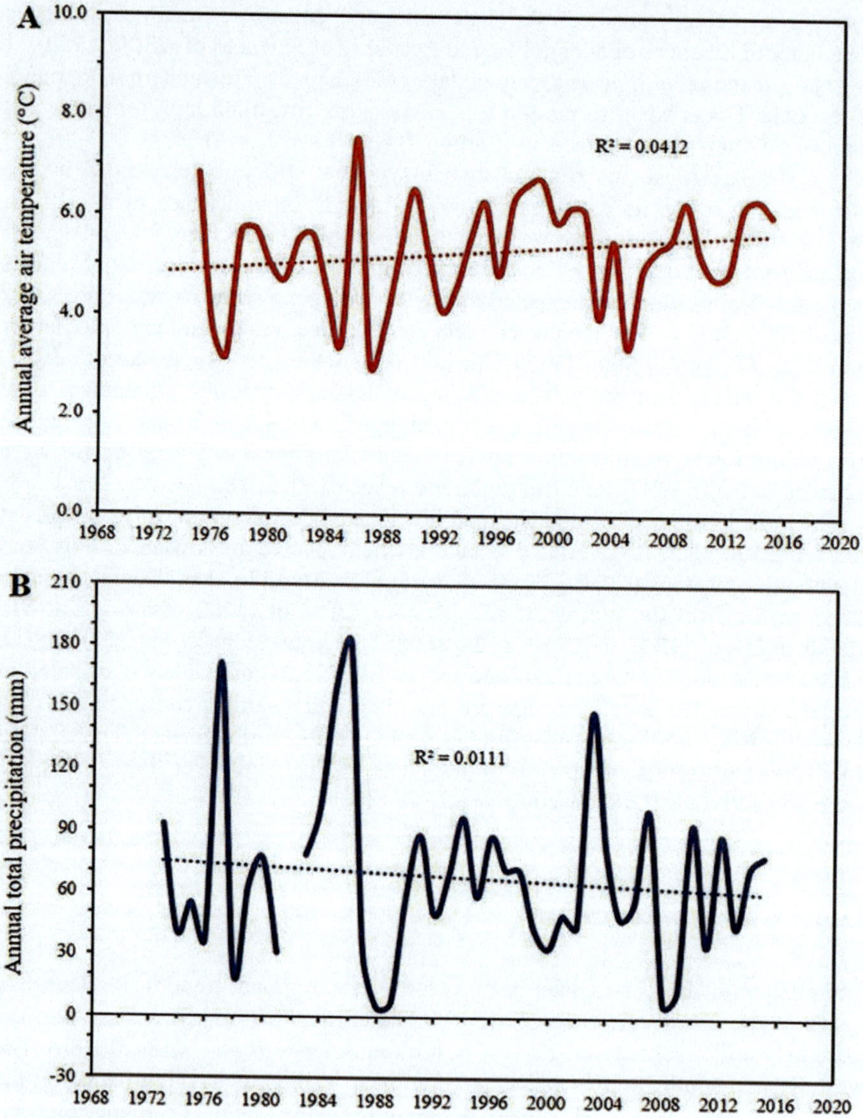

Fig. 11.3 Annual average (**a**) air temperature and (**b**) precipitation recorded at the Mandal Ovoo meteorological station (Figs. 9.1 and 9.3) during 1973–2015

7.5 °C, and the other warmer years with the annual average air temperature above 5.5 °C were found in 1975, 1978–1979, 1982, 1983, 1990, 1994–1995, 1997–2002, 2004, 2008–2009, and 2013–2015 (Fig. 11.3a), whereas the colder years with the annual average air temperature below 3.5 °C were recorded in 1977, 1985, 1987, and 2005. The drier years with the annual precipitation below 50 mm were found in 1974, 1976, 1978, 1981, 1987–1989, 1992, 1999–2002, 2005, 2009, 2011, and 2013 (Fig. 11.3b). The humid years with the annual precipitation above 100 mm were recorded in 1977, 1984–1986, and 2003. The warm and dry years occurred in 1978, 1999–2002, 2009, and 2013, while the cooler and humid years occurred in 1977 and 1985 (Fig. 11.3).

11.5 Hydrological Condition

Lake Ulaan is fed by Ongi River (Figs. 11.1 and 11.2b) draining the southern Khangai Mountain Range (Fig. 1.1). In addition, Sukhait, Leg, and Tsagaan Rivers and small streams draining the Govi Altai Mountain Range in the southwest and the west periodically feed the lake (Tsegmid, 1969). The intermittent meltwater inflow that feeds Lake Ulaan was indicated by fluctuation of the mean grain size (Lee et al., 2011). The Ongi River has shrunk and has not supplied water to the lake continuously since the mid-1990s (Lee et al., 2011), and the river did not feed the lake in 2015 (Orkhonselenge et al., 2018b). Today, Ongi River cannot perennially reach the lake (Fig. 11.2b) because the river temporarily flows depending on the annual precipitation (Orkhonselenge et al., 2018a, 2018b) and human activity at its drainage basin. The water level and area of the lake are unstable (Orkhonselenge et al., 2018a, 2018b), and they often fluctuate depending on the precipitation in the lake basin (Fig. 11.3b) and the intermittent inflow discharge of Ongi River (Figs. 11.1 and 11.2b).

Lake Ulaan almost completely dried out in 1952–1953 when the lake floor was exposed, and salty marshes were bleached at the surface (Tsegmid, 1969). Although the lake covers an area of over 10–100 km^2 when the annual precipitation rises, it is almost dried out during the low precipitation (Orkhonselenge et al., 2018a, 2018b). Lake Ulaan is characterized with a shallow flat floor (Tsegmid, 1969), and the lake shrank in 1986–1989 (Batnasan, 1998), 2015 (Orkhonselenge et al., 2018b), and 2018 (Fig. 11.2a).

Lake Ulaan is a freshwater lake even though it has no outlet (Figs. 11.1 and 11.4) unlike other lakes in the Govi region (Tsegmid, 1969). The freshness of the lake water is related to the fact that the water in the unconsolidated soil around the lake feeds the lake under a highly evaporative climate of the Govi region (Tsegmid, 1969). Moreover, a large part of the lake water infiltrates to the subsurface (Sternberg & Paillou, 2015; Holguin & Sternberg, 2016). According to Tsend (1965), in the lake water, anion HCO_3^- is dominant (Table 11.1). A cation ratio of Na^++K^+ > Ca^{2+} > Mg^{2+} in a stream near the Mandal bag in the east of Lake Ulaan was observed in 2009 (Ariunbileg et al., 2020; Table 11.2).

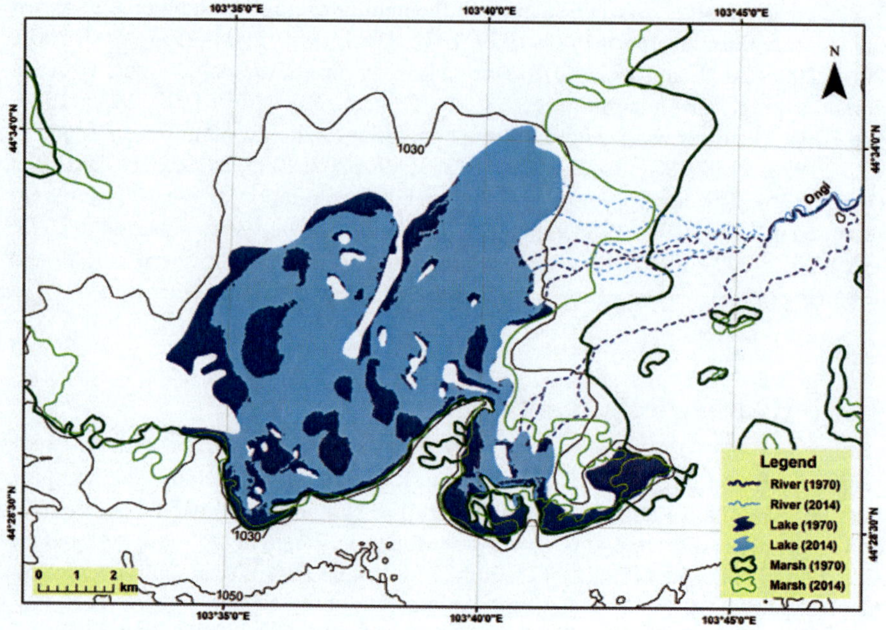

Fig. 11.4 Spatial distributions of areas and marshes of Lake Ulaan in 1970 and 2014. Modified from Orkhonselenge et al. (2018a)

Table 11.1 Hydrochemical composition of Lake Ulaan

Ions (mg/L)	Na^{+}+K^{+}	Ca^{2+}	Mg^{2+}	HCO_3^{-}	SO_4^{2-}	Cl^{-}
384.1	30.4	42.0	20.1	262.3	8.0	21.3

Table 11.2 Hydrochemical composition of groundwater in the east of Lake Ulaan near the Mandal bag

Ions	Na^{+}+K^{+}	Ca^{2+}	Mg^{2+}	Cl^{-}
(mg/L)	152.5	88.98	17.7	1480

11.6 Changes in Area

Lake Ulaan was a stable large lake until the 1950s, but it experienced abrupt fluctuations during the last half century (Table 11.3). In the past, Lake Ulaan was a large paleolake covering an area greater than 19,500 km^2, a medium-sized lake covering a surface area of about 6900 km^2, and a small-sized lake covering an area close to 1700 km^2 (Sternberg & Paillou, 2015; Table 11.3). The ages of the first two large and medium-sized lakes are unknown, but the last one is estimated to be of a Holocene age (Table 11.3).

Table 11.3 Temporal changes in area of Lake Ulaan

#	Year	Area (km^2)	Elevation (m a.s.l.)	Level (m)	Authors
1	162 ± 13 ka	–	–	~43 m	Lehmkuhl et al. (2018)
2	Unknown	19,500.0	1285	–	Sternberg and Paillou (2015)
3	Unknown	6900.0	1150	–	Sternberg and Paillou (2015)
4	Holocene	1700.0	1070	–	Sternberg and Paillou (2015)
5	Middle Holocene	~500.0	1027	–	Lehmkuhl et al. (2018)
6	1948	175.0	–	–	Murzaev (1952)
7	1949	135.0	–	–	Bespalov (1951)
8	1952–1953	0.0	1008	–	Tsegmid (1969)
9	<1960	–	–	2–3	Lehmkuhl et al. (2018)
10	1960	65.0	–	–	Lee et al. (2011)
11	1970	62.38	1024	–	Orkhonselenge et al. (2018a)
12	>1980	–	–	1–2	Batnasan (1998)
13	1986–1989	0.0	–	–	Batnasan (1998)
14	2000	0.0	–	–	Kang et al. (2015)
15	2009	0.0	–	–	Ariunbileg et al. (2020)
16	2010	0.0	–	–	Davaa (2015)
17	2013	20.9	–	–	Davaa (2015)
18	2014	44.16	–	–	Orkhonselenge et al. (2018a)
19	2015	0.0	–	–	Orkhonselenge et al. (2018b)
20	2018	0.0	–	–	Figure 11.2a

The lake shrank in 1952–1953, 1986–1989, 2000, 2010, 2015, and 2018 (Table 11.3). The lake shrinkage in 1986–1989 coincided with the dry years with the annual precipitation below 50 mm during 1987–1989 (see Sect. 11.4), whereas the shrinkage in 2015 coincided with the warm year with the annual average air temperature above 5.5 °C (or 6.0 °C in 2015). The disappearances in 2000 and 2009 coincided with the warm and also dry years of 2000 and 2009 (see Sect. 11.4), but the disappearance in 2010 coincided with the year with the annual average air temperature of 5.2 °C and the annual precipitation of 93.7 mm (Fig. 11.3). The most recent condition of Lake Ulaan shows that the lake is shifting into a playa lake (Orkhonselenge et al., in press).

The recent rapid decrease of the Lake Ulaan area shows that the lake responded considerably to climate change resulting in the intense evaporation and the consequent shrinkage. This reduction of Lake Ulaan area affected by the climate factors agrees with those of other lakes in the Valley of Lakes (Orkhonselenge et al., 2018a) and Lake Yakhi in eastern Mongolia (Orkhonselenge & Bulgan, 2021). The decrease in area of Lake Ulaan by 18.2 km^2 during the past 44 years between 1970 and 2014 (Table 11.3; Fig. 11.4) may have been related to the annual average air temperature rise (Fig. 11.3a) and the annual precipitation drop (Fig. 11.3b), especially to both

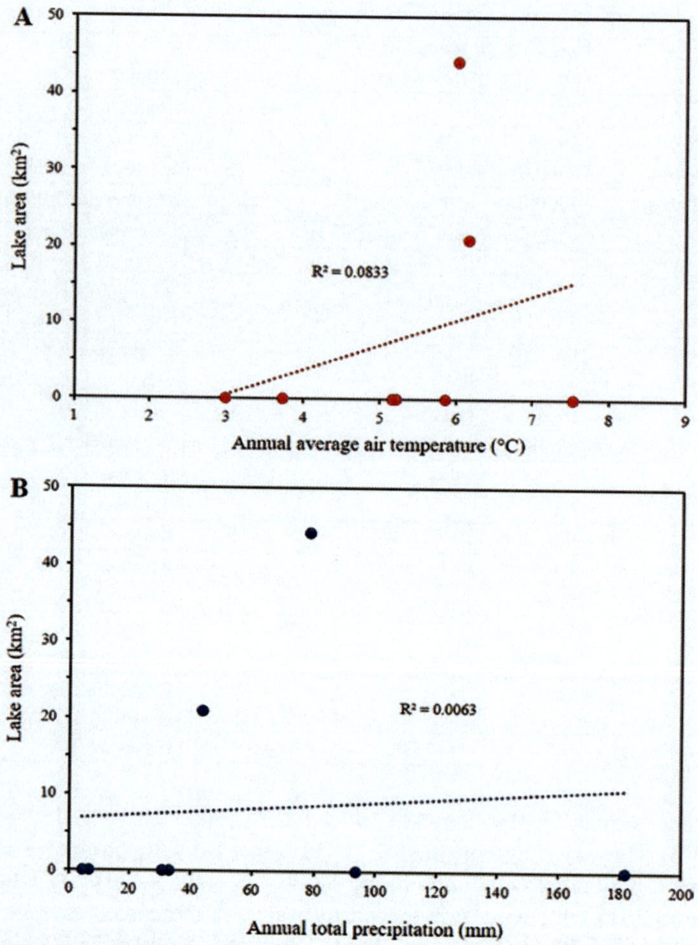

Fig. 11.5 Relationships between area of Lake Ulaan and annual average (**a**) air temperature and (**b**) precipitation during 1986–2014

warmer and drier years with the annual average air temperature above 5.5 °C and the annual precipitation below 50 mm in 1978, 1999–2002, 2009, and 2013 (see Sect. 11.4).

The linear regression analysis (Fig. 11.5) shows that the lake area during 1986–2014 was not clearly related to the annual average air temperature ($R^2 = 0.0833$) and precipitation ($R^2 = 0.0063$). The correlation analysis shows that the lake area during the same period of 1986–2014 had a weak positive correlation with air temperature ($r = 0.288$) but no clear correlation with precipitation ($r = 0.07$).

11.7 Geochemical Review

In the Lake Ulaan sediments, major elements contain large amounts of alkali and alkaline earth metals besides the transition metals and semimetals. The result coincides with the dominant presences of alkali and alkaline earth metals in surface sediments of Lake Ulaan (Orkhonselenge et al., 2018b). The discrimination plots of the major element oxides SiO_2, TiO_2, Fe_2O_3, CaO, Na_2O, K_2O, and MgO vs. Al_2O_3 and CaO vs. loss on ignition (LOI) of the Lake Ulaan sediments show their linear relationships for the cores LU18-1 to LU18-4 (Orkhonselenge et al., in press; Figs. 11.1 and 11.6). TiO_2, Fe_2O_3, and MgO are highly related to Al_2O_3, whereas SiO_2, CaO, Na_2O, and K_2O vs. Al_2O_3 and CaO vs. LOI are randomly related (Fig. 11.6) depending on the core sites (Fig. 11.1).

Although SiO_2 has negative relationships with Al_2O_3 (R^2 = 0.0538–0.8955), terrigenous-derived TiO_2 shows strong positive relationships with Al_2O_3 (R^2 = 0.873–0.9687), implying its presence in phyllosilicates during sedimentary processes, especially illites deriving from the source rock, because aluminum (Al) is a good measure of detrital flux. According to Ferdous and Farazi (2016), the relatively enriched Al_2O_3 is related to the elevated levels of mica, K-feldspar, and clay minerals. Moreover, Fe_2O_3 (R^2 = 0.9113–0.9797) and MgO (R^2 = 0.9299–0.9621) have strong positive relationships with Al_2O_3 (Fig. 11.6). CaO have negative (R^2 = 0.2467–0.8642 for the cores LU18-2 and LU18-3) to positive (R^2 = 0.4504–0.7062 for the cores LU18-1 and LU18-4) relationships with Al_2O_3 (Fig. 11.6). Following Hofer et al. (2013), the weak correlation of CaO with Al_2O_3 means that Ca is not influenced significantly by a detrital source (e.g., feldspars). The linear relationships between CaO and LOI (R^2 = 0.1896–0.9698) in Fig. 11.6 allows us to assume that CaO is largely derived from carbonates. The high content of CaO in the lake sediments may have been related to the evaporated and precipitated cations Ca, Na, and K of the lake water since the 1950s during which the lake has been frequently exposed aerially (see Sect. 11.6), and its brines have been often bleached at the surface. It could also be related to the transported and deposited Ca derived from the Silurian limestone and dolomite in the south and west, and the Cenozoic limestone in the southeast (see Sect. 11.3). The major elements in the lake sediments (Fig. 11.6) are significantly contributed by the middle to late Paleozoic oceanic ophiolites, tholeiitic to calc-alkaline volcanic and volcanoclastic rocks (Badarch et al., 2002) prevailing in the Lake Ulaan basin. This coincides with a conclusion by Lee et al. (2013) and Orkhonselenge et al. (in press) about those elements derived from the bedrocks in an oceanic-arc setting with a mafic igneous provenance.

The measured Chemical Index of Alteration (CIA) proposed by Nesbitt and Young (1982) and Weathering Potential Index (WPI) proposed by Reiche (1943) indicate that the marginal sites (LU18-3, LU18-4) within the Lake Ulaan (Fig. 11.1) are more vulnerable to weathering process than the central sites (LU18-1, LU18-2) (Fig. 11.7a). According to Nesbitt and Young (1982), the weathering with the chemical alteration changes the concentrations of the alkali and alkaline earth metals.

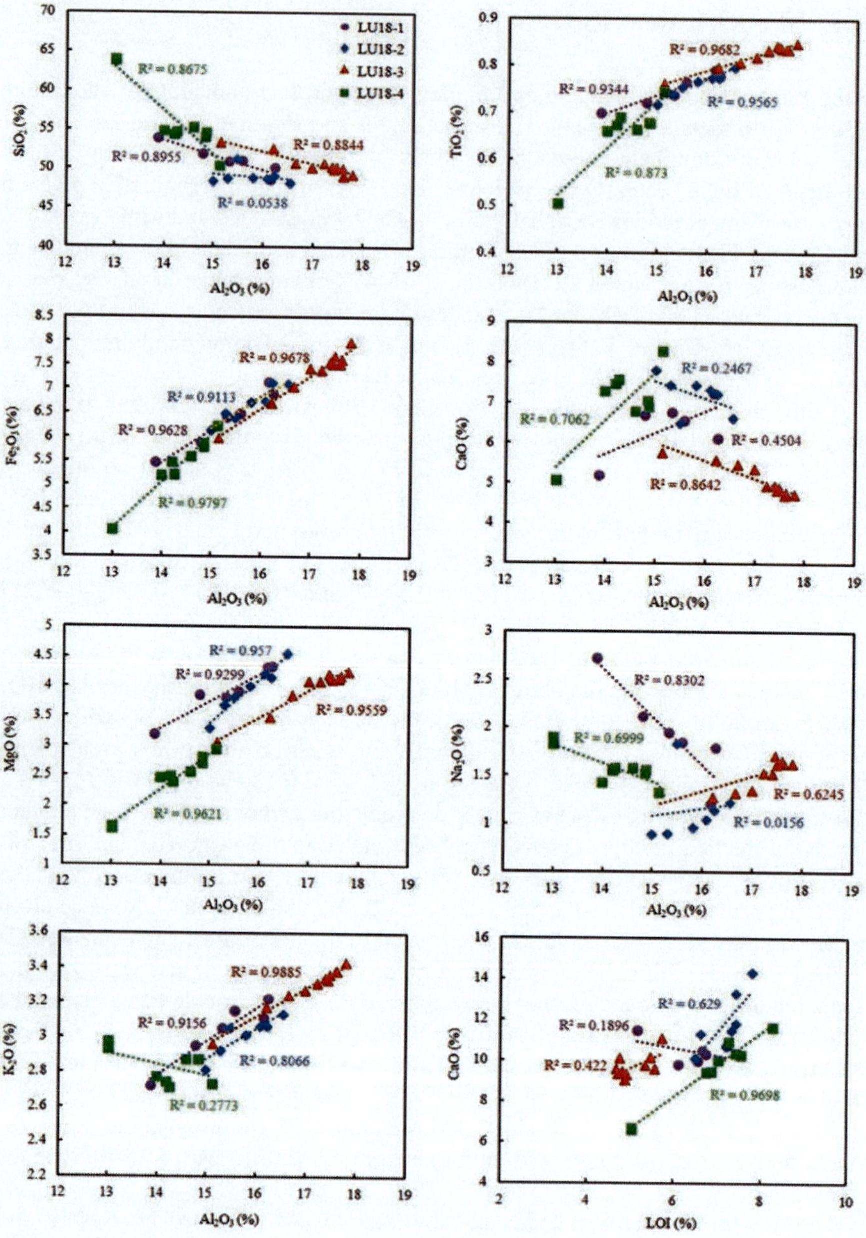

Fig. 11.6 Major element oxides vs. Al_2O_3 and CaO vs. LOI in Lake Ulaan sediments. After Orkhonselenge et al. (in press)

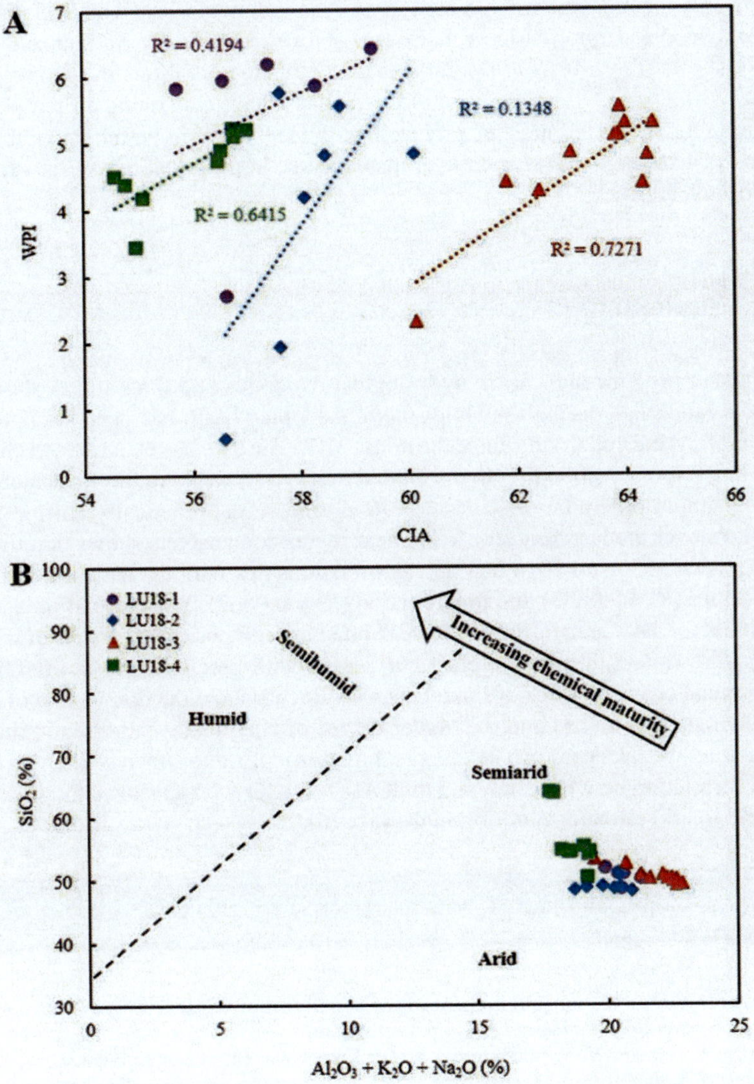

Fig. 11.7 (**a**) Degree of chemical weathering and (**b**) chemical maturity of Lake Ulaan sediments. After Orkhonselenge et al. (in press)

The weathering trend in the Lake Ulaan sediments implies that the lake surface consisting of unconsolidated sediments is easily eroded and/or deflated by wind because the lake often shrinks (see Sect. 11.6) during the low precipitation and intermittent discharge of Ongi River (see Sect. 11.3). The erosion rate associated with the high degree of weathering may have been high in the lake basin, whereas the deposition rate was 1.8 cm/ka in the late Holocene (Orkhonselenge et al.,

2018b). The chemical maturity of the Lake Ulaan sediments, based on a relationship between SiO_2 and total Al_2O_3 + K_2O + Na_2O as proposed by Suttner and Dutta (1986), shows the shift from semiarid to arid climate conditions in the lake basin (Fig. 11.7b). Overall, the shrinkage of Lake Ulaan due to the rising air temperature and dropping precipitation is in agreement with the weathering intensity of the sites within the lake (Fig. 11.7a) and the climate change in the lake basin (Fig. 11.7b) in the recent years.

11.8 Summary

Lake Ulaan presents significant hydrological dynamics and dramatic evolution of the Govi landscape during the Holocene. Lake Ulaan basin has been strengthened with the warming and drying since the mid-1970s. The lake has experienced changes of its area directly responding to the annual average air temperature rise since 1975 and the annual precipitation fall since 1974, and it has practically shifted into a playa lake over the last 50 years. The linear regression analysis shows that the lake area between 1986 and 2014 had no clear relationships with the annual average air temperature ($R^2 = 0.0833$) and precipitation ($R^2 = 0.0063$). The correlation analysis shows that the lake area during 1986–2014 had a weak positive correlation with air temperature ($r = 0.288$) but no clear correlation with precipitation ($r = 0.07$). The geochemical characteristics of Lake Ulaan sediments show the dominance of alkali and alkaline earth metals and the higher degree of chemical weathering in the lake margin than the lake center. The chemical maturity of Lake Ulaan sediments based on the correlation between SiO_2 and total $Al_2O_3 + K_2O + Na_2O$ shows the shift from semiarid to arid climate condition in the lake basin.

References

Academy of Sciences of Mongolian and Academy of Sciences of USSR. (1990). *National Atlas of the Mongolian People's Republic*. 144 p. [In Russian].

Ariunbileg, S., Isupov, V. P., Vladimirov, A. G., Kolpakova, M. N., & Kuibida, L. V. (2020). *Chemical evolution and geochemistry of mineralized lakes of Mongolia*. Admon Printing House, 132 p. [In Mongolian].

Badarch, G., Cunningham, W. D., & Windley, B. F. (2002). A new terrane subdivision for Mongolia: Implications for the Phanerozoic crustal growth of Central Asia. *Journal of Asian Earth Sciences, 21*, 87–110.

Badarch, G., & Tumurtogoo, O. (2001). Tectonostratigraphic Terranes of Mongolia. *Gondwana Research, 4*(2), 143–144.

Batnasan, N. (1998). *Hydrological system, water regime, and evolution of large lakes in the Govi region*. Doctorate Thesis in Geography: Hydrology, Institute of Geoecology, Mongolian Academy of Sciences, Ulaanbaatar. 163 p. [In Mongolian].

Berkey, P. C., & Morris, K. F. (1927). *Geology of Mongolia. A reconnaissance report based on the investigations of the years 1922–1923. Natural History of Central Asia* (Vol. 2). The American Museum of Natural History, 475 p.

Bespalov, N. D. (1951). *Soil of Mongolia*. AH (vol. 41). [In Russian].

Davaa, G. (2015). *Resource and regime of surface water of Mongolia*. 408 p. [In Mongolian].

Ferdous, N., & Farazi, A. H. (2016). Geochemistry of Tertiary sandstones from southwest Sarawak, Malaysia: Implications for provenance and tectonic setting. *Acta Geochemistry, 35*(3), 294–308.

Hofer, G., Wagreich, M., & Neuhuber, S. (2013). Geochemistry of fine-grained sediments of the upper Cretaceous to Paleogene Gosau Group (Austria, Slovakia): Implications for paleoenvironmental and provenance studies. *Geoscience Frontiers, 4*, 449–468.

Holguin, L. R., & Sternberg, T. (2016). A GIS based approach to Holocene hydrology and social connectivity in the Gobi Desert, Mongolia. *Archaeological Research in Asia, 39*, 1–9.

Kang, S., Lee, G., Togtokh, C., & Jang, K. (2015). Characterizing regional precipitation-driven lake area change in Mongolia. *Journal of Arid Land, 7*(2), 146–158.

Lee, M. K., Lee, Y. I., Lim, H. S., Lee, J. I., Choi, J. H., & Yoon, H. I. (2011). Comparison of radiocarbon and OSL dating methods for a Late Quaternary sediment core from Lake Ulaan, Mongolia. *Journal of Paleolimnology, 45*, 127–135.

Lee, M. K., Lee, Y. I., Lim, H. S., Lee, J. I., & Yoon, H. I. (2013). Late Pleistocene–Holocene records from Lake Ulaan, southern Mongolia: Implications for east Asian palaeomonsoonal climate changes. *Journal of Quaternary Science, 28*(4), 370–378.

Lehmkuhl, F., Grunert, J., Hülle, D., Batkhishig, O., & Stauch, G. (2018). Paleolakes in the Gobi region of southern Mongolia. *Quaternary Science Reviews, 179*, 1–23.

Mischke, S., Lee, M. K., & Lee, Y. I. (2020). Climate history of Southern Mongolia since 17 ka: The Ostracod, Gastropod and Charophyte Record from Lake Ulaan. *Frontiers in Earth Science, 8*(221), 1–15.

Murzaev, E. M. (1952). *Paleogeography of the northern Govi region*. [In Russian].

Narantsetseg, T., Badamgarav, J., Ariunbileg, S., Badamgarav, D., Oyunchimeg, T., Idermunkh, T., Uugantsetseg, B., Tuvshinjargal, B., & Dolgorsuren, K. (2011). *Geology of the Meso-Cenozoic depressions*. Institute of Geology and Mineral Resources, 146 p.

Nesbitt, H. W., & Young, G. M. (1982). Early Proterozoic climates and plate motions inferred from major element chemistry of lutites. *Nature, 299*, 715–717.

Orkhonselenge, A., & Bulgan, O. (2021). Geochemical studies and lacustrine geomorphology of Lake Yakhi basin in eastern Mongolia. *Géomorphologie: Relief, Processes, Environnement, 27*(3), 231–242. https://doi.org/10.4000/geomorphologie.15873

Orkhonselenge, A., Komatsu, G., & Uuganzaya, M. (2018a). Climate-driven changes in lake areas for the last half century in the Valley of Lakes, Govi Region, Southern Mongolia. *Natural Science, 10*(7), 263–277.

Orkhonselenge, A., Komatsu, G., & Uuganzaya, M. (2018b). Middle to late Holocene sedimentation dynamics and paleoclimatic conditions in the Lake Ulaan basin, southern Mongolia. *Géomorphologie: Relief, Processus, Environnement, 24*(4), 351–363.

Orkhonselenge, A., Uuganzaya, M., & Davaagatan, T. (in press). Late Holocene sedimentation dynamics in the Lake Ulaan basin, southern Mongolia. *Environmental Earth Science*.

Reiche, P. (1943). Graphic representation of chemical weathering. *Journal of Sedimentary Petrology., 13*(2), 58–68.

Sternberg, T., & Paillou, P. (2015). Mapping potential shallow groundwater in the Gobi Desert using remote sensing: Lake Ulaan Nuur. *Journal of Arid Environments, 118*, 21–27.

Suttner, L. J., & Dutta, P. K. (1986). Alluvial sandstone composition and paleoclimate, I. Framework mineralogy. *Journal of Sedimentary Petrology, 56*, 329–345.

Tsegmid, S. (1969). *Physical geography of Mongolia*. State Press, 405 p. [In Mongolian].

Tsend, N. (1965). *Water chemical composition of lakes in the Govi region*. Institute of Livestock.

Tserensodnom, J. (1971). *Lakes of Mongolia*. State Publishing, 192 p. [In Mongolian].

Tserensodnom, J. (2000). *A catalog of lakes in Mongolia*. Shuvuun Saaral Publishing, 141 p. [In Mongolian].

Chapter 12
Landscape, Lake Distribution, and Evolution in Western Mongolia

Abstract Western Mongolia is a special region with high mountains, modern glaciers, large tectonic depressions and intermontane valleys, and large rivers and lakes. The spectacular landscape of western Mongolia is characterized by the Mongolian Altai Mountain Range and the Depression of Great Lakes. This range occupies a physiographically important place in Central Asia because it hosts glaciated high mountains towering over 4000 m a.s.l., one of the global interfluves, and it supports immense surface water resources in the Mongolian Plateau. The Depression of Great Lakes characterized by the Govi landscape contains many remnant lakes of paleolakes and also modern lakes. Rivers in western Mongolia feed large rivers and lakes belonging to the North Arctic Ocean and Central Asian internal drainage basins. In western Mongolia, the lake area occupies 59.7% of the total lake area, and tectonic, glacial, and aeolian lakes are abundant. Landscape and lacustrine archives of western Mongolia can provide much needed information for reconstructing regional paleoglaciers, paleoclimate, and paleoenvironments in Central Asia and Eurasia during geological time scales.

Keywords Mongolian Altai · Depression of Great Lakes · Paleolakes · Govi · Western Mongolia

12.1 Introduction

Western Mongolia is an important climatic conjunction between the Pacific-influenced and the Atlantic-influenced climates (Sun et al., 2013). The western Mongolia has been primarily influenced by the East Asian Summer Monsoon (EASM) during the Holocene, and the influence was at its maximum during the middle Holocene (e.g., Harrison et al., 1996; Tarasov et al., 2000). Local and regional climates in the Mongolian Altai Mountain Range since 450 a BP were associated with cyclonic versus anticyclonic conditions and the westerlies (Inceoglu et al., 2016). A negative correlation between temperature and effective moisture

A. Orkhonselenge et al., *Lakes of Mongolia*, Syntheses in Limnogeology,
https://doi.org/10.1007/978-3-030-99120-3_12

(warm-dry and cool-wet conditions) was found in the Depression of Great Lakes for the period between the years 200 and 2000 (Shinneman et al., 2010).

This region (Fig. 12.1) hosts the Mongolian Altai Mountain Range, one of the major Central Asian interfluves, and it feeds large rivers (e.g., Bulgan, Khovd, Buyant, and Chigertei Rivers), headwaters of Erchis River (or Irtysh River) in the northwest flowing to the Arctic Ocean (Jigj, 1975). The Mongolian Altai Mountain Range (see Sect. 12.2) is continuous in the east to other parts of the Altai such as the Russian Altai in the north, the Chinese Altai in the south, and the Kazakhstan Altai in the west. The Mongolian Altai is an anticlinal ridge with intermontane valleys and depressions, and it was glaciated twice (Tsegmid, 1969) or three to four times (Jigj, 1976) during the Pleistocene (Blomdin et al., 2018). In the Mongolian Altai, the linkage between glacial and hydrological variations based on dynamic changes of modern glacier and glacial lake, compared to climate and hydrological changes

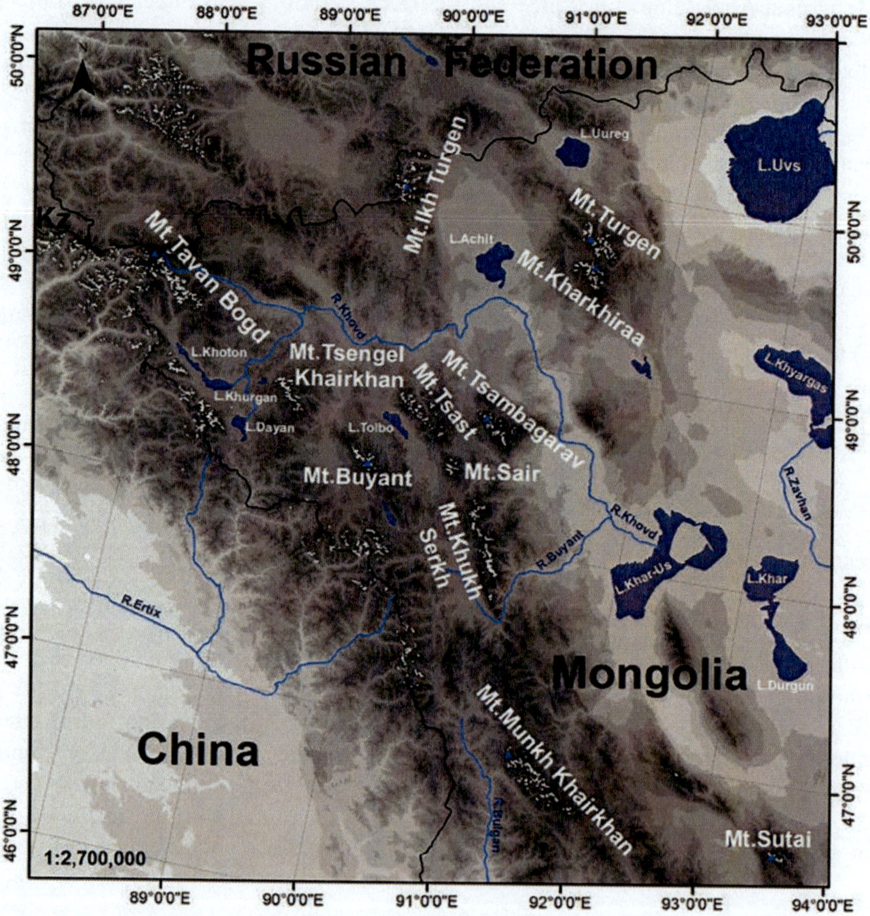

Fig. 12.1 Physiography of the Mongolian Altai Mountain Range and the Depression of Great Lakes. Modified from Orkhonselenge and Harbor (2018)

in the region, was considered by Orkhonselenge and Harbor (2018). Moreover, the aeolian mantles over the Mongolian Altai occur at the elevations of ~1500–2500 m a.s.l., and sand dunes found at some places in the Mongolian Altai provide evidence for the aeolian deposition during the late glacial and the early Holocene (Lehmkuhl, 2014).

The Depression of Great Lakes in the region is the northwestern extension of the unique Govi landscape[1] (see Chap. 9). There are many large and small lakes in the depression (see Chap. 5; Sect. 12.3). For instance, the middle Holocene (5.5 ka BP) shallow lake and the late Eocene (39.0 ka BP) large deep lake at the eastern rim of the Mongol Els sand dunefield (see Sect. 12.2) in western Mongolia were reconstructed by Grunert et al. (2009). The fluviolacustrine and aeolian geomorphological processes of the lakes' formations in the Mongol Els were investigated by Stolz et al. (2012). In the early Holocene, lake levels in the Depression of Great Lakes were high (Lehmkuhl et al., 2018). Moreover, paleolake sediments in the Depression of Great Lakes indicate mainly a humid period from the early to middle Holocene (Lehmkuhl, 2014).

The landscape in western Mongolia has attracted much scientific attention over the last three decades. To date, numerous studies have been conducted in terms of geomorphology, glaciology, limnology, and paleoclimatology. Recently, lake studies conducted in western Mongolia overwhelm those of glacial studies for the purpose of reconstructing regional paleoclimate and paleoenvironmental changes in Mongolia, Central Asia, and Eurasia. For instance, large lakes in the early Holocene were formed by a strong EASM that reached western Mongolia (Lehmkuhl et al., 2018). The Holocene vegetation and climate dynamics in the Mongolian Altai Mountain Range were quantitatively reconstructed from Lake Khoton (Rudaya et al., 2009). Moreover, the endorheic Lake Khar basin formed by a glacier in the Mongolian Altai during the local Last Glacial Maximum (LGM) or Marine Oxygen Isotope Stage (MIS) 2 (Walther et al., 2017).

Hydrogeochemical properties are also investigated for lakes of western Mongolia. For example, the greatest uranium (U) resources (about 6000 tons) are stored in Lake Khyargas (see Sect. 12.3), one of the hypersaline soda closed-basin lakes highly enriched in ^{238}U (<1.0 mg/L) (Isupov et al., 2011, 2015). The U-bearing soda Lake Shaazgai contains high concentrations of U (~1 mg/L) (Isupov et al., 2013; Gaskova et al., 2015). Differences in evaporation-crystallization processes are presented in three different types of lakes, Lakes Shaazgai (soda), Davsan (chloride), and Tonkhil (sulfate) with a wide range of brine composition (Kolpakova & Gaskova, 2018). Over 30 mineralized lakes at the elevations of 1050–1830 m a.s.l. in the Depression of Great Lakes formed during the Neogene and Pleistocene were estimated hydrochemically and geochemically by Ariunbileg et al. (2020).

Western Mongolia has been a crossroad of peoples and cultures from the eastern Mongolia (see Chap. 6) to the western Europe during the thirteenth century when Mongolians, over the next eight centuries, have left unique footprints of their

[1] *Govi* has been often spelled as *Gobi* in the international literature. *Govi* is the correct English transliteration from Mongolian *Говь*.

history and culture throughout the Eurasia. This region includes a natural site and a cultural site registered in the World Heritage List (UNESCO, 2020). The natural site is the Lake Uvs basin (see Chap. 13), an International Geosphere-Biosphere Program (IGBP) site for monitoring global warming, whereas the cultural site is the Petroglyphic Complexes of the Mongolian Altai Mountain Range illustrating the development of culture in Mongolia over a period of 12 ka.

In this chapter, landscape features and lake evolution in western Mongolia are discussed in detail. Tectonic saline and glacial lakes representative in the region are discussed in Chaps. 13 and 14. Paleoclimate implications from lakes in western Mongolia are discussed in Chap. 21.

12.2 Landscape

12.2.1 Geology

In western Mongolia, various tectonostratigraphical terrains occur (e.g., the Altai back-arc/forearc basin, the Khovd and Turgen accretionary wedges, the Tsagaan Shiveet and Tseel metamorphic rocks, and the Baitag and lake island arcs) (Badarch et al., 2002). In the region the middle to late Cambrian, early Ordovician, and Devonian granites, granodiorites, and granosyenites are dominant among the geological structures (Academy of Sciences of Mongolia and Academy of Sciences of USSR, 1990). Regarding region's geological formation, the Devonian terrigenic tuff graywacke is dominant in the mountains, while a continental platform consisting of the Cenozoic red and gray terrigenic sands, gravels, and pebbles with the Paleozoic and Mesozoic granites and basalts is dominant in the Depression of Great Lakes (Academy of Sciences of Mongolia and Academy of Sciences of USSR, 1990).

The high mountains in the Mongolian Altai Mountain Range consist of the Precambrian stratigraphic sequences dislocated during the late Neogene and early Quaternary (Tsegmid, 1969). The Neogene deposits are found on the foothills of the mountains where steeply tilted peneplains are found in the western Mongolian Altai. The Mongolian Altai Mountain Range consists mainly of the Archean and Proterozoic rocks and the Paleozoic deposits underlain by the Precambrian metamorphic rocks (e.g., gneiss, limestones, schists, etc.) in the Altai-Sayan anticline belt during the Caledonian orogeny (Tsegmid, 1969).

The geology of western Mongolia consists of the Ordovician and Silurian mobile belt reworked as the Devonian and Carboniferous mobile belt in the Mongolian Altai, and of the late Neoproterozoic and Ordovician mobile belt and middle to late Paleozoic sedimentary basins in the Depression of Great Lakes over a predominantly juvenile crust (Kröner & Rojas-Agramonte, 2014). There are three major geological complexes of back-arc/forearc basins in the westernmost margin, an accretionary wedge in the middle, and an island arc in the Depression of Great Lakes in the east of the region, and these are dissected by a metamorphic complex in the south (Soejono et al., 2016).

12.2.2 Climate

Western Mongolia is climatologically an important region because it is located under the three atmospheric circulations, the Siberian anticyclone, westerlies, and Central Asian cyclone, and it is also controlled by the North Arctic and Atlantic Oceans. Nevertheless, the air masses coming from these two oceans into the Altai Mountain Range falls in the northern (or Russian) Altai and western (or Kazakhstan) Altai, not in the eastern (or Mongolian) Altai. This phenomenon contributes to the aridity prevailing throughout the region and also to the development of the Govi landscape in the Depression of Great Lakes. Potential changes in the precipitation-to-evaporation balance in western Mongolia are demonstrated with rising air temperatures interpreted from tree-ring records obtained in the areas of Lakes Kholboo, Duruu, Zagas, Baga, Takhilt, and Bayan in the Depression of Great Lakes (Shinneman et al., 2010).

The average air temperature is −30 to −34 °C in the Depression of Great Lakes and -20 °C in the Mongolian Altai Mountain Range in January, while it is 38–41 °C in the Depression of Great Lakes and less than 15 °C in the Mongolian Altai Mountain Range in July (Tsegmid, 1969). The annual air temperature amplitude is 53 °C recorded at the Ulaangom meteorological station (see details in Chap. 13). The annual average precipitation is 100–150 mm in the Depression of Great Lakes and over 300 mm in the Mongolian Altai Mountain Range (Tsegmid, 1969). Another source says that the annual average precipitation is 150–300 mm and the open water evaporation is 300–400 mm in the Mongolian Altai mountainous region (Ministry of Environment and Green Development, 2013).

The annual average air temperature increased by 1.9–2.28 °C in the high mountains of Mongolia including the Mongolian Altai since the 1940s, and the winter average air temperature and precipitation have noticeably increased in comparison to the summer temperature and precipitation since the 1940s (Batima et al., 2005). These temperature increases and precipitation changes are likely to have driven the observed changes in glacier extent, river discharge, and lake area (Sarantuya & Natsagdorj, 2010; Orkhonselenge & Harbor, 2018).

At the Ulgii and Khovd meteorological stations in the Mongolian Altai, the annual average air temperature is recorded to be 1.8 °C, with the summer monthly average air temperature that goes up to 19 °C in Khovd and the winter monthly average air temperature that goes down to −25.5 °C. The annual average precipitation is 56 mm in Ulgii and 210 mm in Khovd (Orkhonselenge & Harbor, 2018).

12.2.3 Landform

The great variety of landscapes makes western Mongolia a very significant region from the geomorphological point of view. The western Mongolia consists of mountainous area in the west and a wide depression in the east (Fig. 12.1).

The mountains in the west were uplifted due to neotectonics, and they have steep slopes and sharp peaks. The mountainous region has an average elevation at 3200–3500 m a.s.l., and there are numerous peaks elevated over 4000 m a.s.l. in the Mongolian Altai Mountain Range. The Mongolian Altai is connected to the Russian Altai in the northwest and north, and the Siilkhem Mountain Range in the northeast, and it extends over 1000 km from Mt. Maanit in the Altai Tavan Bogd Mts. at 4653 m a.s.l. (Tsegmid, 1969; Jigj, 1975) or 4374 m a.s.l. (Academy of Sciences of Mongolia and Academy of Sciences of USSR, 1990) and Mt. Munkh Khairkhan at 4362 m a.s.l. (Jigj, 1975) in the northwest (Fig. 12.1) to the Gichgene Mountain Range (Murzaev, 1952) and Lake Alag basin (Jigj, 1975) in the southeast.

There are numerous branches and isolated mountains over 4000 m a.s.l. in the Mongolian Altai, for example, Mt. Tsengel Khairkhan (4362 m a.s.l.), Mt. Munkh Khairkhan (4231 m a.s.l.), Mt. Sutai (4226 m a.s.l.), Mt. Tsast Bogd (4208 m a.s.l.), Mt. Bugat (4041 m a.s.l.), Mt. Kharkhiraa (4040 m a.s.l.), Mt. Ikh Turgen (4030 m a.s.l.), and Mt. Turgen (4029 m a.s.l.) and the Baruun Mountain Range (4025 m a.s.l.) and the Altan Khukhii Mountain Range (4003 m a.s.l.) (Tsegmid, 1969; Fig. 12.1). Today, some of these mountains (e.g., Mt. Tavan Bogd, Mt. Must, Mt. Tsengel, Mt. Tsast Bogd, Mt. Kharkhiraa, Mt. Turgen, Mt. Tsambagarav (Fig. 12.2)) still host modern glaciers. The mountains are isolated by tectonic depressions of Lakes Achit and Uureg (Jigj, 1975; Fig. 12.1). In the mountainous region, physical weathering is dominant and it forms steep cliffs and gorges.

The Depression of Great Lakes is the second most important landscape in the region, and it consists of numerous large lake basins (Fig. 12.1; see Sect. 12.3) and sand dunefields. The Depression of Great Lakes, having a tectonic origin, extends from the Tagna Mountain Range in the northwest to the Sharga depression (see Chap. 9) in the southeast, and it has an average elevation at 2700 m a.s.l. with the lowest value in the Lake Uvs basin (Fig. 12.1; see Chap. 13) at 743 m a.s.l. (Jigj, 1975). There are large sand dunefields (e.g., Mongol Els, Buurug Del Els, Altan Els, Bor Khyar Els, etc.) in the east of the Depression of Great Lakes. The Mongol Els (Fig. 12.3) is the largest sand dunefield of Mongolia. The crescent-shaped Buurug Del Els sand dunefield occupying over an area of 4000 km^2 along the valleys of Nariin and Baruun Turuun Rivers contours Lake Uvs (see Chap. 13) in the east and southeast (Tsegmid, 1969), and it hosts the aeolian freshwater Lake Bayan covering an area of 30.5 km^2.

12.2.4 Glaciers and Permafrost

In the Mongolian Altai, large ice advances occurred during MIS 4 (74–71 ka) and MIS 2 (25–20 and 18–17 ka) coinciding with the global temperature minima (Lehmkuhl et al., 2016). The distribution of abundant glacial erosional and depositional landforms indicates that the Altai Mountains have experienced predominantly alpine-style glaciations with some small ice caps centered on the higher mountain peaks (Blomdin et al., 2016; Orkhonselenge & Harbor, 2018). Mt. Ikh

Fig. 12.2 Mt. Tsambagarav with modern glaciers (Fig. 12.1) viewing toward the (**a**) southeast and (**b**) northeast. Photos by A. Orkhonselenge on July 30, 2013, for (**a**) and August 1, 2013, for (**b**)

Turgen paleoglaciers in the Mongolian Altai indicate that cold and wet conditions at MIS 3, and cold and dry conditions at MIS 2 in western Mongolia (Blomdin et al., 2018). Humid conditions prevailed in the Mongolian Altai region at the onset of the neoglaciation and the end of the Holocene Climate Optimum approximately 6.0 ka BP, followed by a long dry period from 5.0 ka BP until 250 a BP (Herren et al., 2013).

In the Depression of Great Lakes, the late Pleistocene ice- and the early Holocene moraine-dammed Lake Ak-Hol formed in the catchment of the Mogen-Buren River

Fig. 12.3 Google Earth Pro image of the Mongol Els sand dunefield on July 25, 2020

due to the last catastrophic flood contributing to the sudden drainage into Lake Achit (Agatova et al., 2020).

In the Altai mountainous region, permafrost was investigated by V. Sapojnikov in 1906, the Mongolian and Russian permafrost research team in 1968, and national permafrost scientists in 1974 (e.g., D. Tumurbaatar, N. Sharkhuu, T. Dorj) and 2001 (e.g., D. Tumurbaatar, Ya. Jambaljav, D. Undarmaa) (Tumurbaatar, 2004). Permafrost in the Khan Khukhii Mountain Range caused extensive peat formation in the Lake Uvs and Tes River basins (Minayeva et al., 2004, 2005). Permafrost data and maps about depth, temperature, texture, humidity, and geographical distribution in western Mongolia were provided by Tumurbaatar (2004). The seasonal freezing and thawing depths in the Mongolia Altai and Depression of Great Lakes range at around 2.9–4.5 m (Tumurbaatar, 2004; Table 12.1).

Perennial, intermittent, and occasional permafrost are distributed in the Mongolian Altai Mountain Range, while the seasonal permafrost is distributed in the Depression of Great Lakes (Jambaljav et al., 2016). According to Jambaljav et al. (2016), in the Mongolian Altai, the seasonal freezing and thawing depth is about 3.5–6.5 m under the annual average air temperature of −3.3 to 0.2 °C, while in the Depression of Great Lakes, the seasonal freezing and thawing depth is 3.0 m under the annual average air temperature of −0.2 °C. The permafrost in lake and river drainage basins in the Mongolian Altai and Depression of Great Lakes is defined as cold and warm depending on annual mean ground temperature (AMGT) by Zorigt et al. (2020), and it is shown with its active layer thickness (ALT) in Table 12.2.

Table 12.1 Seasonal freezing and thawing depths in western Mongolia

#	Places	Materials	Depth (m)
1	Lake Tolbo basin in the Mongolian Altai Mountain Range	Clays	2.9
		Sands	3.4–3.9
		Gravels	3.9–4.5
2	Lake Achit basin in the Depression of Great Lakes	Clays	2.9–3.0
		Sands	3.4–3.9
		Gravels	3.9–4.5

Table 12.2 Cold and warm permafrost in the lake and river drainage basins of western Mongolia

Places	Basins	Area of the basins (km²)	Permafrost in the basins (%)		AMGT (°C)	ALT (m)
			Cold permafrost (AMGT<-2 °C)	Warm permafrost (−2 °C < AMGT <0 °C)		
Mongolian Altai	Lake Khar-Khovd River	87,787	22.6	19.5	−2.5	0.9–4.6
	Uyench River-Bodonch River	34,037	4.7	11.1	−1.7	0.8–3.5
Depression of Great Lakes	Lake Uvs-Tes River	53,510	21.7	56.2	−1.4	2.7–4.9
	Lake Khyargas-Zavkhan River	120,707	17.6	17.6	−1.2	0.9–4.1

12.2.5 Groundwater

Groundwater in the region occurs in the metamorphic rocks, granites, sandstones, and limestones along tectonic faults and the borders between igneous and metamorphic rocks (Tsegmid, 1969). Artesian groundwater emerges in the Cretaceous and Neogene deposits in the Depression of Great Lakes, while the groundwater occurs in the Precambrian and Paleozoic metamorphic rocks within the Mongolian Altai Mountain Range (Tsegmid, 1969). The groundwater in the Neogene deposits is abundant in the region and it occurs at around 60 m depth (Tsegmid, 1969).

Table 12.3 shows surface water and groundwater resources of major drainage basins in western Mongolia (Ministry of Environment and Green Development, 2013).

Table 12.3 Water resources of the lake and river drainage basins in western Mongolia

#	Drainage basins	Area (km^2)	Type	Surface water (million m^3/year)			Groundwater (million m^3/year)	
				Total resource	Environmental	Possible use	Potential exploitable	Exploitable resource
1	Lake Khyargas-Zavkhan River	122,315	SW/GW	599	554	44.9	892	10.0
2	Khuis Govi-Lake Tsetseg	43,024	SW/GW	0	0	0.0	493	8.1
3	Uyench River-Bodonch River	34,491	SW/GW	66	64	2.7	237	11.3
4	Bulgan River	10,155	SW	207	199	8.3	86	0.0
5	Lake Khar-Khovd River	88,936	SW/GW	2317	2201	115.8	684	12.7
6	Lake Uvs-Tes River	54,223	SW/GW	1578	1514	63.1	405	6.1
	Total	353,144		4767	4532	234.8	2797	48.2

SW surface water; *GW* groundwater

12.2.6 Surface Water

Surface water resources of major drainage basins in the region are shown in Table 12.3. In the Mongolian Altai, Khurimt River feeds the North Arctic Ocean drainage basin via Ob River, while Khovd,[2] Buyant (Fig. 12.1), Tsenkher, Bulgan, Uyench, Bodonch, Barlag, and Bij Rivers feed the Central Asian internal drainage basin. Moreover, glacial meltwater in the Mongolian Altai also feeds large rivers, such as Katun and Biya Rivers, headwaters of the Ob River in the Russian Altai, and Khovd River (Figs. 12.1 and 12.4), an outflow of Lake Khurgan (see Chap. 14) in the Mongolian Altai and a headwater of Lake Khar Us in the Depression of Great Lakes (Fig. 12.1).

Rivers in the region are mainly fed by rainfall (25–40%), snow meltwater (35–55%), groundwater (15–25%), and glaciers (<5%) (Tsegmid, 1969). The discharge rate is estimated to be about 8–12 L/s in Khovd River and its tributaries Tsagaan and Sogoot Rivers in the northern Mongolian Altai and 1.0–1.5 L/s in

[2] *Khovd* has been often misspelled as *Kobdo* based on the Russian pronunciation. Khovd is the right English transliteration from Mongolian *Ховд*.

Fig. 12.4 Khovd River (Fig. 12.1), an outlet of Lake Khurgan (Fig. 14.3) in the Mongolian Altai Mountain Range. Photo by G. Miimaa near Ulaan Khus soum, Bayan Ulgii aimag, on October 26, 2019

Bulgan, Uyench, and Bodonch Rivers in the southern Mongolian Altai (Tsegmid, 1969).

Lakes (see Sect. 12.3) are one of the major surface water resources in western Mongolia. There are large lakes covering an area of over 10, 50, 100, 500, and 1000 km^2 (Table 12.4; see Chap. 5).

Glaciers in the Mongolian Altai Mountain Range (Fig. 12.2) are also one of the major surface water resources in the region (Orkhonselenge & Harbor, 2018). The glaciers in the Mongolian Altai have shrunk by 6% over the last 40 years (Baast, 1998) or 22% during the last 50–60 years (Davaa, 2010). This is a major concern in the region because of its heavy dependency on glacial meltwater to sustain water supplies for agriculture and other ecosystem services in summer months (Orkhonselenge & Harbor, 2018).

12.3 Lakes

Lakes in western Mongolia (Fig. 1.1) are medium in number but the largest in area (Table 1.2). Tectonic and glacial lakes are the dominant lake types distributed in western Mongolia (see Chap. 4). In this region lakes elevated at the ranges of 2001–2500 m a.s.l. and 2501–3000 m a.s.l. are abundant among the total 567 lakes (Table 4.1; see Chap. 5). A total of 135 lakes at 2001–2500 m a.s.l. constitute 50.0% of the total 270 lakes with the same elevation range in Mongolia, while 189 lakes at 2501–3000 m a.s.l. constitute ~92.2% of the total 205 lakes with the same elevation

Table 12.4 Morphometric components of lakes in western Mongolia

#	Lakes	Origin/type	Elevation (m a.s.l.)	Area (km^2)	Depth (m)	Description
1	Achit	Tectonic	1435	291.41	2.3; 5.0	Section 12.4.4
2	Airag	Freshwater	1030	143	6.0; 10.0	–
3	Bayan	Freshwater	1491	64.2	22.0; 50.0	–
4	Bayan	Aeolian	932	30.8	11.9; 26.5	–
5	Bust	Tectonic	2041	22.1	5.7; 10.0	–
6	Dayan	Glacial	2232.4	67.2	2.7; 4.5	–
7	Durgun	Tectonic	1132	305	14.0; 27.0	–
8	Duruu	Tectonic	2395.0	14.6	5.2; 14.5	–
9	Duruu	Aeolian	1143	15.0	7.0; 12.0	–
10	Khar	Tectonic	1132	561.96	4.5; 7.0	Section 12.4.3
11	Khar	Tectonic	2494	14.2	17.0; 38.0	–
12	Khar Us	Tectonic	1157	946.49	2.3; 5.0	Section 12.4.2
13	Kholboo	Tectonic	1941	14.0	4.0; 6.0	–
14	Khoton	Glacial	2083.6	50.1	26.6; 58.0	Chapter 14
15	Khurgan	Glacial	2073	70.8	7.6; 28.0	Chapter 14
16	Khyargas	Tectonic	1028	1348.2	47.0; 80.0	Section 12.4.1
17	Namiriin Khar	Tectonic	1574	62.7	5.2; 8.2	–
18	Nogoon	Tectonic	1132	26.7	0.8; 2.5	–
19	Oigon	Tectonic	1664	61.3	3.5; 8.0	–
20	Zeeren	Saline	1043	12.1	0.5; 0.9	–
21	Tal	Glacial	2576	32.0	3.4; 8.9	–
22	Tolbo	Glacial	2080	83.5	9.0; 12.7	–
23	Tsetseg	Saline	1687	17.7	0.4; 1.5	Figure 12.6b
24	Ulaagchnii Khar	Freshwater	1980	84.5	20.0; 47.0	Figure 12.6a
25	Uvs	Tectonic	743	3451.6	11.9; 20.0	Chapter 13
26	Uureg	Tectonic	1425	249.12	27.0; 42.0	Section 12.4.5

range in Mongolia (see Chap. 5). In addition, a total of 33 lakes are located above 3001 m a.s.l. and they are distributed only in this region (Table 4.1; see Chap. 5). The greatest number of lakes of each class of areas (Table 4.1; see Chap. 5) in the whole country are distributed in this region.

In the region, there are 567 lakes each covering an area of over 0.1 km^2 (see Chaps. 1 and 5), and they have diverse origins (see Chap. 4). There are 3 largest lakes, Lakes Uvs (Fig. 12.1; see Chap. 13), Khyargas (Figs. 12.1 and 12.5a), and Khar Us (Figs. 12.1 and 12.5b) among 17 lakes each covering an area of over 50 km^2 and 8905.8 km^2 in total. Two of these lakes occupy an area of over 1000 km^2 each and 4799.8 km^2 in total (Table 12.4), but Lake Khar Us is excluded from the list because it has lost its area rapidly in recent years (Orkhonselenge and Uuganzaya, unpublished; see Sect. 12.4, Table 12.5). Lake Uvs situated in this region is the largest lake (~3500 km^2) in the entire Mongolia (Tables 12.4 and 12.5; see Chap. 13).

Fig. 12.5 Landsat 8 images of large lakes in the Depression of Great Lakes (Fig. 12.1): (**a**) Lake Khyargas on July 28, 2015 (path of 140 and row of 26), (**b**) Lake Khar Us on July 25, 2014 (path of 140 and row of 27), (**c**) Lake Khar on August 20, 2015 (path of 141 and row of 25), (**d**) Lake Achit on September 12, 2015 (path of 142 and row of 26), and (**e**) Lake Uureg on August 19, 2015 (path of 142 and row of 25). Modified from Orkhonselenge and Uuganzaya (unpublished)

12.3.1 Lakes in the Mongolian Altai Mountain Range

Lakes in the Mongolian Altai Mountain Range are situated at 1500–3500 m a.s.l. and they do not have large areas. These lakes originally formed in glacial cirques and intermontane valleys by thawing of paleoglaciers and modern glaciers and by damming of moraines. In the Mongolian Altai, the lakes occupying an area up to 50 km^2 constitute 40% among all the mountainous lakes, whereas the lakes occupying an area close to 100 km^2 are few (e.g., Lakes Uureg and Achit). They are mostly freshwater lakes (see Chap. 5).

The lakes in the region are filled with glacial freshwater such as in the cases of Lakes Khoton (see Chap. 14), Tolbo, Dayan, Khurgan (Figs. 12.1 and 14.3), Khotgor, Bort, Tsunkhel, Khukh, Bayan, Buyant, and Chichirtei. The glacial Lakes Tolbo, Khoton, Khurgan, and Dayan are distributed in the Mongolian Altai Mountain Range (e.g., Jigj, 1976; Blomdin et al., 2018; Orkhonselenge & Harbor, 2018).

12.3.2 *Lakes in the Depression of Great Lakes*

The lakes in the Depression of Great Lakes are the sinks for water and sediment transported by the rivers draining the eastern slopes of the Mongolian Altai, northern slopes of the Govi Altai, and western slopes of the Khangai Mountain Ranges, and they are included in the Central Asian internal drainage basin. Tectonic lakes such as Lakes Khyargas, Khar Us, Uureg, Achit, Uvs (see Chap. 13), Tsetseg (Fig. 12.6b), and Tonkhil are distributed in the Depression of Great Lakes (see Chap. 4). Thermokarst Lakes Eleet and Asgat occur in the Tes River valley (see

Fig. 12.6 Lakes (**a**) Ulaagchnii Khar and (**b**) Tsetseg in the Mongolian Altai Mountain Range. Photos by G. Miimaa for (**a**) and B. Chimgee for (**b**)

Chap. 4). There are abundant aeolian lakes within the areas of transverse, crescent, and longitudinal dunes in the Depression of Great Lakes. For example, aeolian Lakes Khunt, Khunguin Khar, Mukhar Khungui, Ulaagchnii Khar (Fig. 12.6a), Bayan, and Thookhor are located in the Zavkhan, Khungui, Tes, and Nariin Rivers' valleys (see Chap. 4).

Lakes in the Depression of Great Lakes are saline (Tsegmid, 1969; see Chap. 5). For example, the lakes such as Lakes Zergiin Tsagaan, Tsetseg (Fig. 12.5b), Tonkhil, Ikhes, and Alag are all saline. Lakes in the region contain larger amounts of brines and minerals than the lakes in other regions of Mongolia (see Chap. 23). For instance, the highly mineralized soda lakes located close to U deposits within western Mongolia contain high concentrations of U (1–3 mg/L), B (<250 mg/L), Br (<1.1 g/L), Li (<600 mg/L), and Sr (<8 mg/L) (e.g., Isupov et al., 2015). Lake Shaazgai is effective in sorption of U and arsenic (As) speciation only at a high concentration of FeOOH sorbent at pH 9.4 (Gaskova et al., 2015). Ariunbileg et al. (2020) found soda lakes (48%), saline lakes (41%), and sulfate lakes (11%) among the studied 38 mineralized lakes in the Depression of Great Lakes.

12.4 Lake Evolution

Lakes in western Mongolia have experienced frequent disappearance, reappearance, expansion, shrinkage, and mineralization during various geological time scales. Lakes in the Depression of Great Lakes mostly disappeared during the MIS 6, but large paleolakes existed from the last interglacial MIS 5e to the beginning of the LGM (Lehmkuhl et al., 2018). At the MIS 3 (or 39.0 ka BP), a large and deep Lake Ereen covered the floodplain formed by converging Shurag and Zavkhan Rivers at the eastern rim of the Mongol Els sand dunefield (Grunert et al., 2009; Stolz et al., 2012; Fig. 12.3) at the eastern Depression of Great Lakes. The lakes reached the highest levels in the early Holocene, but the lake levels were generally low in the late Holocene (Lehmkuhl et al., 2018). The endorheic Lake Khar formed as a result of the blockage of Kharganat River valley by a glacier at MIS 2 showed a more or less stable lake level since 1948/1970 despite the increasing precipitation and melting glaciers and permafrost in high mountains of the Mongolian Altai Mountain Range (Walther et al., 2017). Orkhonselenge and Harbor (2018) showed an increased area from 10.37 km^2 in 2007 to 10.41 km^2 in 2011 for a glacial lake in Mt. Ikh Turgen in the northeastern Mongolian Altai due to the recent rise of the annual average air temperature.

Three large lakes (i.e., Lakes Uvs, Khyargas, and Khar) exhibited distinct drawdowns in the late 2000s (Davaa, 2015). In contrast, the Lake Khar Us level remained more or less constant, although the summer lake level increased in the late 2000s (Kang et al., 2015), but a continuous decrease in area since 1966 has been found (Table 12.5; Fig. 12.7). In the last 50 years, many lakes completely disappeared due to the strong human usage of the water resources (Lehmkuhl et al., 2018) or the local climate changes (see Chaps. 13 and 14). Six largest lakes in the Depression of

Table 12.5 Temporal changes in areas (km^2) of large lake in the Depression of Great Lakes

		Tserensodnom (1971)	Orkhonselenge and Uuganzaya (unpublished)	Davaa (2015)		Orkhonselenge and Uuganzaya (unpublished)	
#	Lakes	1966	1970	2010	2013	2014	2015
1	Uvs	3518.3	3354.47	3663.6	3646.6		3451.6
2	Khyargas	1406.8	1380.96	1396.1	1373.1		1348.2
3	Khar Us	1859.2	1101.45	1002.1	1030.3	946.49	
4	Khar	574.8	584.48	574.9	568.8		561.96
5	Achit	296.8	305.54	301.5	298.3		291.41
6	Uureg	237.6	250.6	252.3	252.4		249.12

Great Lakes (Figs. 1.1 and 12.1) show decreases in their areas between 1966 and 2014–2015, except for the Lake Uvs (Table 12.5). Areas of Lake Uvs is described in detail in Chap. 13, but the areas of other five lakes are described in brief here (Fig. 12.7).

12.4.1 Lake Khyargas

Lake Khyargas, a tectonic saline lake, is located at 1028 m a.s.l. in the Depression of Great Lakes (Figs. 1.1 and 12.1). The lake extends for 75 km in the maximum length, 19 km on average width or 31 km in the maximum width, and 253.2 km along the shoreline (Tserensodnom, 1971). The water depth of the lake is 47 m on average and 80 m in the maximum (Tserensodnom, 1971). Lake Khyargas is fed by Galba River, an outflow of Lake Airag (Figs. 12.5a and 12.7a), draining from an area of 115,500 km^2 (Tserensodnom, 1971) including Khovd, Buyant, Khoid Tsenkher, and Urd Tsenkher Rivers in the Mongolian Altai Mountain Range, and Zavkhan and Khungui Rivers in the Khangai Mountain Range, but the lake has no outlet.

Water level of Lake Khyargas is highly dependent on the inflow river discharge, and it varies 22.0 cm annually and 127.5 cm seasonally (Batnasan, 1998). The water temperature of Lake Khyargas is 18–20 °C at the surface in July and 0.1–0.2 °C at the surface and 2–3 °C at the bottom in March (Tserensodnom, 1971). The mineralization value of Lake Khyargas is 2.55 g/L, in which 1.02 g/L NaCl and 0.99 g/L Na_2SO_4 are dominant (Tserensodnom, 1971), or 8.4 g/L (Batnasan, 1998; Tserensodnom, 2000). The mineralization value of the lake reached 7.46 g/L in 1926 and 7.27–8.37 g/L in 1941–1967 (Tserensodnom, 1971) or 7.34–9.45 g/L in 2010 dominated by Na^+, Cl^-, SO_4^{2-}, and HCO_3^- (Ariunbileg et al., 2020). Lake Khyargas occupies an area of 1406.8 km^2 (Tserensodnom, 1971), but the lake lost its area for 32.75 km^2 or ~2.4% from 1380.96 km^2 in 1970 to 1348.2 km^2 in 2015 (Orkhonselenge and Uuganzaya, unpublished; Table 12.5; Fig. 12.7a).

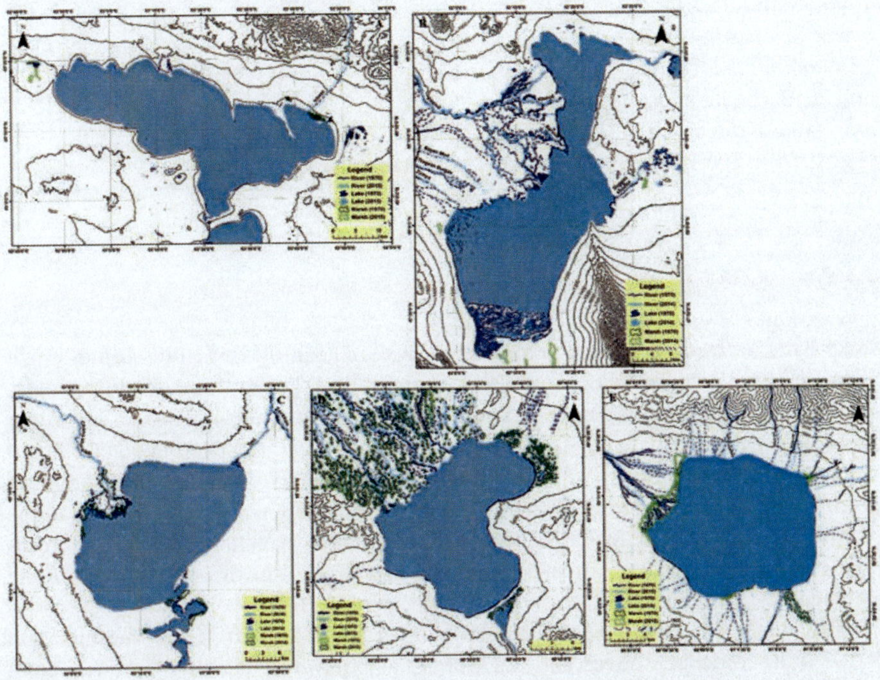

Fig. 12.7 Spatial distributions of areas and marshes of Lakes (**a**) Khyargas, (**b**) Khar Us, (**c**) Khar, (**d**) Achit, and (**e**) Uureg in 1970 and 2014–2015. Modified from Orkhonselenge and Uuganzaya (unpublished)

12.4.2 Lake Khar Us

Lake Khar Us, a tectonic freshwater lake, is located at 1157 m a.s.l. in the Depression of Great Lakes (Figs. 1.1 and 12.1). Lake Khar Us extends for 72.2 km in the maximum length, 25.8 km on average width or 26.2 km in the maximum width, and 306.8 km along the shoreline, and it is 4.5 m deep (Tserensodnom, 1971). The lake is fed by rivers and streams draining from an area of 70,450 km^2 including Khovd, Buyant, Tsenkher, Baruun, Tsagaan Burgas, and Zumya Rivers (Figs. 12.5b and 12.7b), and the lake feeds only Chono Kharaikh River (Figs. 12.5c and 12.7c), a headwater of Lake Khar. The water level of Lake Khar Us is highly dependent on Khovd River, and it fluctuates with 69.3 cm annually and 30.4 cm seasonally (Batnasan, 1998). The mineralization value of the lake is 242.9 mg/L in the west and 348.0 mg/L in the east (Tserensodnom, 1971) and 107 mg/L (Batnasan, 1998) or 5.12–5.5 g/L in 2010 dominated with Cl^-, Na^+, SO_4^{2-}, and HCO_3^- (Ariunbileg et al., 2020).

The lake once covered an area of 1859.2 km^2 (Tserensodnom, 1971) or 946.49 km^2 (Orkhonselenge and Uuganzaya, unpublished; Table 12.5; Fig. 12.7b). The lake area is decreased by 83.9 km^2 from 1607.3 km^2 in 1992 to 1523.33 km^2 in

1996 (Batnasan, 1998). The lake area dropped from 1596.73 km^2 in 1994 with the water level at 1160.36 m a.s.l. to 1409.1 km^2 in 2007 with the water level at 1159.64 m a.s.l. (Davaa, 2015). Although the lake rose in area again between 2010 and 2013, the lake lost its area for 154.96 km^2 or ~14.1% from 1101.45 km^2 in 1970 and 946.49 km^2 in 2014 (Orkhonselenge and Uuganzaya, unpublished; Table 12.5; Fig. 12.7b). The lake continuously decreases in its area.

12.4.3 Lake Khar

Lake Khar, a freshwater lake, is located at 1132 m a.s.l. in the Depression of Great Lakes (Figs. 1.1 and 12.1). The lake extends for 36.8 km in the maximum length, 15.7 km on average width or 23.6 km in the maximum width, and 158.4 km along the shoreline, and it is 7.0 m deep in the center and 4.0–5.0 m deep in other parts of the lake (Tserensodnom, 1971). Lake Khar is fed by rivers and streams draining from an area of 72,000 km^2 including a major inflow Chono Kharaikh River in the northwest, and the lake feeds Teel River, a tributary of Zavkhan River in the northeast (Figs. 12.5c and 12.7c), and Khom River, a headwater of Lake Durgun in the southeast.

The water temperature of the lake ranges at 20.0–25.0 °C in summer and 1.5–2.0 °C under ice cover and 3.0–5.0 °C at bottom in winter (Tserensodnom, 2000). The mineralization value of the lake water is 387.7 mg/L in the east and 261.1 mg/L in the west (Tserensodnom, 1971). The water level of Lake Khar is highly dependent on inflow discharge of Chono Kharaikh River, and it fluctuates 69.3 cm annually and 30.4 cm seasonally (Batnasan, 1998). The lake occupies an area of 574.8 km^2 (Tserensodnom, 1971), but the lake lost its area for 22.52 km^2 or ~3.9% from 584.48 km^2 in 1970 to 561.96 km^2 in 2015 (Orkhonselenge and Uuganzaya, unpublished; Table 12.5; Fig. 12.7c).

12.4.4 Lake Achit

Lake Achit, a tectonic freshwater lake, is located at the elevation of 1435 m a.s.l. in the northwestern corner of the Depression of Great Lakes (Figs. 1.1 and 12.1). The lake extends for 30.0 km in the maximum length and 16.0 km in width (Tsegmid, 1969). Another source states that it is 24.0 km long, 18.0 km wide in the maximum and 12.3 km on average, and 93.0 km along the shoreline (Tserensodnom, 1971). The lake water depth is 2.3 m on average and 5.0 m at its maximum (Table 12.4). The lake occupies a relatively low volume of 0.665 km^3 although the lake covers a large area of 291.41 km^2 (Orkhonselenge and Uuganzaya, unpublished; Table 12.6).

Lake Achit, a terminal lake, is fed by Tsagaan Nuur, Khatuu, Altan Gadas, Bukh Murun, and Uliastai Rivers (Figs. 12.5d and 12.7d) and streams draining from an area of 10,500 km^2 of glaciers and snow in the Mt. Kharkhiraa, Mt. Turgen, and

Siilkhem Mountain Range (Tsegmid, 1969; Tserensodnom, 1971). This lake has also inflows by Khuv Us and Mogen Buren Rivers (Sun et al., 2013; Agatova et al., 2020), and the lake feeds Khovd River and Lake Khar Us through a channel called *Usnii Khooloi* (Tsegmid, 1969; Figs. 12.5d and 12.7d) or *Nuuriin Khooloi* (Tserensodnom, 1971). Lake Achit connects with Lakes Khoton and Khurgan (see Chap. 14) in the Mongolian Altai Mountain Range and Lake Khar Us in the Depression of Great Lakes through the outlet channel Usnii (or Nuuriin) Khooloi and Khovd River.

Lake Achit has fluctuated its area during the last 46 years (Table 12.6). However, these changes were small. The lake lost an area by 14.2 km^2 between 1969 and 1971, gained by 4.7 km^2 between 1971 and 2010, and lost by 10.1 km^2 between 2010 and 2015. Overall, the lake lost an area of 19.6 km^2 between 1969 and 2015 and of 14.13 km^2 between 1970 and 2015 (Orkhonselenge and Uuganzaya, unpublished; Table 12.6, Fig. 12.7d) except for a minor gain. In total, Lake Achit has lost an area by ~20 km^2 during the last half century, and the reduction of lake area has coincided with the gradually rising air temperature and fluctuating precipitation in the lake basin since 1985. The linear regression analysis shows that the lake area between 2010 and 2015 is strongly related to both the annual average precipitation ($R^2 = 0.9992$) and the annual average air temperature ($R^2 = 0.9514$). The correlation analysis shows that the lake area during 2010–2015 had a strong positive correlation with precipitation ($r = 0.99$) and a strong negative correlation with temperature ($r = -0.97$).

12.4.5 Lake Uureg

Lake Uureg, a tectonic saline lake, is located at 1425 m a.s.l. in a depression between Mt. Tsagaan Shuvuut (3496 m a.s.l.) and Mt. Turgen (3979 m a.s.l.) at the northernmost corner of the Depression of Great Lakes (Figs. 1.1 and 12.1). The lake extends for 19.5 km in length, 18.0 km in width, and 65.6 km along the shoreline, and it reaches the maximum depth of 42.0 m at the center of the lake (Tserensodnom, 1971). Lake Uureg (Figs. 12.5e and 12.7e) is fed by rivers and streams draining from an area of 3360 km^2 including Tsagaan and Kharig Rivers,

Table 12.6 Temporal changes in area of Lake Achit

#	Year	Elevation (m a.s.l.)	Area (km^2)	Authors
1	1969	1464	311.0	Tsegmid (1969)
2	1970	–	305.54	Orkhonselenge and Uuganzaya, unpublished
3	1971	1435	296.8	Tserensodnom (1971)
4	2010	–	301.5	Davaa (2015)
5	2013	–	298.3	Davaa (2015)
6	2013	1444 ± 5	–	Sun et al. (2013)
7	2015	–	291.41	Orkhonselenge and Uuganzaya, unpublished

and the last one draining from the Siilkhem Mountain Range in the northwest is the largest one among the inflows.

The water temperature of Lake Uureg is 14–15 °C at the beach and 12.1–13 °C at the center in July (Tserensodnom, 1971). The mineralization value is 2.88–5.1 g/L in 1926–1959 (Tserensodnom, 1971) or 3.95 g/l in 2008 dominated with SO_4^{2-}, HCO_3^-, Cl^-, and Na^+ (Ariunbileg et al., 2020). Lake Uureg covers an area of 237.6 km^2 (Tserensodnom, 1971), but the lake lost its area for 1.48 km^2 from 250.6 km^2 in 1970 to 249.12 km^2 in 2015 (Orkhonselenge and Uuganzaya, unpublished; Table 12.5; Fig. 12.7e).

12.5 Summary

Western Mongolia is an important region in Central Asia because of the presence of the Mongolian Altai Mountain Range, preserving paleoglacier remnants and modern glaciers, feeding large rivers and lakes in the North Arctic Ocean and Central Asian internal drainage basins, and large remnant and modern lakes in the Depression of Great Lakes (e.g., the three largest lakes of Mongolia cover a total area of 5746.29 km^2). In western Mongolia, lakes cover the largest areas of the total lake area in the country (59.7%), and tectonic, glacial, aeolian, and thermokarst lakes are mainly distributed. Lakes in the region have diverse origins, but most of them have tectonic origins due to the presence of mountains and the large depression linked to faults extending from the northwest to the southeast. Large lakes in the Depression of Great Lakes have dramatically lost their areas during the last 45 years, and they are highly mineralized, while lakes in the Mongolian Altai have stabilized or increased their areas in the period. The region archives long-term records of paleoglaciers, paleolakes, paleoclimates, and paleoenvironments related to the internal processes that occurred in the western tip of the Central Asian Orogenic Belt.

References

Academy of Sciences of Mongolia and Academy of Sciences of USSR. (1990). *National Atlas of the Mongolian People's Republic*. [In Russian].

Agatova, A. R., Nepop, R. K., & Glebova, A. N. (2020). Lake systems in mountain borders of the Great Lakes Basin of Mongolia: From Pleistocene catastrophes to Holocene gradual evolution (case study from southwestern Tuva). *Quaternary International, 538*, 53–62.

Ariunbileg, S., Isupov, V. P., Vladimirov, A. G., Kolpakova, M. N., & Kuibida, L. V. (2020). *Chemical evolution and geochemistry of mineralized lakes of Mongolia*. Admon Printing House, 132 p. [In Mongolian].

Baast, P. (1998). *Modern glaciers and hydrology of Mongolia*. A dissertation for Philosophical doctor of Geography. Institute of Geoecology, Mongolian Academy of Sciences, Ulaanbaatar. Mongolia, 162 p. [In Russian with English abstract].

Badarch, G., Cunningham, W. D., & Windley, B. F. (2002). A new terrane subdivision for Mongolia: Implications for the Phanerozoic crustal growth of Central Asia. *Journal of Asian Earth Sciences, 21*, 87–110.

Batima, P., Natsagdorj, L., Gomboluudev, P., & Erdenetsetseg, B. (2005). Observed climate change in Mongolia. *AIACC Working Paper No. 12*, 26 p.

Batnasan, N. (1998). *Hydrological system, water regime, and evolution of large lakes in the Govi region.* Doctorate Thesis in Geography: Hydrology, Institute of Geoecology, Mongolian Academy of Sciences, Ulaanbaatar, 163 p. [In Mongolian].

Blomdin, R., Heyman, J., Stroeven, A. P., Hättestrand, C., Harbor, J. M., Gribenski, N., Jansson, K. N., Petrakov, D. A., Ivanov, Orkhonselenge, A., Rudoy, A. N., & Walther, M. (2016). Glacial geomorphology of the Altai and western Sayan mountains. *Central Asia. Journal of Maps, 12*(1), 123–136.

Blomdin, R., Stroeven, A. P., Harbor, J. M., Gribenski, N., Caffee, M. W., Heyman, J., Rogozhina, I., Ivanov, M. N., Petrakov, D. A., Walther, M., Rudoy, A. N., Zhang, W., Orkhonselenge, A., Hättestrand, C., Lifton, N. A., & Jansson, K. N. (2018). Timing and dynamics of glaciation in the Ikh Turgen mountains, Altai Region, High Asia. *Quaternary Geochronology, 47*, 54–71.

Davaa, G. (2010). Climate change impacts on water resources in Mongolia. In *Proceedings of Consultative Meeting on Integration of Climate Change Adaptation into Sustainable Development in Mongolia.* Institute for Global Environmental Strategies (IGES), pp. 30–36.

Davaa, G. (2015). *Resource and regime of surface water of Mongolia.* 408 p. [In Mongolian].

Gaskova, O. L., Isupov, V. P., Vladimirov, A. G., Shvartsev, S. L., & Kolpakova, M. N. (2015). Thermodynamic Modeling of the Behavior of Uranium and Arsenic in Mineralized Shaazgai_ Nuur Lake (Northwest Mongolia). *Doklady Earth Sciences, 465*(1), 1159–1163.

Grunert, J., Stolz, C., Hempelmann, N., Hilgers, A., Hulle, D., Lehmkul, F., Felauer, T., & Dasch, D. (2009). The evolution of small lake basin in the Gobi desert in Mongolia. *Quaternary Sciences, 7410*, 04-678-09.

Harrison, S. P., Yu, G., & Tarasov, P. E. (1996). Late Quaternary lake-level record from Northern Eurasia. *Quaternary Research, 45*, 138–159.

Herren, P., Eichler, A., Machguth, H., Papina, T., Tobler, L., Zapf, A., & Schwikowski, M. (2013). The onset of Neoglaciation 6000 years ago in western Mongolia revealed by an ice core from the Tsambagarav mountain range. *Quaternary Science Reviews, 69*, 59–68.

Inceoglu, F., Knudsen, M. F., Olsen, J., Karoff, C., Herren, P. A., Schwikowski, M., Aldahan, A., & Possnert, G. (2016). A continuous ice-core ^{10}Be record from Mongolian mid-latitudes: Influences of solar variability and local climate. *Earth and Planetary Science Letters, 437*, 47–56.

Isupov, V. P., Ariunbileg, S., Razvorotneva, L. I., Lyakhov, N. Z., Shvartsev, S. L., Vladimirov, A. G., Kolpakova, M. N., Shatskaya, S. S., Chupakhina, L. E., Moroz, E. N., & Kuibida, L. V. (2013). Geochemical Model of Uranium Accumulation in Shaazgai-Nuur Lake (Northwestern Mongolia). *Doklady Earth Sciences, 448*(1), 143–148.

Isupov, V. P., Vladimirov, A. G., Lyakhov, N. Z., Shvartsev, S. L., Ariunbileg, S., Kolpakova, M. N., Shatskaya, S. S., Chupakhina, L. E., Kuibida, L. V., & Moroz, E. N. (2011). Uranium in Saline Lakes of Northwestern Mongolia. *Doklady Earth Sciences, 437*(1), 352–356.

Isupov, V., Vladimirov, A., Sodov, A., Kolpakova, M., Shvartsev, S., & Volkova, N. (2015). Hydromineral Resources of Saline Lakes of Mongolia and Russian Altai. *Advanced Materials Research, 1085*, 166–170.

Jambaljav, Y., Gansukh, Y., Temuujin, K., Tsogt-Erdene, G., Undrakhtsetseg, T., Saruulzaya, A., Amarbayasgalan, Y., Dashtseren, A., & Narangerel, S. (2016). *Report of New Map of Permafrost Distribution in Mongolia: Scale 1:1000000.* Institute of Geography and Geoecology, Mongolian Academy of Sciences, 90 p.

Jigj, S. (1975). *Primary feature of Mongolian landforms.* Institute of Geography & Permafrost. Mongolian Academy of Sciences (MAS), 126 p. [In Mongolian].

Jigj, S. (1976). *A brief description of paleoglaciations and paleoglaciers.* 51 p. [In Mongolian].

Kang, S., Lee, G., Togtokh, C., & Jang, K. (2015). Characterizing regional precipitation-driven lake area change in Mongolia. *Journal of Arid Land, 7*(2), 146–158.

Kolpakova, M. N., & Gaskova, O. L. (2018). Major ions behaviour during evaporation of different saline type water of Western Mongolian lakes (geochemical modelling). *Hydrology Research, 49*(1), 163–176.

Kröner, A., & Rojas-Agramonte, Y. (2014). The Altaids as seen by Eduard Suess, and Present Thinking on the Late Mesoproterozoic to Palaeozoic Evolution of Central Asia. *Austrian Journal of Earth Sciences, 107*(1), 156–168.

Lehmkuhl, F. (2014). Aeolian dust in mountain areas of Tibet and Mongolia. *ProScience, 1*, 44–50.

Lehmkuhl, F., Grunert, J., Hülle, D., Batkhishig, O., & Stauch, G. (2018). Paleolakes in the Gobi region of southern Mongolia. *Quaternary Science Reviews, 179*, 1–23.

Lehmkuhl, F., Klinge, M., Rother, H., & Daniela Hulle, D. (2016). Distribution and timing of Holocene and Pleistocene glacier fluctuations in western Mongolia. *Annals of Glaciology, 57*(71), 1–10.

Minayeva, T., Gunin, P., Sirin, A., Dugardzhav, C., & Bazha, S. (2004). Peatlands in Mongolia: The typical and disappearing landscape. *Peatlands International, 2*, 44–47.

Minayeva, T., Sirin, A., Dorofeyuk, N., Smagin, V., Bayasgalan, D., Gunin, P., Dugardjav, C., Bazha, S., Tsedendash, G., & Zoyo, D. (2005). Mongolia mires: From taiga to desert. In G. M. Steiner (Ed.), *Moore von Sibirien bis Feuerland* (Vol. 85, pp. 335–352). Stapfia.

Ministry of Environment and Green Development. (2013). *Integrated water management plan of Mongolia*. 339 p.

Murzaev, E. M. (1952). *The description of physical geography of Mongolia* (2nd ed.). [In Russian].

Orkhonselenge, A., & Harbor, J. M. (2018). Impacts of Modern Glacier Changes on Surface Water Resources in Western and Northern Mongolia. *Journal of Water Resource and Protection, 10*(6), 559–576.

Rudaya, N., Tarasov, P., Dorofeyuk, N., Solovieva, N., Kalugin, I., Andreev, A., Daryin, A., Diekmann, B., Riedel, F., Tserendash, N., & Wagner, M. (2009). Holocene environments and climate in the Mongolian Altai reconstructed from the Hoton-Nur pollen and diatom records: A step towards better understanding climate dynamics in Central Asia. *Quaternary Science Reviews, 28*, 540–554.

Sarantuya, C., & Natsagdorj, L. (2010). Climate change in Mongolia. In *Proceedings of Consultative Meeting on Integration of Climate Change Adaptation into Sustainable Development in Mongolia. Institute for Global Environmental Strategies (IGES)*, pp. 23–28.

Shinneman, A. L. C., Umbanhowar, C. E., Edlund, M. B., & Soninkhishig, N. (2010). Late-Holocene moisture balance inferred from diatom and lake sediment records in western Mongolia. *The Holocene*, 123–138.

Soejono, I., Buriánek, D., Svojtka, M., Žáček, V., Čáp, P., & Janoušek, V. (2016). Mid-Ordovician and Late Devonian magmatism in the Togtokhinshil Complex: New insight into the formation and accretionary evolution of the Lake Zone (western Mongolia). *Journal of Geosciences, 61*, 5–23.

Stolz, C., Hülle, D., Hilgers, A., Grunert, J., Lehmkuhl, F., & Dasch, D. (2012). Reconstructing fluvial, lacustrine and aeolian process dynamics in Western Mongolia. *Zeitschrift für Geomorphologie, 56*(3), 267–300.

Sun, A., Feng, Z., Ran, M., & Zhang, C. (2013). Pollen-recorded bioclimatic variations of the last ~22,600 years retrieved from Achit Nuur core in the western Mongolian Plateau. *Quaternary International, 311*, 36–43.

Tarasov, P., Dorofeyuk, N., & Metel'tseva, E. (2000). Holocene vegetation and climate changes in Hoton-Nur basin, northwest Mongolia. *Boreas, 29*(2), 117–126.

Tsegmid, S. (1969). Physical geography of Mongolia. 759 p. [In Mongolian].

Tserensodnom, J. (1971). *Lakes of Mongolia*. State Publishing, 202 p. [In Mongolian].

Tserensodnom, J. (2000). *A catalog of lakes in Mongolia*. Shuvuun Saaral Publishing, 141 p. [In Mongolian].

Tumurbaatar, D. (2004). *Seasonal and Permanent Permafrost of Mongolia*. Mongolian Academy of Sciences, 147 p. [In Mongolian].

UNESCO. (2020). *World Heritage List*. Retrieved from https://whc.unesco.org/en/list/&order=country#alphaM.

Walther, M., Dashtseren, A., Kamp, U., Temuujin, K., Meixner, F., Pan, C. G., & Gansukh, Y. (2017). Glaciers, permafrost and lake levels at the Tsengel Khairkhan Massif, Mongolian Altai, during the late Pleistocene and Holocene. *Geosciences, 7*(73), 1–20.

Zorigt, M., Myagmar, K., Orkhonselenge, A., van Beek, E., Kwadijk, J., Tsogtbayar, J., Yamkhin, J., & Dechinlkhundev, D. (2020). Modelling permafrost distribution over the river basins of Mongolia using remote sensing and analytical approaches. *Environmental Earth Sciences, 79*(308), 1–11.

Chapter 13
Lake Uvs

Abstract Lake Uvs is the largest tectonic saline lake situated at the northern corner of the Depression of Great Lakes in western Mongolia. Lake Uvs has experienced large fluctuations of its area during 1951–2015 depending on the annual average air temperature and precipitation during the last 80 years. The lake gained an area by 240.6 km^2 between 1951 and 2010 and lost its area by 145.3 km^2 between 2010 and 2015. The linear regression analysis shows that the lake area between 1951 and 2015 was weakly related to the annual temperature ($R^2 = 0.3449$), but it had no clear relationship with the annual precipitation ($R^2 = 0.0243$). The correlation analysis shows that the lake area during 1951–2015 had a moderately positive correlation with air temperature ($r = 0.58$) and a weak negative correlation with precipitation ($r = -0.15$). Lake Uvs would provide information on the regional paleoclimatic and paleoenvironmental conditions linked to histories of glacial, fluvial, aeolian, lacustrine, and Govi environments in the Depression of Great Lakes and the Mongolian Altai Mountain Range, which are major physiographic components in western Mongolia, Eurasia, and Central Asia.

Keywords Lake Uvs · Tes River · Depression of Great Lakes · Govi region · Climate change · Western Mongolia

13.1 Introduction

Lake Uvs occupies a very large area at the northern corner of the Depression of Great Lakes between the Mongolian Altai and Khangai Mountain Ranges in western Mongolia (Figs. 1.1 and 12.1). The lake preserves a historical archive regarding glacial, fluvial, aeolian, and lacustrine environments in the Govi[1] landscape of the Depression of Great Lakes.

[1] *Govi* has been often spelled as *Gobi* in the international literature. Govi is the correct English transliteration from Mongolian *Говь*.

A. Orkhonselenge et al., *Lakes of Mongolia*, Syntheses in Limnogeology,
https://doi.org/10.1007/978-3-030-99120-3_13

Lake Uvs, the largest lake (over 3500 km^2) in Mongolia, is one of the lakes which attracted considerable attention by scientists. For instance, Lake Uvs was first noted by the Russian tourist V. Tumenets in 1916 (Jigj, 1975), and it was initially described by Bespalov (1951), Murzaev (1952), Tsegmid (1969), Tserensodnom (1971), and Sevastyanov (1994) providing morphometric components (e.g., water regime, area, volume, mineralization, etc.). The original formation of Lake Uvs basin is estimated to be in the early Mesozoic (Tserensodnom, 2000). The Lake Uvs basin likely occupied an area of 16,000 km^2, extending 220 km in length, 110 km in width, and 300 m in depth during the early and middle Pleistocene, and the lake water level probably reached 1180 m a.s.l. during the late Pleistocene (Murzaev, 1952). Moreover, the presence of paleolake deposits and terraces isolated 20–30 km away from the present shoreline implies much higher lake levels than the present and a hydrological connection of the lake with Lakes Khyargas and Khar Us in the past (Tsegmid, 1969).

The first meteorological and hydrological stations installed in the Lake Uvs basin in 1942 and 1962 have been measuring the local climate of the lake basin and the water level and thermal regime of the lake to date. Recent studies showed that annual and seasonal fluctuations of the lake level are not high (e.g., Tsegmid, 1969; Tserensodnom, 1971). Since 1995 the Mongolian and German joint expedition conducted lacustrine geomorphological and paleolimnological investigations of the Lake Uvs basin. The lake's hydrological system, spatial and temporal changes in water regime, characteristics of bottom deposits, and mineralization value of the lake water were investigated by Batnasan (1998). Lake Uvs changed in its water level during the Holocene with a major drop of about 45.0 m from 10.0 to 5.0 ka with several small, superimposed transgressions, and it has risen since 2.8 ka (Grunert et al., 2000).

Lake Uvs is a saline lake, and it is included in the World Heritage List by UNESCO in 2003. The lake basin has been an International Geosphere-Biosphere Program (IGBP) site since 1997 for monitoring global warming on the basis of natural criteria of (1) international scientific importance on the unchanging nature of the nomadic pastoral use of its grasslands because of its climatic and hydrological regimes and (2) the major biomes of eastern Eurasia with a number of endemic plants, extremely important habitats for a wide range of wild animals, many of them threatened or endangered, and the mountains at the western end of the basin, which are important refuges for the globally threatened snow leopard, mountain sheep (argali), and the Asiatic ibex (UNESCO, 2020). Lake Uvs itself is an important habitat for waterfowl as well as for birds migrating from Siberia toward south.

In this chapter, research results from Lake Uvs are synthesized, and changes of its area are discussed in detail. Paleoclimatic implications from Lake Uvs basin are considered in Chap. 21.

13.2 Physiographic Condition

Lake Uvs is located at the elevation of 759 m a.s.l. (Murzaev, 1952) or 743 m a.s.l. (Tserensodnom, 1971) (see Sect. 13.6) at the northern end of the Depression of Great Lakes (Figs. 1.1 and 12.1). In the Lake Uvs basin, a Govi landscape dominates although the lake lies at the higher northern latitudes within Mongolia. The Govi region extends toward the north through the Depression of Great Lakes between the Mongolian Altai and Khangai Mountain Ranges (see Chaps. 9 and 12).

Lake Uvs has been listed as the 42nd largest lake in the world. The lake (Fig. 13.1) extends 84.0 km in the maximum length and 79.0 km in the maximum width (Tsegmid, 1969) as shown in Table 1.3. The lake's water volume is 39.6 km^3 (Tserensodnom, 1971) or 35.7 km^3 (Davaa, 2015). The water volume increased by 1.5 km^3 in 1966 and 1.1 km^3 in 1986 due to the flooding in those years (Davaa, 2015). Lake Uvs (Fig. 13.1) covers an area of 3423 km^2 (Bespalov, 1951), 3518.3 km^2 (Davaa, 2015), or 3451.62 km^2 (Orkhonselenge and Uuganzaya, unpublished; see Sect. 13.6).

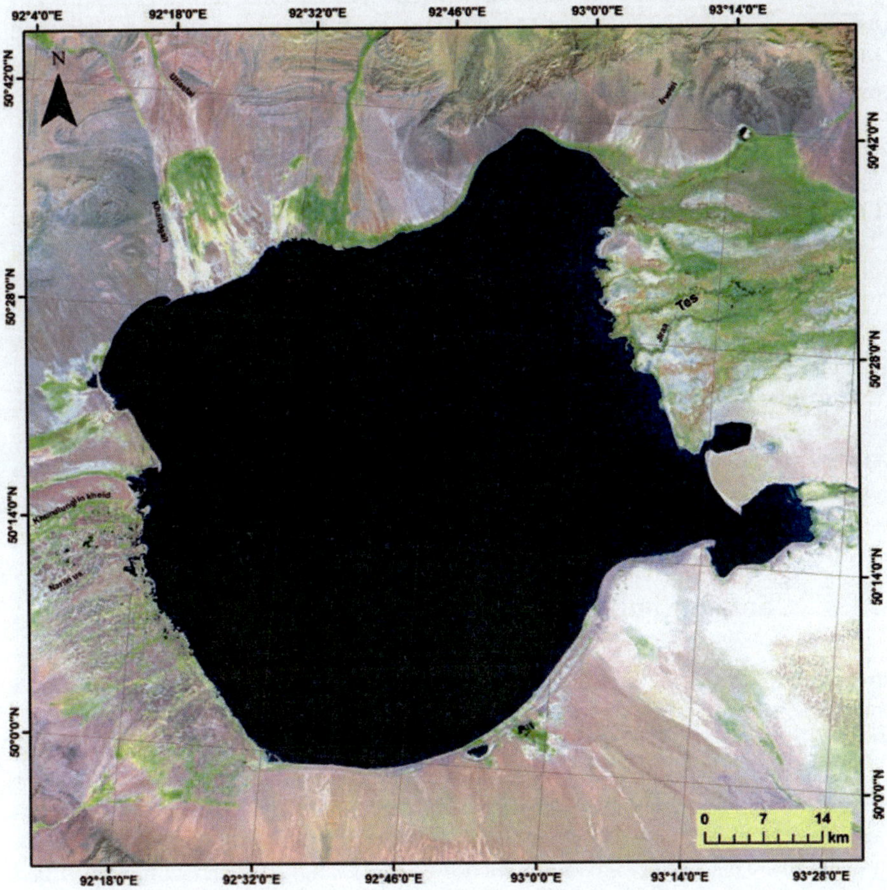

Fig. 13.1 Landsat 8 image of Lake Uvs on June 6, 2015 (path of 133 and row of 29). Modified from Orkhonselenge and Uuganzaya (unpublished)

13.3 Geological and Geomorphological Settings

Lake Uvs is situated over the Neogene and Quaternary marine molassic deposits in the Cenozoic continental platform structure (Academy of Sciences of Mongolian and Academy of Sciences of USSR, 1990) consisting of red and gray terrigenic sands, gravels, and pebbles. The lake bottom is filled by the late Quaternary and modern-day deposits (Academy of Sciences of Mongolian and Academy of Sciences of USSR, 1990) and the underlying Mesozoic and Cenozoic sediments (Kröner & Rojas-Agramonte, 2014). Lake Uvs lies tectonostratigraphically on the Cenozoic alluvial basin (Badarch et al., 2002).

Lake Uvs, classified as a tectonic lake (see Chap. 5), is surrounded by Mt. Kharkhiraa (4037 m a.s.l.) and Mt. Turgen (3979 m a.s.l.) with modern glaciers in the southwest, Mt. Tsagaan Shuvuut (3496 m a.s.l.) in the west, Tagna Mountain Range (3081 m a.s.l.) and Mt. Davst (1810 m a.s.l.) in the north, Bulnai (2645 m a.s.l.) and Khan Khukhii (2928 m a.s.l.) Mountain Ranges and Buurug Del and Altan sand dunefields in the east and southeast, and Mt. Togtokhiin Shil (2356 m a.s.l.) in the south (Fig. 12.1).

Lake Uvs is almost circle shaped (Fig. 13.1) and surrounded by the Govi landscape. However, the lake floor is trench shaped (Batnasan, 1999). There is a 5.0–10.0-km-long peninsula consisting of coarse sands, gravels, and boulders (Tsegmid, 1969) at the mouths or deltas of Kharkhiraa, Turgen, Khundlun, and Torkhilog Rivers in the west and north (Fig. 13.1). The peninsula is often inundated during the heavy rainfall, and it is frequently modified during the transgressions (Tserensodnom, 1971).

Rocky plains, gravelly channels, paleolake deposits, and terraces are widely distributed around the lake beaches. Swamps, marshes, and streams are formed in the downstreams of the inflow rivers (Fig. 13.1) draining the glaciers on the Kharkhiraa, Turgen, and Tagna Mountain Ranges (Tsegmid, 1969). Dense reeds and broom grasses are abundant in the mouths of Kharkhiraa, Khundlun, Sagil, Tes, and Nariin Rivers in the west and northeast.

13.4 Climate Condition

In the Lake Uvs basin, an arid period in the 1970s lasted to the early 1980s, and a humid period continued in the late 1980s to 1997 (Batnasan, 1999).

The local climate data recorded at the Ulaangom meteorological station located 27.2 km southwest of the lake (Fig. 13.1) show variations in the annual average air temperature and precipitation during the last 80 years (Fig. 13.2). The annual average air temperature is −1.53 °C, and it is rapidly rising since the 1990s (Figs. 13.2a and 13.3b). This rise was also recorded by Kang et al. (2015) showing that it had been the highest rate of temperature rising observed at the Ulaangom meteorological station. The average temperature is −32.3 °C in January and 19.5 °C in July.

The annual precipitation incoming into the lake is 140 mm (Tserensodnom et al., 1986) or 100–150 mm (Batnasan, 1999). In the lake basin, the maximum rainfall occurs in June and July; however the maximum runoff occurs in May because of the flooding caused by snow melting (Batnasan, 1998). The annual average precipitation is 138.9 mm, and it shows a steady rate of fluctuations (Fig. 13.2b). The monthly average precipitation has been 24.4 mm in June, 31.6 mm in July, and 29.8 mm in August during the last eight decades.

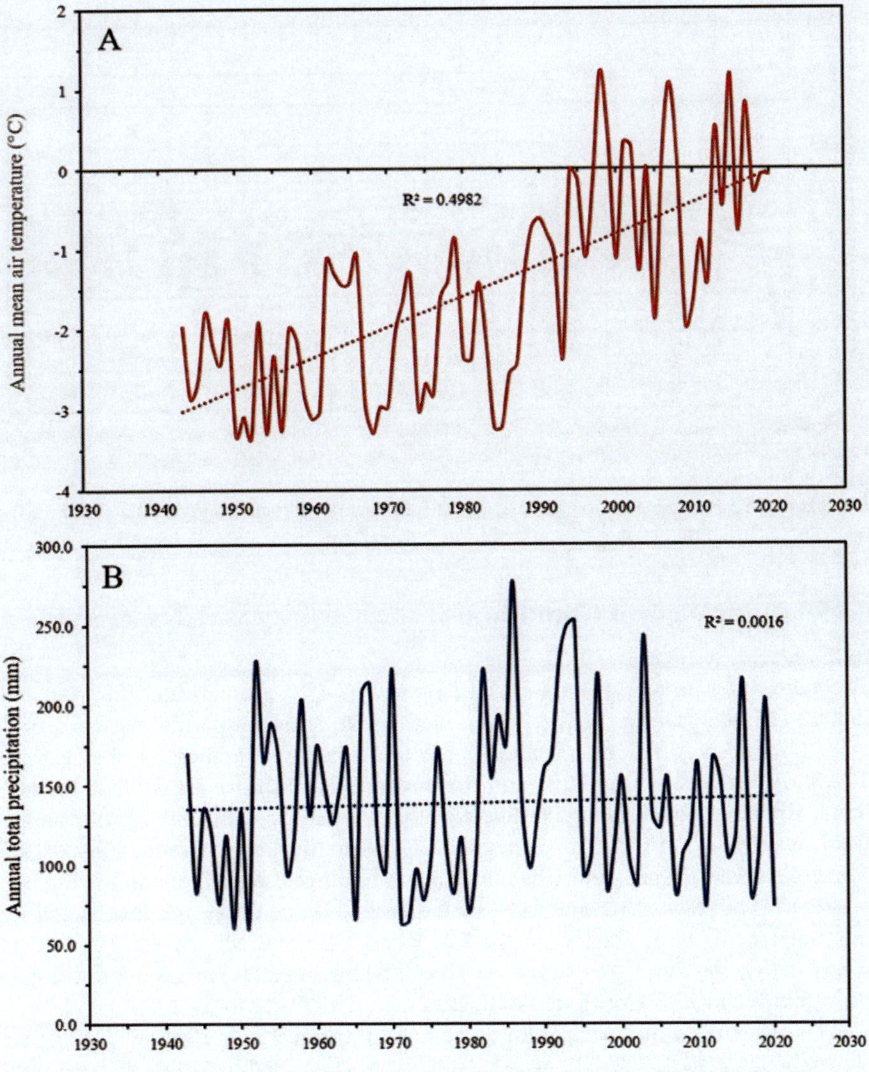

Fig. 13.2 Annual average (**a**) air temperature and (**b**) precipitation recorded at the Ulaangom meteorological station during 1943–2020

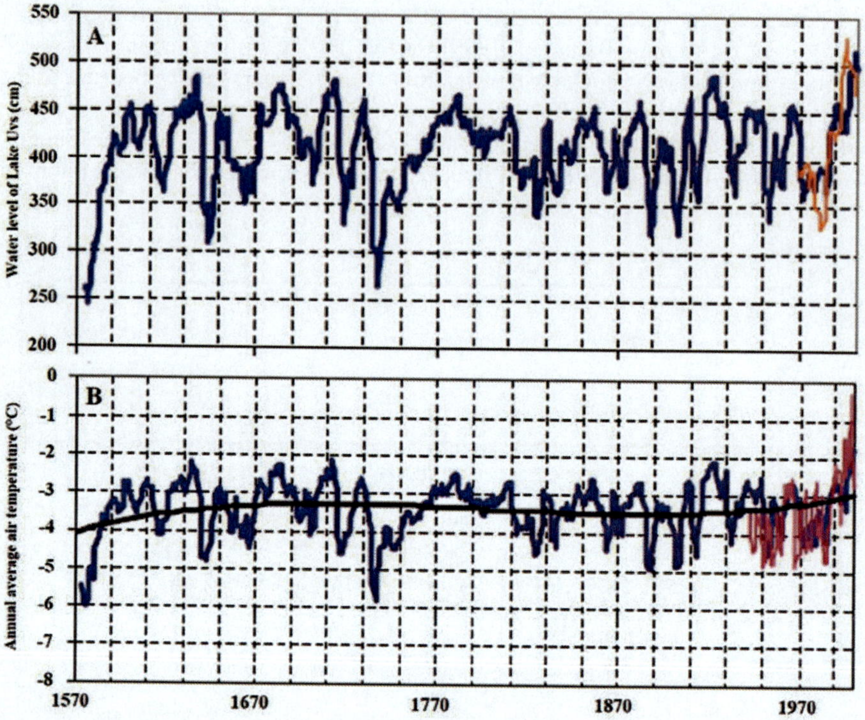

Fig. 13.3 (**a**) Water level of Lake Uvs and (**b**) annual average air temperature recorded at the Ulaangom meteorological station since 1570 (bluish black, dendrochronological data; red for (**a**) and pink for (**b**), observed data). Modified from Davaa (2003)

13.5 Hydrological Condition

The main water sources of Lake Uvs are runoff (57.4%), groundwater (28.6%), and rainfall (14.0%) (Davaa, 2015). In this lake basin, seasonal permafrost is distributed. Lake Uvs is fed by over ten large and small rivers such as Tes, Nariin, Torkhilog, and Guramsan Rivers in the east and Kharkhiraa, Khundlun, Turgen, Sagil, Borshoo, and Khandgait Rivers in the west (Figs. 13.1 and 13.4) draining from an area of 71,100 km^2 (Tsegmid, 1969) in the branch mountains of the Mongolian Altai, Tagna, and Khangai Mountain Ranges. Among the inflowing rivers, Tes River (Figs. 13.1 and 13.4) is the largest. In addition, the lake is fed by meltwater of modern glaciers in the Mt. Kharkhiraa and Mt. Turgen (Fig. 12.1) where the glacier areas are estimated to be 62.0 km^2 and 52.0 km^2 (Batnasan, 1998) or 39.5 km^2 and 33.8 km^2 (Lehmkuhl, 2012).

The measured water temperatures of Lake Uvs for 1964, 1965, and 1971 are shown in Table 13.1. The lake water temperature decreased by −0.07 °C/year until

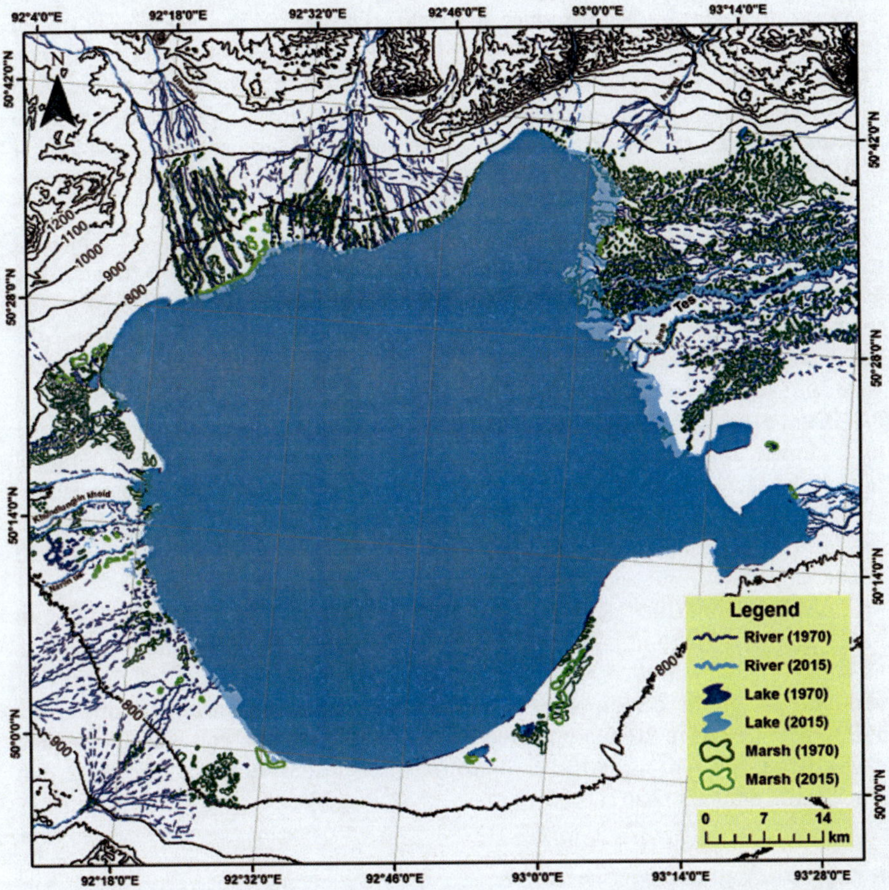

Fig. 13.4 Spatial distributions of areas and marshes of Lake Uvs in 1970 and 2015. Modified from Orkhonselenge and Uuganzaya (unpublished)

Table 13.1 Water temperature of Lake Uvs

	Temperature range (°C)[1]		Temperature range (°C)[2]
Month	1964	1965	1971
May	7.4	10.4	5.8–10.4
June	16.0	–	15.1–16.5
July	–	–	25.0–27.3
August	23.6	25.0	15.9–23.6
September	–	–	10.7–16.8
October	–	–	5.8–9.5
November	–	–	2.3–3.1

Source: [1] Tsegmid (1969), [2] Tserensodnom (1971)

1989 and rose to 0.27 °C/year since the 1990s (Davaa, 2015). Lake Uvs is frozen from middle November to middle May, and the mean ice thickness is ~1.0 m (Tsegmid, 1969) or 1.2–1.5 m (Tserensodnom, 2000). The annual evaporation from the lake surface is about 800–1000 mm (Tsegmid, 1969), 850–900 mm (Tserensodnom, 1971), 913.0 mm (Batnasan & Tserensodnom, 1987), or 900–1000 mm (Batnasan, 1999). The transparency of the lake is 6 m in the northwest and 0.3 m in the east (Horn et al., 1999).

The water depth of Lake Uvs is 13.5 m on average, and it reaches ~23.0 m at its maximum (Horn et al., 1999). The maximum depth of 21.1 m was also found in the west of the lake (Batnasan, 1999). The water level rises from April to May and reaches its maximum during July to August because the annual fluctuation of the lake level strongly depends on precipitation, evaporation, and discharges of groundwater and surface water (Tsegmid, 1969). The annual fluctuation of the lake level is 0.7–0.8 m (Tsegmid, 1969) or 0.5–0.9 m (Tserensodnom, 1971). Depending on the local climate, high waves up to 2 m form on the surface of the lake. The water level of the lake increased by 1.0 m from the 1960s to the mid-1970s, in the late 1980s, and also during 1994–1996 (Fig. 13.3a) and dropped in the early 1980s (Batnasan, 1998, 1999). The lake water level fluctuates with the average rate of 23.7 cm/year (Batnasan, 1998).

Lake Uvs is a saline lake. The lake is saturated with salts (Table 13.2) due to halite deposits (Tsegmid, 1969) and salt deposits in the lake basin, and the lake has no outlet (Tserensodnom, 1971). The mineralization of the lake is 9.2 g/L in the east and 12.6 g/L in the center, and Na^+ and Cl^- are the dominant ions (Horn et al., 1999), or 17.3 g/L in 2010 dominated by Cl^-, SO_4^{2-}, HCO_3^-, and Na^+ (Ariunbileg et al., 2020). The hydrochemical composition of the lake is shown in Table 13.2, and NaCl is the dominant salt.

Table 13.2 Hydrochemical composition of Lake Uvs

Salts (g/L)	NaCl	Na_2SO_4	$MgSO_4$	Na_2CO_3	$MgCO_3$	Authors	Year
11.94	5.75	4.16	0.43	–	–	G.N. Potanin	1879
18.53	10.44	4.16	2.59	–	–	V.A. Smirnov	1929
15.85–17.57	+	+	+	–	–	S. Lamozav	1948
19.7	+	+	+	2.08	–	Bespalov	1951
18.41–18.78	+	–	+	–	+	Davaasuren	1961
–	+	+	–	–	–	Tserensodnom	1971
18.7	–	–	–	–	–	Batnasan	1998
12.6	+	+	+	+	+	Horn et al.	1999
12.0	–	–	–	–	–	Tserensodnom	2000
17.3	10.78	9.53	4.78	5.32	0.57	Ariunbileg et al.	2020

Source data between 1879 and 1948 were found in Tserensodnom (1971)

13.6 Changes in Area

Table 13.3 shows changes in area of Lake Uvs during the last 65 years. The lake gained an area of 240.6 km^2 from 1951 to 2010 and lost an area of ~145.3 km^2 from 2010 to 2015 (Table 13.3). Lake Uvs increased in area by ~97.2 km^2 between 1970 and 2015 when the gain was significantly observed on the northeastern shore of the lake (Orkhonselenge and Uuganzaya, unpublished; Fig. 13.4). The increase in area of the lake until 2010 coincided with the rise in the annual precipitation during 1975–1977 and 1982–1986 (Fig. 13.2b), with the continuous increase of the lake water volume for 7.2 km^3 during the period of 1945–1971 (Table 13.3) and with the increased lake water level since the 1970s (Fig. 13.3a). Lake Uvs rose by 2 m in water level during 1964–1995, but it shows a falling trend ever since (Ministry of Environment and Green Development, 2013).

The linear regression analysis (Fig. 13.5) for the period of 1951–2015 shows that the area of Lake Uvs was weakly related to the annual average air temperature ($R^2 = 0.3449$), but there was no clear relationship between the lake area and the precipitation ($R^2 = 0.0243$). The correlation analysis shows that the lake area had a moderate positive correlation with the air temperature ($r = 0.58$) and a weak negative correlation with the precipitation ($r = -0.15$). These may coincide with Fig. 13.3 showing the water level of the lake increased since the 1990s following the rise of air temperature in the same period and another result given by Kang

Table 13.3 Temporal changes in area and volume of Lake Uvs

#	Year	Elevation (m a.s.l.)	Area (km^2)	Volume (km^3)	Authors
1	1945	–	–	32.4	Davaa (2015)
2	1951	–	3423	–	Bespalov (1951)
3	1952	759	3350	–	Murzaev (1952)
4	1966	–	–	33.9	Davaa (2015)
5	1970	–	3354.47	–	Orkhonselenge and Uuganzaya, unpublished
6	1971	743	–	39.6	Tserensodnom (1971)
7	1986	759.94	3518.3	35.7	Davaa (2015)
8	1988	–	–	38.88	Davaa (2015)
9	1992	–	3599.25	–	Batnasan (1998)
10	1996	–	3641.34	–	Batnasan (1998)
11	1999	761	3650	49.3	Horn et al. (1999)
12	2010	–	3663.6	–	Davaa (2015)
13	2013	–	3646.6	–	Davaa (2015)
14	2015	–	3451.62	–	Orkhonselenge and Uuganzaya, unpublished

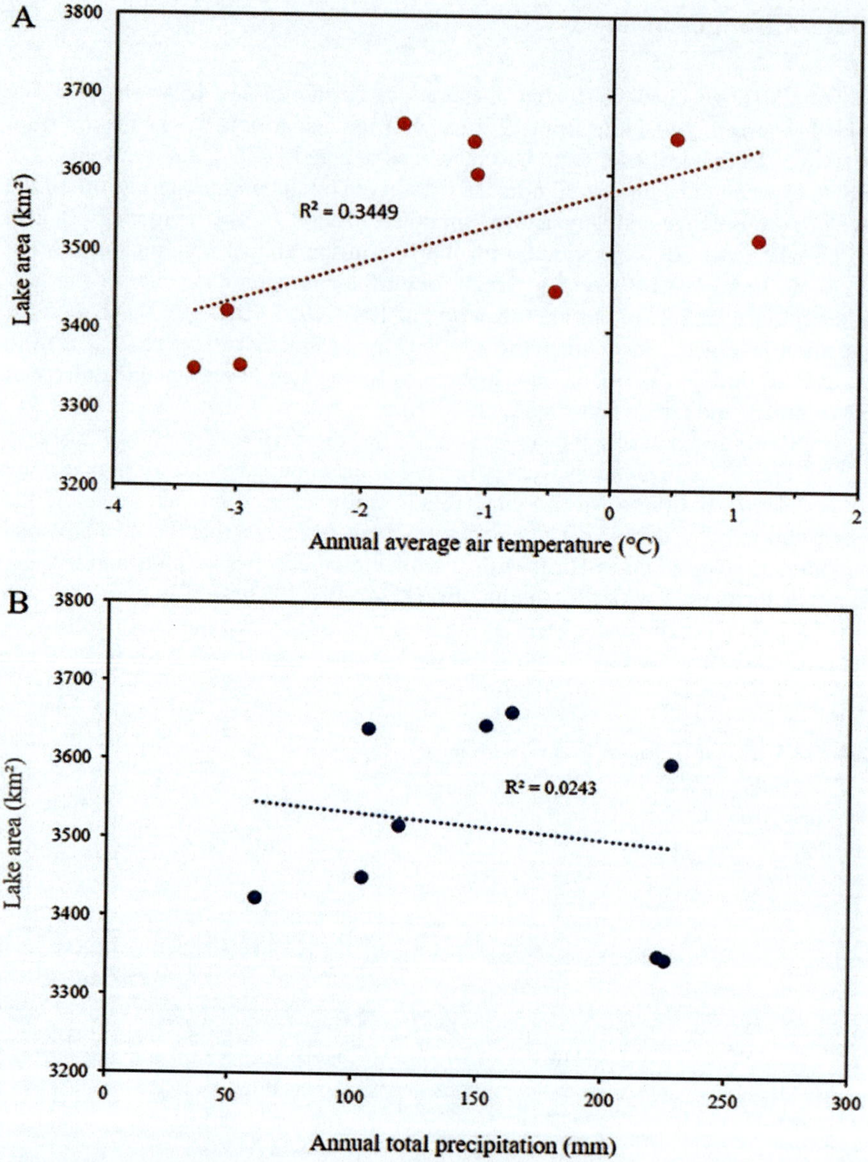

Fig. 13.5 Relationships between area of Lake Uvs and annual average (**a**) air temperature and (**b**) precipitation during 1951–2015

et al. (2015) showing a significant negative ($r = 0.99$) correlation of the lake area change with the annual precipitation in western Mongolia. Moreover, a recent rise in the lake level was observed, and this was confirmed by the flooding of sandbars and the lake's encroachment on the dunefield at the eastern side of Lake Uvs (Grunert et al., 2000).

13.7 Summary

Lake Uvs, the largest saline lake in Mongolia, is one of representative lakes in the Govi landscape of the Depression of Great Lakes in western Mongolia. The lake shows local climatological and hydrological changes in the region during the last eight decades. Lake Uvs experienced significant fluctuations of its area during the last 65 years gaining 240.6 km^2 during 1951–2010 and losing ~97.2 km^2 during 2010–2015. The linear regression analysis shows that the lake area between 1951 and 2015 was weakly related to the annual average air temperature ($R^2 = 0.3449$), but not clearly related to the annual precipitation ($R^2 = 0.0243$). Moreover, the correlation analysis shows that the lake area during 1951–2015 had a moderately positive correlation with air temperature ($r = 0.58$) and a weak negative correlation with precipitation ($r = -0.15$). Sedimentological, geochemical, and geochronological records from Lake Uvs would lead to the understanding of the lake evolution and regional paleoclimate changes in the Depression of Great Lakes in western Mongolia, Eurasia, and Central Asia.

References

Academy of Sciences of Mongolian and Academy of Sciences of USSR. (1990). *National Atlas of the Mongolian People's Republic*. 156 p. [In Russian].

Ariunbileg, S., Isupov, V. P., Vladimirov, A. G., Kolpakova, M. N., & Kuibida, L. V. (2020). *Chemical evolution and geochemistry of mineralized lakes of Mongolia*. Admon Printing House, 132 p. [In Mongolian].

Badarch, G., Cunningham, W. D., & Windley, B. F. (2002). A new terrane subdivision for Mongolia: Implications for the Phanerozoic crustal growth of Central Asia. *Journal of Asian Earth Sciences, 21*, 87–110.

Batnasan, N. (1998). *Hydrological system, water regime, and evolution of large lakes in the Govi region*. Doctorate Thesis in Geography: Hydrology, Institute of Geoecology, Mongolian Academy of Sciences, Ulaanbaatar. 163 p. [In Mongolian].

Batnasan, N. (1999). Morphology and long-term water level fluctuation of Lake Uvs. In *Proceedings of International Conference of Global Change and Uvs Nuur on Sustainable Development of the Altai-Sayan Ecoregion and Transboundary Nature Conservation Issues*, Ulaangom, Uvs aimag, pp. 94–108. [In Mongolian with English and Russian abstract].

Batnasan, N., & Tserensodnom, J. (1987). Aspects in water regime of Lake Buun Tsagaan. *Mongolian Geographical Issues, 27*, 18–24.

Bespalov, N. D. (1951). *Soils of the Mongolian People's Republic*. Issue of AN SSSR, 41. Moscow. [In Russian].

Davaa, G. (2003). *Changes in surface water regime and resources in Mongolia. Oral Presentation.*

Davaa, G. (2015). *Resource and regime of surface Water of Mongolia*. 408 p. [In Mongolian].

Davaasuren, D. (1961). *Physical and chemical characteristics of Lakes Uvs and Duruu and their brine chemical compositions*. Scientific Report of National University of Mongolia.

Grunert, J., Lehmkuhl, F., & Walther, M. (2000). Paleoclimatic evolution of the Uvs Nuur basin and adjacent areas (Western Mongolia). *Quaternary International, 65–66*, 171–192.

Horn, W., Paul, M., Davaa, G., Dulmaa, A., & Tseveendorj, N. (1999). Water quality of several lakes and its influencing factors. In *Proceedings of International Conference of Global Change and Uvs Nuur on Sustainable Development of the Altai-Sayan Ecoregion and Transboundary Nature Conservation Issues*, Ulaangom, Uvs aimag, pp. 120–130.

Jigj, S. (1975). *Primary feature of Mongolian landforms*. Institute of Geography & Permafrost. Mongolian Academy of Sciences (MAS), 126 p. [In Mongolian].

Kang, S., Lee, G., Togtock, C., & Jang, K. (2015). Characterizing regional precipitation-driven lake area change in Mongolia. *Journal of Arid Land, 7*(2), 146–158.

Kröner, A., & Rojas-Agramonte, Y. (2014). The Altaids as seen by Eduard Suess, and present thinking on the late Mesoproterozoic to Palaeozoic evolution of Central Asia. *Austrian Journal of Earth Sciences, 107*(1), 156–168.

Lehmkuhl, F. (2012). Holocene glaciers in the Mongolian Altai: An example from the Turgen-Kharkhiraa Mountains. *Journal of Asian Earth Sciences, 52*, 12–20.

Ministry of Environment and Green Development. (2013). *Integrated Water Management Plan of Mongolia*. 339 p.

Murzaev. E. M. (1952). *The description of Physical Geography of Mongolia* (2nd ed.). [In Russian].

Sevastyanov, D. V. (1994). *Limnology and paleolimnology of Mongolia*. Science, 302 p. [In Russian].

Tsegmid, S. (1969). *Physical geography of Mongolia*. 759 p. [In Mongolian].

Tserensodnom, J. (1971). *Lakes of Mongolia*. State Publishing. [In Mongolian].

Tserensodnom, J. (2000). *A catalog of lakes in Mongolia*. Shuvuun Saaral Publishing, 141 p. [In Mongolian].

Tserensodnom, J., Sugar, T., & Batnasan, N. (1986). *Results from Lake Uvs*. Institute of Geography, Mongolian Academy of Sciences, 128 p. [In Mongolian].

UNESCO. (2020). *World Heritage List*. Retrieved from https://whc.unesco.org/en/list/&order=country#alphaM.

Chapter 14
Lake Khoton

Abstract Lake Khoton, a glacial lake, constitutes a valuable archive useful for reconstructing paleoglaciations in the Mongolian Altai Mountain Range during the Pleistocene and also for understanding regional paleoclimate changes in western Mongolia and Eurasia. Lake Khoton plays an important role in surface water resources of the region, and it is a major headwater of Khovd River through Lake Khurgan. Lake Khoton records the hydrological and modern glacial dynamic changes caused by the present global warming. The lake has gained an area by 0.67 km^2 during the past 45 years. This phenomenon is associated with the recent rapid melting of glaciers on high mountains of the Mongolian Altai Mountain Range due to the gradually rising annual average air temperature and precipitation. The area of Lake Khoton between 2010 and 2015 showed no clear linear relationships with precipitation ($R^2 = 0.0839$) and air temperature ($R^2 = 0.0004$). The correlation analysis shows that the lake area during 2010–2015 had a weak positive correlation with precipitation ($r = 0.289$) and a weak negative correlation with air temperature ($r = -0.021$). Nevertheless, short-term data are still insufficient to infer these relationships further. The high mountain region around Lake Khoton in the Mongolian Altai Mountain Range has been warmed up during the last 35 years, and this results in the recent changes of surface water resources within western Mongolia and Eurasia.

Keywords Lake Khoton · Lake Khurgan · Khovd River · Mongolian Altai · Climate change · Western Mongolia

14.1 Introduction

Lake Khoton is classified as a glacial lake with a marvelous scenery in the Mongolia Altai Mountain Range at the westernmost corner of Mongolia (Fig. 1.1). It attracts attention by tourists, but not much so by scientists in comparison with other spectacular lakes in western Mongolia. Lake Khoton plays a valuable role in the surface

A. Orkhonselenge et al., *Lakes of Mongolia*, Syntheses in Limnogeology,
https://doi.org/10.1007/978-3-030-99120-3_14

water dynamics of the region because it is a headwater of Khovd[1] River feeding Lake Khar Us in the Great Lakes Depression (Fig. 12.1).

During the Pleistocene glaciations in the Mongolian Altai, which occurred twice (Tsegmid, 1969) or three to four times (Jigj, 1976), Lake Khoton may have developed and evolved. The extents of paleo- and modern glaciations in Mongolia were the greatest and broadest in the Mongolian Altai, comparing to those in the Khangai, Khuvsgul, and Khentii mountain ranges. There, the dominant style of glaciations recorded in erosional and depositional landforms was intermontane valley glaciers (e.g., Tsegmid, 1969; Jigj, 1976; Blomdin et al., 2016; Orkhonselenge, 2016; Orkhonselenge & Uuganzaya, 2018).

Today, modern glaciers are preserved only in a few mountains of the Mongolian Altai and Khangai and Khuvsgul mountain ranges (Orkhonselenge & Harbor, 2018). Nevertheless, the modern glaciers in these mountain ranges play a vital role in water resources of Central Asia and also in regional ecological, social, and economic developments (Orkhonselenge & Harbor, 2018). The modern glaciers in these mountain ranges provide over 70% of freshwater resources in Mongolia (Davaa, 2010).

To date, numerous glacial studies were conducted regarding the extents of the Pleistocene and modern glaciers (including of the Little Ice Age) in the Mongolian Altai Mountain Range (e.g., Devjatkin, 1981; Florensov & Korzhnev, 1982; Lehmkuhl et al., 2011; Blomdin et al., 2016, 2018). For instance, Blomdin et al. (2016, 2018) compiled a glacial geomorphological map containing detailed individual glacial landforms in Mt. Tsambagarav, and they also provided the Pleistocene glacier extension in Mt. Ikh Turgen (Fig. 12.1). Moreover, Krumwiede et al. (2014) showed recent glacier changes of Mt. Munkh Khairkhan and Mt. Tavan Bogd (Fig. 12.1).

The modern glaciers occupy small areas, and the glaciated areas have been decreasing rapidly (Orkhonselenge & Harbor, 2018). The intermontane valley glaciers are sensitive indicators of regional and local climates (Tronov, 1966), and over a glacial cycle they respond to variations in the position and strength of major climate systems (Blomdin et al., 2016). The current global climatic shifts are having a pronounced effect on glaciers in Mongolia (Kamp et al., 2013). Glaciers in the Mongolian Altai shrank by 22% in size during the last 50–60 years (Davaa, 2010). Over the last 40 years, the glacier areas decreased by 6% (Baast, 1998). However, there is no precise work linking fluctuations of glaciers and surface waters in Mongolia except for a work presented by Orkhonselenge and Harbor (2018) who considered the linkage between glacial and hydrological variations based on dynamic changes of the modern glaciers and glacial lakes in the Mongolian Altai and Khuvsgul Mountain Ranges.

To date, there is no comprehensive study regarding the formation and evolution histories of glacial lakes although the paleo- and modern glaciers in the Mongolian Altai have been relatively well documented. Lake Khoton is one of the glacial lakes

[1] *Khovd* has been often misspelled as *Kobdo* based on the Russian pronunciation. Khovd is the right English transliteration from Mongolian *Ховд*.

occurring in the center of the Mongolian Altai, and it has attracted only limited scientific attention. There is still poor scientific knowledge about the glaciers around this lake and the lake's formation. Nevertheless, Rudaya et al. (2009) suggested the oligotrophic/mesotrophic type of Lake Khoton prior to 11.5 ka BP. Recent study by Klinge et al. (2021) provided the development of Lake Khoton by the terminal moraine damming during or before MIS 2 and the late glacial period.

Rudaya and Li (2013) worked on pollen and diatom records obtained from Lake Khoton sediments, and they reconstructed the Holocene climate and environments of the Mongolian Altai. Moreover, Slynko et al. (2010) investigated the distribution and morphoecological forms of species *Thymallus brevirostris* in Lake Khoton with analyses of plastic and meristic characters, distribution of fish, and feeding. Aside from these results, there has been a major lack of work on geomorphology, hydrology, and paleoglaciology of Lake Khoton.

In this chapter, research results from Lake Khoton are synthesized, and changes in area of the lake are considered in detail. Paleoclimate implications recorded in the Lake Khoton basin are discussed in Chap. 21.

14.2 Physiographic Condition

Lake Khoton is located at the elevation of 2084 m a.s.l. (Tsegmid, 1969) or 2083.6 m a.s.l. (Tserensodnom, 1971) in the center of the Mongolian Altai Mountain Range in western Mongolia (Figs. 1.1 and 12.1).

Lake Khoton has a glacial origin (see Chap. 4). The lake (Fig. 14.1) extends for 28.5 km in length, 3–6 km in width, and 37.7 m in depth (Murzaev, 1952). Another source indicates that the lake extends for 21.5 km in length and 2.3 km on average width but reaches 4.0 km at the maximum width and 48.8 km along the shoreline length (Tserensodnom, 1971) as shown in Table 1.3. The bottom deposits of the lake consist of glacial sediments of muds, clays, sands, boulders, and moraines (Tserensodnom, 1971). The lake occupies an area of 60.0 km^2 (Tsegmid, 1969) or 53.79 km^2 (Orkhonselenge and Uuganzaya, unpublished; see Sect. 19.6).

14.3 Geological and Geomorphological Settings

Lake Khoton resides in the early Paleozoic geosynclinal zone (Academy of Sciences of Mongolia and Academy of Sciences of USSR, 1990). The lake basin's geological formation is a continental platform consisting of the Cenozoic red and gray terrigenic sands, gravels, and pebbles in the south, southwest, and west and the Devonian terrigenic tuff graywacke in the northwest, north, and northeast (Academy of Sciences of Mongolia and Academy of Sciences of USSR, 1990).

Lake Khoton is underlain by the middle to late Quaternary deposits in the south, southwest, and west and the middle to late Cambrian and the early Ordovician system in the northwest, north, and northeast (Academy of Sciences of Mongolia and

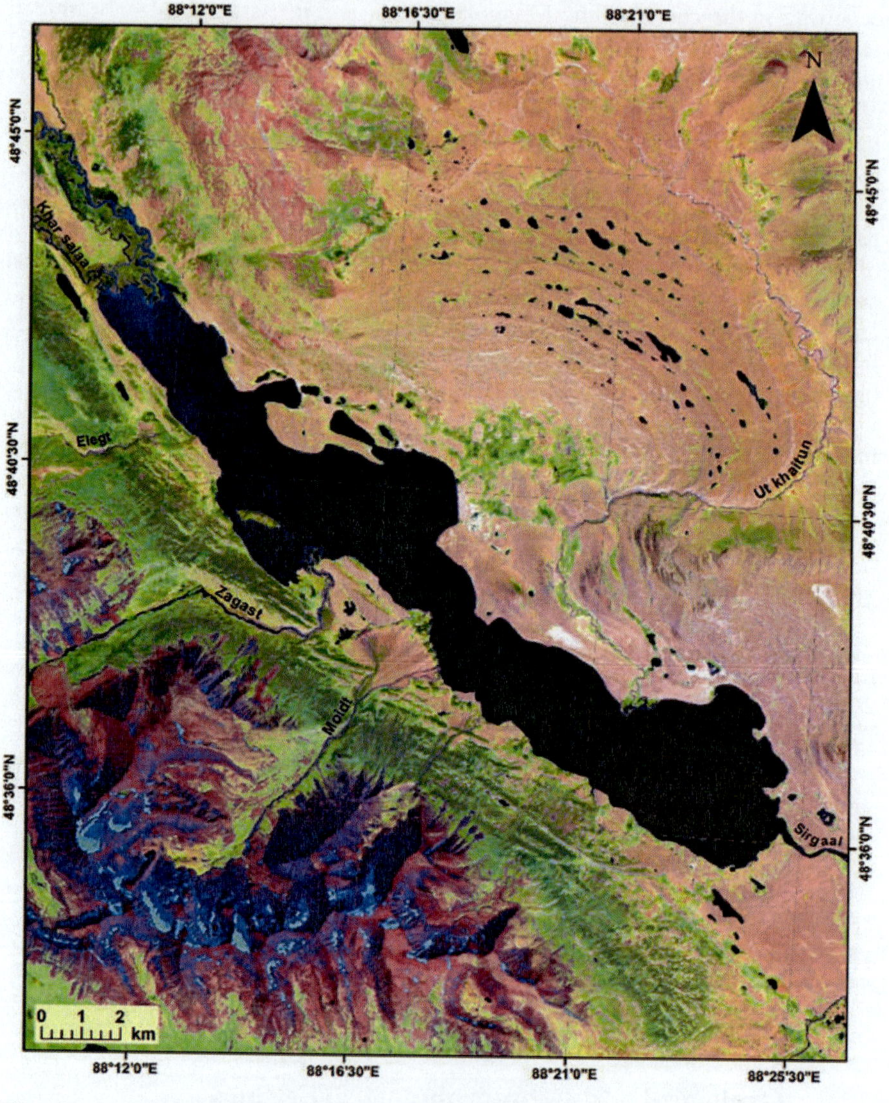

Fig. 14.1 Landsat 8 image of Lake Khoton on August 2, 2015 (path of 143 and row of 26). Modified from Orkhonselenge and Uuganzaya (unpublished)

Academy of Sciences of USSR, 1990). The lake lies tectonostratigraphically on the Altai backarc/forearc basin terrain consisting of nonmarine clastic sedimentary rocks, the Cambrian and Ordovician granite rocks, and the Devonian felsic and alkaline volcanic and volcaniclastic rocks (Badarch et al., 2002).

Lake Khoton is an elongated lake extending from the northwest to the southeast in the Mongolian Altai Mountain Range (Figs. 14.1 and 14.2). The lake is connected with the glacial Lake Khurgan (Fig. 14.3) in the southeast through Sirgaal River (Fig. 14.1; see Sect. 14.5). Beaches of the lake encroach deeply into the Altai

Fig. 14.2 Lake Khoton viewing toward the (**a**) southeast and (**b**) northeast. Photos by D. Enkhzaya on June 18, 2017

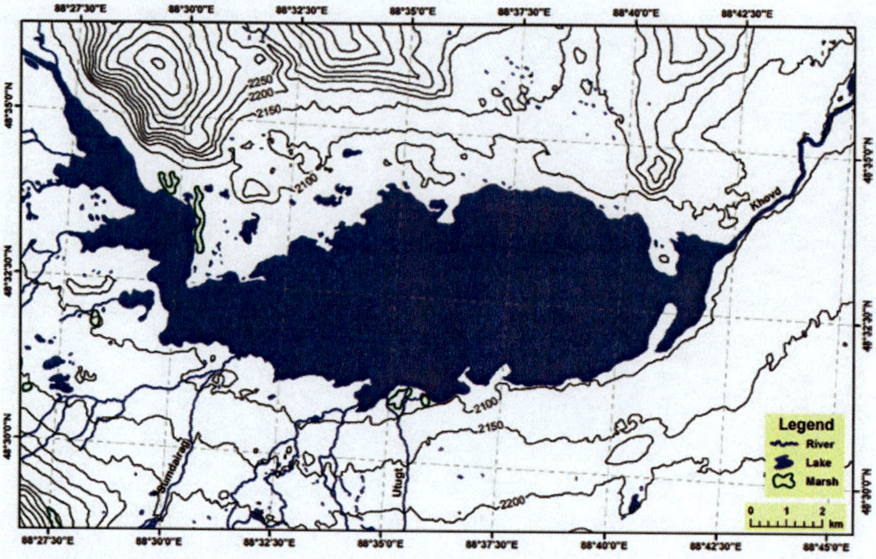

Fig. 14.3 Physiography of Lake Khurgan and its outlet Khovd River. Modified from Orkhonselenge and Uuganzaya (unpublished)

mountains where timberline is at 2400–2500 m a.s.l. The lake contains an island in the west occupying an area of 0.5 km^2 near the downstream of Zagast River (Fig. 14.1), and the island is fully covered by forest (Tserensodnom, 1971).

Lake Khoton is dammed by a cross-valley terminal moraine of a large Pleistocene valley glacier which itself is connected to a glaciofluvial outwash fan that extends into the glaciolacustrine sediments of Lake Khurgan (Lehmkuhl et al., 2016; Klinge et al., 2021; Fig. 14.3) that is 17.1 m deep (Murzaev, 1952). Morainic sediments in a hummocky terrain around Lake Khurgan incorporate large amounts of laminated glaciolacustrine clay and silt, which provide evidence for a proglacial lake which may have existed at 88.5 ± 14.5 ka (silt) and 85.5 ± 10.4 ka (clay) during the late Pleistocene ice advances and glacial oscillations (Klinge et al., 2021). At Lake Khoton, damming of terminal moraines occurred at 16.2 ± 2.0 ka and 18.3 ± 2.2 ka during or before the MIS 2 (Klinge et al., 2021).

There are approximately 200 lakes in the vicinity of Lakes Khoton and Khurgan; however, most of them are located in glacial cirques at above 3500 m a.s.l. (Tserensodnom, 1971). These lakes are connected to each other by rivers and streams. For instance, in the southwest of Lake Khoton, there are many small glacial lakes (e.g., Lakes Turgen, Zagast, Khelkhee, etc.; see Chap. 4).

14.4 Climate Condition

Local climate data recorded at the Tsengel meteorological station located 65.6 km northeast of the lake show the annual average air temperature of −2.1 °C and the annual average precipitation of 89.1 mm (Fig. 14.4). The average air temperature is −23.0 °C in January and 15.6 °C in July, and the average precipitation is 23.9 mm from June to August.

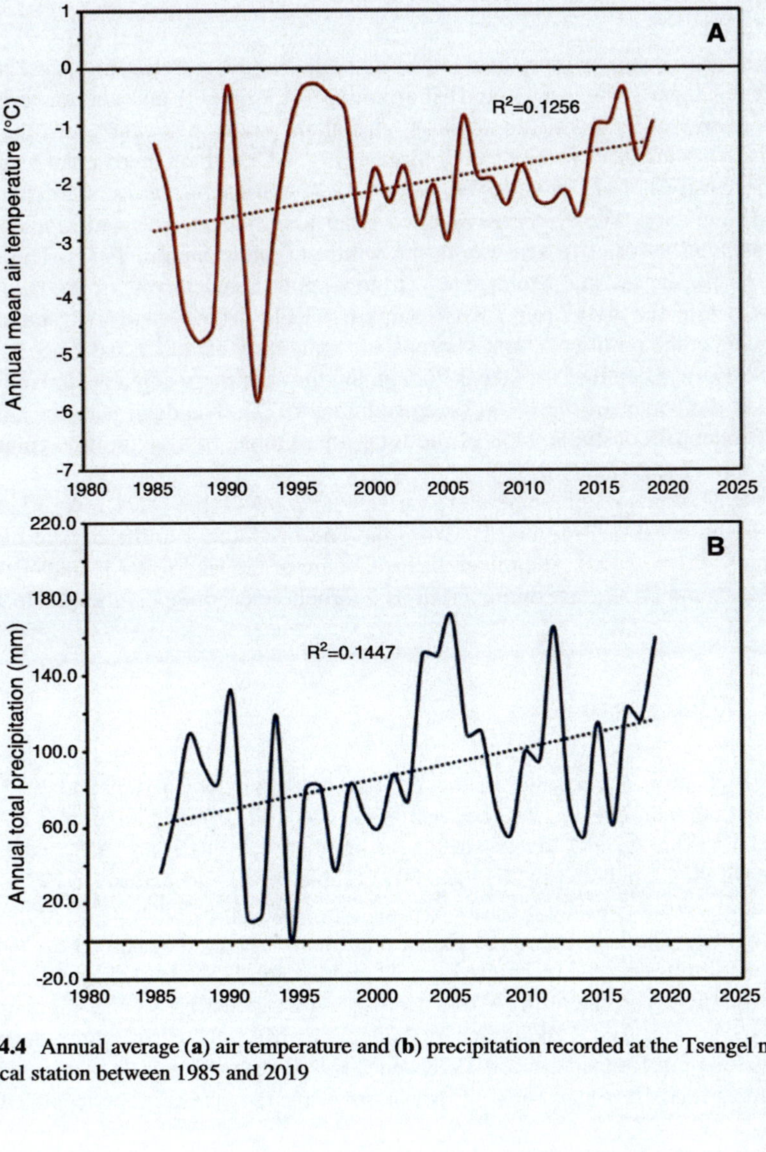

Fig. 14.4 Annual average (**a**) air temperature and (**b**) precipitation recorded at the Tsengel meteorological station between 1985 and 2019

Warm years were recorded in 1990, 1995–1998, 2006, 2016–2017, and 2019 with the annual average air temperature range of −0.28 °C to −0.97 °C (Fig. 14.4a). Humid years were recorded in 2003–2007, 2012, and 2017–2019 with the annual precipitation of 110.7–165.4 mm (Fig. 14.4b). The data show that the air temperature and precipitation in the Mongolia Altai have risen during the last 35 years (Fig. 14.4).

14.5 Hydrological Condition

Lake Khoton is fed by more than ten rivers including Khar Salaa (or Khar Khovd), Tsagaan, Zagast (Fig. 14.1), and Khoromdog (or Chuluut) and streams draining from an area of 2670 km^2. In addition, glacial and snow meltwaters feed the lake. The lake has an outflow river called Sirgaal (Fig. 14.1) which feeds Lake Khurgan, a major headwater of Khovd River (Fig. 14.3). The Sirgaal River is 3.0 km long and 100–150 m wide. The discharge rate in Sirgaal River is 23.0 m^3/s with a velocity of 1.2 m/s, and it is not frozen during the winter (Tserensodnom, 1971). The rivers (e.g., Elegt, Zagast, and Moidt rivers) inflowing at the southwest of the lake form deltas, while the rivers (e.g., Khar Salaa (or Khar Khovd) and Tsagaan rivers) inflowing at the northwest form swamps and marshes (Figs. 14.1 and 14.5).

The average depth of the lake is 26.6 m, but the maximum depth is 58.0 m (Table 12.4) at the center of the lake (Tserensodnom, 1971). The deep parts of the lake (>10 m depth) constitute 74% of the total area (Table 14.1). The lake volume is 1.341 km^3 (Tserensodnom, 1971).

Lake Khoton's water temperature varies with depth (Table 14.2). Lake Khoton is covered by a 1.3–1.6-m-thick ice from the end of October to the middle of June (Tserensodnom, 2000). The mineralization value of the lake water is 0.09–0.1 g/L, and SO_4^{2-} and HCO_{3-} are dominant in its hydrochemical composition (Table 14.3).

14.6 Changes in Area

Table 14.4 shows the changes in area of Lake Khoton between 1969 and 2015. The lake lost an area of ~10.0 km^2 between 1969 and 1971, gained an area of 6.4 km^2 between 1971 and 2010, lost an area of 3.1 km^2 between 2010 and 2013, and gained an area of ~0.4 km^2 between 2013 and 2015 (Table 14.4). The increase in 1971–2010 may have been contributed by the rise of the annual precipitation in 2003–2007 (Fig. 14.4b). The lake expanded for an additional area of 0.67 km^2 from 1970 to 2015 (Orkhonselenge and Uuganzaya, Unpublished; Table 14.4; Fig. 14.5). This observation is consistent with the decreases of glacier areas and the increase observed for one glacial lake's area by 0.04 km^2 on Mt. Ikh Turgen in the Mongolian Altai Mountain Range between 1970 and 2011 (Orkhonselenge & Harbor, 2018). The linear regression analysis (Fig. 14.6) shows that the lake area between 2010 and

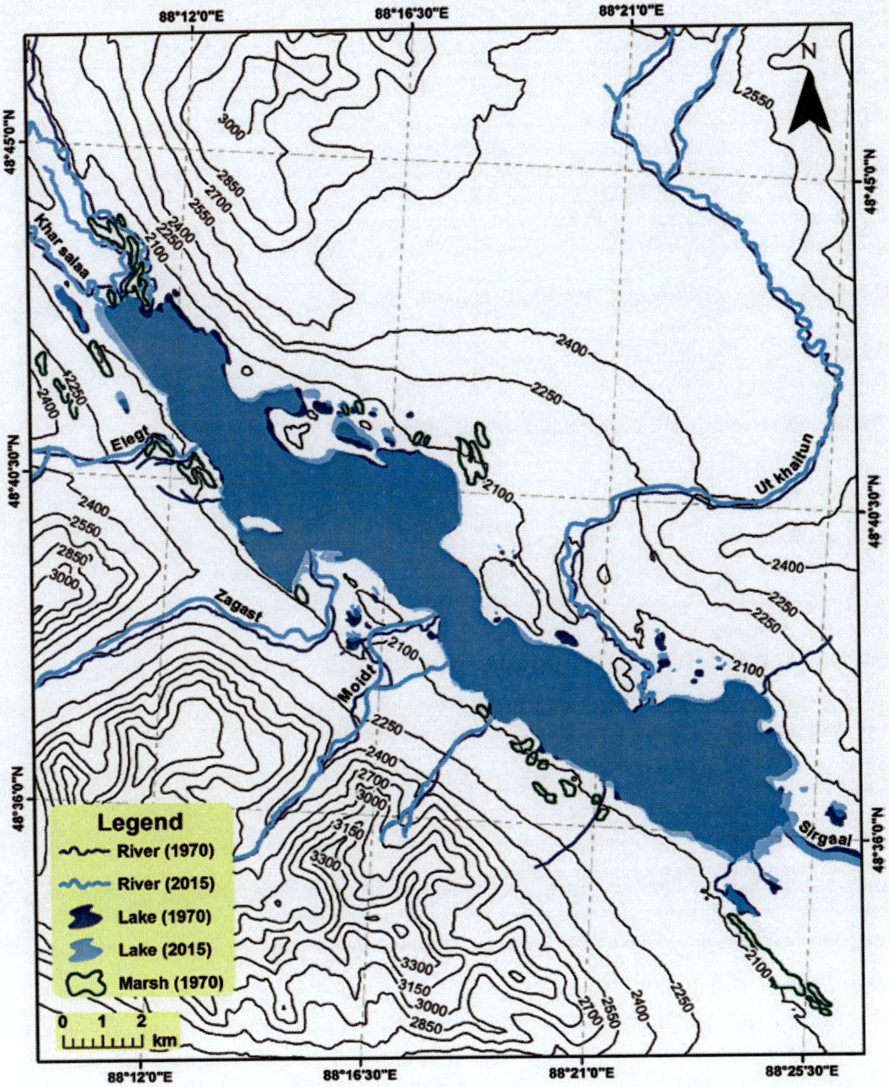

Fig. 14.5 Spatial distributions of areas and marshes of Lake Khoton in 1970 and 2015. Modified from Orkhonselenge and Uuganzaya (unpublished)

Table 14.1 Bathymetric data with corresponding area and volume of Lake Khoton

#	Depth (m)	Area (%)	Volume (km^3)
1	4.0<	46.9	0.301
2	12.0<	10.4	0.067
3	20.0<	3.0	0.014

Table 14.2 Water temperature of Lake Khoton

#	Depth (m)	June (°C)	July (°C)
1	0.0	4.6–9.0	13.3–22.3
2	3.0	–	20.1
3	6.0–10.0	6.3–5.2	–
4	12.0	5.0–5.1	15.3
5	25.0	–	10.3
6	30.0	–	13.3
	Position		
7	Beach	–	22.3
8	Center	–	13.0–14.0
9	Bottom	–	7.9

Table 14.3 Hydrochemical composition of Lake Khoton

	Ions (g/L)						
Depth (m)	Na^+	Mg^{2+}	Ca^{2+}	Cl^-	SO_4^{2-}	HCO_3^-	Total
0.0	0.01	0.007	0.005	0.007	0.03	0.03	0.09
25.0	0.02	0.005	0.005	0.007	0.01	0.06	0.1
45.0	0.01	0.008	0.005	0.007	0.03	0.03	0.09

Table 14.4 Temporal changes in area of Lake Khoton

#	Year	Area (km^2)	Authors
1	1969	60.0	Tsegmid (1969)
2	1970	53.12	Orkhonselenge and Uuganzaya, unpublished
3	1971	50.1	Tserensodnom (1971)
4	2010	56.5	Davaa (2015)
5	2013	53.4	Davaa (2015)
6	2015	53.79	Orkhonselenge and Uuganzaya, unpublished

2015 was not clearly related to the annual precipitation ($R^2 = 0.0839$) nor air temperature ($R^2 = 0.0004$). The correlation analysis shows that the lake area during 2010–2015 had a weak positive correlation with the precipitation ($r = 0.289$) and a weak negative correlation with the air temperature ($r = -0.021$). Nevertheless, the short-term data are still insufficient to infer these relationships further.

The increase of Lake Khoton's area between 1970 and 2015 (Orkhonselenge and Uuganzaya, unpublished; Table 14.4; Fig. 14.5) may have been correlated to the gradually rising air temperature and precipitation during the last 35 years (Fig. 14.4). The air temperature rising is contributing to the recent rapid melting of glaciers on high mountains in the Mongolian Altai Mountain Range, and the glacial meltwater supports various surface water resources (Orkhonselenge & Harbor, 2018) including lakes, rivers, streams, springs, and even flooding. The changes of Lake Khoton's area show that the lake is also sensitive to the rising air temperature contributing to melting of glaciers for other lakes (Table 12.4). The short-term (since 2010)

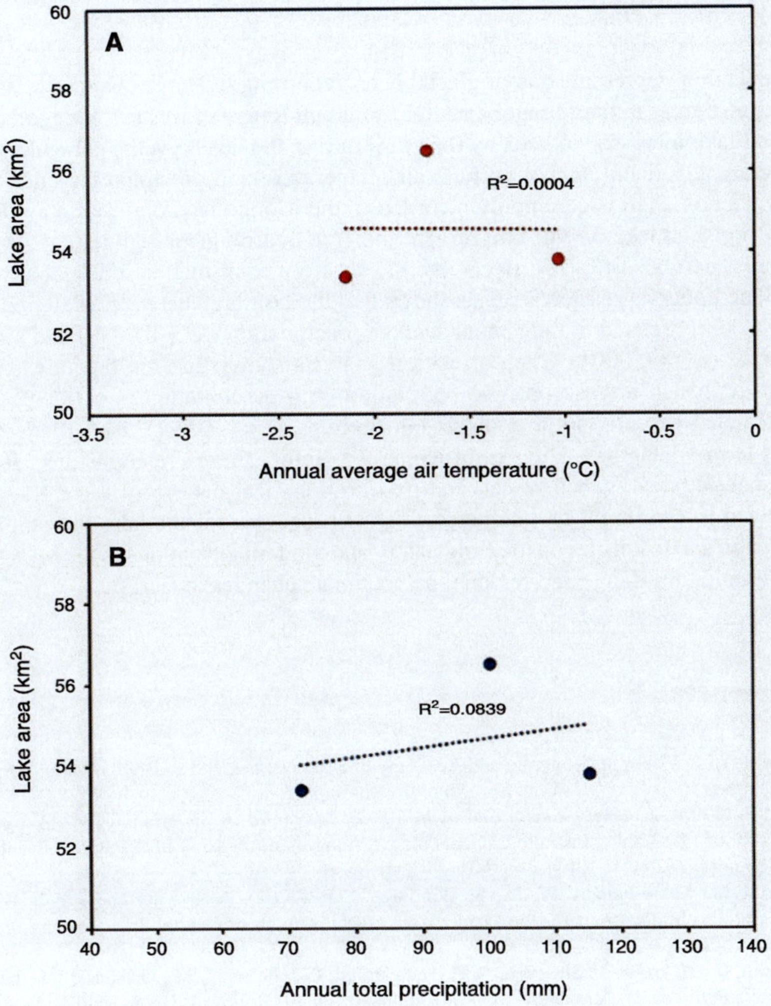

Fig. 14.6 Relationships between area of Lake Khoton and annual average (**a**) air temperature and (**b**) precipitation during 2010–2015

phenomenon for Lake Khoton having a positive correlation with the precipitation and a negative correlation with the temperature is also found for Lake Achit (see Chap. 12) in the Great Lakes Depression. In contrast, the long-term (since 1950) phenomenon for Lake Uvs (see Chap. 13) in the Great Lakes Depression was found, having a moderate positive correlation with the temperature and a weak negative correlation with the precipitation.

14.7 Summary

Lake Khoton represents one of glacial lakes recording dynamic changes of surface water resources in the Mongolian Altai Mountain Range in western Mongolia. This glacial lake increased its area by 0.67 km^2 during the last 45 years coinciding with the gradually rising annual average air temperature and precipitation since 1985. This phenomenon is commonly recorded in the Mongolian Altai Mountain Range (e.g., a glacial lake on Mt. Ikh Turgen), i.e., the hydrological regime (e.g., modern glaciers, permafrost, lakes, rivers, etc.) is strongly controlled by the global warming. The linear regression analysis shows that the lake area between 2010 and 2015 is not clearly related to the annual average precipitation ($R^2 = 0.0839$) and air temperature ($R^2 = 0.0004$). The correlation analysis shows that the lake area during 2010–2015 had a weak positive correlation with precipitation ($r = 0.289$) and a weak negative correlation with air temperature ($r = -0.021$). Nevertheless, the short-term data are still insufficient to infer these relationships further. Sedimentological, geochemical, and geochronological data from Lake Khoton is expected to shed light on geomorphological processes for the lake formation and evolution and to help reconstructing paleo- and modern glaciations in the Mongolian Altai Mountain Range and regional paleoclimate changes in Eurasia.

References

Academy of Sciences of Mongolia and Academy of Sciences of USSR. (1990). *National Atlas of the Mongolian People's Republic*. 156 p. [In Russian].

Baast, P. (1998). *Modern glaciers and hydrology of Mongolia*. A dissertation for Philosophical doctor of Geography, Institute of Geoecology, Mongolian Academy of Sciences, Ulaanbaatar, Mongolia, 162 p. [In Russian with English abstract].

Badarch, G., Cunningham, W. D., & Windley, B. F. (2002). A new terrane subdivision for Mongolia: Implications for the Phanerozoic crustal growth of Central Asia. *Journal of Asian Earth Sciences, 21*, 87–110.

Blomdin, R., Heyman, J., Stroeven, A. P., Hattestrand, C., Harbor, J. M., Gribenski, N., Jansson, K. N., Petrakov, D. A., Ivanov, M. N., Orkhonselenge, A., Rudoy, A. N., & Walther, M. (2016). Glacial geomorphology of the Altai and Western Sayan Mountains. *Central Asia. Journal of Maps, 12*(1), 123–136.

Blomdin, R., Stroeven, A. P., Harbor, J. M., Gribenski, N., Caffee, M. W., Heyman, J., Rogozhina, I., Ivanov, M. N., Petrakov, D. A., Walther, M., Rudoy, A. N., Zhang, W., Orkhonselenge, A., Hättestrand, C., Lifton, N. A., & Jansson, K. N. (2018). Timing and dynamics of glaciation in the Ikh Turgen mountains, Altai Region, High Asia. *Quaternary Geochronology, 47*, 54–71.

Davaa, G. (2010). Climate change impacts on water resources in Mongolia. In *Proceedings of Consultative Meeting on Integration of Climate Change Adaptation into Sustainable Development in Mongolia*. Institute for Global Environmental Strategies (IGES), pp. 30–36.

Davaa, G. (2015). *Resource and regime of surface water of Mongolia*. 408 p. [In Mongolian].

Devjatkin, E. V. (1981). *Cenozoic of Inner Asia*. Nauka, 196 p. [In Russian].

Florensov, N. N., & Korzhnev, S. S. (1982). Geomorphology of Mongolian People Republic. *Joint Soviet-Mongolian scientific research geologic expeditions. Transactions, 28*. Moscow. [In Russian].

Jigj, S. (1976). *A brief description of paleoglaciations and paleoglaciers.* 51 p. [In Mongolian].

Kamp, U., McManigal, K. G., Dashtseren, A., & Walther, M. (2013). Documenting glacial changes between 1910, 1970, 1992 and 2010 in the Turgen Mountains, Mongolian Altai, using repeat photographs, topographic maps, and satellite imagery. *The Geographical Journal, 179*(3), 248–263.

Klinge, M., Schlütz, F., Zander, A., Hülle, D., Batkhishig, O., & Lehmkuhl, F. (2021). Late Pleistocene lake level, glaciation and climate change in the Mongolian Altai deduced from sedimentological and palynological archives. *Quaternary Research, 99*, 168–189. https://doi.org/10.1017/qua.2020.67

Krumwiede, B. S., Kamp, U., Leonard, G. J., Kargel, J. S., Dashtseren, A., & Walther, M. (2014). Chapter 22: Recent glacier changes in the Mongolian Altai Mountains: Case studies from Munkh Khairkhan and Tavan Bogd. In J. S. Kargel et al. (Eds.), *Global land ice measurements from space* (pp. 481–508). Springer Praxis Books, Springer-Verlag.

Lehmkuhl, F., Klinge, M., Rother, H., & Daniela Hulle, D. (2016). Distribution and timing of Holocene and Pleistocene glacier fluctuations in western Mongolia. *Annals of Glaciology, 57*(71), 1–10.

Lehmkuhl, F., Klinge, M., & Stauch, G. (2011). The extent and timing of Late Pleistocene Glaciations in the Altai and neighboring mountains systems. In J. Ehlers & P. L. Gibbards (Eds.), *Development in quaternary science—Extent and chronology—A closer look* (Vol. 15, pp. 967–979). Elsevier.

Murzaev, E. M. (1952). *The description of Physical Geography of Mongolia* (2nd ed.). [In Russian].

Orkhonselenge, A. (2016). Glacial Geomorphology of Mt. Munkh Saridag in the Khuvsgul Mountain Range, Northern Mongolia. *Géomorphologie: Relief, Processes, Environnement, 22*(4), 387–396.

Orkhonselenge, A., & Harbor, J. M. (2018). Impacts of modern glacier changes on surface water resources in western and northern Mongolia. *Journal of Water Resource and Protection, 10*(6), 559–576.

Orkhonselenge, A., & Uuganzaya, M. (2018). Glacial Geomorphology of Mt. Asralt Khairkhan and Mt. Baga Khentii Saridag in Khentii Mountain Range, Northeastern Mongolia. *International Journal of Geosciences, 9*(6), 308–319.

Rudaya, N., & Li, H. (2013). A new approach for reconstruction of the Holocene climate in the Mongolian Altai: The high-resolution ^{13}C records of TOC and pollen complexes in Hoton-Nur Lake sediments. *Journal of Asian Earth Sciences, 69*, 185–195.

Rudaya, N., Tarasov, P., Dorofeyuk, N., Solovieva, N., Kalugin, I., Andreev, A., Daryin, A., Diekmann, B., Riedel, F., Tserendash, N., & Wagner, M. (2009). Holocene environments and climate in the Mongolian Altai reconstructed from the Hoton-Nur pollen and diatom records: A step towards better understanding climate dynamics in Central Asia. *Quaternary Science Reviews, 28*, 540–554.

Slynko, Y. V., Mendsaykhan, B., & Kasanov, A. N. (2010). On Intraspecies Forms of the Mongolian Grayling (*Thymallus brevirostris* Kessl.) from Hoton Nur Lake (Western Mongolia). *Journal of Ichthyology, 50*(1), 28–37.

Tronov, M. V. (1966). *Glaciers and climate*. Gidrometeoizdat, 407 p. [In Russian].

Tsegmid, S. (1969). *Physical geography of Mongolia.* 759 p. [In Mongolian].

Tserensodnom, J. (1971). *Lakes of Mongolia*. State Publishing, 202 p. [In Mongolian].

Tserensodnom, J. (2000). *A catalog of lakes in Mongolia.* Shuvuun Saaral Publishing, 141 p. [In Mongolian].

Chapter 15
Landscape, Lake Distribution, and Evolution in Northern Mongolia

Abstract Northern Mongolia plays a vital role in climatology, hydrology, and geomorphology of Eurasia. The region reflects the climate change and landscape evolution of Eurasia because it is situated at the southern border of eastern Siberia under the influences of the East Asian winter monsoon and westerlies. The region is unique with the presence of the high Khuvsgul mountains, perennial permafrost, large tectonic depressions (e.g., Darkhad), large rivers (e.g., Shishkhid, Selenge rivers), and Lake Khuvsgul, the deepest in Mongolia. Rivers in northern Mongolia feed large rivers (e.g., Yenisei, Ob rivers) and lakes (e.g., Lake Baikal) in the North Arctic Ocean drainage basin. The rivers and lakes in northern Mongolia constitute the maximum resource of surface water in Mongolia. Lakes in northern Mongolia occupy 20.6% of the total lake area, but the number of lakes is the lowest (7.7%), and they have primarily tectonic, glacial, fluvial, and thermokarst origins. Geochemical, geochronological, and hydrological records obtained in northern Mongolia would potentially lead to understanding of the geomorphological and sedimentological processes in the region, and they could help in reconstructing paleoclimate and paleoenvironmental changes of Eurasia. Thus, such reconstruction in northern Mongolia is essential for the prediction of future climate change impacts in Eurasia and Central Asia.

Keywords Khuvsgul mountains · Darkhad basin · Lake Khuvsgul · Selenge River · Northern Mongolia · Eg River

15.1 Introduction

Northern Mongolia is situated in the Precambrian Tuva-Mongolian terrain (Fig. 15.1) at the important geotectonic conjunctions in the Central Asian Orogenic Belt (CAOB) between the Siberian and Sino-Korean cratons. Northern Mongolia lies on the geosynclines and tectonic faults of the southern Baikal Rift Zone (BRZ) in southeastern Siberia. The Paleozoic oceanic environment in northern Mongolia shifted to the continental environment since the late Paleozoic and early Mesozoic

A. Orkhonselenge et al., *Lakes of Mongolia*, Syntheses in Limnogeology,
https://doi.org/10.1007/978-3-030-99120-3_15

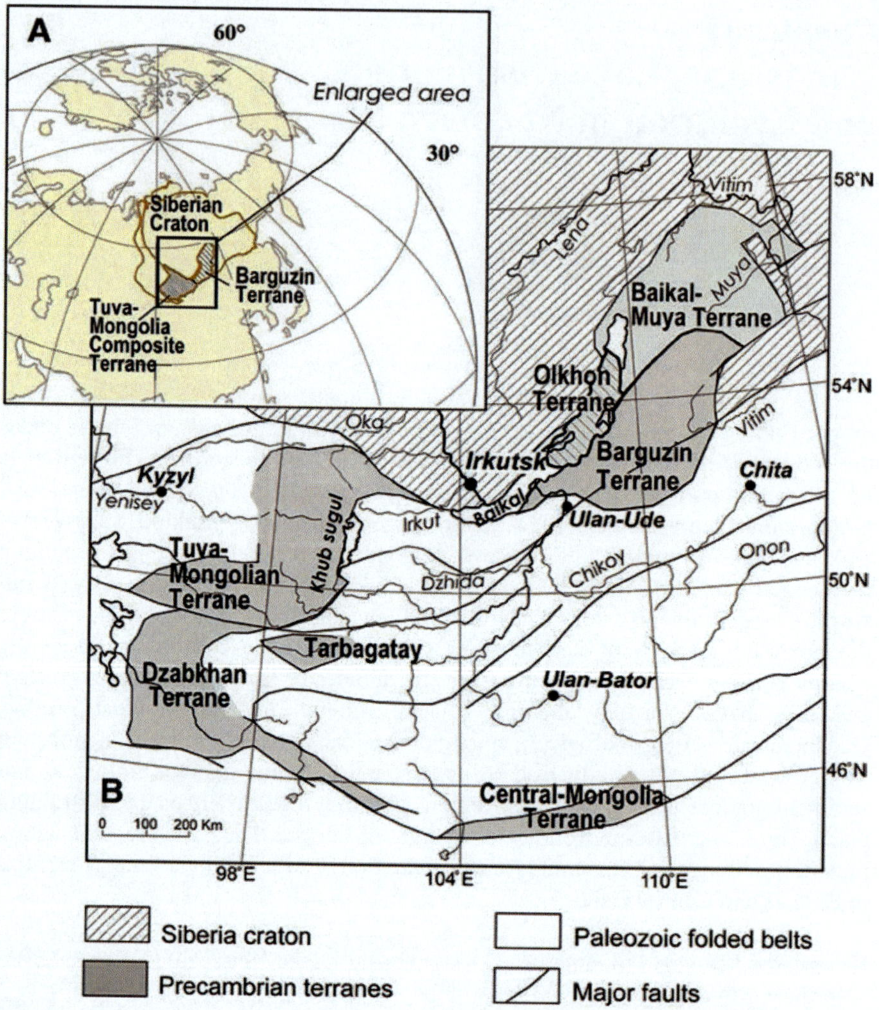

Fig. 15.1 Tectonic terrain of northern Mongolia. Modified from Kravchinsky et al. (2010)

Eras (Jigj, 1975). Numerous Mesozoic and Cenozoic uplifts and volcanic eruptions (Batrom et al., 2006; Table 15.1) result in the present landforms existing in the region.

Northern Mongolia (Fig. 1.1) encompasses regional landscape peculiarities of geology, geomorphology, permafrost, climate, and hydrology. It hosts high mountains of Khuvsgul[1] and northern Khangai[2] Mountain Ranges, feeding

[1] *Khuvsgul* has been misspelled as *Hovsgol* and *Khubsugul* in publications. Khuvsgul is the right English transliteration from Mongolian *Хөвсгөл*.

[2] *Khangai* has been misspelled as *Khangay* (or *Hangay*) and *Hangai* in many publications. Khangai is the right English transliteration from Mongolian *Хангай*.

Table 15.1 Evolutionary histories of paleogeography and geology of northern Mongolia and dynamic changes of landforms and geology around the Lake Khargal basin (see Chap. 17). Modified from Orkhonselenge and Gerelsaikhan (unpublished)

Geological timescale				Paleogeography		Geology		Geomorphology	
Erathem/era	System/period	Series/epoch	Stage/age	Jigj (1975)	Batrom et al. (2006)	Jigj (1975)	Batrom et al. (2006)	Batrom et al. (2006)	Orkhonselenge and Gerelsaikhan, unpublished
Cenozoic	Quaternary	Holocene	Megallian	Weathering, incision of rivers, solifluction and erosion	Modern uplifts, erosion and ephemeral floods	Lacustrine and alluvial deposits	Floodplain deposits	Floodplains, peneplain and pediments formed	Lacustrine and fluvial environments
			Northgrippian						
			Greenlandian				Alluvial fan deposits		
		Pleistocene	Late	Neotectonics strengthened, Alpine orogeny and glaciations	Uplift strengthened and dominance of humid climate	Alluvial fan deposits	Terrace deposits	Paleolake divided into small lakes such as Lakes Khargal, Olon, and Airkhan, and Eg River changed its channel	Initiation of development of independent Lake Khargal basin
			Chibanian						
			Calabrian		Dominance of arid climate		Loess deposits		
			Gelasian	Formation of the third peneplains and glaciations	Uplift restarted				
	Neogene	Pliocene	Piacenzian	Formation of the second peneplains			Basalts		
			Zanclean			Sedimentary rocks			
		Miocene	Late	Formation of the first peneplains	Volcanic eruption occurred and faults formed				
			Early			Basalts exposed			
	Paleogene	Oligocene		Landform erosion and initial formation of peneplains	Continental uplift occurred and graben formed	Terrigenic sedimentary rocks	Uilgan River sedimentary rock formation		
		Eocene							
		Paleocene							

(continued)

Table 15.1 (continued)

Geological timescale				Paleogeography		Geology		Geomorphology	
Erathem/era	System/period	Series/epoch	Stage/age	Jigj (1975)	Batrom et al. (2006)	Jigj (1975)	Batrom et al. (2006)	Batrom et al. (2006)	Orkhonselenge and Gerelsaikhan, unpublished
Mesozoic	Cretaceous	Late							
		Early		Orogeny strengthened	Continental divergence and volcanism	Igneous rocks			
	Jurassic	Late				Sedimentary rocks	Selenge igneous rock complex	Paleolake	A large paleolake
		Middle		Hercynian orogeny and formation of Khanui basin and faults					
		Early				Igneous rocks			
	Triassic	Late		Neotectonism		Sedimentary rocks			
		Middle							
		Early							

Paleozoic	Permian	Lopingian	Predominance of terrigenic plateau		Terrigenic and oceanic sedimentary rocks	Tulbur formation		
		Guadalupian				Khargana formation		
		Cisuralian	Transition to plateau and Hercynian orogeny			Khustai formation		
	Carboniferous	Pennsylvanian		Oceanic basin shifted into continental terrain	Dominance of igneous rocks			
		Mississippian	Predominance of continental terrain		Dominance of oceanic deposits			
	Devonian	Late	Presence of oceanic environment		Dominance of igneous rocks with mollassic deposits			
		Middle						
		Early						
	Silurian	Pridoli	Caledonian orogeny		Sedimentary rocks	Urkhtei formation		
		Ludlow						
		Wenlock						
		Liandovery						
	Ordovician	Late						
		Middle						
		Early						
	Cambrian	Furongian	Subsidence of the continent	Collision occurrence and mantle diapers happened	Oceanic deposits with igneous rock in Selenge River basin	Zed igneous rock complex		
		Series 3						
		Series 2						
		Terreneuvian		Presence of island arc complex		Badar River sedimentary and metamorphic rock formation		
Proterozoic	Neoproterozoic	Ediacaran	Orogenic uplifts	Paleooceanic environment	Metamorphic rocks			
		Cryogenian						
		Tonian						

numerous large rivers such as Tengis, Shishkhid, Delger, Ider, Bugsei, Eg, and Selenge[3] rivers (Fig. 1.1), which are headwaters of the Lake Baikal[4], Yenisei River, and Ob River in the North Arctic Ocean drainage basin. This region hosts one of Central Asian interfluves, Khuvsgul Mountain Range (Jigj, 1975), which feeds large rivers such as Tengis and Shishkhid rivers, major headwaters of Yenisei River in the northwest of the region, and Delger River, one of major headwaters of Selenge River that is a major headwater of Lake Baikal in the northeast of the region. The region contains three large N-S trend rift grabens, which are the southwestern continuous extension of the BRZ (see Sect. 15.2.3).

Moreover, northern Mongolia is a climatologically important region because variations in intensity of the East Asian winter monsoon (EAWM) and westerlies due to their collision have a strong control on the local climate there (see Sect. 15.2.2), i.e., northern Mongolia becomes an influential barrier on atmospheric circulations. Northern Mongolia also plays an important role in the regional climatic system because it isolates the Siberian high-pressure and Central Asian low-pressure atmospheric circulations. The westerly atmospheric transport influences northern Mongolia (Lydolph, 1977) where summers are cool and winters are extremely cold, i.e., the annual average air temperature amplitude is high. The climate in this region is an East Siberian ultra-continental climate with high aridity and long-lasting, harsh winters with scarce snowfall, a significant difference between winter and summer air temperatures, and large amplitudes of absolute maxima and minima (Kitov et al., 2015). These severe climatic conditions are responsible for the widespread occurrence of permafrost and associated cryogenic landforms in the region. To this date, the interrelationships between the westerlies and the East Asian summer monsoon (EASM) and the EAWM are not well understood in northern Mongolia (see Chap. 21).

In the region, the Khuvsgul basin has attracted much attention by scientists since 1830 when Russian botanists N.S. Turchaninov and N.V. Kirillov conducted the first research (Bogoyavlensky et al., 1989). The Khuvsgul Mountain Range has been also studied in glaciology recently. For example, morphometric changes in extension, age of paleoglaciers (e.g., Krivonogov et al., 2005; Gillespie et al., 2008; Komatsu et al., 2009; Batbaatar et al., 2018), modern glaciers (Kitov et al., 2015), a glacial geomorphology (Orkhonselenge, 2016), and dynamic changes of modern glaciers and of a glacial lake area (Orkhonselenge & Harbor, 2018) have been provided (see Sect. 15.2.4). In northern Mongolia, hydrology is an integrated system with dynamics in snow cover, perennial and seasonal permafrost, and large rivers and lakes. The rivers and lakes in the region (see Sect. 15.2.6) are the most important surface water resources of Mongolia and Siberia, particularly for the North Arctic Ocean drainage basin in Eurasia.

---

[3] *Selenge* has been often incorrectly written as *Selenga* internationally based on the Russians' pronunciation. Selenge is the correct English transliteration from Mongolian *Сэлэнгэ*.

[4] *Baigali* meaning *Nature* has been misspelled as *Baikal* by the Russians. Baigali is the right English transliteration from Mongolian *Байгаль*.

Lakes in northern Mongolia have been intensely studied in recent years. In the region, Lake Khuvsgul (see Chap. 16) has been a focus of attention in terms of limnology, geomorphology, hydrology, sedimentology, and paleoclimatology because it is a headwater of Lake Baikal through Eg and Selenge rivers (Fig. 1.1). The investigation results from Lake Khuvsgul and adjacent areas were shown in the Atlas of Lake Khuvsgul produced by Irkutsk and Mongolian national universities after the Soviet and Mongolian joint Khuvsgul expedition in 1971–1986 (Bogoyavlensky et al., 1989). Moreover, information on geology, climate, hydrophysics, hydrochemistry, permafrost, pedology, biodiversity, and ecology of Lake Khuvsgul was provided by Goulden et al. (2006). The international Khuvsgul Drilling Project (KDP-01) revealed the warm marine isotope stage (MIS) 3, MIS 5, MIS 7, and MIS 9 with the increases in $\delta^{18}O$ value, high contents of Br, low amounts of clay particles, and high bio-productivities in Lake Khuvsgul, whereas positive $\delta^{13}C$ values linked with high inflowing dissolved inorganic carbonates and forest-steppe vegetation were evidenced during cold MIS 2, MIS 4, MIS 6, and MIS 8 (Fedotov et al., 2015). The international Khuvsgul (or Hovsgol) Drilling Project (HDP-04) proposed a provisional age model on the basis of two procedures with some dates (^{14}C, ^{238}U-^{234}U-^{230}Th and paleomagnetic events) and auto-orbital tuning for sedimentary records from Lake Khuvsgul (Kashiwaya et al., 2010). Besides the Lake Khuvsgul, other lakes have also been studied to date. For instance, in the paleolake Darkhad or the modern Darkhad basin, Lake Dood revealed climatic, lake level, and environmental changes during the late glacial and the early to middle Holocene as presented in Narantsetseg et al. (2013). The international Darkhad Drilling Project (DDP-2010) provided geomorphological, sedimentological, and geochronological records showing that the Darkhad sedimentary sequence had been formed since the Pliocene (Krivonogov et al., 2012). Pechersky et al. (2013) determined the cosmic origin of native iron (Fe) in the Quaternary sediment from the paleolake Darkhad basin, which is similar to the native irons of Eurasia and the Atlantic.

In this chapter, landscape feature and lake evolution in northern Mongolia are considered in detail. Geomorphological and geochemical implications from two tectonic and fluvial lakes representing specific landscapes in this region are discussed in detail in Chaps. 16 and 17.

15.2 Landscape

15.2.1 Geology

Tectonostratigraphic terrains such as the Sangil metamorphic rock, Khug accretionary wedge, Cambrian shelf carbonate rock in the Khuvsgul basin, Darkhad island arc, Targan cratonal rock, Permian-Triassic volcanic-plutonic belt, Ilchir ophiolite,

Khamar Davaa metamorphic rock, Chita island arc (Fig. 15.1), Buteel metamorphic rock, Selenge belt Devonian-Carboniferous turbidite basin, Bayangol island arc, Kharaa backarc/forearc basin, and Khangai-Khentii[5] Devonian-Carboniferous turbidite basin (Badarch et al., 2002) are distributed from the west to the east of the region. The Proterozoic metamorphic rocks in the region are overlain by the Paleozoic and Mesozoic igneous rocks and the Cenozoic sedimentary rocks (Table 15.1).

The geology of northern Mongolia consists of the late Neoproterozoic and Ordovician mobile belt with the Paleozoic and Mesozoic granitoids intruded in the predominantly juvenile crust and the Neoproterozoic and early Paleozoic sedimentary basins in the west, and the Ordovician and Silurian mobile belt reworked as the Devonian and Carboniferous mobile belt in the east (Kröner & Rojas-Agramonte, 2014).

Quaternary deposits in the Darkhad basin, Lakes Khuvsgul and Erkhel in northern Mongolia, were studied in detail by Narantsetseg and Khishigsuren (2017). The Darkhad basin is mainly underlain by the Proterozoic metamorphic and plutonic rocks with spotty occurrences of the Paleozoic marine sedimentary deposits (Nishida & Jamsran, 2009). In the east of the Selenge River basin, zircon dating of footwall rocks within the Buteel metamorphic complex establishes the emplacement of granitoid orthogneiss at 240–211 Ma, a felsic metavolcanic rock at 265.0 ± 1.2 Ma, a syenite at 265.5 ± 1.2 Ma, and a metarhyolite of the pregranitoid basement at 553.6 ± 2.9 Ma, while $^{40}Ar/^{39}Ar$ dating indicates ages of 133.5 ± 1.8 Ma of hornblende from amphibolite and 122.6 ± 1.8 Ma of biotite from mylonitized gabbro-dolerite, and a per-alkaline granite intruding orthogneisses of the Zag metamorphic complex indicates 151.6 ± 0.7 Ma (Donskaya et al., 2008).

15.2.2 Climate

Climates in the mountainous regions and depressions are variable within northern Mongolia. The annual average precipitation is 190–300 mm overall and 400–500 mm in the high mountains with the maximum recorded in July (Tsegmid, 1969; Tserensodnom, 1971). There, the average air temperature is −25 to −34 °C in January and 15–20 °C in July (Tserensodnom, 1971).

According to the local climate data recorded at the Khatgal meteorological station (Fig. 15.3), the annual average air temperature is −1.9 °C with its amplitude of 70–80 °C and the diurnal air temperature amplitude of 20–25 °C (Orkhonselenge, 2016). However, the climate data at the station during the last almost 60 years show the annual average air temperature of −3.36 °C with a rising trend (Fig. 15.2a) and the annual average precipitation of 302.9 mm with an equally fluctuating trend

[5] *Khentii* has been misspelled as *Khentey* (or *Hentey*) and *Khentei* (or *Hentei*) in many publications. Khentii is the right English transliteration from Mongolian *Хэнтий*.

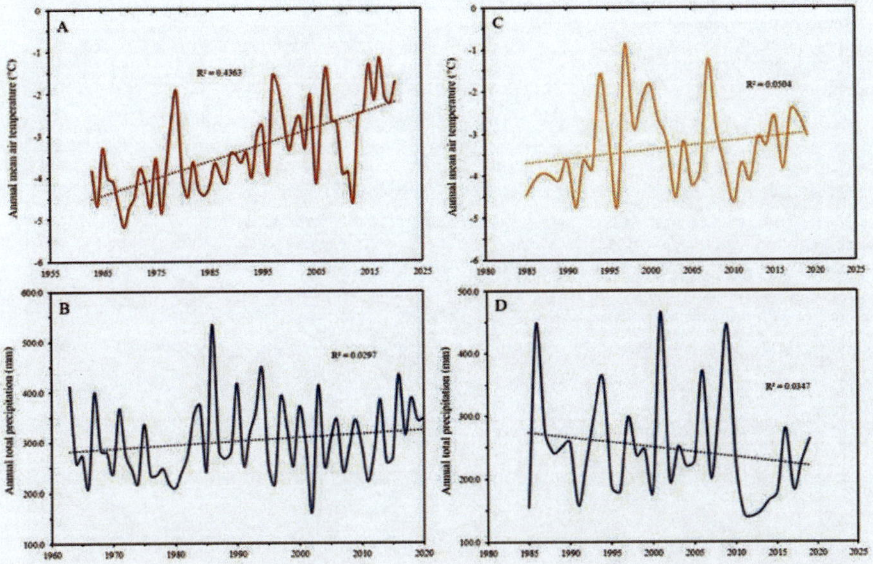

Fig. 15.2 Annual average (**a**) air temperature and (**b**) precipitation recorded at the Khatgal meteorological station within the Khuvsgul basin (Fig. 15.3) during 1963–2020. Annual average (**c**) air temperature and (**d**) precipitation recorded at the Khankh meteorological station within the Khuvsgul basin (Fig. 15.3) during 1985–2019

(Fig. 15.2b). The annual average air temperature at the Khankh meteorological station (Fig. 15.3) located at the elevation of 1650 m a.s.l. has increased by 2.17°C during the last 40 years (Orkhonselenge & Harbor, 2018). However, the local climate data at the station during the last 35 years show the annual average air temperature of −3.33 °C with a rising trend (Fig. 15.2c) and the annual average precipitation of 247.7 mm with a dropping trend (Fig. 15.2d). In addition, at the Teshig meteorological station in the east of the region, the annual average air temperature is −1.87 °C and the annual average precipitation is 285 mm (see Chap. 17).

The Darkhad basin in the west of the region is characterized by cool summers and extremely cold winters due to the large annual air temperature amplitude with the average air temperatures of −32.3 °C in January and 15.7 °C in July. The climate of the Darkhad basin (Fig. 15.3) is extremely continental with an annual average air temperature range of 40 °C and an annual average precipitation of 250–400 mm (Peck et al., 2001). Another source states that the average air temperature is -30 °C in January and 15 °C in July, and the annual average precipitation is 300–350 mm (Bogoyavlensky et al., 1989). The annual average air temperature in the basin was −6.43 °C between August 2002 and August 2003 with the maximum of 38.1 °C and the minimum of −45.1 °C (Suzuki & Natsagdorj, 2009). The annual and monthly air temperatures in the Darkhad basin between 1974 and 2003 (Fuji et al., 2009) rose by 3.7 °C (Table 15.2). The meteorological stations in the basin are located as Renchin Lkhumbe soum in the east, Ulaan Uul soum in the south, and Tsagaan Nuur soum in the northwest.

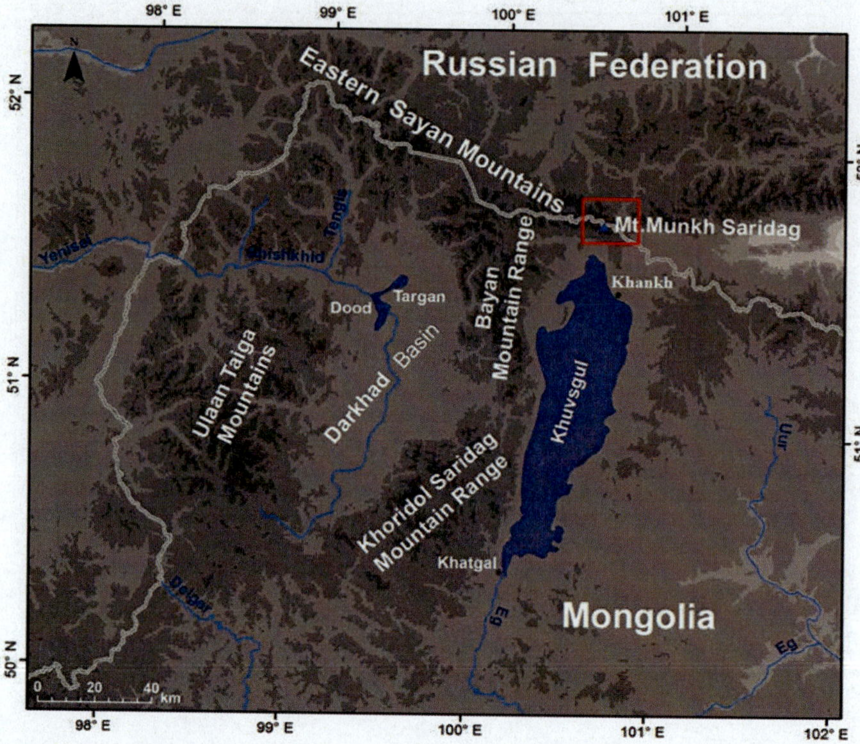

Fig. 15.3 Physiography of Khuvsgul Mountain Range and depressions. A box shows location of Mt. Munkh Saridag (Fig. 15.4). Modified from Orkhonselenge (2016)

15.2.3 Landform

The landscapes of this region are represented primarily by the Khuvsgul Mountain Range, Darkhad basin, and Eg River and eastern Selenge River drainage basins (Table 1.2). Northern Mongolia (Fig. 1.1) hosts high mountains of the Sayan Mountain Range, lakes in the BRZ, and large rivers in the Selenge River basin that is a part of the North Arctic Ocean drainage basin. Moreover, there are large intermontane valleys between the Khuvsgul and northern Khangai mountain ranges, and these intermontane valleys are dominated by the forest landscape. The region is surrounded by the Sayan Mountain Range in the north (Fig. 15.3), Lake Baikal basin in the northeast, Khangai Mountain Range in the south (Fig. 1.1), and Siilkhem Mountain Range in the west. In other words, this region mostly consists of the high mountainous zone in the west and the wide intermontane valley zone in the east and south (Fig. 15.3).

Table 15.2 Annual and monthly average air temperatures recorded in the Darkhad basin between 1974 and 2003

#	Year	Meteorological stations	January (°C)	July (°C)	Annual (°C)
1	1974	Renchin Lkhumbe	−33.7	12.9	−7.0
		Ulaan Uul	−29.9	14.4	−7.7
2	1978	Renchin Lkhumbe	−27.9	12.9	−5.4
		Ulaan Uul	−24.1	14.2	−5.9
3	1983	Renchin Lkhumbe	−30.5	11.7	−8.3
		Ulaan Uul	−27.5	12.6	−5.3
4	1988	Renchin Lkhumbe	−34.0	11.9	−7.7
		Ulaan Uul	−30.5	11.3	–
5	2002	Tsagaan Nuur	−27.2	18.2	−5.7
		Renchin Lkhumbe	−26.9	16.3	−5.8
		Ulaan Uul	−30.9	17.9	−3.3
6	2003	Tsagaan Nuur	−30.7	15.6	−6.7
		Renchin Lkhumbe	−30.4	14.3	−6.7
		Ulaan Uul	−29.0	15.9	−3.3

Khuvsgul Mountain Range extends from the north to the south, and it consists of isolated massifs at the elevation range of 2500–3000 m a.s.l. with the highest peak Mt. Munkh[6] Saridag[7] (3491 m a.s.l.) (Orkhonselenge, 2016; Fig. 15.4). There are abundant paleoglacier traces in the Khuvsgul Mountain Range (see Sect. 15.2.4). The high mountains in the Khuvsgul Mountain Range, located at the northern corner of Mongolia, serve as watersheds including large lakes and rivers of the North Arctic Ocean drainage basin (Orkhonselenge, 2016). The Burekht (2156 m a.s.l.), Khantai (2150 m a.s.l.), and Buteel (1814) mountain ranges in the eastern Selenge River drainage basin are included in this region, and these ranges extend from the southwest to the northeast.

There are three large tectonic rift depressions, namely, *Khuvsgul*, *Darkhad*, and *Bus* from the east to the west, which are the southwestern end of the BRZ. The first depression *Khuvsgul* is the easternmost (Fig. 15.3). The second depression *Darkhad*, lying at 1550–1650 m a.s.l., is isolated from the Khuvsgul depression by the Bayan (or Bayan Zurkh) in the northeast and Khoridol Saridag mountain ranges (2600–3200 m a.s.l.) in the southeast, and from the third depression *Bus* by the Ulaan Taiga mountains (2500–3100 m a.s.l.) in the west (Fig. 15.3). The Darkhad basin, extending for about 140.0 km in length from the north to the south and 70 km

[6] *Munkh* has been often misspelled as *Munku* or *Munh* based on the Russian pronunciation in the international literature. Munkh is the correct English transliteration from Mongolian *Мөнх*.

[7] *Saridag* has been also misspelled as *Sardyk* or *Sardig* by the Russians internationally. Saridag is the correct English transliteration from Mongolian *Сарьдаг*.

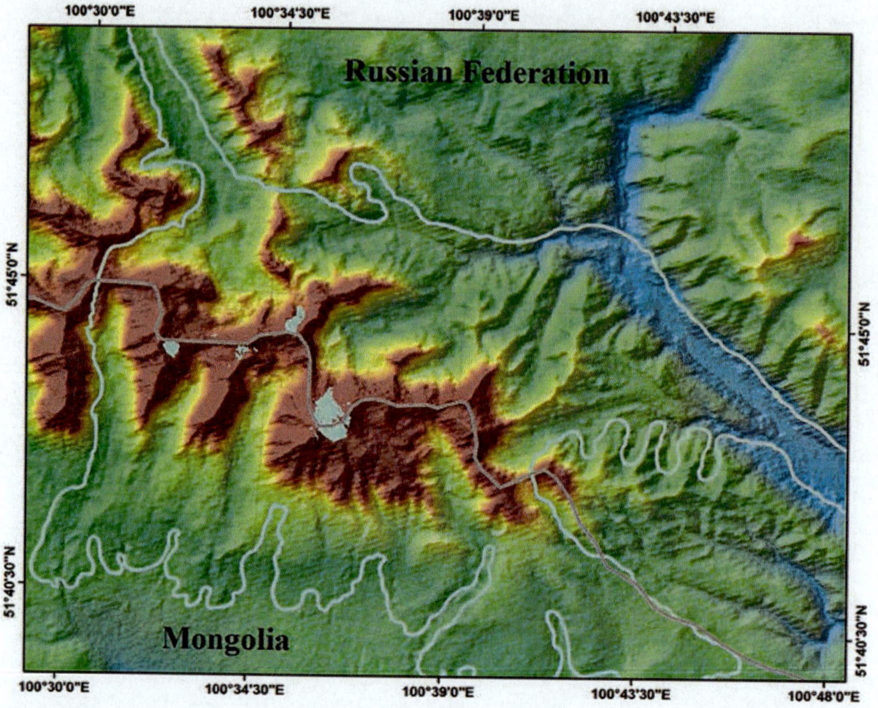

Fig. 15.4 ASTER image of Mt. Munkh Saridag (Fig. 15.3) with the maximum paleoglacier reconstruction based on the spatial distributions of glacial landforms. Modified from Orkhonselenge (2016)

in width from the west to the east, is a wide plain dotted with many lakes, paleoshorelines, paleolake terraces, rivers, swamps, marshes, sand dunefields, and pingos. This basin is bound by the Sayan Mountain Range in the north, Bayan (or Bayan Zurkh) Mountain Range in the northeast, Khoridol Saridag Mountain Range in the southeast, and Mt. Ulaan Taiga in the west (Fig. 15.3). The basin is divided into two areas: Deed Tsagaan in the south and Dood Tsagaan in the north. The Deed Tsagaan area includes only small lakes, swamps, and bogs, but the Dood Tsagaan area contains large remnant lakes (e.g., Dood, Dood Tsagaan, Targan), rivers, swamps, and pingos.

15.2.4 Glaciers and Permafrost

Evidence for paleo- and modern glaciers is observed in the landforms of the region. The glaciers in northern Mongolia during MIS 2 were the smallest in sizes due to the high regional aridity (Fedotov et al., 2015). In the Khuvsgul Mountain Range, there are abundant paleoglacier remnants such as postglaciers pediments, wide glacial depressions, and large glacial lakes between Khoridol Saridag Mountain Range, Bayan (or Bayan Zurkh) Mountain Range, Ulaan Taiga Mountain Range, Shishkhid River basin, and the head of Uur River (Orkhonselenge, 2016; Fig. 15.3). The Khuvsgul Mountain Range experienced glaciations twice (Tsegmid, 1969) centered on Mt. Munkh Saridag (Jigj, 1976). The Mt. Munkh Saridag (Fig. 15.4) still hosts remnants of the paleoglaciers indicating the modern glacier dynamics with glacial cirques at 3000 m a.s.l. and moraines at 2705 m a.s.l. (Orkhonselenge, 2016). The modern glacier complex in Mt. Munkh Saridag has formed only during the last 4–5 ka (Kitov et al., 2015). For example, there is a small glacier (0.12 km^2) on the Mongolian side of Mt. Munkh Saridag (Osipov et al., 2013). This modern glacier has melted and decreased its size from 0.68 to 0.34 km^2 since the end of the Little Ice Age, the tendency accelerating especially in recent years (Kitov et al., 2015). The modern glaciers on Mt. Munkh Saridag have decreased by melting from 0.95 km^2 in 1940 to 0.86 km^2 in 2010 (Davaa, 2015), and the changes of the modern glacier areas in Mt. Munkh Saridag progressed as follows: 0.9 km^2 in 1970, 0.85 km^2 in 1986, 0.55 km^2 in 2000, and 0.38 km^2 in 2007 (Orkhonselenge, 2016).

In the Darkhad basin, there are glacial remnant lakes and paleolake terraces. It is hypothesized that an ice field spanning over the Sayan Mountain Range and Ulaan Taiga Mountain Range or extensive intermontane valley glaciers during the different ice ages blocked the Shishkhid River that is part of the upper Yenisei River system, and these conditions formed a great palaeolake within the Darkhad basin (e.g., Grosswald & Rudoy, 1996; Krivonogov et al., 2005; Gillespie et al., 2008; Komatsu et al., 2009). This paleolake Darkhad has drained catastrophically westward though the Yenisei River drainages probably more than once during the late Pleistocene and possibly during the Holocene, leaving many geomorphological and sedimentological traces along the river valley (e.g., Grosswald & Rudoy, 1996; Krivonogov et al., 2005; Komatsu et al., 2009, 2016; Batbaatar & Gillespie, 2016a, 2016b). The cause of the draining is considered to be collapse of the ice or sedimentary dam. The remains of end, medial, and lateral moraines in the valleys of Tengis and Shishkhid Rivers (Fig. 15.3) and intermontane valleys reveal that large glaciers advanced in the late Pleistocene (Krivonogov et al., 2005). The late Pleistocene glaciers around the Darkhad basin advanced twice at 35–53 ka and 17–19 ka BP and at least once earlier (Gillespie et al., 2008).

Perennial permafrost of Mongolia is widely distributed in northern Mongolia where seasonal freezing and thawing depth of permafrost ranges around 1.3–4.5 m on average (Tumurbaatar, 2004; Table 15.3). The seasonal freezing and thawing depth of permafrost in the Khatgal (Fig. 15.3) was 3.6 m in 1969, 4.0 m in 1983, 4.7 m in 2004 (Sharkhuu et al., 2007), and 5.96 m in 2010 (Davaa, 2015). This shows that permafrost is rapidly melting due to the recent climate change. Permafrost and swamps are widely developed around Lake Khuvsgul (Orkhonselenge et al., 2014). Perennial, intermittent, and spotted types of permafrost are distributed in northern Mongolia with predominance of perennial in the center, intermittent permafrost in the west, and spotted permafrost in the east (Jambaljav et al., 2016). In the Khuvsgul Mountain Range, the seasonal freezing and thawing depth is 2.0–4.5 m under the annual average air temperature of −3.0 to −0.1 °C, while in the

Table 15.3 Seasonal freezing and thawing depths in northern Mongolia

#	Places	Materials	Depth (m)	Year	Depth (m)	Year
1	Khuvsgul region	Clays	1.3–1.5			
		Sands and gravels	2.5–4.5			
2	Khatgal region	Gravels	3.2–4.3	1978	1.1	1998
3	Darkhad basin	Sands	1.3–2.2			
4	High mountainous region	Near river and lake	3.5–4.5			
		Limestone, granite, and phosphorite	2.8–3.2			

Table 15.4 Cold and warm permafrost in the lake and river drainage basins of northern Mongolia

Places	Basins	Area of the basins (km²)	Permafrost in the basins (%)		AMGT (°C)	ALT (m)
			Cold permafrost (AMGT < −2 °C)	Warm permafrost (−2 °C < AMGT < 0 °C)		
Khuvsgul	Delger-Bugsei rivers	23,018	35.8	37.9	−2.3	2.6–5.2
	Lake Khuvsgul-Eg River	41,321	23.7	61.4	−1.7	1.5–4.4
	Shishkhid River	20,096	46.0	51.7	−4.1	1.7–7.2
Khangai	Selenge River	30,983	0.6	21.7	−0.7	1.4–3.6

intermontane valleys the seasonal freezing and thawing depth is 4.0 m under the annual average air temperature of −0.7 to −0.1 °C (Jambaljav et al., 2016).

The active layer within the permafrost area of the Darkhad basin ranges between 1.45 m and 1.74 m in depth depending on the vegetation type, while it is about 1.5 m in the marsh (Takeda et al., 2009). In the periglacial areas of the Darkhad basin, actions of frost and ice are shaping landforms and landscapes (Fuji et al., 2009). The permafrost in lake and river drainage basins in the Khuvsgul and northern Khangai mountain ranges is defined as cold and warm depending on annual mean ground temperature (AMGT) by Zorigt et al. (2020), and it is shown with its active layer thickness (ALT) of 1.4–7.2 m in Table 15.4.

Table 15.5 Water resource of drainage basins in northern Mongolia

#	Drainage basin	Area (km^2)	Surface water resource (million m^3/year)			Groundwater resource (million m^3/year)	
			Total resource	Environmental	Possible use	Potential exploitable	Exploitable resource
1	Selenge River	31,395	2133	1856	277.3	697	90.3
2	Lake Khuvsgul-Eg River	41,871	2971	2570	401.1	432	0.2
3	Shishkhid River	20,362	519	481	39.0	206	0.2
4	Delger River	23,324	1080	999	81.0	229	2.7
	Total	116,952	6703	5906	798.4	1564	93.4

15.2.5 Groundwater

Groundwater in the Precambrian and Paleozoic lithology dominantly resides in metamorphic, igneous, and intrusive rocks within the Khuvsgul Mountain Range (Tsegmid, 1969). The groundwater in intrusive rocks of the Khuvsgul Mountain Range has an average discharge rate of 0.15–10.0 l/s and a maximum rate of 100.0 l/s at the depth of 14.8–152.0 m, whereas the groundwater in volcanic rocks has a discharge rate of 0.01–12.0 l/s (Tserensodnom, 1971).

Table 15.5 shows surface water and groundwater resources in large lake and river drainage basins in northern Mongolia (Ministry of Environment and Green Development, 2013). The groundwater resources are annually recharged, mainly through infiltration of surface water. In recent years, a decrease in the groundwater level has been observed in Murun, and the range of the decrease reached 0.63 m in 2000–2013 (Institute of Meteorology, Hydrology and Environment, 2013).

The groundwater in the Darkhad basin is hydrochemically dominated by Ca^{2+} and HCO_3^-, and the average value of mineralization in the groundwater varies from 103.5 mg/L to 327.9 mg/L (Narantsogt et al., 2009).

15.2.6 Surface Water

Major surface water resources in northern Mongolia (Table 15.5) are rivers, lakes (see Sect. 15.3), and meltwaters of snow cover, modern glaciers, and permafrost. The region has numerous large rivers (e.g., Tengis, Shishkhid, Delger, Eg (Fig. 15.3), Azarga, Zelter, Ider, and Selenge rivers). The rivers draining the Khuvsgul and northern Khangai mountain ranges in northern Mongolia are the headwater of large rivers and lakes (e.g., Lake Baikal, Yenisei River, Ob River) in the North Arctic

Fig. 15.5 Selenge River flowing under the bridge in the Ulaan Ud City in southeastern Russia viewing toward the east. Photo by A. Orkhonselenge on July 24, 2016

Ocean drainage basin. For example, in the Khuvsgul Mountain Range, the density of rivers is high, and most of the rivers drain into Tengis and Shishkhid rivers in the northwest (Fig. 15.3), Lake Khuvsgul in the center, and Eg and Selenge rivers in the southeast of the region.

Selenge River (Fig. 15.5), occupying a large drainage area (Tables 15.4 and 15.5) in high mountainous regions of northern Mongolia, feeds Lake Baikal, the largest freshwater lake by volume in the world. Moreover, Shishkhid River (Figs. 15.3 and 15.6), the largest in the Darkhad basin, feeds Yenisei River in the eastern Sayan Mountain Range through various drainages in the basin (e.g., Mungarag, Khug, Sharga, Tengis, Bayan, Alkhan, Ongolog, Jarai, Khodon, Arsai, Jargalant, and Khogorog rivers in Fig. 15.6). The rivers in the Darkhad basin are primarily dominated by Ca^{2+} in the range of 15–28 mg/L and HCO_3^- of 130.9–170.5 mg/L (Narantsogt et al., 2009). For instance, the average value of mineralization is 63.2 mg/L in Shishkhid River, 25.6 mg/L in Deed Turuu River, and 103.4 mg/L in Sharga River (Narantsogt et al., 2009).

Lakes in northern Mongolia (see Sect. 15.3) play very important role for the surface water in southeastern Siberia and hydrological dynamic regime in the North Arctic Ocean drainage basin. The lakes have dominantly tectonic, glacial, fluvial, and thermokarst origins (see Chap. 4), and they are deep and freshwater (see Chap. 5).

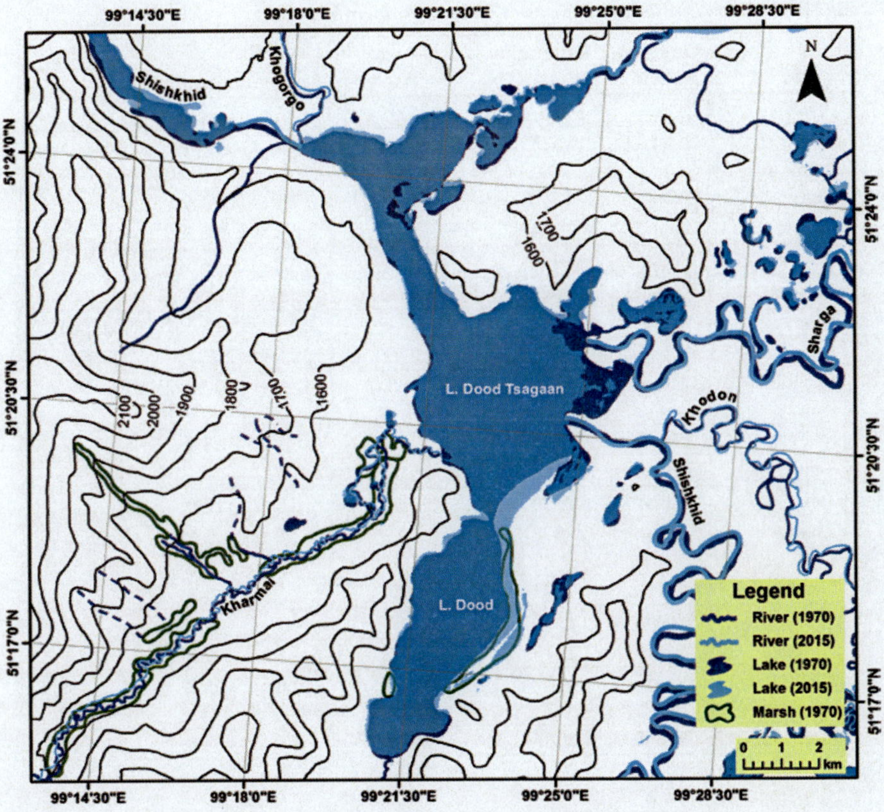

Fig. 15.6 Spatial distributions of areas and marshes of Lakes Dood and Dood Tsagaan in 1970 and 2015. Modified from Orkhonselenge and Uuganzaya (unpublished)

15.3 Lakes

Lakes in northern Mongolia (Fig. 1.1) are the lowest in number but the second largest in area (Table 1.2). The large lakes distributed in high mountains of the region are correlated with a high precipitation range of 250–500 mm/year and associated with dense river networks (see Chap. 5). There are 236 lakes each covering an area larger than 0.1 km^2 within northern Mongolia, and they constitute 7.7% of the total number of lakes in Mongolia (see Chaps. 1 and 5). The lakes occupy an area of 3297.4 km^2 in total, and it constitutes 20.6% of the total area of lakes in Mongolia. Table 15.6 lists large lakes in northern Mongolia covering an area over 5.0 km^2. These lakes cover 95% of total lakes in the region, i.e., they cover an area of 3132.3 km^2 in total. In the region, lakes elevated at ranges of 1501–2000 m a.s.l. and

Table 15.6 Morphometric components of lakes in northern Mongolia

#	Lakes	Origin/ genesis	Elevation (m a.s.l.)	Area (km^2)	Depth (m)		Details
					Average	Maximum	
1	Airkhan	Oxbow	936	5.0	1.3	2.0	–
2	Dood Tsagaan	Glacial	1538	31.3	4.7	17.0	Section 15.4.2
3	Emt	Oxbow	1540	5.0	1.1	2.5	–
4	Erkhel	Tectonic	1544	13.2	0.9	4.5	–
5	Gandan	Tectonic	1988	9.6	4.8	9.5	–
6	Jugnai	Tectonic	1997	27.0	5.4	9.5	–
7	Khargal	Oxbow	1071	13.6	9.3	15.6	Chapter 17
8	Khuvsgul	Tectonic	1645	2760	138.0	262.4	Chapter 16
9	Sangiin Dalai	Tectonic	1888	165.0	12.0	30.0	–
10	Sharga	Volcanic	1335	12.4	3.7	5.9	–
11	Shar Gert	Glacial	2160	5.0	1.7	3.5	–
12	Taij	Glacial	2330	7.5	6.0	12.0	–
13	Targan	Glacial	1538	19.9	2.8	5.5	Section 15.4.3
14	Tsegeen	Tectonic	1523	7.2	2.5	4.8	–
15	Tsoitson	Oxbow	1539	7.5	1.4	3.0	–
16	Tunamal	Tectonic	1874	20.2	5.5	9.9	–
17	Tushig Tsagaan	Tectonic	680	5.8	0.6	1.5	–
18	Zuun	Tectonic	2006	17.5	8.3	17.6	–

2001–2500 m a.s.l. are abundant among the total 236 lakes (Table 4.1), and 84 lakes at 1501–2000 m a.s.l. constitute 30.9% of the total 271 lakes with the same elevation range in Mongolia, while 89 lakes at 2001–2500 m a.s.l. constitute 32.9% of the total 270 lakes with the same elevation range in Mongolia (see Chap. 5). Most lakes in the region cover an area less than 5.0 km^2 (Table 4.1); however, their areas are ranked at the lowest level in the country, i.e., they constitute only 7.5% of the total lakes in the whole country (see Chap. 5).

All the lakes in the Khuvsgul mountainous region are found in two large depressions: (1) Darkhad (or Shishkhid) and (2) Khuvsgul (Tserensodnom, 1971). The lakes in these two depressions have tectonic and glacial origins, and they record different development histories during the Pleistocene and Holocene. For instance, the paleolakes in the Khuvsgul stood at lower levels than the present level, while the lakes in the Darkhad were very large during the Pleistocene glaciations. Moreover, the lakes in the paleolake Darkhad basin have shrunk or decreased in the Holocene, but Lake Khuvsgul is deep and large today. In the region, the largest lakes are Lake

Khuvsgul in the Khuvsgul basin and Lake Dood Tsagaan in the Darkhad (or Shishkhid) basin. The lakes in the Khuvsgul basin are tectonically originated, whereas the lakes in the Darkhad (or Shishkhid) basin are paleoglacially formed in thermokarst and permafrost zones (see Chap. 4).

Large tectonic lakes such as Lake Khuvsgul (see Chap. 16) and remnants of the paleolake Darkhad are distributed in this region. Glacial Lakes Semsee, Pine, Shar, and Taij are found in the Khuvsgul Mountain Range (Tserensodnom, 1971; Orkhonselenge, 2016). Thermokarst lakes are distributed in the Sharga River and Ar Sair River valleys of the Darkhad basin (see Chap. 4). For example, Lake Dood Tsagaan (Fig. 15.6) in the Darkhad basin is a glacial lake (e.g., Krivonogov et al., 2005; Narantsetseg et al., 2013). Lakes Erkhel and Khargal (see Chap. 17) are oxbow lakes. In the region, many lakes are located along or nearby river channels. Moreover, riverside lakes in the Eg and Selenge River drainage basins depend on annual and seasonal variations of Selenge River discharge (Kang et al., 2015).

Lakes in the Khuvsgul Mountain Range are different from other mountainous lakes regarding water regime and mineralization level. For example, several tens of lakes including Lakes Semsee, Pine, Taij, and Shar lying above 2000 m a.s.l. are ice-free for only 2–3 months in summer, and their hydrothermal resource potentials and mineralization levels are low. In addition, because snow cover, permafrost, and groundwater contributed to water inputs into the lakes, they are characterized by extremely freshwater.

In the Khuvsgul mountainous region, there are numerous lakes that have similar origins, water regimes, and water sources, but 75% of them are small and each occupies an area less than 10.0 km^2 and 50–60 m deep (Tserensodnom, 1971). These lakes are connected by channels, rivers, and streams, and they are included in the North Arctic Ocean drainage basin.

Lakes in the Khuvsgul basin (e.g., Lakes Semsee, Taij, Pine, Shar) are located above 2000 m a.s.l., and these lakes have low thermal resources and high snow-water resources. The tectonically originated freshwater Lake Khuvsgul, the deepest lake in Mongolia, holds a large water volume of 383.7 km^3 (see Chap. 16).

In the Darkhad basin, there are remnant glacial lakes (e.g., Lakes Dund, Dood, Dood Tsagaan (Fig. 15.6), Targan (Fig. 15.7)). There are many freshwater lakes and some mineralized lakes in the Darkhad basin, where their depths vary around 5.0–10.0 m on average and reach the maximum of 20.0 m. Lakes (e.g., Deed Tsagaan, Duuren, Tsetsegt, Tovgiin Turuu, Tuvkhun, and Khort) in the basin are dominated by HCO_3^- ranging in concentration of 101.3–175.6 mg/L (Narantsogt et al., 2009; Table 15.7). There are some sodium lakes (e.g., Lake Tuvkhun), and sediments on the exposed shoreline of Lake Tuvkhun show abundant Mg^{2+}, Ca^{2+}, and Na^+ (Nakagawa et al., 2009).

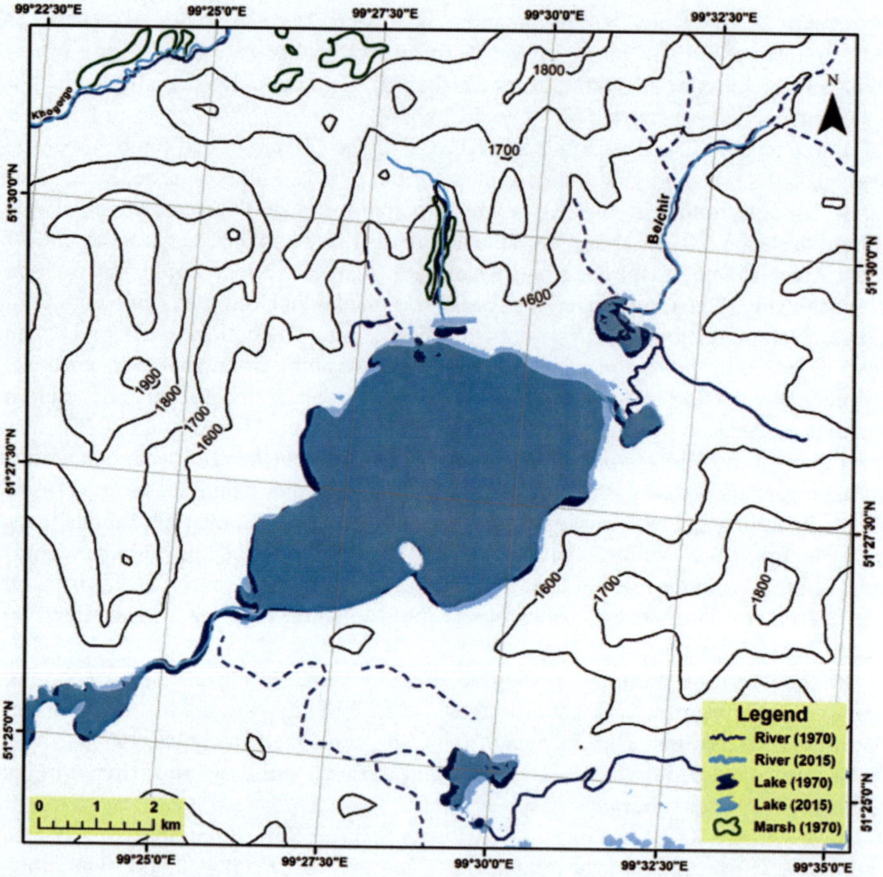

Fig. 15.7 Spatial distribution of areas and marshes of Lake Targan in 1970 and 2015. Modified from Orkhonselenge and Uuganzaya (unpublished)

15.4 Lake Evolution

Lakes in the region rapidly respond to the recent climate change. The rising annual and monthly average air temperatures impact on lacustrine environments. However, each lake depending on its locality or physiographic condition responds to the changes in air temperature and precipitation differently. In northern Mongolia, the temporal variations of lake area in the Bulgan area in the eastern part of the region showed a generally decreasing pattern; however, in the Murun area in the western part of the region, the lake areas were similar across different years with absolute relative changes of only 0.6% (Kang et al., 2015). In the Bulgan area, Lake Khargal shows limited fluctuations in its area (see Chap. 17).

Table 15.7 Hydrochemical composition of lakes in the Darkhad basin

#	Lakes	Mineralization value (mg/L)	Ions (mg/L)									
			Ca^{2+}	Mg^{2+}	Cl^{-}	NO_2^{-}	NH_3^{+}	Fe^{2+}	HNO_3^{-}	F^{-}	HCO_3^{-}	SO_4^{2-}
1	Deed Tsagaan	297.6	3.0	3.6	28.0	0.8	0.08	0.64	0.01	1.23	175.6	32.53
2	Duuren	127.1	12.0	27.6	32.0	0.9	0.06	0.58	0.02	1.12	173.6	39.7
3	Tsetsegt	98.9	4.0	12.0	13.0	0.6	0.2	0.2	0.002	0.73	101.3	24.0
4	Tovgiin Turuu	270.6	11.8	16.9	10.0	0.3	0.15	0.01	0.006	0.21	180.5	37.6
5	Tuvkhun	295.3	60.1	255.1	24.8	3.1	0.31	0.05	0.02	0.91	241.9	139.1
6	Khort	73.2	40.0	28.2	14.5	0.9	0.13	0.13	0.002	0.78	160.9	35.8

In the Khuvsgul Mountain Range, a glacial lake on Mt. Munkh Saridag decreased in area from 0.726 km^2 in 1970 to 0.624 km^2 in 2007 (Orkhonselenge & Harbor, 2018). Moreover, Lake Khuvsgul has lost an area during the past 45 years (see Chap. 16).

In the Darkhad basin, 84 lakes within the perennial permafrost zone with the total water surface area of 3.38 km^2 newly formed between 1962 and 2006, and 173 lakes increased their total areas by 3.84 km^2 (Saruulzaya, 2017). Although the Darkhad basin is drying under the continuous air temperature rise (Table 15.2), Lakes Dood and Targan increased in an area over the last 45 years, while Lake Dood Tsagaan lost its area in the same period (Figs. 15.6 and 15.7). Indeed, the Lakes Targan, Dood, and Dood Tsagaan form a common hydrological system.

15.4.1 *Lake Dood*

Lake Dood is located at the elevation of 1538 m a.s.l. in the northernmost corner of the Darkhad basin (Fig. 15.3). Lake Dood, an oligotrophic freshwater lake (Tserensodnom, 2000), is 13 km long and 1–3 km wide, and it has the maximum depth of about 15 m (Peck et al., 2001). Lake Dood gained an area of 1.12 km^2 changing from 11.87 km^2 in 1970 to 12.99 km^2 in 2015 (Orkhonselenge and Uuganzaya, unpublished) when the northeastern and eastern shores expanded dramatically (Fig. 15.6).

15.4.2 *Lake Dood Tsagaan*

Lake Dood Tsagaan is located at the elevation of 1538 m a.s.l. in the northernmost corner of the Darkhad basin. Lake Dood Tsagaan occupies an area of 31.3 km^2 (Tserensodnom, 2000; Table. 15.6) or 30.9 km^2 (Orkhonselenge and Uuganzaya, unpublished; Fig. 15.6). Lake Dood Tsagaan is fed by large rivers such as Deed Shishkhid, Khodon, Sharga, Kharmai, and Arsai rivers (see Sect. 15.2.6) draining from an area of 1110 km^2 (Tserensodnom, 1971) or 171.3 km^2 (Tserensodnom, 2000), and it feeds Yenisei River through Dood Shishkhid River in the west. Lake Dood Tsagaan decreased its area of 1.08 km^2 changing from 30.99 km^2 in 1970 to 29.91 km^2 in 2015 (Orkhonselenge and Uuganzaya, unpublished), and it lost its area on the northeastern and eastern shores, although the lake gained some area on the southeastern shore (Fig. 15.6).

A history of Lakes Dood and Dood Tsagaan in the paleolake Darkhad basin during the late glacial and the early to middle Holocene was presented by Narantsetseg et al. (2013) (see Chap. 21). Moreover, pronounced weathering/sediment influx changes and water level fluctuations in the Lakes Dood and Dood Tsagaan during the Holocene were identified (Peck et al., 2001).

15.4.3 Lake Targan

Lake Targan is located at the elevation of 1538 m a.s.l. in the northernmost corner of the Darkhad basin (Figs. 1.1 and 15.3). Lake Targan is one of the tectonic and glacial remnant lakes of the great paleolake Darkhad, and the lake extends for 7.5 km in length, 4.6 km in width, and 26.9 km along the shoreline (Tserensodnom, 2000). Lake Targan's water depth ranges up to 10.0 m (Tserensodnom, 1971) or 2.8 m on average and 5.5 m at its maximum (Tserensodnom, 2000). Lake Targan is fed by small rivers and streams in the north and northeast, and it is connected with Lakes Dood Tsagaan and Dood (Fig. 15.6) through a wide canal in the southwest (Fig. 15.7). The lake water is classified into the hydrocarbonate and sulfate type (Tserensodnom, 2000). The mineralization value of the lake is about 187.9–366.0 mg/L with the dominant ion of HCO_3^- (Tserensodnom, 1971) and 149.7 mg/L (Tserensodnom, 2000).

Lake Targan occupies an area of 19.9 km^2 (Tserensodnom, 2000; Tables 15.6 and 15.8) or 20.7 km^2 (Orkhonselenge and Uuganzaya, unpublished; Table. 15.8). The lake increased its area by 0.6 km^2 between 1970 and 2015 (Fig. 15.7). The linear regression analysis shows that the lake area between 2000 and 2015 was moderately related to the annual average air temperature (R^2 = 0.6653), but there is no clear relationship between the lake area and the annual precipitation (R^2 = 0.0039). The correlation analysis shows that the lake area during 2000–2015 had a strong positive correlation with the annual average air temperature (r = 0.815), but no clear correlation with the annual precipitation (r = 0.062). The gaining of the lake area during the last 45 years coincided with the increase of the annual precipitation, possibly linked to the recent rapid melting of permafrost in the Darkhad basin due to the gradual rise of the annual average air temperature.

Table 15.8 Temporal changes in area of Lake Targan

#	Year	Area (km^2)	Authors
1	1970	20.1	Orkhonselenge and Uuganzaya, unpublished
2	2000	19.9	Tserensodnom (2000)
3	2010	20.3	Davaa (2015)
4	2013	21.4	Davaa (2015)
5	2015	20.7	Orkhonselenge and Uuganzaya, unpublished

15.5 Summary

Northern Mongolia shows a spectacular landscape consisting of high mountains, tectonic rift depressions, modern glacier, permafrost, large rivers, and lakes. In northern Mongolia, tectonic, glacial, fluvial, and thermokarst lakes are dominantly distributed, and lakes in this region are the lowest in number (7.7%) but the second largest in area (20.6%) in the country. The recent lake evolution in the region provides a wider context for Mongolia, Eurasia, and Central Asia. Because the geographical location of northern Mongolia is under the dominant presence of the Siberian high-pressure cell or the East Asian winter monsoon and the westerlies, the region is an archive of dynamic changes of atmospheric, cryospheric, and hydrospheric systems in Eurasia and associated landforms. Due to its high elevations and intercontinental location on the Central Asian Orogenic Belt and the southern Siberian taiga landscape, northern Mongolia is particularly important as a geosphere of Eurasia and Central Asia. The information on the paleo- and modern-day climate changes in Eurasia and Central Asia is expressed in geomorphology, sedimentology, geochemistry, and hydrology of northern Mongolia.

References

Badarch, G., Cunningham, W. D., & Windley, B. F. (2002). A new terrane subdivision for Mongolia: Implications for the Phanerozoic crustal growth of Central Asia. *Journal of Asian Earth Sciences, 21*, 87–110.

Batbaatar, J., & Gillespie, A. R. (2016a). Outburst floods of the Maly Yenisei. Part I. *International Geology Review, 58*, 1723–1752.

Batbaatar, J., & Gillespie, A. R. (2016b). Outburst floods of the Maly Yenisei. Part II—New age constraints from Darhad basin. *International Geology Review, 58*, 1753–1779.

Batbaatar, J., Gillespie, A. R., Fink, D., Matmon, A., & Fujioka, T. (2018). Asynchronous glaciations in arid continental climate. *Quaternary Science Reviews, 182*, 1–19.

Batrom, B., Tulgaa, U., Batbayar, N., Tumursukh, N., Adiyasuren, D., & Sukkhbaatar, B. (2006). *Report: General exploring and geological mapping in a scale of 1:50,000 of Olon Nuur area in Zed metallogenic region*. Geological Information Center Press. [In Mongolian].

Bogoyavlensky, B.A., Batjargal, B., Martinov, V.P., Kozlov, U.P., Borobiev, V.V., & Batsuuri, D. (1989). Atlas of Lake Khuvsgul: Mongolian People Republic. State Printing Industry of Geodesy and Cartography, SSSR, 118 p. [In Russian with an abstract in English, German and Mongolian].

Davaa, G. (2015). *Resource and regime of surface water of Mongolia*. 408 p. [In Mongolian].

Donskaya, T. V., Windley, B. F., Mazukabzov, A. M., Kröner, A., Sklyarov, E. V., Gladkochub, D. P., Ponomarchuk, V. A., Badarch, G., Reichow, M. K., & Hegner, E. (2008). Age and evolution of late Mesozoic metamorphic core complexes in southern Siberia and northern Mongolia. *Journal of the Geological Society, 165*, 405–421.

Fedotov, A. P., Ignat'ev, A. V., & Velivetskaya, T. A. (2015). Reconstruction of deglaciation of Northern Mongolia for the last 330 ka BP, inferred from ostracod stable isotope records from Lake Khubsugul. *Environmental Earth Science, 74*, 2041–2054.

Fuji, N., Khadbaatar, S., Ulambayar, N., & Nasanjargal, E. (2009). Geoscientific and Ecosystematic Researches of the Darhadyn Wetland Basin, Hovsgol Province, Northern Mongolia. In H. Nishida & T. Jamsran (Eds.), *Darhadyn Wetland in Mongolia: Synthesis investigation on ecosystems* (pp. 165–201). NPO Mongolia Ecology Information Center/Alex Company Press.

Gillespie, A. R., Burke, R. M., Komatsu, G., & Bayasgalan, A. (2008). Late Pleistocene glaciers in Darhad Basin, northern Mongolia. *Quaternary Research, 69*(2), 169–187.

Goulden, C. E., Sitnikova, T., Gelhaus, J., & Bazartseren, B. (2006). *The Geology, Biodiversity and Ecology of Lake Hövsgöl (Mongolia)*. Backhuys Publishers, 525 p.

Grosswald, M. G., & Rudoy, A. N. (1996). Quaternary glacier-dammed lakes in the mountains of Siberia. *Polar Geography, 20*(3), 180–198.

Institute of Meteorology, Hydrology and Environment. (2013). *Water and weather of Mongolia*. 27 p. [In Mongolian].

Jambaljav, Y., Gansukh, Y., Temuujin, K., Tsogt-Erdene, G., Undrakhtsetseg, T., Saruulzaya, A., Amarbayasgalan, Y., Dashtseren, A., & Narangerel, S. (2016). *Report of New Map of Permafrost Distribution in Mongolia: Scale 1:1000000*. Institute of Geography and Geoecology, Mongolian Academy of Sciences, 90 p.

Jigj, S. (1975). *Primary feature of Mongolian landforms*. Institute of Geography & Permafrost. Mongolian Academy of Sciences (MAS), 126 p. [In Mongolian].

Jigj, S. (1976). *A brief description of paleoglaciations and paleoglaciers*. 51 p. [In Mongolian].

Kang, S., Lee, G., Togtokh, C., & Jang, K. (2015). Characterizing regional precipitation-driven lake area change in Mongolia. *Journal of Arid Land, 7*(2), 146–158.

Kashiwaya, K., Ochiai, S., Sumino, G., Tsukamoto, T., Szyniszewska, A., Yamamoto, M., Sakaguchi, A., Hasebe, N., Sakai, H., Watanabe, T., & Kawai, T. (2010). Climato-hydrological fluctuations printed in long lacustrine records in Lake Hövsgöl. *Mongolia Quaternary International, 219*, 178–187.

Kitov, A. D., Kovalenko, S. N., Plyusnin, V. M., & Suvorov, E. G. (2015). Modern changes of the high-mountain landscapes and glaciation in Southern Siberia (Russia) by the example of the Eastern Sayan mountains. *Environmental Earth Sciences, 74*(3), 1931–1946.

Komatsu, G., Arzhannikov, S. G., Gillespie, A. R., Burke, R. M., Miyamoto, H., & Baker, V. R. (2009). Quaternary paleolake formation and cataclysmic flooding along the upper Yenisei River. *Geomorphology, 104*, 143–164.

Komatsu, G., Baker, V. R., Arzhannikov, S. G., Gallagher, R., Arzhannikova, A. V., Murana, A., & Oguchi, T. (2016). Catastrophic flooding, palaeolakes and late Quaternary drainage reorganization in northern Eurasia. *International Geology Review, 58*, 1693–1722.

Kravchinsky, V. A., Sklyarov, E. V., Gladkochub, D. P., & Harbert, W. P. (2010). Paleomagnetism of the Precambrian Eastern Sayan rocks: Implications for the Ediacaran–Early Cambrian paleogeography of the Tuva-Mongolian composite terrane. *Tectonophysics, 486*, 65–80.

Krivonogov, S. K., Sheinkman, V. S., & Mistruykov, A. A. (2005). Stages in the development of the Darhad dammed lake (Northern Mongolia) during the Late Pleistocene and Holocene. *Quaternary International, 136*(1), 83–94.

Krivonogov, S. K., Yi, S., Kashiwaya, K., Kim, J. C., Narantsetseg, T., Oyunchimeg, T., Safonova, I. Y., Kazansky, A. Y., Sitnikova, T., Kim, J. Y., & Hasebe, N. (2012). Solved and unsolved problems of sedimentation, glaciation and paleolakes of the Darhad Basin, Northern Mongolia. *Quaternary Science Reviews, 56*, 142–163.

Kröner, A., & Rojas-Agramonte, Y. (2014). The Altaids as seen by Eduard Suess, and Present Thinking on the Late Mesoproterozoic to Palaeozoic Evolution of Central Asia. *Austrian Journal of Earth Sciences, 107*(1), 156–168.

Lydolph, P. E. (1977). Climates of the Soviet Union. In H. E. Landsberg (Ed.), *World survey of climatology* (Vol. 7)., 443 p.

Ministry of Environment and Green Development. (2013). *Integrated Water management plan of Mongolia.* 339 p.

Nakagawa, M., Suzuki, M., & Nishida, H. (2009). Ingredient of soil at four points in Darhadyn Wetland. In H. Nishida & T. Jamsran (Eds.), *Darhadyn Wetland in Mongolia: Synthesis investigation on ecosystems* (pp. 203–208). NPO Mongolia Ecology Information Center/Alex Company Press.

Narantsetseg, T., & Khishigsuren, S. (2017). Classification and description of Quaternary deposits in Mongolia. *Khaiguulchin, 58*, 65–70.

Narantsetseg, T., Krivonogov, S. K., Oyunchimeg, T., Uugantsetseg, B., Burr, G. S., Tomurhuu, D., & Dolgorsuren, K. (2013). Late Glacial to Middle Holocene climate and environmental changes as recorded in Lake Dood sediments, Darhad Basin, northern Mongolia. *Quaternary International, 311*, 12–24.

Narantsogt, N., Suzuki, Y., & Yunden, M. (2009). Physical components and Chemical compositions of Surface water and Groundwater in Darhadyn Hotgor Wetland. In H. Nishida & T. Jamsran (Eds.), *Darhadyn Wetland in Mongolia: Synthesis investigation on ecosystems* (pp. 227–234). NPO Mongolia Ecology Information Center/Alex Company Press.

Nishida, H., & Jamsran, T. (2009). *Darhadyn Wetland in Mongolia: Synthesis investigation on ecosystems*. NPO Mongolia Ecology Information Center/Alex Company Press, 539 p.

Orkhonselenge, A. (2016). Glacial geomorphology of Mt. Munkh Saridag in the Khuvsgul Mountain Range, Northern Mongolia. *Géomorphologie: Relief, Processes, Environnement, 22*(4), 387–396.

Orkhonselenge, A., & Harbor, J. M. (2018). Impacts of modern glacier changes on surface water resources in western and northern Mongolia. *Journal of Water Resource and Protection, 10*(6), 559–576.

Orkhonselenge, A., Krivonogov, S. K., Mino, K., Kashiwaya, K., Yamamoto, M., & Nakamura, T. (2014). Holocene landform evolution of lake Khuvsgul basin, Mongolia. *Géomorphologie: Relief, Processus, Environnement, 20*(4), 343–354.

Osipov, E. Y., Ashmetiev, A. Y., Osipova, O. P., & Klevczov, E. V. (2013). Novaiy inventarizacziy lednikov v ugovostochnoy chasti Vostochnogo Sayna. *Led i sneg (Ice and Snow), 3*, 45–54. [In Russian].

Peck, J. A., Khosbayar, P., King, J. W., Fowell, S. J., Ariunbileg, S., & Soninkhishig, N. (2001). *The latest Holocene sedimentary environmental magnetic record from Lake Dood, Mongolia.* American Geophysical Union, Fall Meeting, Abstract Id. GP41A-0243.

Pechersky, D. M., Gil'manova, D. M., Kazansky, A. Y., Krivonogov, S. K., Nurgaliev, D. K., & Tsel'movich, V. A. (2013). Native iron in Quaternary deposits of the Darhad Basin (*northern Mongolia*). *Russian Geology and Geophysics, 54*(12), 1499–1514.

Saruulzaya, A. (2017). *Thermokarst lake dynamics in the continuous and isolated permafrost zones, Mongolia.* Ph.D. Thesis, Hokkaido University, Sapporo, Japan. 76 p.

Sharkhuu, A., Sharkhuu, N., Etzelmüller, B., Heggem, E. S. F., Nelson, F. E., Shiklomanov, N. I., Goulden, C. E., & Brown, J. (2007). Permafrost monitoring in the Hovsgol mountain region, Mongolia. *Journal of Geophysical Research, 112*, 1–11.

Suzuki, Y., & Natsagdorj, N. (2009). Air temperature over span at Darhadyn Hotgor. In H. Nishida & T. Jamsran (Eds.), *Darhadyn Wetland in Mongolia: Synthesis investigation on ecosystems* (pp. 213–219). NPO Mongolia Ecology Information Center/Alex Company Press.

Takeda, K., Tamir, J., & Nishida, H. (2009). Influence of Permafrost to Natural Environment at Marsh in Darkhad Valley. In H. Nishida & T. Jamsran (Eds.), *Darhadyn Wetland in Mongolia: Synthesis investigation on ecosystems* (pp. 153–157). NPO Mongolia Ecology Information Center/Alex Company Press.

Tsegmid, S. (1969). *Physical geography of Mongolia.* 759 p. [In Mongolian].

Tserensodnom, J. (1971). *Lakes of Mongolia.* State Publishing, 202 p. [In Mongolian].

Tserensodnom, J. (2000). *A catalog of lakes in Mongolia*. Shuvuun Saaral Publishing, 141 p. [In Mongolian].

Tumurbaatar, D. (2004). *Seasonal and permanent permafrost of Mongolia*. Mongolian Academy of Sciences, 147 p. [In Mongolian].

Zorigt, M., Myagmar, K., Orkhonselenge, A., van Beek, E., Kwadijk, J., Tsogtbayar, J., Yamkhin, J., & Dechinlkhundev, D. (2020). Modelling permafrost distribution over the river basins of Mongolia using remote sensing and analytical approaches. *Environmental Earth Sciences, 79*(308), 1–11.

Chapter 16
Lake Khuvsgul

Abstract Lake Khuvsgul, a tectonic lake, plays a vital role in surface water resources of northern Mongolia and Eurasia. The lake stores three quarter (~384 km^3) of the total volume of freshwater resources in Mongolia, and it is a major headwater of Lake Baikal in eastern Siberia through Eg and Selenge rivers. The water resource of Lake Khuvsgul is the largest not only in Mongolia but also in Central Asia. Lake Khuvsgul responses to the present climate change with gradual hydrological fluctuations. The lake gained an area by 15.6 km^2 between 1970 and 2010, and it lost by 24.6 km^2 between 2010 and 2015. The area increase coincided with the rise in lake level and increase in precipitation, whereas the area decrease was consistent with the air temperature rise since 2010. The Lake Khuvsgul area during 1989–2015 had weak linear relationships with the annual average air temperature ($R^2 = 0.1431$) and precipitation ($R^2 = 0.1059$) recorded at the northern shore of the lake, and here the correlation analysis shows that the lake area had weak negative correlations with both air temperature ($r = -0.378$) and precipitation ($r = -0.325$). The lake area during 1970–2015 had weak linear relationships with the annual average air temperature ($R^2 = 0.1467$) and precipitation ($R^2 = 0.1089$) recorded at the southern shore of the lake, and here the correlation analysis shows that the lake area had a weak positive correlation with precipitation ($r = 0.330$) and a weak negative correlation with air temperature ($r = -0.382$).

Keywords Lake Khuvsgul · Eg River · Selenge River · Lake Baikal · Climate change · Northern Mongolia

16.1 Introduction

Lake Khuvsgul[1] constitutes an important part of surface water resources in northern Mongolia because it is the largest lake in the region and the deepest lake in the country. Lake Khuvsgul may serve as a sensitive water gauge in continental interior

[1] *Khuvsgul has been misspelled as Hovsgol and Khubsugul in many publications. Khuvsgul is the right English transliteration from Mongolian* Хөвсгөл.

A. Orkhonselenge et al., *Lakes of Mongolia*, Syntheses in Limnogeology,
https://doi.org/10.1007/978-3-030-99120-3_16

Asia, and Khuvsgul's sedimentary record preserves reliable proxies for past changes of effective moisture balance in the Lake Baikal[2] catchment area (Prokopenko et al., 2005). Lake Khuvsgul has attracted much scientific attention in terms of geology, seismology, geomorphology, sedimentology, geochemistry, pedology, paleoclimatology, hydrology, cryology, and ecology. Lake Khuvsgul, one of the great lakes in Central Asia, has been known to Russian and Chinese scholars for a long time (Meybeck, 1995). This lake has been investigated by Russian scientists since the 1830s (Bogoyavlensky et al., 1989), Mongolian scientists since the 1960s, and international scientists since the late 1990s (see Chap. 2). For instance, in 1830, botanists N.S. Turchaninov and N.V. Kirillov initiated the first investigation of Lake Khuvsgul, in 1859 G.E. Radde measured water level of the lake, and in 1897–1902 S.P. Peretolchin and E.V. De-Genning-Michelis collected bathymetric data along six cross sections (Bogoyavlensky et al., 1989).

In early studies of the Lake Khuvsgul basin, the Russian scientists (e.g., physical geographer E.M. Murzaev; geologists E.P. Pachkovsky, A.K. Ivanov, N.L. Cudryavtseva, and N.A. Marinov; pedologist N.D. Bespalov; and botanist A.A. Yunatov) played important roles (Bogoyavlensky et al., 1989; see Chap. 2). Geology and geomorphology of the Lake Khuvsgul basin were initially studied by G.N. Potanin who emphasized a wide distribution of igneous rocks in his publication entitled *Notes About the Northwestern Mongolia* in 1883 (Kozlov, 1989). Moreover, geomorphological features in the lake basin were described also by the Russian scientists (e.g., G.M. Permikin, E.S. Selisky, G.E. Radde, A.L. Chekanovsky, P.A. Kropotkin, E.V. De-Genning-Michelis, S.P. Peretolchin, V.S. Elpatievsky, V.L. Komarov, V.C. Dorogostaisky, N.V. Mikheev). For instance, V.L. Komarov investigated the origin of Lake Khuvsgul and changes of lake water level in 1902, and he noted that some terraces along its shores were of tectonic origins and others were erosive in origin (Tserensodnom, 1971). Moreover, S.P. Peretolchin studied water regime, discharge, depth, and deposit of Lake Khuvsgul in 1897–1902, and he provided their data in the book entitled *Physical Geography of Lake Khuvsgul* in 1903 (Table 16.1). In addition, based on the terraces 3, 10, 15, and 26 m higher than the present-day lake level, he noted that Lake Khuvsgul's water level had decreased gradually due to glacier retreats at Mt. Munkh[3] Saridag[4] in the north of the lake and intensive erosion of Eg River, which is one of the major tributaries of Selenge[5] River, in the south of the lake. V.S. Elpatievsky noted that the Lake Khuvsgul basin had formed during the uplift of Bayan (or Bayan Zurkh) Mountain Range in the west and northwest of the lake (Fig. 15.3); however, dating of the uplift remains unclear.

[2] *Baigali* meaning *nature* has been misspelled as *Baikal* by the Russians. Baigali is the right English transliteration from Mongolian Байгаль.

[3] *Munkh* has been often misspelled as *Munku* or Munh based on the Russian pronunciation in the international literature. Munkh is the correct English transliteration from Mongolian *Мөнх*.

[4] *Saridag* has been also misspelled as *Sardyk* or Sardig by the Russians internationally. Saridag is the correct English transliteration from Mongolian *Сарьдаг*.

[5] *Selenge* has been often incorrectly written as *Selenga* based on the Russian pronunciation. Selenge is the correct English transliteration from Mongolian Сэлэнгэ.

Table 16.1 Morphometric components of Lake Khuvsgul

#	Elevation (m a.s.l.)	Length (km)	Width (km)		Depth (m)		Number of inflow rivers and streams	Authors
			Average	Maximum	Average	Maximum		
1	–	–	–	–	–	170.0	–	Fedotov et al. (2004)
2	–	133.4	39.5	–	–	238.3	49	S.P. Peretolchin (1903) in Tsegmid (1969)
3	1554	139.0	45.0	–	–	245.7	–	V.S. Elpatievsky (1903) in Tserensodnom (1971)
4	1645	–	–	–	–	262.0	46	Tsegmid (1969)
5	–	136.0	20.3	36.5	139.0	262.0	69	Bogoyavlensky et al. (1989)
6	–	–	–	–	–	262.4	–	Goulden et al. (2006)
7	–	–	–	–	–	252.6	–	Kumagai et al. (2006)

More recently, some Mongolian scientists (e.g., physical geographers Sh. Tsegmid, B. Natsag, Ch. Duvjir, J. Tserensodnom, and T. Tsengel and hydrobiologist A. Dulmaa, etc.) conducted geomorphological and hydrological studies in the Lake Khuvsgul basin (Bogoyavlensky et al., 1989). Geology and sedimentology of Lake Khuvsgul (see Sect. 16.3) have been studied by geologists (e.g., O. Tumurtogoo, D. Tumurkhuu, Ts. Narantsetseg). In addition, A. Dashdorj, A.A. Tomilov, P.P. Bochkarov, and E.P. Nikolaeva completed hydrochemical and hydrobiological investigations of Lake Khuvsgul in 1959–1960. The age of Lake Khuvsgul is estimated to be 25 ka BP (Tserensodnom, 2000). During 1971–1986, the Mongolian and Russian joint expedition revealed the basic characteristics of the Lake Khuvsgul basin, such as shoreline features and dynamics, wind direction, water temperature of the lake, hydrochemical composition, geochemical composition of its bottom deposits, and the lake biosphere in *Atlas of Khuvsgul* (Bogoyavlensky et al., 1989). More complete data on geology, geomorphology, geodynamics, and bathymetry of Lake Khuvsgul have been provided since the 1970s (Bogoyavlensky et al., 1989). The annual precipitation, evaporation, and changes in water volume of Lake Khuvsgul for the 47 years spanning between 1964 and 2010 and the Eg River discharge recorded at the Khatgal and Alag Erdene hydrological stations for 25 years were provided by T. Tsengel in 2013 (Davaa, 2015; see Sect. 16.5).

Since the late 1990s, international teams (e.g., Mongolian, Russian, American, Japanese, Chinese, Korean) led many research projects in the Lake Khuvsgul basin. For instance, the international teams conducted the Khuvsgul (or Hovsgol) Drilling Project (HDP) in 2001, 2004, 2006, and 2009, and it provided detailed data from the lake regarding regional and local paleoclimate cycles of the middle to late Pleistocene (see Chap. 21). The international teams (e.g., HDP Members, 2009; Watanabe et al., 2009) recovered the age dataset for Lake Khuvsgul sediments. For example, the

HDP-04 core spans sedimentary sequences for about 1.0 Ma (Kuzmin et al., 2008; Prokopenko et al., 2009), and it reconstructs a deep lake basin during the cold stage comparable to glacial periods at the marine isotope stage (MIS) 5d and another lake during the interglacial period at the MIS 11 (Kashiwaya et al., 2010). However, in the proposed age model, the upper 24 m of 81 m HDP-04 drill core suggests the sedimentary records of late MIS 7 to MIS 1, while its lower 24 m represents the stratigraphic records of MIS 27 to MIS 13 interval (Prokopenko et al., 2009). Diatom and biogenic silica proxies in the core sediments reconstructed Holocene climate changes within the Lake Khuvsgul basin with high precipitation/runoff events from 11.0 to 7.0 ka and warmer summers between 6.0 and 3.5 ka as the result of higher soil temperatures and lower moisture availability (Prokopenko et al., 2007). Hydro-geomorphological fluctuations in the Lake Khuvsgul catchment system during the past 10,000 years were presented by Orkhonselenge et al. (2011) suggesting three periods of large discharges from outside related to meltwater and/or heavy rainfall. The three short events of higher-rate sedimentation were found at 7.4–7.1, 4.8–4.5, and 1.0–0.9 cal ka BP, which are possibly related to drops in lake level, reduction of lake area, progradation of the Borsog River delta, and high erosion of the exposed lake shores (Orkhonselenge et al., 2013).

In this chapter, previous studies in the fields of geomorphology, sedimentology, geochemistry, and hydrology on Lake Khuvsgul are synthesized, and changes in area of the lake are discussed in detail together with local climate data. Paleoclimatic records from the lake are considered in Chap. 21.

16.2 Physiographic Condition

Lake Khuvsgul is located at the elevation of 1645 m a.s.l. in an intermontane basin within the Khuvsgul Mountain Range of northern Mongolia (Figs. 1.1 and 15.3). Morphometric components of Lake Khuvsgul are shown in Tables 1.3, 15.6, and 16.1. The lake extends for 133.4 km in length from the north to the south and 39.5 km in width from the east to the west (Table 16.1). The lake belongs to the vast Lake Baikal watershed (see Sect. 16.5) because it is located in the Baikal Rift Zone (BRZ) (see Chap. 15).

Lake Khuvsgul, the second largest lake in area within Mongolia, occupies an area of 2612 km^2 excluding islands (Tsegmid, 1969), 2760 km^2 (Bogoyavlensky et al., 1989), or 2773.2 km^2 including islands (Orkhonselenge and Uuganzaya, unpublished; see Sect. 16.6). The depth at the deepest point is measured to be 238.3 m or 245.7 m (Table 16.1) at the southwest of the island Dalain Khui[6] (see Sect. 16.3; Fig. 16.2). Lake Khuvsgul's water volume is estimated to be 317.5 km^3 (Tsegmid, 1969), 380.7 km^3 (Tserensodnom, 1971; Goulden et al., 2006) or 383.7 km^3 (Davaa et al., 2007) (see Sect. 16.6).

[6] *Khui*, meaning *Navel*, has been misspelled as *Khuu* and/or *Huu* and mistranslated into English as *Boy* in Wikipedia. Khui is the right English transliteration from Mongolian *Хүй*.

The Khuvsgul basin, filled by <500 m thick deposits, constitutes the southwestern end of the BRZ (Bogoyavlensky et al., 1989), and so then the lake is neotectonically similar with the Lake Baikal (see Sect. 16.3). Lake Khuvsgul sediments are well documented. For example, the tentative correlation shows that sediment accumulation rates in Lake Khuvsgul were relatively slow (ca. 18 m = 200 ka) during the MIS 24–MIS 17 interval and high during the MIS 17–MIS 16 interval (ca. 16 m = 100 ka) (Prokopenko et al., 2009). The channel Ulkhen is the major source of transportation of sediments to be deposited at the outlet of the lake or at the headwater of Eg River (Orkhonselenge et al., 2013; Fig. 16.1).

16.3 Geological and Geomorphological Settings

Geological descriptions of Lake Khuvsgul were given by G.N. Potanin (1883), the Mongolian and Tuva joint expedition (1924–1933), Mongolian and Russian geologists (1959–1965), and the Mongolian and Irkutsk joint expedition (1971–1986).

In the Khuvsgul region, the Archean, early Proterozoic, Rhipheic, Cambrian, Jurassic, Neogene, and Quaternary rocks are widely distributed. The 5.5-km-thick Archean bedrock, consisting of gneiss, amphibolites, marbles, calcites, granulites, and magmatites, is overlain by the 3.0-km-thick early Proterozoic sequence consisting of limestones, marbles, shales, quartzites, gneiss, and amphibolites, while the Neogene and Quaternary igneous and modern-day deposits are hundreds of meters thick consisting of deluvial, deluvial-proluvial, lacustrine, glacial, and aeolian sediments (Bogoyavlensky et al., 1989). Tectonically, Lake Khuvsgul is located at the northern border of the Central Asian Orogenic Belt which surrounds the southern Siberia that is made of the Precambrian accumulation center (Bogoyavlensky et al., 1989; Fig. 15.1). The floor of the Precambrian accumulation is exposed at the end of the Precambrian Khamar Davaa and the eastern Tuva structures. During the Cenozoic, particularly in the middle Pliocene and Pleistocene, new uplifting (orogeny) and subsidence (e.g., Khuvsgul, Darkhad, Bus basins) events occurred (Table 15.1), and a rework of paleofault formed and a new transform plate movement also took place (Bogoyavlensky et al., 1989). The tectonics of Lake Khuvsgul are expressed by geosynclinals, orogeny stages, and tectonic activities (Table 15.1).

Lake Khuvsgul, an elliptically shaped lake, resides in a tectonic graben within the BRZ. The elliptic shape of the lake becomes narrow at the headwater of Eg River (Fig. 16.1). The Lake Khuvsgul basin formed during the uplift of the Bayan (or Bayan Zurkh) Mountain Range in the northwest (Fig. 15.3). Due to the formation of Lake Khuvsgul in the deep tectonic graben with a rugged surrounding topography, its shores are deeply dissected and include abundant bays, gulfs, spurs, islands, and cliffs (Tsegmid, 1969). Lake Khuvsgul is bound by high mountains such as steep Khoridol Saridag and Bayan (or Bayan Zurkh) mountain ranges, elevated at 3000–3200 m a.s.l. in the southwest and northwest (Fig. 15.3), which have dome-blocked structures (Orkhonselenge et al., 2014); by Mt. Munkh Saridag (Fig. 15.3) and Mt. Buren Khaan in the north; by gentle Mt. Khivan Zaluu Uuriin Saridag,

Fig. 16.1 Landsat 8 image of Lake Khuvsgul on September 18, 2015 (path of 136 and row of 25). Modified from Orkhonselenge and Uuganzaya (unpublished)

elevated up to 2367 m a.s.l.; and by basaltic flows (Zolotarev et al., 1982) in the east. The pediment of Mt. Munkh Saridag (Orkhonselenge, 2016) encounters the lake and forms 40–50 m high basaltic terrace (Tsegmid, 1969). Among the numerous peninsulas extending from the Bayan (or Bayan Zurkh) Mountain Range into the lake, Dulaan Uul (2174 m a.s.l.) in the northwest is the largest one standing 200 m higher than the lake level. S.P. Peretolchin observed four paleoterraces, and he noted that the Doloon Uul on the northwestern shore of the lake had been an island. He suggested that at that time, the lake level was 40 m higher than the recent level (Tserensodnom, 1971). There are numerous terraces, implying drops of the lake level, on the eastern slope of the paleo-peninsula Doloon Uul (Tsegmid, 1969).

Lake Khuvsgul has four major islands, namely, Dalain Khui, Khadan Khui, Modon Khui, and Baga Khui (Fig. 16.2). The island Dalain Khui is the largest one, lying at center of the lake (Figs. 16.1 and 16.2a), and it is 126 m high above the lake level and occupies an area of 8.4 km^2, extending for 3.0 km in length, 2.0 km in width, and 8.5 km along the shoreline (Tsegmid, 1969). The island Baga Khui has disappeared since 1971 due to the rise of the lake level (Davaa, 2015; see Sect. 16.5). At the estuaries of inflowing rivers and streams (Table 16.1), swampy bays are densely developed. For example, there are six large bays (e.g., Dalbai, Sevsuul, Noyon) developed on the eastern shore of the lake (Fig. 16.1). The presence of the paleoterraces found at diverse of elevations indicates various regressions and decrease of precipitation due to climate change (see Chap. 21) and tectonic movements (see Chap. 15) occurring in the lake basin.

16.4 Climate Condition

The climate in the Lake Khuvsgul basin is harsh continental, and it is characterized by extremely cold and dry winter and mild and cool summer. The average air temperature in the lake basin is 10–12 °C in July, and the annual average precipitation is 300–350 mm (Bogoyavlensky et al., 1989). The Lake Khuvsgul region has warmed up by ~1.7 °C in the annual average air temperature since 1963, and the maximum and minimum air temperatures have been in the ranges of 27–30 °C and −30 to −40 °C (Namkhaijantsan, 2006). The summer precipitation in the Dalbai bay on the eastern shore of Lake Khuvsgul (Fig. 16.1) had 40 short heavy rainfall events (≥0.8 mm/5 min) from 2008 to 2012, but these rainfalls lasted for no more than 30–40 min each (Goulden et al., 2016).

Local climate data in the Lake Khuvsgul basin have been recorded at the Khankh meteorological station on the northeastern shore of the lake (Fig. 15.3) during the last 35 years and at the Khatgal meteorological station on the southernmost tip of the lake (Fig. 15.3) during the last 60 years (Fig. 15.2). The annual variations of air temperature and precipitation are shown in Chap. 15. On the northeastern shore of the lake, the average air temperature is 10.6 °C in summer (12.6 °C in July) and −17.7 °C in winter (−21.0 °C in January) (Fig. 16.3a), and the average precipitation is 197.9 mm in summer (Fig. 16.3b) with 88.6 mm falling in July. On the southern

Fig. 16.2 (**a**) Island Dalain Khui in the center of Lake Khuvsgul (Fig. 16.1) viewing toward the northwest. (**b**) Lake Khuvsgul viewing toward the south. Photo B by A. Orkhonselenge on August 25, 2008, from the Sukhbaatar cruise during the excursion for the seventh International Symposium on *Environmental Changes in East Eurasia and Adjacent Areas—High Resolution Records of Terrestrial Sediments in Lake Khuvsgul*

shore of the lake, the average air temperature is 11.2 °C in summer (12.5 °C in July) and −21.2 °C in winter (−23.2 °C in January) (Fig. 16.3c), and the average precipitation is 216.4 mm in summer (Fig. 16.3d) with 87.3 mm falling in July.

The summer and spring temperatures have risen on the northern shore during the last 35 years (Fig. 16.3a) and on the southern shore during the last 60 years (Fig. 16.3c). The seasonal average precipitation is decreasing on the northern shore (Fig. 16.3b), but it is gradually increasing on the southern shore except for only the autumn season with a decreasing trend (Fig. 16.3d).

16.5 Hydrological Condition

Lake Khuvsgul, a freshwater lake, is fed by more than 46 (Table 16.1) permanent rivers and streams (17 in the west and 29 in the east) draining from an area of 5300 km^2 in the surrounding mountains (Tsegmid, 1969), and the lake feeds only Eg

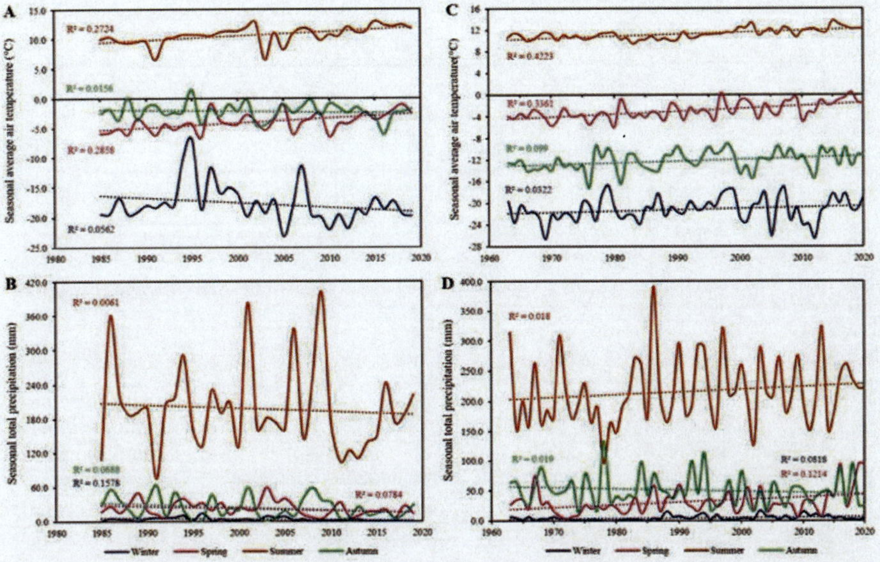

Fig. 16.3 The seasonal average (**a**) air temperature and (**b**) precipitation recorded at the Khankh meteorological station on the northern shore of Lake Khuvsgul (Fig. 15.3) during 1985–2019 and the seasonal average (**c**) air temperature and (**d**) precipitation recorded at the Khatgal meteorological station on the southern shore of Lake Khuvsgul (Fig. 15.3) during 1963–2020

River in the south (Fig. 16.1). According to Bogoyavlensky et al. (1989), 96 rivers and creeks feed the lake in total, and 69 of them (Table 16.1) are permanent. The main source of Lake Khuvsgul is runoff, precipitation, and meltwaters of glacier, snow, and permafrost. Over 69% of the lake's total area are deeper than 100 m (Tsegmid, 1969). The depth of the lake reaches 100–200 m within the 1–2 km distance from the beach toward the center of the lake because the lake's graben wall is steep (Tsegmid, 1969). By the beginning of the Holocene, Lake Khuvsgul evolved from a low-level closed basin lake to an overflowing freshwater lake (Prokopenko et al., 2005).

Lake Khuvsgul's water level increased from 1960 to the mid-1970s and in the late 1980s, and it dropped in the early 1980s (Batnasan, 1998). Another source says that the lake water level increased in the mids-1970s, 1980s, 1990s, and the early 2000s (Kumagai et al., 2006). Recent studies inform inundated 50–80 years old trees and low terraces around the lake implying that the lake level increased in the last few decades (Tsegmid, 1969). The water level of Lake Khuvsgul seasonally fluctuates in the range of 0.5–1.2 m (Tsegmid, 1969) or 0.45–0.58 m (Tserensodnom, 1971). The lake level increases with rainfall in July and August, and it reaches the maximum level in September and October. The water level of the lake rose in 1976–1977, 1986–1988, 1993–1995, and 2000 possibly due to the sedimentary dam formed at the headwater of Eg River in 1971–1972 (Orkhonselenge et al., 2013; Fig. 16.4). As shown in Table 16.2, the precipitation trend in the lake basin is almost

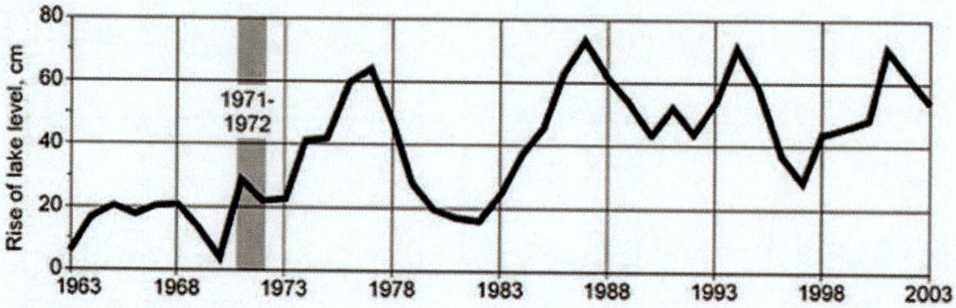

Fig. 16.4 Water level of Lake Khuvsgul between 1963 and 2003. After Orkhonselenge et al. (2013)

Table 16.2 Water balance of Lake Khuvsgul

Period	Eg River's discharge		Evaporation		Precipitation		Volume		Amplitude of inflowing and outflowing groundwaters	
	km³	mm	km³	mm	km³	mm	km³	mm	km³	mm
1964–1989	0.574	208	1754	635	0.784	284	-0,038	−14	1506	545
1964–2010	–	–	1730	626	0,815	295	0,071	25	–	–

Table 16.3 Water temperature with corresponding depth of Lake Khuvsgul

#	Water depth (m)	July (°C)	August (°C)
1	0.0	12.0–13.0	9.6
2	10.0	–	8.9
3	25.0	–	7.2
4	50.0	–	4.6
5	100.0	–	3.9
6	200.0	–	3.7
7	262.0	3.5–4.0	–

consistent with the discharge of Eg River, while discharge of inflowing surface water and groundwater coincides with the evaporation (Davaa, 2015).

Lake Khuvsgul's water temperature ranges with the depth (Table 16.3). The measured water temperature is about 11.1–13.9 °C near the beaches and 5.6–8.0 °C at the lake center (Tserensodnom, 1971). The water temperature decreased at the rate of −0.02 °C/year until 1989, and it rose at the rate of 0.32 °C/year since 1990 (Davaa, 2015). The temperature at the surface of the lake does not exceed 18.0 °C (Table 16.4). Lake Khuvsgul is covered by a 1.0–1.5-m-thick ice from the end of September to the middle of June, but it melts between the end of June and the beginning of September (Tsegmid, 1969). The freezing and thawing periods of the lake ice have been delayed for about 20 days in comparison with previous three decades (Batima et al., 2004). The mineralization value of the lake is low (Tsegmid, 1969), and HCO_3 is dominant with the total ions of about 190–210 mg/L (Table 16.5) or 227 mg/L (Davaa, 2015).

Table 16.4 Water temperature at the surface of Lake Khuvsgul

#	Year	Water temperature (°C)									Authors
		April	May	June	July	August	September	October	November	Maximum	
1	1963	–	–	7.4	12.0	13.2	7.9	2.5	0.4	17.8	Tsegmid (1969)
2	1964	–	2.7	6.8	12.3	13.0	9.1	3.2	–	18.0	
3	1965	–	–	8.3	11.4	13.1	8.3	–	–	16.4	Tserensodnom (1971)
4	1966	0.4	2.8	7.9	9.9	11.6	9.9	5.1	1.2	14.6	
5	1967	–	1.6	6.9	12.0	11.0	8.3	5.7	–	15.7	
6	1968	0.5	2.1	7.6	11.4	11.1	7.2	3.5	–	16.6	

Table 16.5 Hydrochemical composition of Lake Khuvsgul

Ions (mg/L)	HCO_3^-	Ca^{2+}	Cl^-	SO_4^{2-}	Mg^{2+}	Na^++K^+
190–210	134.2–151.9	24.6–34.0	1.1–5.0	0.7–15.8	8.2–13.8	0.2–10.9

16.6 Changes in Area

The Russian and Mongolian hydrobiological expedition led by A.A. Tomilov and A. Dashdorj during 1959–1960 and the hydrological team led by A.E. Cherkasov, V.P. Shumyev, and N. Batsukh with the Khuvsgul Mongolian and Russian joint expedition in 1972 measured the surface area of Lake Khuvsgul and determined it as 2760 km^2 (Bogoyavlensky et al., 1989; Table 16.6). Here, we should note that the area in 1969 is a measurement result excluding its islands (see Sect. 16.2).

Lake Khuvsgul lost an area of 9.0 km^2 between 1970 and 2015 (Orkhonselenge and Uuganzaya, unpublished; Fig. 16.5) while experiencing fluctuations (Table 16.6). There were two periods of losing area and a period of gaining area, i.e., the lake area decreased by 22.2 km^2 in 1970–1989, increased by 37.8 km^2 in 1989–2010, and decreased again by 24.6 km^2 in 2010–2015 (Table 16.6). The changes in the lake area in 1970–2010 coincided with the dropped water level of the lake in 1963–1983 except for the peaks of increased water level in 1976–1977 and with the general rising level in 1983–2003 except for the drop in 1997 (Fig. 16.4). The increase in area of 15.6 km^2 during the 40 years between 1970 and 2010 (Table 16.6) is an agreement with the rises of the lake level in 1975–1978, 1985–1988, 1993–1996, and 2000–2003 (Kumagai et al., 2006).

The humidifying trend in the Lake Khuvsgul region with the increase of the annual average precipitation by 15–30 mm since 1963 and the increase of the annual average precipitation (partially prediction) from 30 to 107 mm between 2000 and 2050 (Namkhaijantsan, 2006) may have contributed to the changes in area of the lake. This is consistent with the melting of modern glaciers on Mt. Munkh Saridag by 9.5% between 1940 and 2010 (Davaa, 2015) and by 42.6% between 1970 and 2007 (Orkhonselenge, 2016; Orkhonselenge & Harbor, 2018). In addition to permafrost and groundwater, such glacier meltwater supplies Lake Khuvsgul with a large quantity of water resource. The water level of Lake Khuvsgul has steadily increased for about 1.0 m since 1963, and it has been attributed to the melting of the permafrost or an increase of rainfall to the obstruction of outflow due to sand banks in the Eg River and to the reduction of evaporation from the water surface due to the lower water temperatures (Ministry of Environment and Green Development, 2013).

Figure 16.5 shows the lake decreased in area of 9 km^2 along the eastern and southern shores between 1970 and 2015 (Orkhonselenge and Uuganzaya, unpublished). The recent decrease in area of Lake Khuvsgul since 2010 may have been related to the rising annual average air temperature on the southern shore ($R^2 = 0.4363$ in Fig. 15.2a). The trend is contributed by the predicted annual average air temperature increase from 0.7 to 3.2 °C between 2000 and 2050 (Namkhaijantsan,

Table 16.6 Temporal changes in area and volume of Lake Khuvsgul

#	Year	Area (km²)	Volume (km³)	Authors
1	1969	2612.0	317.5	Tsegmid (1969)
2	1970	2782.2	–	Orkhonselenge and Uuganzaya, unpublished
3	1971	–	380.7	Tserensodnom (1971)
4	1989	2760.0	383.3	Bogoyavlensky et al. (1989)
5	2006	–	380.7	Goulden et al. (2006)
6	2007	2770.0	383.7	Davaa et al. (2007)
7	2010	2797.8	–	Davaa (2015)
8	2013	2785.8	–	Davaa (2015)
9	2015	2773.2	–	Orkhonselenge and Uuganzaya, unpublished

2006) on the southern shore of the lake, where soil temperature rose by 2.6 °C or 0.06 °C/year ($R^2 = 0.39$) between 1970 and 2008 (Davaa, 2015). The incision of Eg River contributes to the lowering level of Lake Khuvsgul (Orkhonselenge et al., 2013), and it plays an important role in reduction of the lake area.

The linear regression analysis shows that the area of Lake Khuvsgul for 1989–2015 had weak relationships with the annual average air temperature ($R^2 = 0.1431$) and precipitation ($R^2 = 0.1059$) on the northern shore of the lake (Fig. 16.6a, b). Here, the correlation analysis shows that the lake area had weak negative correlations with both air temperature ($r = -0.378$) and precipitation ($r = -0.325$). However, on the southern shore of the lake, the lake area for 1970–2015 shows weak linear relationships with the annual average air temperature ($R^2 = 0.1467$) and precipitation ($R^2 = 0.1089$) (Fig. 16.6c, d). Here, the correlation analysis shows that the lake area had a weak positive correlation with precipitation ($r = 0.330$) and a weak negative correlation with air temperature ($r = -0.382$). It implies that the annual average precipitation and air temperature on the southern shore of the lake provide more impact on the lake area changes than in the case of the northern shore.

16.7 Summary

Lake Khuvsgul, lying in the Baikal Rift Zone, is the largest freshwater lake by volume in Mongolia, and it plays an important role in surface water resources of Mongolia, Eurasia, and Central Asia. The lake represents dynamic changes in the hydrological regime of northern Mongolia, and it has gradually responded to climate change during the last 60 years. Lake Khuvsgul enlarged its area from 1970 to 2010 coinciding with the rise of its water level in the mids-1970s, 1980s, 1990s, and early 2000s due to the 50% increased groundwater inflow caused by melting of glaciers and permafrost under the rising annual and seasonal air temperatures. However, the lake has lost its area since 2010 coinciding with the decreasing

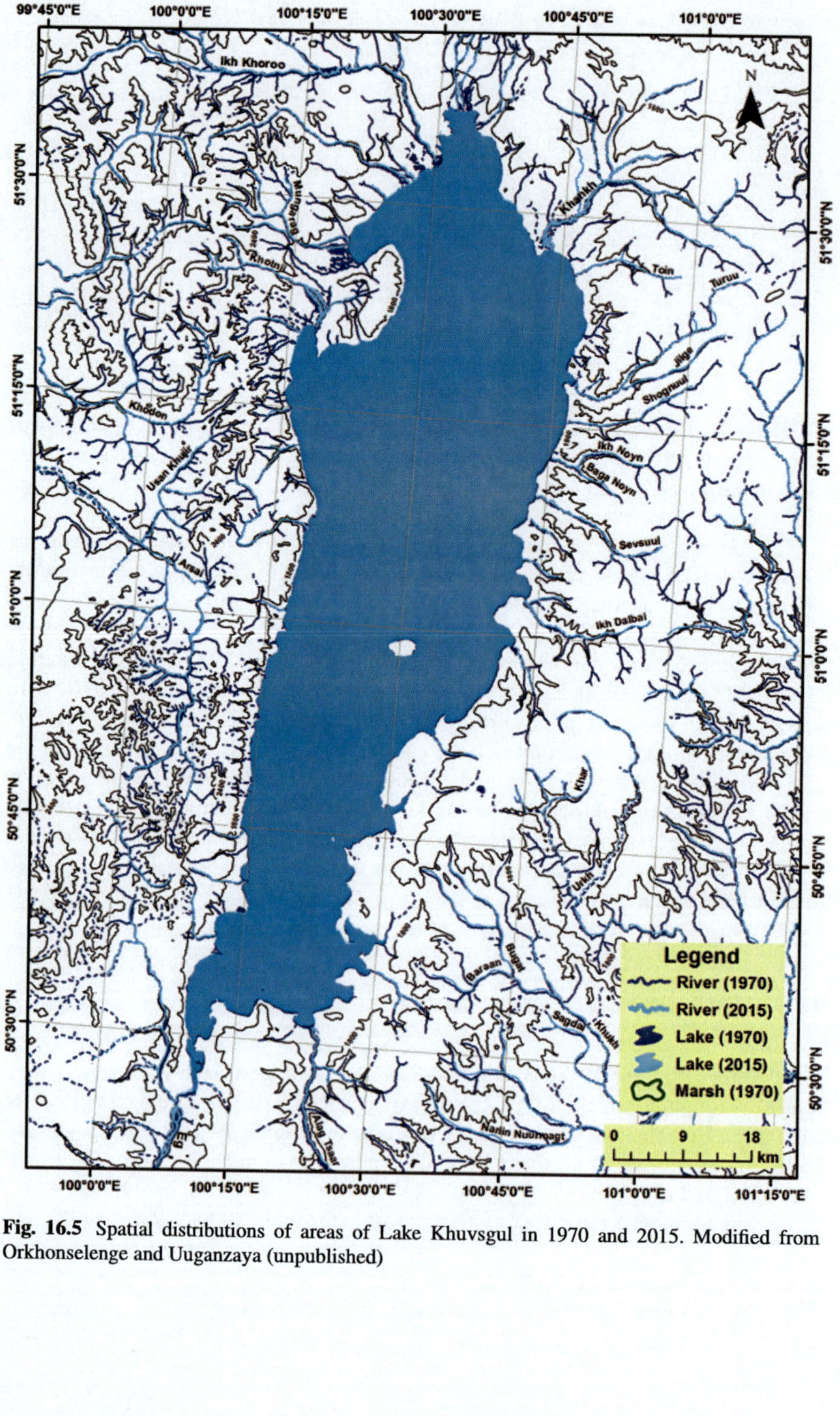

Fig. 16.5 Spatial distributions of areas of Lake Khuvsgul in 1970 and 2015. Modified from Orkhonselenge and Uuganzaya (unpublished)

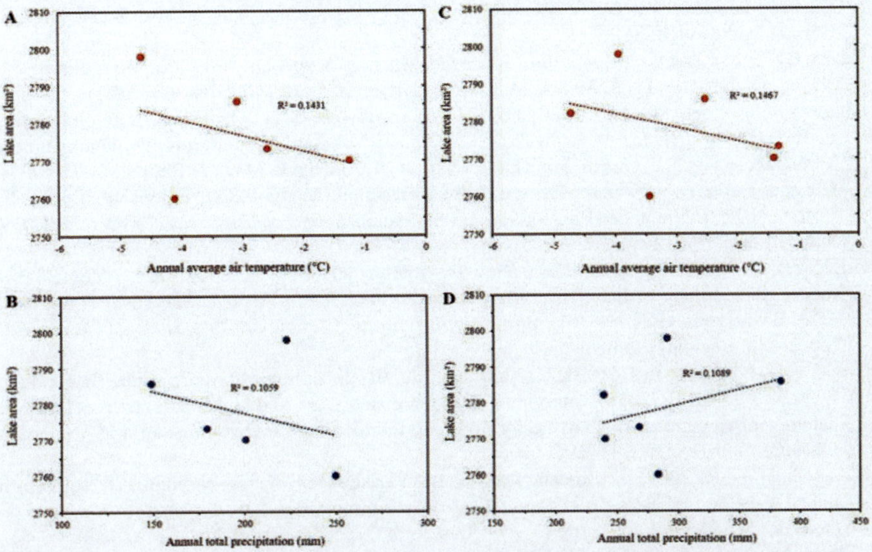

Fig. 16.6 Relationships between area of Lake Khuvsgul and annual average (**a**) air temperature and (**b**) precipitation recorded on the northern shore during 1989–2015 and annual average (**c**) air temperature and (**d**) precipitation recorded on the southern shore during 1970–2015

precipitation and intensifying incision of Eg River, an outlet of Lake Khuvsgul. The Lake Khuvsgul area for 1989–2015 had weak linear relationships with the annual average air temperature ($R^2 = 0.1431$) and precipitation ($R^2 = 0.1059$) measured on the northern shore of the lake. The correlation analysis shows that the lake area had weak negative correlations with both air temperature ($r = -0.378$) and precipitation ($r = -0.325$). Although the lake area for 1970–2015 had weak linear relationships with the annual average air temperature ($R^2 = 0.1467$) and precipitation ($R^2 = 0.1089$) measured on the southern shore of the lake, the correlation analysis shows that the lake area had a weak positive correlation with precipitation ($r = 0.330$) and a weak negative correlation with air temperature ($r = -0.382$).

References

Batima, P., Batnasan, N., & Bolormaa, B. (2004). Trends in river and lake ice in Mongolia. *AIACC Working Paper, 4*, 3–12.

Batnasan, N. (1998). *Hydrological system, water regime, and evolution of large lakes in the Govi region*. Doctorate Thesis in Geography: Hydrology, Institute of Geoecology, Mongolian Academy of Sciences, Ulaanbaatar. 163 p. [In Mongolian].

Bogoyavlensky, B. A., Batjargal, B., Martinov, V. P., Kozlov, U. P., Borobiev, V. V., & Batsuuri, D. (1989). *Atlas of Lake Khuvsgul: Mongolian People Republic*. State Printing Industry of

Geodesy and Cartography, SSSR. Moscow. 118 p. [In Russian with an abstract in English, German and Mongolian].

Davaa, G. (2015). *Resource and regime of surface water of Mongolia*. 408 p. [In Mongolian].

Davaa, G., Oyunbaatar, D., & Sugita, M. (2007). Surface water of Mongolia. *Hydrology.*, 55–68.

Fedotov, A. P., Chebykin, E. P., Semenov, M. Y., Vorobyova, S. S., Osipov, E. Y., Golobokova, L. P., Pogodaeva, T. V., Zheleznyakova, T. O., Grachev, M. A., Tumurhuu, D., Oyunchimeg, T., Narantsetseg, T., Tumurtogoo, O., Dolgikh, P. T., Arsenyuk, M. I., & Batist, M. D. (2004). Changes in the volume and salinity of Lake Khubsugul (Mongolia) in response to global climate changes in the upper Pleistocene and the Holocene. *Paleogeography, Paleoclimatology, Paleoecology, 209*, 245–257.

Goulden, C. E., Tomurtogoo, O., Karabanov, E., & Munguntsetseg, A. (2006). The geological history and geography of Lake Hövsgöl, Mongolia. In C. E. Goulden, T. Sitnikova, J. Gelhaus, & B. Bazartseren (Eds.), *The geology, biodiversity and ecology of lake Hövsgöl (Mongolia)* (pp. 1–19). Backhuys Publishers.

Goulden, C. E., Mead, J., Horwitz, R., Goulden, M., Nandintsetseg, B., McCormick, S., Boldgiv, B., & Petraitis, P. S. (2016). Interviews of Mongolian herders and high resolution precipitation data reveal an increase in short heavy rains and thunderstorm activity in semi-arid Mongolia. *Climatic Change, 136*, 281–295.

Members, H. D. P. (2009). Sedimentary record from Lake Hovsgol, NW Mongolia: Results from the HDP-04 and HDP-06 drill cores. *Quaternary International, 205*, 21–37.

Kashiwaya, K., Ochiai, S., Sumino, G., Tsukamoto, T., Szyniszewska, A., Yamamoto, M., Sakaguchi, A., Hasebe, N., Sakai, H., Watanabe, T., & Kawai, T. (2010). Climato-hydrological fluctuations printed in long lacustrine records in Lake Hövsgöl, Mongolia. *Quaternary International, 219*, 178–187.

Kozlov, J. P. (1989). *Atlas ozera Chubsugul*. 119 p.

Kumagai, M., Urabe, J., Goulden, C. E., Soninkhishig, N., Hadbaatar, D., Tsujimura, S., Hayami, Y., Sekino, T., & Maruo, M. (2006). Recent rise in water level at Lake Hövsgöl, Mongolia. In C. E. Goulden, T. Sitnikova, J. Gelhaus, & B. Bazartseren (Eds.), *The geology, biodiversity and ecology of lake Hövsgöl (Mongolia)* (pp. 77–92). Backhuys Publishers.

Kuzmin, M. I., Bezrukova, E. V., Kulagina, N. V., Khomutova, M. Y., Osukhovskaya, Y. N., Letunova, P. N., & Abzaeva, A. A. (2008). Comparison of paleoclimatic records from Lakes Baikal and Hovsgol. In *Proceedings of the 7th International Symposium on Environmental Changes in East Eurasia and Adjacent Areas—High resolution records of terrestrial sediments in Lake Khuvsgul, Khatgal, Mongolia*. 20 p.

Meybeck, M. (1995). Global distribution of lakes. In A. Lerman, D. M. Imboden, & J. R. Gat (Eds.), *physics and chemistry of lakes* (pp. 1–35). Springer-Verlag.

Ministry of Environment and Green Development. (2013). *Integrated water management plan of Mongolia*. 339 p.

Namkhaijantsan, G. (2006). Climate of Hövsgöl Lake region. In C. E. Goulden, T. Sitnikova, J. Gelhaus, & B. Bazartseren (Eds.), *The geology, biodiversity and ecology of lake Hövsgöl (Mongolia)* (pp. 63–76). Backhuys Publishers.

Orkhonselenge, A. (2016). Glacial Geomorphology of Mt. Munkh Saridag in the Khuvsgul Mountain Range, Northern Mongolia. *Géomorphologie: Relief, Processes, Environnement, 22*(4), 387–396.

Orkhonselenge, A., & Harbor, J. M. (2018). Impacts of modern glacier changes on surface water resources in western and northern Mongolia. *Journal of Water Resource and Protection, 10*(6), 559–576.

Orkhonselenge, A., Krivonogov, S. K., Mino, K., Kashiwaya, K., Safonova, I. Y., Yamamoto, M., Kashima, K., Nakamura, T., & Kim, J. Y. (2013). Holocene sedimentary records from Lake Borsog, eastern shore of Lake Khuvsgul, Mongolia, and their paleoenvironmental implications. *Quaternary International, 290–291*, 95–109.

Orkhonselenge, A., Krivonogov, S. K., Mino, K., Kashiwaya, K., Yamamoto, M., & Nakamura, T. (2014). Holocene landform evolution of lake Khuvsgul basin, Mongolia. *Géomorphologie: Relief, Processus, Environnement, 20*(4), 343–354.

Orkhonselenge, A., Mino, K., Kashiwaya, K., Krivonogov, S. K., Yamamoto, T., Sakguchi, A., & Nakamura, T. (2011). Holocene hydro-environmental changes in northern Mongolia inferred from lacustrine sediments in Borsog Bay of Lake Khuvsgul. *Journal of Earth Environment, 2*(3), 457–461.

Prokopenko, A. A., Khursevich, G. K., Bezrukova, E. V., Kuzmin, M. I., Boes, X., Williams, D. F., Fedenya, S. A., Kulagina, N. V., Letunova, P. P., & Abzaeva, A. A. (2007). Paleoenvironmental proxy records from Lake Hovsgol, Mongolia, and a synthesis of Holocene climate change in the Lake Baikal watershed. *Quaternary Research, 68*, 2–17.

Prokopenko, A. A., Khursevich, G. K., Kuzmin, M. I., & Kawai, T. (2009). Productivity cycles in Lake Hovsgol, NW Mongolia, during the last 1 Ma and the age model of the HDP-04 drill core record. *Quaternary International, 205*, 111–125.

Prokopenko, A. A., Kuzmin, M. I., Williams, D. F., Gelety, V. F., Kalmychkov, G. V., Gvozdkov, A. N., & Solotchin, P. A. (2005). Basin-wide sedimentation changes and deglacial lake-level rise in Lake Hovsgol basin, NW Mongolia. *Quaternary International, 136*, 59–69.

Tsegmid, S. (1969). *Physical geography of Mongolia.* 759 p. [In Mongolian].

Tserensodnom, J. (1971). *Lakes of Mongolia.* State Publishing, 202 p. [In Mongolian].

Tserensodnom, J. (2000). *A catalog of lakes in Mongolia.* Shuvuun Saaral Publishing, 141 p. [In Mongolian].

Watanabe, T., Nakamura, T., Nara, F. W., Kakegawa, T., Horiuchi, K., Senda, R., Oda, T., Nishimura, M., & Matsumoto, G. I. (2009). High-time resolution AMS ^{14}C data sets for Lake Baikal and Lake Hovsgol sediment cores: Changes in radiocarbon age and sedimentation rates during the transition from the last glacial to the Holocene. *Quaternary International, 205*, 12–20.

Zolotarev, A. G., Kulakov, V. S., & Khilko, S. D. (1982). Khubsugul upland. In *Geomorphology of Mongolian People Republic. Proceedings of Join Soviet-Mongolian Scientific-Research Geological Expedition* (Vol. 28, pp. 109–122). Nauka. [In Russian].

Chapter 17
Lake Khargal

Abstract Lake Khargal is one of geotectonically, geomorphologically, and geochemically important lakes in the taiga forest region of northern Mongolia. Lake Khargal represents dynamic changes of surface water in the region. The lake also plays an important role in the hydrological system of the North Arctic Ocean drainage basin. This is because it is a paleo-oxbow lake of Eg River, one of tributaries of Selenge River, a major headwater of Lake Baikal. Lake Khargal enlarged its area by 1.7 km^2 during 1970–2010 and reduced by 0.3 km^2 during 2010–2018. The changes in lake area during the last 50 years imply that Lake Khargal is sensitive to the present climate change indicated by the gradually rising annual average air temperature and fluctuating annual average precipitation since 1985. The linear regression analysis shows that the lake area between 1986 and 2018 was weakly related to the annual precipitation ($R^2 = 0.139$), but it was not clearly related to the annual average air temperature ($R^2 = 0.0187$). The correlation analysis shows that the lake area during 1986–2018 had a weak positive correlation with air temperature ($r = 0.136$) and a weak negative correlation with precipitation ($r = -0.372$). The A-CN-K ternary plot shows that Lake Khargal sediments were derived from the source rocks enriched in plagioclase, gabbro, tonalite, and granodiorite under a low degree of chemical weathering condition in the lake basin suggesting that a cool and arid climate dominated in the source area. The climate condition of the lake basin is shown in the chemical maturity plot, shifting from semiarid to arid. Geochronological records from Lake Khargal sediments show Holocene paleoclimate changes in northern Mongolia and Eurasia.

Keywords Lake Khargal · Eg River · Selenge River · Lake Baikal · Climate change · Northern Mongolia

A. Orkhonselenge et al., *Lakes of Mongolia*, Syntheses in Limnogeology,
https://doi.org/10.1007/978-3-030-99120-3_17

17.1 Introduction

Lake sediments are preferentially used to identify weathering intensity in source area and depositional conditions in lake basin because they record long-term continuous local erosional, transportation, and depositional processes under the local and regional climatic systems (Orkhonselenge et al., 2021b). The sediments archived in terrestrial lakes contain geochemical proxies of the surface processes over various geological timescales. Thus, sediments obtained from Lake Khargal in northern Mongolia (Fig. 1.1) can be used to decipher local and regional climate changes. Understanding and reconstructing past and present climate and environmental changes in northern Mongolia is essential in predicting the impacts of future climate change scenarios and landform evolution in the region. The climate change in Mongolia can reflect the great environmental changes of Eurasia and Central Asia.

Due to the highly elevated intercontinental location of Mongolia on the Central Asian Orogenic Belt (CAOB) and in the transitional region between the Siberian taiga forest in the north and Govi[1] desert in the south, northern Mongolia is particularly sensitive to changes in air mass circulation, moisture balance, and air temperature due to the collision of different atmospheric circulations of the East Asian winter monsoon (EAWM) and westerlies (Orkhonselenge et al., 2019, 2021a). The westerly atmospheric transport influences the northern Mongolia (Lydolph, 1977) where summers are cool and winters are extremely cold, i.e., the annual average air temperature amplitude is high. In northern Mongolia, geochemical investigation results from that the Lake Khargal basin may thus enhance interpretations of data obtained from other surrounding regions, including Siberia and northern China (see Chap. 21).

To this date, there have been numerous studies from Lake Khuvsgul[2] regarding paleoclimate changes of northern Mongolia (e.g., Fedotov et al., 2004; Prokopenko et al., 2007; Orkhonselenge et al., 2013) and from Lakes Dood (e.g., Peck et al., 2001; Narantsetseg et al., 2013) and Erkhel (e.g., Khosbayar, 2005; Watanabe et al., 2007) (see Chap. 21). For instance, Prokopenko et al. (2007) reconstructed the Holocene climate changes in the Lake Khuvsgul basin. They revealed high precipitation/runoff from 11.0 ka to 7.0 ka and also warmer summers between 6.0 ka and 3.5 ka resulting from higher soil temperatures and lower moisture availability interpreted based on diatom and biogenic silica proxies. However, Lake Khargal has not attracted much attention by scholars until today. Some general descriptions were given, for instance, Tserensodnom (1971, 2000) provided morphometric values of Lake Khargal. The Lake Khargal basin experienced the evolution of paleooceanic and continental terrains, and in the early Cretaceous, a 30-km-long depression formed between the Eg River and Lake Olon faults due to the intensive uplift of surrounding landforms (Batrom et al., 2006; Table 15.1). Orkhonselenge et al.

[1] *Govi* has been often spelled as *Gobi* in the international literature. Govi is the correct English transliteration from Mongolian *Говь*.

[2] *Khuvsgul* has been misspelled as *Hovsgol* and *Khubsugul* in publications. Khuvsgul is the right English transliteration from Mongolian *Хөвсгөл*.

(2019, 2021a) noted that reconstructing the impact of paleoclimate and paleoenvironmental changes in the Lake Khargal basin would provide us future climate forecasts in Eurasia and Central Asia in the regional scale through comparison with data from surrounding sites. Recently, Barna et al. (2021) reconstructed the Holocene vegetation and climate changes in northern Mongolia based on palynological, macrofossil, and geochemical analyses of livingstone piston core sediments from Lake Khargal (see Chap. 21).

Lake Khargal (Fig. 1.1) archives information not only about surface erosional, transportation, and depositional processes in the Eg and Selenge[3] River drainage basins and in the Lake Baikal[4] basin but also about historical atmospheric circulations and geochemical characteristics of Eurasia due to its unique position in the downwind direction of the EAWM and westerlies. As a recorder of regional paleoclimate and paleoenvironmental changes, Lake Khargal sediments can shed light on the history of geomorphological evolution of the lake basin (see Sect. 17.3). In this chapter, the changes in area of Lake Khargal during the last 50 years and the geochemical nature of its sediments are considered in detail. Paleoclimatic implications from Lake Khargal basin are considered in Chap. 21.

17.2 Physiographic Condition

Lake Khargal is located at the elevation of 1070.8 m a.s.l. (Tserensodnom, 1971) or 1057 m a.s.l. (Orkhonselenge et al., 2019) in a tectonic rift in the southwest of Kharlan and the west of Burekht mountain ranges (Fig. 17.1), branches of Zed-Khantai and Buteel mountain ranges in northern Mongolia (Fig. 1.1). Lake Khargal extends for 4.9 km in length, 4.2 km (Tserensodnom, 1971) or 4.4 km (Tserensodnom, 2000) in width, and 14.6 km (Tserensodnom, 1971) or 14.8 km (Tserensodnom, 2000) along shoreline length. The lake covers an area of 13.6 km^2 (Tserensodnom, 1971) or 13.83 km^2 (Gerelsaikhan & Orkhonselenge, 2019; see Sect. 17.6).

The lake resides in the Eg and Selenge River drainage basins belonging to the North Arctic Ocean drainage basin. Water depth of the lake is 15.6 m, whereas the water volume is 0.127 km^3 (Tserensodnom, 2000). The Lake Khargal water depth (Table 17.1) increases from the east to the southwest (Tserensodnom, 2000) where the maximum depth is 16.8 m (Orkhonselenge et al., 2019).

Lake Khargal is filled with the Holocene lacustrine deposits. In the lake, clays and muds are abundant on its bottom, while silts and sands are distributed on its shorelines (Tserensodnom, 2000). Mean grain size of the Lake Khargal sediments fluctuates between fine silt and clay to fine sand (Barna et al., 2021). The lake is situated in the taiga forest region of northern Mongolia, which is the southern border of the Siberian taiga zone.

[3] *Selenge* has been often incorrectly written as *Selenga* internationally based on the Russian pronunciation. Selenge is the correct English transliteration from Mongolian *Сэлэнгэ*.

[4] *Baigali* meaning *nature* has been misspelled as *Baikal* by the Russians. Baigali is the right English transliteration from Mongolian *Байгаль*.

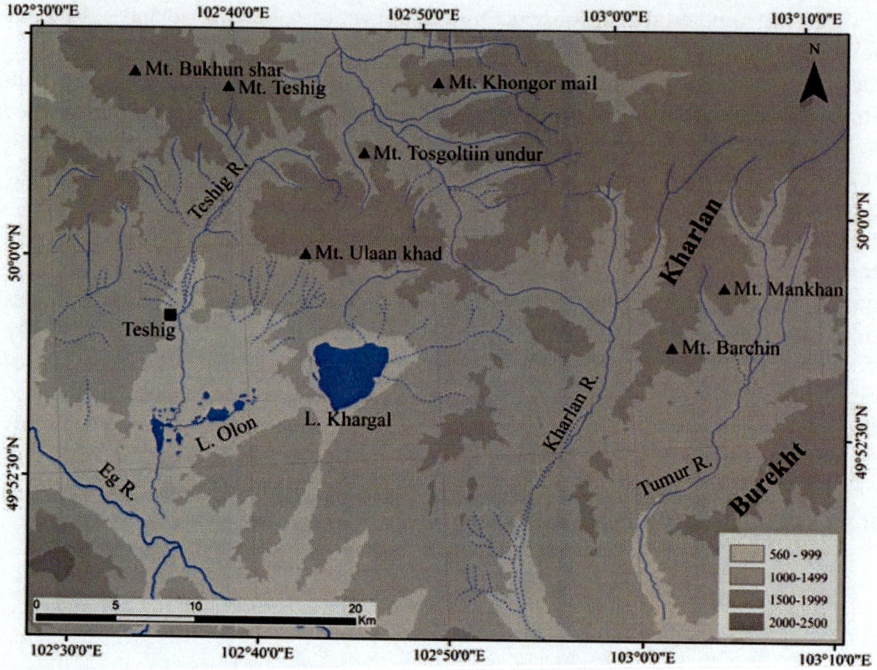

Fig. 17.1 Physiography of Lake Khargal basin with location of the Teshig meteorological station. Modified from Orkhonselenge and Gerelsaikhan (unpublished)

Table 17.1 Bathymetric data with corresponding area of Lake Khargal

#	Water depth (m)	Area (%)
1	0.0–4.0	14.7
2	4.0–6.0	12.6
3	6.0–8.0	11.0
4	8.0<	61.7

17.3 Geological and Geomorphological Settings

Tectonically, Lake Khargal lies at the border of steeply dipping faults between the late Riphean and Cambrian amphibolic and granite-carbonate complex and the late Paleozoic volcanic formation complex (Academy of Sciences of Mongolia and Academy of Sciences of USSR, 1990). Lake Khargal is tectonostratigraphically underlain by the Chita island arc terrain composed of nonmarine clastic sedimentary rocks and mafic-volcanic and volcaniclastic rocks (Badarch et al., 2002) and the Paleozoic and Mesozoic granitoids of volcanic-plutonic associations (Kröner & Rojas-Agramonte, 2014).

Lake Khargal is located at the confluence between the Eg River island arc and Orkhon-Selenge continental rift volcanic terrain in the northern Mongolian Orogenic Belt. The lithospheric stratigraphy of the Lake Khargal basin consists of the Neoproterozoic and early Cambrian Badar River meta-andesite and sedimentary-metamorphic rock formation; the middle to late Cambrian Zed igneous rock complex; the Paleozoic Khanui group made of Urkhtei, Khustai, Khargana, and Tulbur formations; the early to middle Mesozoic Selenge igneous rock complex; the late Mesozoic and early Cenozoic Uilgan River formation; and the middle to late Cenozoic deposits (Batrom et al., 2006, Table 15.1).

Modern landforms of the Lake Khargal basin are directly related to the paleogeographical history and geological evolution of northern Mongolia (Table 15.1). In general, the relict landforms formed during the times earlier than the Cretaceous are not preserved; however, the paleolake shorelines since the early Cretaceous period have been preserved until today. A large paleolake existed and it was divided into numerous lakes such as Lakes Khargal, Olon (Fig. 17.1), and Airkhan due to the uplifting and graben formation throughout the basin during the early Cretaceous (Batrom et al., 2006). Evidence for modern uplifts and erosion are revealed by formations of floodplains, peneplain, and pediments (Table 15.1).

Lake Khargal joins the Eg River valley and its low terraces distributed on the alluvial plains in the south and southwest; high mountains in the west, northwest, and north; and surface alignment in the northeast, east, and southeast (Academy of Sciences of Mongolia and Academy of Sciences of USSR, 1990). The lake is surrounded by Mt. Mankhan (1743 m a.s.l.) in the northeast, Mt. Barchin (1794 m a.s.l.) in the east (Fig. 17.1), Mt. Zurkh (1717 m a.s.l.) in the southeast, Mt. Uliin Undur (1393 m a.s.l.) in the south, Dulaan Khan Mountain Range in the southwest, Mt. Ovoot (1230 m a.s.l.) in the northwest (Orshikh et al., 2014), and Mt. Ulaan Khad, Mt. Tosgoltiin Undur, and Mt. Teshig in the north (Fig. 17.1). The surrounding mountains are as high as 1600–1930 m a.s.l., whereas the surrounding lowlands are at the range of 1200–1600 m a.s.l., (Fig. 17.1). Lake Khargal is quasi triangle-shaped (Fig. 17.1), and the paleo- and modern shores are distinctly observed (Fig. 17.2a, b).

17.4 Climate Condition

The local climate data were recorded at the Teshig meteorological station located at 9 km northwest of Lake Khargal (Fig. 17.1) between 1985 and 2019. The data show that the annual average air temperature ranges between −3.7°C in 1985 and 2.1 °C in 2019 with an average value of −1.9 °C (Fig. 17.3a). The annual average precipitation varied from 76.5 mm in 1985 to 317 mm in 2019 with an average value of 285.5 mm/year (Fig. 17.3b). The average air temperature was −27.7 °C in January and 18.2 °C in July (Fig. 17.3c).

The most humid years were observed with the annual precipitation ranging between 353.3 mm in 1989 and 452 mm in 1994 (Fig. 17.3b). The summer and

Fig. 17.2 (**a**) Lake Khargal viewing toward the southeast. (**b**) Lake Khargal on the right and a separated lake on the left viewing toward the southwest. (**c**) The separated lake viewing toward the southeast. Photos by A. Orkhonselenge on June 23, 2019, for (**a**) and June 29, 2019, for (**b** and **c**)

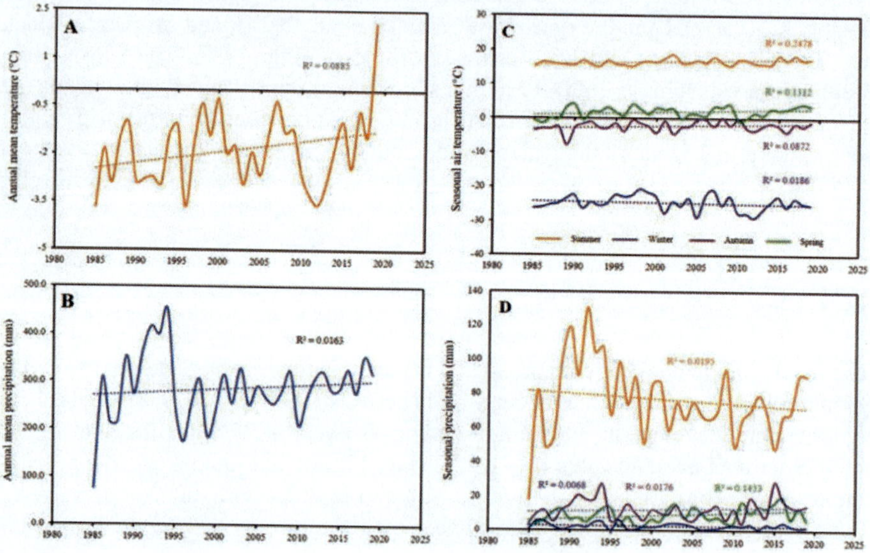

Fig. 17.3 Annual average (**a**) air temperature and (**b**) precipitation and seasonal average (**c**) air temperature and (**d**) precipitation recorded at the Teshig meteorological station (Fig. 17.1) between 1985 and 2019

winter precipitations slightly decreased; however, the spring and autumn precipitations gradually increased (Fig. 17.3d). The warm years were observed in 1988–1989, 1994–1995, 1998–2000, 2007, and 2017 except for 2019 (Fig. 17.3a). The summer and spring air temperatures were gradually rising (Fig. 17.3c).

17.5 Hydrological Condition

Lake Khargal is located in the drainage basin of Eg River (Fig. 17.1), running between Khuvsgul and Khantai mountain ranges, and it is one of major inflows of Selenge River running between Buteel and Buren mountain ranges in the Siberian geomorphological region (Fig. 1.1). Hydrogeologically, Lake Khargal is included in the Khuvsgul-Zed Orogenic Belt.

Lake Khargal is fed by Khavchraanga River in the north and Khoid and Dund streams in the east, and it has no outlet (Fig. 17.4). In the drainage basin of Lake

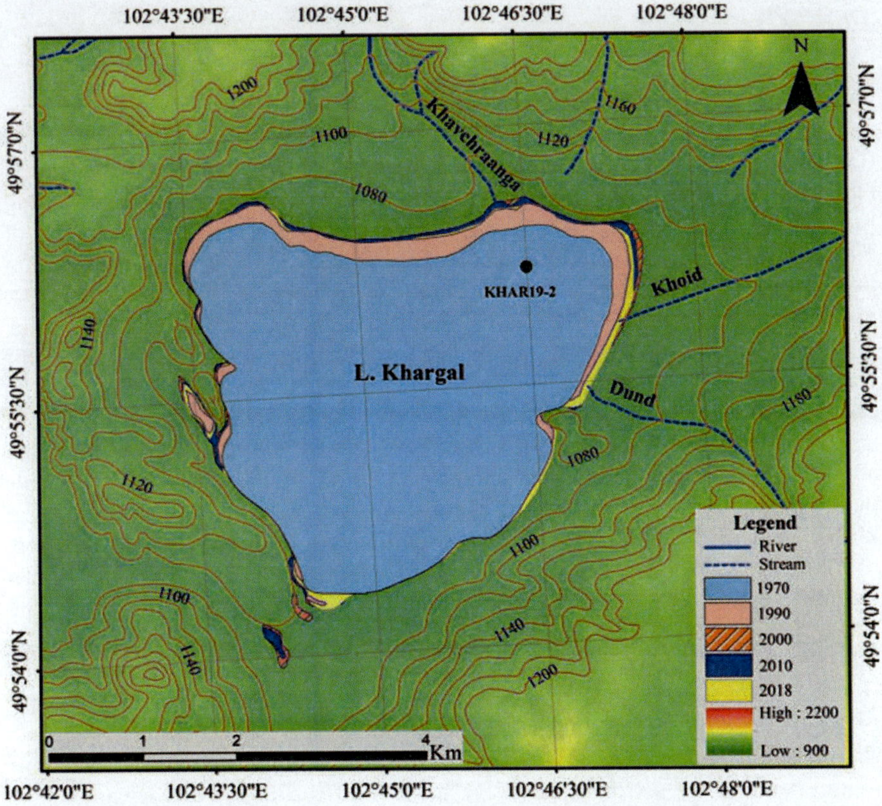

Fig. 17.4 Spatial distributions of areas of Lake Khargal and a separated lake (Fig. 17.2c) in 1970, 1990, 2000, 2010, and 2018 with the location of the core KHAR19–2 for geochemical review (see Sect. 17.7). After Orkhonselenge et al. (2019, 2021a)

Khargal, there are widely distributed perennial and ephemeral rivers and streams (Fig. 17.1), and their water levels are highly dependent on the local climate condition. For instance, there are Eg River, an outlet of Lake Khuvsgul in the northwest, large inflows of the Eg River such as Erin and Ilenkh rivers, and numerous small rivers such as Uramt, Tsalam, Evert, and Kharlan rivers (Fig. 17.1). Eg, Kharlan, Erin, Tsalam, Evert, Ikh Ilenkh, Baga Ilenkh, Urint, Deed Nariin, Dund Nariin, Dood Nariin, and Teshig rivers pass at the elevation range of 939–1530 m a.s.l. Today, rivers such as Eg, Kharlan, Teshig, and Erin are dropping their water levels or tend to shrink due to aridness, and lakes such as Lake Olon and other small lakes (Fig. 17.2c) are decreasing their areas (see Sect. 17.6).

The lake water temperature is 19.5–20.8 °C at its surface and 17.5 °C at 10 m depth during June to August (Tserensodnom, 2000). The lake is frozen between November and April, and the lake transparency is 3.5–4.0 m. The lake is saline with pH of 8.0 (Davaa, 2015), and the mineralization value of the lake is 9.15–9.85 g/L (Tserensodnom, 2000).

17.6 Changes in Area

Table 17.2 shows the changes in area of Lake Khargal during 1970–2018. The lake increased its area by 1.02 km^2 in 1970–2000 and by 0.73 km^2 in 1990–2000 (Orkhonselenge et al., 2021a; Fig. 17.4). There were two major periods of long-term area increasing in 1970–1995 and 2000–2010 and short-term area decreasing in 1995–2000 and 2010–2018 (Table 17.2). The first period of area gaining of 1.85 km^2 in 1970–1995 may have been caused by the rises in the annual precipitation during 1989–1994 (Fig. 17.3b), summer precipitation in 1986 and 1988–1994, and autumn precipitation in 1991–1994 (Fig. 17.3d). The second period of area gaining of 0.22 km^2 in 2000–2010 coincided with the maximum annual average air temperatures in 2000 and 2007 (Fig. 17.3a) which may have contributed to thawing of snow and permafrost meltwater. However, the decreased lake areas of 0.1 km^2 in 1995–2000 and of 0.15 km^2 in 2010–2018 were in agreement with the drops in the annual and summer precipitations and the rises in the annual and seasonal air temperatures with the maximum annual average air temperature recorded in 1998 and 2017 (Fig. 17.3a), which may have strengthened evaporation from the lake surface.

The changes in area of the lake estimated by Normalized Difference Water Index (NDWI), which is additionally calculated for validating the estimation by the Landsat data, show increases of the lake area by 0.7 km^2 in 1990–2000 and by 0.2 km^2 in 2000–2010 (Orkhonselenge et al., 2021a; Table 17.2). The continuous rise of the lake area between 1990 and 2010 is from 13.45 km^2 in 1990 to 14.13 km^2 or 14.40 km^2 in 2010 (Table 17.2). The comparative estimates are distinctly demonstrated by the area increase of a separated lake situating on the western shore of Lake Khargal (Figs. 17.2b, c, and 17.4) in 1995–2010 although this separated lake decreased its area in 1987–1995. Overall, the area of Lake Khargal increased by 1.4 km^2 between 1970 and 2018 based on the Landsat estimation and by 0.4 km^2

Table 17.2 Temporal changes in area of Lake Khargal

#	Year	Area (km²)		Area (km²)	Author
		Landsat	NDWI	Separated lake	
1	1970	12.43	–	–	Gerelsaikhan and Orkhonselenge (2019); Orkhonselenge et al. (2021a)
2	1971	13.60	–	–	Tserensodnom (1971)
3	1986	13.32	–	0.02	Gerelsaikhan and Orkhonselenge, unpublished
4	1987	13.69	–	0.07	Gerelsaikhan and Orkhonselenge, unpublished
5	1988	13.70	–	0.04	Gerelsaikhan and Orkhonselenge, unpublished
6	1989	13.41	–	0.03	Gerelsaikhan and Orkhonselenge, unpublished
7	1990	13.45	13.8	–	Gerelsaikhan and Orkhonselenge (2019); Orkhonselenge et al. (2021a)
8	1991	13.08	–	0.04	Gerelsaikhan and Orkhonselenge, unpublished
9	1995	14.28	–	0.01	Gerelsaikhan and Orkhonselenge, unpublished
10	2000	14.18	14.5	–	Gerelsaikhan and Orkhonselenge (2019); Orkhonselenge et al. (2021a)
11	2005	14.31	–	–	Gerelsaikhan and Orkhonselenge, unpublished
12	2010	14.13	14.7	–	Gerelsaikhan and Orkhonselenge (2019); Orkhonselenge et al. (2021a)
13	2010	14.40	–	–	Davaa (2015)
14	2010	13.91	–	0.07	Gerelsaikhan and Orkhonselenge, unpublished
15	2013	14.20	–	–	Davaa (2015)
16	2015	13.57	–	–	Gerelsaikhan and Orkhonselenge, unpublished
17	2018	13.83	14.2	–	Gerelsaikhan and Orkhonselenge (2019); Orkhonselenge et al. (2021a)

between 1990 and 2018 with the NDWI estimation (Orkhonselenge et al., 2021a; Table 17.2). The changes in the lake area match with the results given by Davaa (2015) showing that the lake area decreased from 2010 to 2013 (Table 17.2). Situating in the taiga forest region of northern Mongolia similar to Lake Khargal, the temporal variations of lake area in the Bulgan region showed a generally decreasing pattern (Kang et al., 2015).

The Lake Khargal area has responded directly to the variations in the annual average air temperature and precipitation. The linear regression analysis (Fig. 17.5) shows that the lake area between 1986 and 2018 was weakly related to the annual precipitation ($R^2 = 0.139$) but was not clearly related to the annual average air temperature ($R^2 = 0.0187$). The correlation analysis shows that the lake area during 1986–2018 had a weak positive correlation with the air temperature ($r = 0.136$) and a weak negative correlation with the precipitation ($r = -0.372$). This trend in Lake Khargal (Fig. 17.5b) is consistent with the result provided by Kang et al. (2015), who showed that the relationship between lake area and precipitation was less distinct for the lakes in the Bulgan region, i.e., annual precipitation accounted for 45% of the lake area change ($P < 0.05$) was more important than that ($R^2 = 0.34$, $P = 0.077$) of monthly precipitation, with low sensitivity of lake area to precipitation change ($S = 0.047$ km²/mm). The relationship between the lake area and air

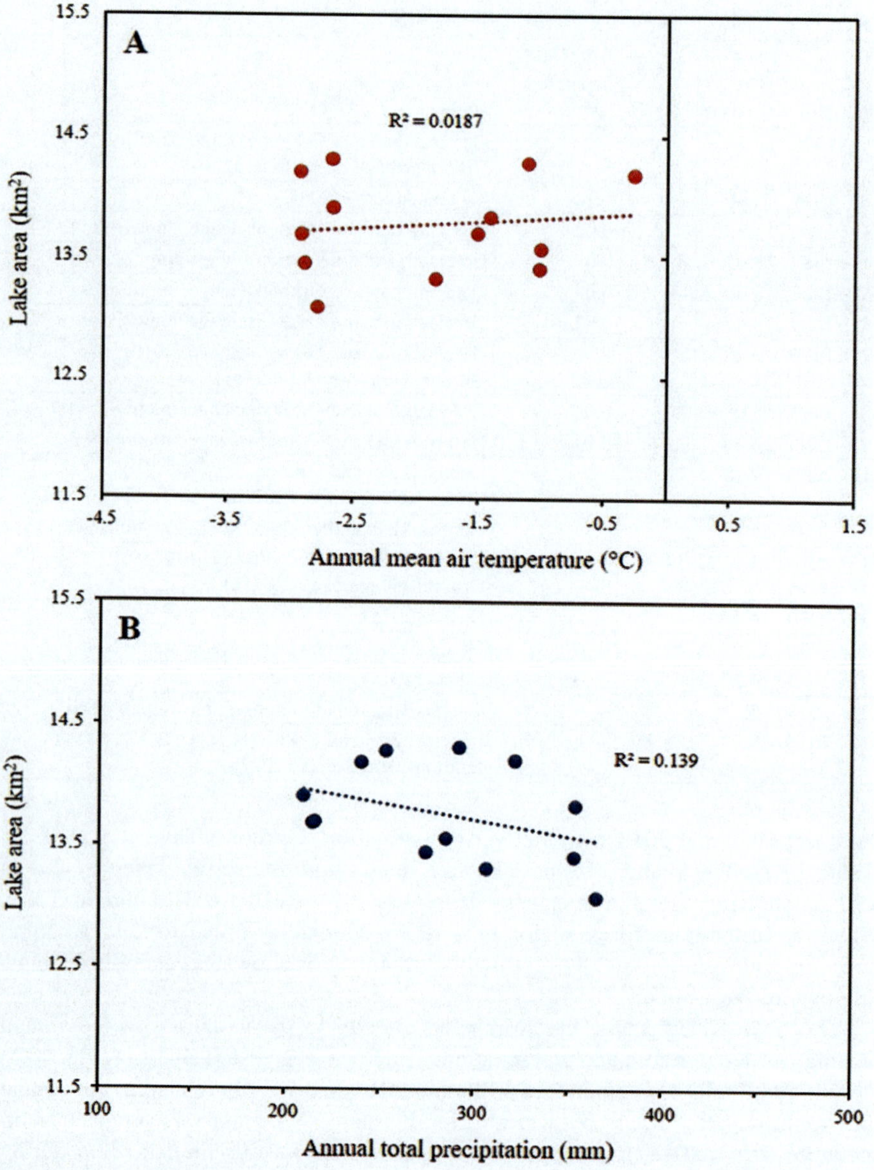

Fig. 17.5 Relationships between area of Lake Khargal and annual average (**a**) air temperature and (**b**) precipitation during 1986–2018

temperature (Fig. 17.5a) implies that the most significant contributions to the increase in lake area are snow meltwater in spring in addition to groundwater and permafrost meltwater from the surrounding mountains. In general, summer and spring air temperatures (Fig. 17.3c) and spring precipitation (Fig. 17.3d) become greater contributions to the lake area gaining than other seasonal parameters.

17.7 Geochemical Review

In the core KHAR19–2 sediments of Lake Khargal (Fig. 17.4), SiO_2 is the major constituent with a value of 30.55% on average sharing with CaO (20.41%), Al_2O_3 (7.27%), MgO (4.05%), Fe_2O_3 (3.72%), Na_2O (1.2%), K_2O (1.1%), TiO_2 (0.54%), P_2O_5 (0.26%), and MnO (0.11%) (Orkhonselenge et al., 2021b). However, loss on ignition (LOI) is large throughout the core sediments with an average value of 30.8%. This is an indication of high biogenic matters in the sediments. The lake sediments contain large amounts of semimetal, alkaline earth metals, and transition metals. In Fig. 17.6, SiO_2 (R^2 = 0.98), TiO_2 (R^2 = 0.96), K_2O (R^2 = 0.94), Na_2O (R^2 = 0.88), and Fe_2O_3 (R^2 = 0.84) show strong relationships with Al_2O_3, whereas MnO (R^2 = 0.29) and P_2O_5 (R^2 = 0.13) indicate relatively weak relationships with Al_2O_3. LOI (R^2 = 0.84) shows a strong relationship with CaO. However, CaO (R^2 = 0.95) shows a strong negative relationship with Al_2O_3, while MgO (R^2 = 0.05) shows a weak negative relationship with Al_2O_3 (Orkhonselenge et al., 2021b).

Terrigenous SiO_2 is the most dominant in the lake sediments. According to Condie (1992), terrigenous-derived TiO_2 showing the strong positive relationship with Al_2O_3 (R^2 = 0.96) implies its presence in phyllosilicates. The terrigenous-indicative elements (e.g., Si, Al) dominant in the lake sediments are consistent with the nonmarine sedimentary rocks underlain by the volcanic and volcaniclastic rocks (Badarch et al., 2002) on the Chita island arc (see Sect. 17.3). Following Hofer et al. (2013), the negative correlation of CaO with Al_2O_3 means that Ca is not significantly influenced by a detrital source (e.g., feldspars). The strong positive correlation between CaO and LOI (Fig. 17.6) allows to assume that CaO is largely derived from carbonates and represents carbonate minerals, largely calcite (Orkhonselenge et al., 2021b).

The Al_2O_3-CaO + Na_2O-K_2O (A-CN-K) ternary diagram (Nesbitt & Young, 1984) shows that the weathered rocks in the Lake Khargal basin were derived from tonalites to granodiorites with plagioclase and gabbro source terrains (Orkhonselenge et al., 2021b; Fig. 17.7a). The granodiorite and gabbro sources are related to the composition of volcanic clastic sedimentary rocks. The Lake Khargal sediments are nearly parallel to the A-CN line (Fig. 17.7a), defining a steady-state weathering trend toward the plagioclase, which is probably due to the presence of Na-rich feldspar rather than K-rich feldspar. The source rocks enriched in plagioclase, gabbro, tonalities, and granodiorites (Orkhonselenge et al., 2021b; Fig. 17.7a) match with the Paleozoic tonalities, granodiorites, plagioclase, and adamellites in the Lake Khargal basin described by the Academy of Sciences of Mongolia and Academy of Sciences of USSR (1990). The chemical maturity of Lake Khargal sediments based on a correlation between SiO_2 and $Al_2O_3 + K_2O + Na_2O$ as proposed by Suttner and Dutta (1986) shows a shift from semiarid to arid climate conditions in the lake basin (Orkhonselenge et al., 2021b; Fig. 17.7b).

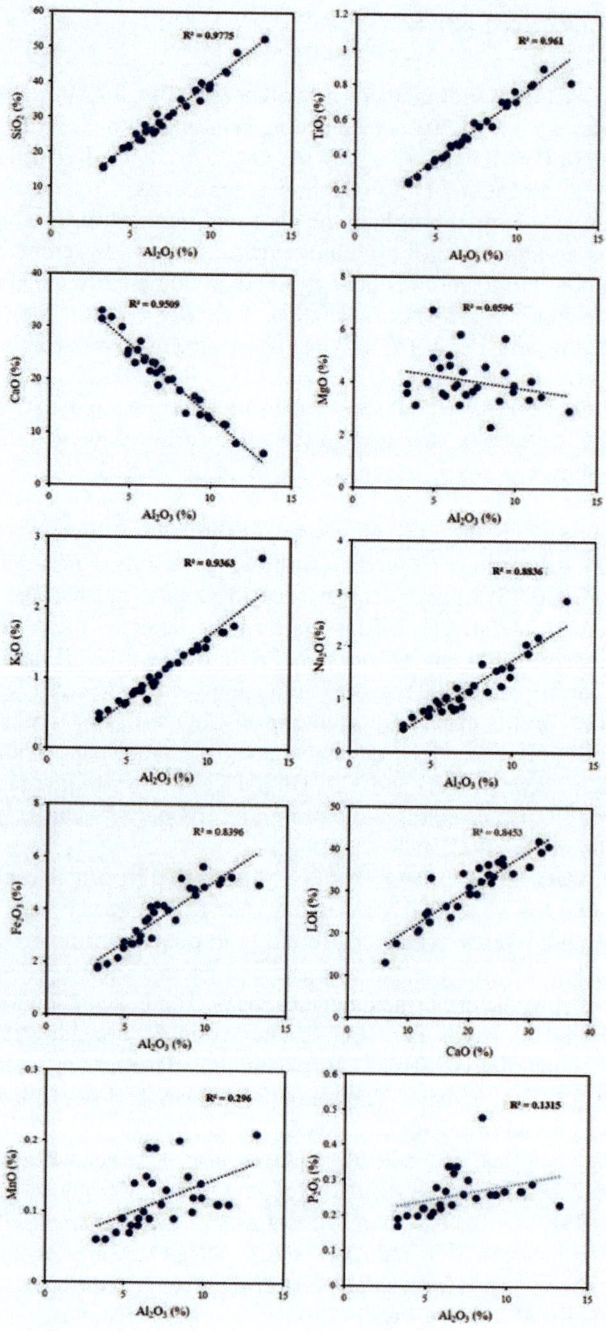

Fig. 17.6 Major element oxides vs. Al_2O_3 and CaO vs. LOI in Lake Khargal sediments. After Orkhonselenge et al. (2021b)

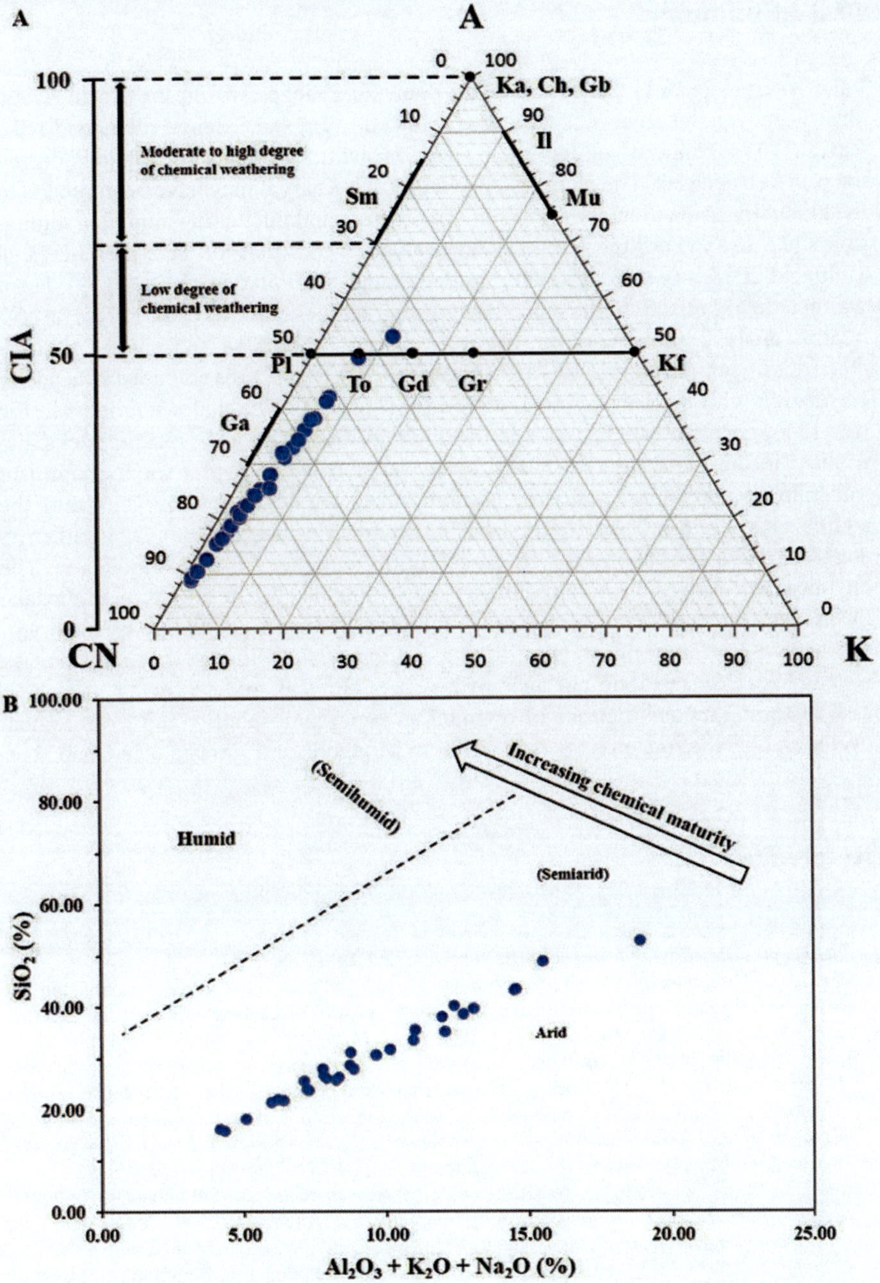

Fig. 17.7 (**a**) Degree of chemical weathering of Lake Khargal sediments in A-CN-K ternary diagram showing source rock compositions of Ka (kaolinite), Ch (chlorite), Gb (gibbsite), Il (illite), Ms (muscovite), Kf (K-feldspar), Sm (smectite), Pl (plagioclase), To (tonalite), Gd (granodiorite), Gr (granite), and Ga (gabbro). (**b**) Chemical maturity of Lake Khargal sediments in a plot of SiO_2 vs. $Al_2O_3 + K_2O + Na_2O$. Modified from Orkhonselenge et al. (2021b)

17.8 Summary

Lake Khargal records the hydrological dynamic changes within the North Arctic Ocean drainage basin because the lake is linked to Eg and Selenge rivers and Lake Baikal. The lake increased its area by 1.4 km^2 with significant fluctuations during the past five decades. The changes in area of Lake Khargal may have been related to the gradually rising annual average air temperature and fluctuating annual precipitation since 1985. The linear regression analysis shows that the lake area between 1986 and 2018 was weakly related to the annual precipitation ($R^2 = 0.139$), but it was not clearly related to the annual average air temperature ($R^2 = 0.0187$). The correlation analysis shows that the lake area during 1986–2018 had a weak positive correlation with the annual average air temperature ($r = 0.136$) and a weak negative correlation with the annual precipitation ($r = -0.372$). The major elements in the lake sediments show their derivation from the sedimentary and volcanic rocks distributed in the Lake Khargal basin. Moreover, the A-CN-K plot shows dominant minerals of plagioclase, gabbro, granodiorites, and tonalities derived from the source rocks. In the Lake Khargal basin, the low degree of chemical weathering suggests that a cool and arid climate condition dominated in the source area. The chemical maturity of the lake sediments shows that the climate condition in the lake basin is shifting from semiarid to arid. For understanding the evolutionary history of the Lake Khargal basin and its driving geologic and climatic mechanisms, we strongly emphasize the importance of geochronological study showing more reliable historical reconstructions of regional tectonic, atmospheric, and hydrological cycles and paleoclimate changes in northern Mongolia and Eurasia (see Chap. 21).

References

Academy of Sciences of Mongolia and Academy of Sciences of USSR. (1990). *National Atlas of the Peoples' Republic of Mongolia.* [In Russian with English abstract].

Badarch, G., Cunningham, W. D., & Windley, B. F. (2002). A new terrane subdivision for Mongolia: Implications for the Phanerozoic crustal growth of Central Asia. *Journal of Asian Earth Sciences, 21*, 87–110.

Barna, J., Bigelow, N., Orkhonselenge, A., Fowell, S., Gardner, W. R. M., Burentogtokh, J., & Honeychurch, W. (2021). Holocene paleoenvironmental changes in the forest-steppe ecotone of northern mongolia based on palynological and sedimentological analyses of cores from Lake Khargal. In *GSA Annual Meeting—The Geological Society of America. Presentation #27–12. Portland, Oregon. October 10–13, 2021*. https://doi.org/10.1130/abs/2021AM-370452.

Batrom, B., Tulgaa, U., Batbayar, N., Tumursukh, N., Adiyasuren, D., & Sukhbaatar, B. (2006). *Report: "General exploring and geological mapping in a scale of 1:50 000 of Olon Nuur area in Zed metallogenic region"*. Geological Information Center Press. [In Mongolian].

Condie, K. C. (1992). Chapter 12: Proterozoic terranes and continental accretion in southwestern North America. In K. C. Condie (Ed.), *Proterozoic crustal evolution* (pp. 447–480). Elsevier.

Davaa, G. (2015). *Resource and regime of surface water of Mongolia*. 408 p. [In Mongolian].

Fedotov, A. P., Chebykin, E. P., Yu, S. M., Vorobyova, S. S., Yu, O. E., Golobokova, L. P., Pogodaeva, T. V., Zheleznyakova, T. O., Grachev, M. A., & Tomurhuu, D. (2004). Changes in the volume and salinity of Lake Khubsugul (Mongolia) in response to global climate changes in the upper Pleistocene and the Holocene. *Palaeogeography, Palaeoclimatology, Palaeoecology, 209*, 245–257.

Gerelsaikhan, D., & Orkhonselenge, A. (2019). Recent lake area changes of Lake Khargal in northern Mongolia. *The Environment, 1*(1), 15–20. [In Mongolian with English abstract].

Hofer, G., Draganits, E., Wagreich, M., Hofmann, C. C., Reischenbacher, D., Neuhuber, S., Grundtner, M. L., & Bottig, M. (2013). Stratigraphy and geochemical characterisation of Upper Cretaceous non-marine e marine cycles (Grünbach Formation, Gosau Group, Austria). *Austrian Journal of Earth Sciences, 104*(2), 90–107.

Kang, S., Lee, G., Togtokh, C., & Jang, K. (2015). Characterizing regional precipitation-driven lake area change in Mongolia. *Journal of Arid Land, 7*(2), 146–158.

Khosbayar, P. (2005). *Mesozoic and Cenozoic paleogeography and paleoclimate of Mongolia*. Institute of Geology and Mineral Resources of Mongolian Academy of Sciences (MAS), Transection, 15, Ulaanbaatar, 184 p. [In Mongolian].

Kröner, A., & Rojas-Agramonte, Y. (2014). The Altaids as seen by Eduard Suess, and Present Thinking on the Late Mesoproterozoic to Palaeozoic Evolution of Central Asia. *Austrian Journal of Earth Sciences, 107*(1), 156–168.

Lydolph, P. E. (1977). Climates of the Soviet Union. In H. E. Landsberg (Ed.), *World survey of climatology* (Vol. 7)., 443 p.

Narantsetseg, T., Krivonogov, S. K., Oyunchimeg, T., Uugantsetseg, B., Burr, G. S., Tomurhuu, D., & Dolgorsuren, K. (2013). Late Glacial to Middle Holocene climate and environmental changes as recorded in Lake Dood sediments, Darhad Basin, northern Mongolia. *Quaternary International, 311*, 12–24.

Nesbitt, H. W., & Young, G. M. (1984). Prediction of some weathering trends of plutonic and volcanic rocks based on thermodynamic and kinetic considerations. *Geochimica et Cosmochimica Acta, 48*, 1523–1534.

Orkhonselenge, A., Gerelsaikhan, D., & Davaagatan, T. (2021a). Spatial and temporal responses of lakes in Northern Mongolia to climate change. In *EGU2021 Virtual Meeting. Presentation #EGU21–9344*. https://doi.org/10.5194/egusphere-egu21-9344.

Orkhonselenge, A., Fowel, S., Nancy, G., Josh, B., Gerelsaikhan, D., & Camillie, T. (2021b). Geochemical properties of Lake Khargal sediments in northern Mongolia. In *Goldschmidt2021 Virtual Conference. Presentation. July 4–9, 2021. Lyon, France.* https://doi.org/10.7185/gold2021.4898.

Orkhonselenge, A., Fowel, S., Josh, B., Nancy, G., Camillie, T., & Gerelsaikhan, D. (2019). *A report from fieldwork measurement and mapping of changes in lake area of Lake Khargal*. Laboratory of Geochemistry and Geomorphology, National University of Mongolia, 15 p. [In Mongolian].

Orkhonselenge, A., Krivonogov, S. K., Mino, K., Kashiwaya, K., Safonova, I. Y., Yamamoto, M., Kashima, K., Nakamura, T., & Kim, J. Y. (2013). Holocene sedimentary records from Lake Borsog, eastern shore of Lake Khuvsgul, Mongolia, and their paleoenvironmental implications. *Quaternary International, 290–291*, 95–109.

Orshikh, N., Enkhbayar, G., Garig, G., Enkhbat, R., & Chantuu, B. (2014). *Atlas of Mongolia with a scale of 1:800,000*. Monsudar Press, 117 p. [In Mongolian].

Peck, J. A., Khosbayar, P., King, J. W., Fowell, S. J., Ariunbileg, S., & Soninkhishig, N. (2001). *The latest Holocene sedimentary environmental magnetic record from lake Dood, Mongolia*. American Geophysical Union, Fall Meeting, Abstract Id. GP41A-0243.

Prokopenko, A. A., Khursevich, G. K., Bezrukova, E. V., Kuzmin, M. I., Boes, X., Williams, D. F., Fedenya, S. A., Kulagina, N. V., Letunova, P. P., & Abzaeva, A. A. (2007). Paleoenvironmental proxy records from Lake Hovsgol, Mongolia, and a synthesis of Holocene climate change in the Lake Baikal watershed. *Quaternary Research, 68*, 2–17.

Suttner, L. J., & Dutta, P. K. (1986). Alluvial sandstone composition and paleoclimate, I. Framework mineralogy. *Journal of Sedimentary Petrology, 56*, 329–345.

Tserensodnom, J. (1971). *Lakes of Mongolia*. State Publishing, 202 p. [In Mongolian].

Tserensodnom, J. (2000). *A catalog of lakes in Mongolia*. Shuvuun Saaral Publishing, 141 p. [In Mongolian].

Watanabe, T., Nakamura, T., & Kawai, T. (2007). Radiocarbon dating of sediments from large continental lakes (Lakes Baikal, Hovsgol and Erhel). *Nuclear Instruments and Methods in Physics Research B, 259*, 565–570.

Chapter 18
Landscape, Lake Distribution, and Evolution in Central Mongolia

Abstract Central Mongolia plays an important role in climatology, hydrology, and geomorphology of Mongolia and Central Asia because it borders the North Arctic Ocean and Central Asian internal drainage basins through the Khangai Mountain Range, one of global interfluves. The extraordinary landscape of central Mongolia is characterized by the Khangai Mountain Range, a wide intermontane valley between Khangai and Khentii mountain ranges, large rivers, and many small lakes. Central Mongolia is physiographically a transition region from the Siberian high-pressure cell to the Central Asian low-pressure cell. The region is geologically diverse with the presence of Quaternary formations with igneous rocks in the north, northeast, and east of the Khangai Mountain Range. Large rivers such as Ider, Chuluut, Khanui, Tamir, and Orkhon draining the northern and northeastern Khangai and Tuul, Kharaa, and Yeruu draining the southwestern and northwestern Khentii feed Selenge River, which is a major headwater of Lake Baikal in the North Arctic Ocean drainage basin. Lakes in central Mongolia occupy only 2.9% of the total lake area (the lowest in the country), and they have primarily tectonic, fluvial, and karst origins. Lakes in this region would provide valuable information regarding the paleogeographical history of the region and paleoclimate changes within Mongolia and Central Asia.

Keywords Khangai Mountain Range · Khentii Mountain Range · Orkhon River · Tuul River · Central Mongolia

18.1 Introduction

Central Mongolia is climatologically, hydrologically, geomorphologically, and historically an important region in Mongolia. Understanding the landscape of central Mongolia is one of essential tasks for filling the gap of knowledge needed for reconstructing paleoclimate changes in Mongolia and Central Asia. Central Mongolia is

A. Orkhonselenge et al., *Lakes of Mongolia*, Syntheses in Limnogeology,
https://doi.org/10.1007/978-3-030-99120-3_18

influenced by numerous atmospheric systems such as the Siberian-Mongolian anti-cyclone, the midlatitude westerlies, and the Central Asian cyclone. Because of its geographical position at the junction of these three large-scale climatic systems, central Mongolia plays an important role in the climate of Mongolia. The climate over the high mountains in the west of the region is cool and humid, whereas it is warm and dry in the depression at the east of the region. The landscape of central Mongolia is represented by Khangai Mountain[1] Range (Fig. 1.1), a wide intermontane valley, large rivers (e.g., Orkhon, Ider, Chuluut, Khanui, Khunui, Tamir, Tuul, Kharaa, and Yeruu rivers), and a limited number of medium-sized Lakes Telmen, Terkhiin Tsagaan, and Ugii and many small lakes.

Physical geography of the Khangai Mountain Range was studied initially by Russian scientists (e.g., M.V. Pevtsov, N.M. Prejevalsky, G.N. Potanin, J.G. Gronno, P.K. Kozlov), and they drew attention to the present belt-like distribution of environmental conditions, occurrence of former intermontane valley glaciations, and lake level oscillations (Klimek et al., 1980). In the Khangai Mountain Range, numerous international research projects have been conducted since the 1960s. For instance, landform, soil, cryogenic process, and vegetation cover in the Tsagaan Turuut River drainage basin and the Sant valley were studied by a German expedition in 1960 (Klimek et al., 1980). Moreover, the Quaternary geomorphology, paleogeography, climatology, hydrology, pedology, and botany in the Ulziit River drainage basin were investigated by a Polish and Mongolian expedition during 1974–1976.

The Khangai Mountain Range is genetically connected with Khentii Mountain[2] Range (Fig. 18.1) in the northeast of Mongolia (Murzaev, 1952; Tsegmid, 1969; Munkhuu, 1979). However, these ranges extend to opposite directions, for example, the Khangai is oriented from the northwest to the southeast, while the Khentii extends from the northeast to the southwest (Fig. 1.1), i.e., the trends of their extensions imply these ranges have different tectonic origins. The Pleistocene glaciations on Mt. Otgon Tenger, the peak of Khangai Mountain Range (Tsegmid, 1969; Jigj, 1976), left moraines dated to be during the large ice advances (Rother et al., 2014).

In central Mongolia, the Khangai Mountain Range is an interfluve bordering the North Arctic Ocean and Central Asian internal drainage basins. In this region, we consider rivers draining from the northern, northeastern, and eastern Khangai and southwestern, western, and northwestern Khentii and their drainage basins and associated landforms. A wide intermontane valley between eastern Khangai and western Khentii is characterized by tributary rivers feeding the Selenge[3] and Orkhon rivers (Fig. 18.1). The valley is filled with the Quaternary basalts (Munkhuu, 1979).

---

[1] *Khangai* has been misspelled as *Khangay* (or *Hangay*) and Hangai in many publications. Khangai is the right English transliteration from Mongolian *Хангай*.

[2] *Khentii* has been misspelled as *Khentey* (or *Hentey*) and *Khentei* (or *Hentei*) in many publications. Khentii is the right English transliteration from Mongolian *Хэнтий*.

[3] *Selenge* has been often incorrectly written as *Selenga* internationally based on the Russian pronunciation. Selenge is the correct English transliteration from Mongolian *Сэлэнгэ*.

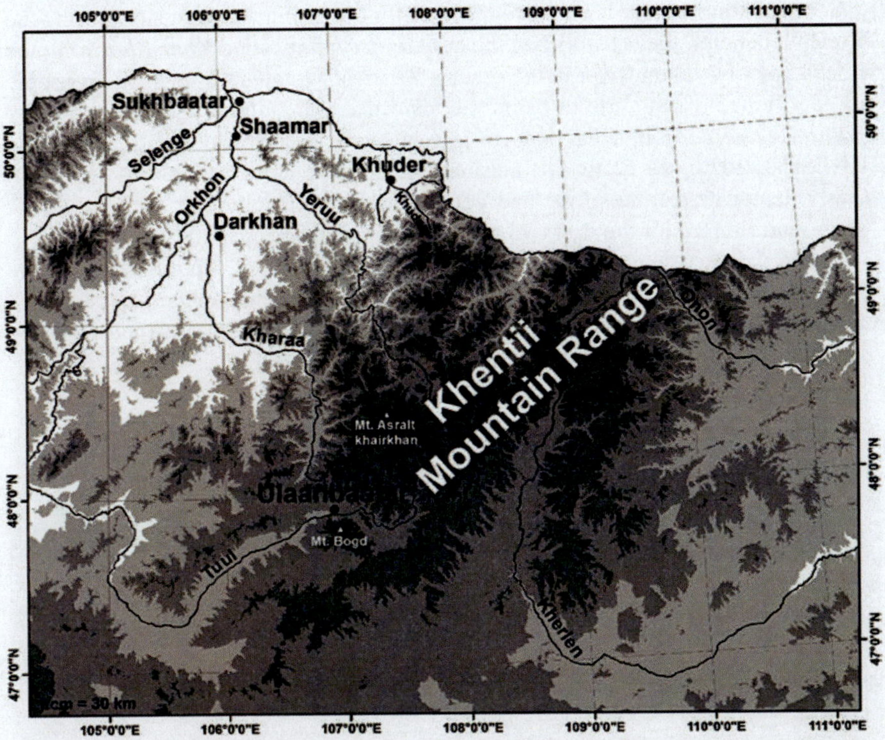

Fig. 18.1 Physiography of Khentii Mountain Range and large rivers in northeastern and central Mongolia. Modified from Orkhonselenge and Uuganzaya (2018)

There are not many studies to have considered central Mongolia (Fig. 1.1) as an independent landscape and a region each. However, there are a few results obtained from lakes in central Mongolia. For instance, for Lake Telmen, its saline level was measured (Peck et al., 2002), and both summer and winter air temperatures of 2 °C greater at ~6.0 ka and >10% lower moisture than today were recorded (Tarasov et al., 1999). In Lake Terkhiin Tsagaan (see Chap. 19), high contents of biogenic silica and organic matters were determined, and the lake was considered as the mountainous type represented with high precipitation, soil moisture, and vegetation cover (Davaagatan et al., 2015). For Lake Ugii (see Chap. 20), geochemical and mineralogical properties of lacustrine sediments (Schwanghart et al., 2008) and a high-resolution pollen- and diatom-based bioclimatic reconstruction (Wang et al., 2011) were presented. Moreover, in Lake Khuis, high-resolution multi-proxy data reviewing vegetation and climate changes in the southeastern Khangai over the last 1.2 ka (Tian et al., 2013) were provided. Lake Gun, at the northernmost corner of the region, was reviewed by Zhang et al. (2012) regarding its Holocene hydrology, and different lake levels were found: low levels at 10.8–5.7, 4.1–3.6, and 3.0–2.5 cal ka BP, intermediate levels at 5.7–4.1 and 3.6–3.0 cal ka BP, and high levels at 2.5–1.6 cal ka BP.

Central Mongolia has been a major center for Mongolian settlements. During the thirteenth century, they established the country's capital called *Khar Khorin*[4] in the wide valley of Orkhon River in the eastern Khangai Mountain Range where people still are living nomadic life to date. The historical and cultural site *Orkhon Valley Cultural Landscape* in Khar Khorin is registered as a World Heritage in 2004 (UNESCO, 2020; see Chap. 24) because the Orkhon River valley (Fig. 18.2a, b) clearly demonstrates the strong and persistent nomadic culture as the core of the Mongolian Empire for the past two millennia (Orkhonselenge, 2021a).

In this chapter, the landscape feature and lake evolution of central Mongolia are discussed, providing valuable insights into paleoclimate change and paleoenvironmental evolution of the region. The paleoclimatic implications obtained from the lakes in this region are discussed further in Chap. 21.

Fig. 18.2 The midstream of Orkhon River in the eastern Khangai viewing toward the (**a**) west and (**b**) southwest and the Khetsuu Els sand dunefield in an intermontane valley between the Khangai and Khentii mountain ranges viewing toward the (**c**) southeast and (**d**) southwest in front of the Mt. Khugnu Khan (1967 m a.s.l.). The arrows show the flow directions of Orkhon River for (**a** and **b**) and extension directions of the sand dunefield for (**c** and **d**). Photos by Orkhonselenge on August 14, 2011

[4] *Khar Khorin* has been often misspelled as *Kara Korum* (or *Khara Khorum*) and *Har Horin* in publications, Wikipedia, etc. *Khar Khorin* is the right English transliteration from Mongolian *Хар Хорин*.

18.2 Landscape

18.2.1 Geology

In central Mongolia, variably aged geosynclines create a complicated geological structure of sedimentary, sedimentary-volcanic, igneous, and intrusive rocks synthesized in the numerous Proterozoic to early Paleozoic, middle to late Paleozoic, Mesozoic, and Cenozoic complexes (Munkhuu, 1979). The Khangai Mountain Range consists of granites with schists, sandstones, and basalts (Tsegmid, 1969), and it contains the thick Devonian and Carboniferous sediments interbedded with continental volcanic and clastic rocks of the Permian age (Klimek, 1980). Regarding the tectonostratigraphical terrain (Badarch et al., 2002), central Mongolia is underlain by the Zavkhan cratonal block and the Tsagaan Olom basin-Cambrian shelf carbonate rocks in the west, the Tarvagatai Permian-Triassic volcanic-plutonic belt and Tarvagatai cratonal block in the north, and the Baidrag and Bayankhongor cratonal block, Zag ophiolite, and the Khangai Devonian-Carboniferous turbidite basin in the south.

The geology of central Mongolia consists of the Ordovician and Silurian mobile belt reworked during the Devonian and Carboniferous in the north, the Mesozoic and Cenozoic sedimentary basins in the south, and the Paleozoic and Mesozoic granitoids in the center (Kröner & Rojas-Agramonte, 2014). The Neogene and Quaternary basalts are widely distributed in the valleys of Orkhon, Suman, and Chuluut rivers (Munkhuu, 1979) in the northeastern Khangai. The Quaternary deposits have been investigated well in Lakes Telmen and Ugii basins (Narantsetseg & Khishigsuren, 2017).

During the Devonian, the continental island arc evolved along the paleo-Asian Ocean convergence zone around the southern margin of the Siberian Craton, producing an extensive felsic volcanic activity in the Tsetserleg region of the central Khangai that ended in the late Devonian period, and subsequently, during the early Carboniferous, plutonic rocks were intruded along the continental island arc (Hara et al., 2013) in the Ulaanbaatar[5] region of the southwestern Khentii (Fig. 18.1).

18.2.2 Climate

Central Mongolia lies under impacts of atmospheric systems of the Siberian-Mongolian anticyclone, the midlatitude westerlies, and Central Asian cyclone. In the Lake Gun basin, Holocene climate changes are summarized with the cold/wet and warm/arid conditions punctuated with a series of abrupt climate shifts linked with those in the North Atlantic Ocean, and the strength and position of the North

[5] *Ulaanbaatar* has been misspelled as *Ulan Bator* or *Ulan Bataar* in publications, Wikipedia, etc. Ulaanbaatar (Fig. 18.1) is the right English transliteration from Mongolian *Улаанбаатар*.

Atlantic Oscillation-affected westerly winds modulated the Siberian-Mongolian high-pressure system (winter monsoon) (Wang et al., 2004).

The climate of central Mongolia is characterized by the long cold winter and dry summer. In the region, the annual average air temperature is −6 to −4 °C and the annual average precipitation is 200–300 mm (Tsegmid, 1969). In the intermontane valley between Khangai and Khentii mountain ranges (Fig. 18.2c, d), the annual average air temperature is −2 °C (−25 to −15 °C in January and 15–20 °C in July), and the annual average precipitation is 200–300 mm (Tserensodnom, 1971) or 300 mm (Munkhuu, 1979).

In the Khangai region, it is cool and humid with the annual average air temperature ranging −0.1 to −6.7 °C and annual average precipitation of 180–275 mm (Tsegmid, 1969). According to Munkhuu (1979), in the Khangai region, the annual average air temperature is −4 to −2 °C (-30 °C in January and 15 °C in July) and the precipitation is 400 mm. In the Khangai mountainous region, the annual average precipitation is 250–350 mm, and the annual open water evaporation is 300–400 mm (Ministry of Environment and Green Development, 2013).

According to Dashkhuu et al. (2015), significant decreases greater than 2 days/decade for cool nights were observed in some parts of central Mongolia, whereas increasing trends of more than 2 days/decade for warm days were observed in the central Mongolia, where warm summer days and nights increased more prominently. Recent trends in local climate of the central Khangai and in the intermontane valley between Khangai and Khentii mountain ranges are given further in Chaps. 19 and 20.

18.2.3 Landform

A variety of landscapes make central Mongolia a very significant region from the geomorphological point of view. Central Mongolia consists of the high Khangai Mountain Range in the west, the medium-high mountainous region in the Selenge River drainage basin in the north, and a wide intermontane valley between the eastern Khangai and the western Khentii mountain ranges in the east. In other words, the landscape of this region is represented by the northern, northeastern, and eastern Khangai Mountain Range; southwestern, western, and northwestern Khentii Mountain Range; western Selenge River drainage basin; and a wide intermontane valley between Khangai and Khentii mountain ranges (Table 1.2). The western Selenge River drainage basin is included in this region, i.e., it is taken into account as the northern border of this region (Fig. 1.1), but the eastern Selenge River drainage basin is included in northern Mongolia (see Chap. 15). To date, the medium-high mountainous region in the Selenge River drainage basin and the intermontane valley between Khangai and Khentii mountain ranges (Fig. 18.2c, d) have not been studied as an independent landscape each.

The Khangai Mountain Range is an elliptical tectonic block located at the western part of the Khangai-Khentii (or Mongolian-Transbaikalian) fold system (Klimek, 1980). The range extends for ~700 km from the northwest to the southeast in central Mongolia (Fig. 1.1). The Khangai Mountain Range is elevated at 1500–2500 m a.s.l. on average with peaks at 3000–4000 m a.s.l. in the southeastern end of the Altai and Sayan mountain ranges (Munkhuu, 1979). The peak of the range, Mt. Otgon Tenger, is elevated at 4031 m a.s.l. (Tsegmid, 1969), 3905 m a.s.l. (Tserensodnom, 1971), or 4021 m a.s.l. (Orshikh et al., 2014). The main Khangai ridge reaches 3000 m a.s.l. (Tsegmid, 1969) and 3000–3500 m a.s.l. on average (Klimek, 1980). Low mountains in the southeast of the Khangai are in the elevation range of 2000–2500 m a.s.l., while high mountains are at 3200–3500 m a.s.l. in the west (Tserensodnom, 1971). In the Khangai, there are numerous dead or dormant volcanic mountains such as Mt. Khorgo (Fig. 18.3; see Chap. 19) and Mt. Uran. The volcanism dammed river valleys and formed closed lakes (Tsegmid, 1969).

There are independent branch ranges of the Khangai Mountain Range such as Khan Khukhii (2928 m a.s.l.), Tarvagatai (3240 m a.s.l.), and Bulnai (2600 m a.s.l.) in the northern Khangai, and these ranges extend from the west to the east lying parallel to the latitudes. The Bulnai Mountain Range is the interfluve bordering the North Arctic Ocean and Central Asian internal drainage basins (Tsegmid, 1969). The summit areas of the Khangai elevated above 2700 m a.s.l. are associated with extensive planation surfaces, where active cryogenic processes produce cryoplanation terraces on the slopes (Klimek & Starkel, 1980). The Buren Mountain Range (1986 m a.s.l.) between Selenge and Orkhon rivers is included in this region, and this range extends from the southwest to the northeast.

The Khentii Mountain Range (Fig. 18.1) extends from the northeast to the southwest in the northeastern Mongolia. The peak of the range is Mt. Asralt Khairkhan (2751 m a.s.l. in Tsegmid, 1969) or 2800 m a.s.l. (Jigj, 1975; see Chap. 6). Most high mountains in the Khentii Mountain Range are at the elevation range of 2200–2400 m a.s.l. (Tserensodnom, 1971), and they contain modern glacial tracers such as cirques, moraines, and lakes (Orkhonselenge & Uuganzaya, 2018) formed during the Last Glacial Maximum (LGM) (Khandsuren et al., 2019). The southwestern, western, and northwestern Khentii feed large rivers (e.g., Tuul, Kharaa, Yeruu in Fig. 18.1) and headwaters of Lake Baikal[6] through Orkhon and Selenge rivers (see Sect. 18.2.6).

There are national parks (e.g., Orkhon valley, Khugnu-Tarna mountains, Khustai Mountain Range, Gorkhi-Terelj mountains, etc.) and natural reserves (e.g., Mt. Batkhaan) in the intermontane valley between the Khangai and Khentii mountain ranges.

18.2.4 Glaciers and Permafrost

In Mongolia, there were two major glaciations in the early and late Pleistocene (Tsegmid, 1969) covering the Mongolian Altai, Khangai, Khentii, and Khuvsgul mountain ranges. Another source states that there are a series of evidence for two to

[6] *Baigali* meaning *nature* has been misspelled as *Baikal* by the Russians. Baigali is the right English transliteration from Mongolian *Байгаль*.

Fig. 18.3 (**a**) Volcanic Mt. Khorgo viewing toward the west. (**b**) The caldera of Mt. Khorgo viewing toward the west. Photos by A. Orkhonselenge on August 15, 2011

three Pleistocene glaciations in the Khangai and Khuvsgul, unlike three to four Pleistocene glaciations in the Mongolian Altai and one Pleistocene glaciation in the Khentii (Jigj, 1976). Today, the modern glaciers occur in only a few mountains of the Mongolian Altai, Khangai, and Khuvsgul mountain ranges (Orkhonselenge & Harbor, 2018).

In the Khangai Mountain Range, paleoglacial cirques and U-shaped valleys are abundantly observed, and moraines and hummocky terrains occur in river valleys; however, at the present time, there are only small modern glaciers on Mt. Otgon Tenger, Mt. Khalzan Burged, Mt. Noyon Khangai, and Mt. Yamaat (Tsegmid, 1969). The last maximum extension of glaciations on Mt. Otgon Tenger occurred at 40.0–35.0 ka during the marine oxygen isotope stage (MIS) 3, and the large ice expansion readvanced at ~23.0 and 17.0–16.0 ka during the MIS 2 (Rother et al., 2014). In the Khangai, intensive frost weathering and solifluction processes were detected by grain size and shape and microelements in soil profiles along the upper and lower slopes of Mt. Ovoot (Davaakhuu & Orkhonselenge, 2015).

The major perennial permafrost of central Mongolia is distributed in the Khangai, Tarvagatai, and Bulnai mountain ranges, where the seasonal freezing and thawing depth of perennial permafrost in sands and silts is 0.9–2.3 m at 3000–3200 m a.s.l., 2.8–3.8 m at 2000–2300 m a.s.l., and 1.8–2.1 m at 1800 m a.s.l., respectively (Tumurbaatar, 2004; Table 18.1). The seasonal freezing and thawing depth of permafrost is 1.5–2.0 m in the Chuluut River valley (see Chap. 19), 1.9–2.0 m in the Lake Telmen basin, 1.1–1.6 m in the Lake Terkhiin Tsagaan basin, 2.0–2.5 m at 2000 m a.s.l. in the Khanui River valley, 1.5–2.0 m in swamps around Erdene Tolgoi and Lun Tolgoi, and 3.8–4.2 m in the Orkhon River valley (Tumurbaatar, 2004; Table 18.1).

The perennial, intermittent, and spotted types of permafrost are distributed in the Khangai Mountain Range in the west of the region, while the seasonal permafrost occurs in the intermontane valley between Khangai and Khentii mountain ranges (Jambaljav et al., 2016) in the east of the region. In the Khangai, the seasonal freezing and thawing depth of perennial permafrost is 2.5–5.5 m under the annual average air temperature of −0.3 to 1.5 °C, while in the intermontane valleys at the head of the Orkhon River basin, the seasonal freezing and thawing depth of perennial permafrost is 2.5 m under the annual average air temperature of −2.8 °C (Jambaljav et al., 2016). In the southeastern Khangai, the freezing and thawing depth of the seasonal permafrost is 3.2 m. The permafrost in the river drainage basins in the northwestern and southwestern Khentii and northeastern Khangai mountain ranges

Table 18.1 Seasonal freezing and thawing depths in central Mongolia

#	Places	Elevation (m a.s.l.)	Depth (m)
1	Khangai Mountain Range	3000–3200	0.9–2.3
2	Tarvagatai Mountain Range	2000–2300	2.8–3.8
3	Bulnai Mountain Range	1800	1.8–2.1
		Materials	
4	Chuluut River valley	Clays	1.5–2.0
5	Lake Terkhiin Tsagaan basin	Clays and silts	1.1–1.6
6	Lake Telmen basin	Clays	1.9–2.0
		Sands	3.0–3.4
7	Khanui River valley	Clays	2.0–2.5
8	Orkhon River valley	Gravels and sands	3.8–4.2

Table 18.2 Cold and warm permafrost in the river drainage basins of central Mongolia

Places	Basins	Area of the basins (km²)	Permafrost in the basins (%)		AMGT (°C)	ALT (m)
			Cold permafrost (AMGT < −2 °C)	Warm permafrost (−2 °C < AMGT < 0 °C)		
Northwestern Khentii	Yeruu-Khuder-Minj rivers	21,986	3.1	61.1	−1.9	2.4–3.7
	Kharaa-Shar rivers	17,463	0.7	21.8	−0.8	2.3–3.6
Southwestern Khentii	Tuul River	49,416	3.7	11.6	−1.1	1.3–3.6
Northeastern Khangai	Ider River	22,757	39.8	40.6	−1.6	1.9–4.2
	Chuluut River	19,813	30.0	41.4	−1.4	1.5–4.0
	Khanui River	15,549	0.5	7.1	−0.2	2.2–3.2
	Orkhon River	52,753	3.8	13.3	−1.6	2.1–3.4

is defined as cold and warm depending on annual mean ground temperature (AMGT) by Zorigt et al. (2020), and it is shown with its active layer thickness (ALT) of 1.3–3.7 m in Table 18.2.

18.2.5 Groundwater

In the Khangai Mountain Range, groundwater resides in the Precambrian and Paleozoic metamorphic, igneous, and intrusive rocks. In the southwestern Khentii Mountain Range, there is artesian groundwater in the Cretaceous and Neogene deposits (Tsegmid, 1969). Table 18.3 shows quantities of groundwater resource belonging to some large river drainage basins in central Mongolia (Ministry of Environment and Green Development, 2013). The groundwater resources are annually renewed, mainly through infiltration of surface water (Ministry of Environment and Green Development, 2013).

The renewable groundwater resource in the Khangai region is 3409.9 million m^3/year, and their usable resource is 3005.1 million m^3/year (Jadambaa & Tserenjav, 2012; Table 18.4). The very high resources of groundwater occur in valleys of large rivers (e.g., Selenge, Ider, Tamir rivers, etc.) and lakes where the hydraulic conductivity in alluvial deposits ranges at 50–400 m^2/day with 150 m^2/day on average. The medium-high resources of groundwater (Table 18.4) occupy headwater basins of large rivers and floodplains of small rivers, while the medium-low resources of groundwater are in valleys of small rivers and around alluvial fans. The low resources of groundwater occur in depressions and valleys with limestones and

Table 18.3 Water resource of river drainage basins in central Mongolia

#	Drainage basin	Area (km^2)	Surface water resource (million m^3/year)			Groundwater resource (million m^3/year)	
			Total resource	Environmental	Possible use	Potential exploitable	Exploitable resource
1	Ider River	23,061	710	657	53.3	129	0.5
2	Khanui River	15,755	231	217	13.9	96	0.2
3	Orkhon River	53,455	2345	2123	221.6	838.3	26.7
4	Tuul River	50,074	1073	1010	63.1	637.7	142.8
5	Kharaa River	17,697	432	406	25.9	182	52.6
6	Yeruu River	22,280	1121	925	196.2	239	0.6
	Total	182,322	5912	5338	574.0	2122	223.4

Table 18.4 Groundwater resource in the Khangai region

#	Groundwater resource	Thickness of alluvial deposits (m)	Discharge (mm/year)	Thousands m^3/year*km^2	Area		Resource to use	
					km^2	%	Million m^3/year	%
1	Very low	1.0–2.0	–	<1.0	49,640	13.2	40.5	1.3
2	Low	1.0–5.0	5.0–20.0	1.0–3.0	144,870	38.2	303.1	10.1
3	Medium-low	2.0–7.0	10.0–20.0	3.0–10.0	142,210	37.4	686.5	22.8
4	Medium-high	2.0–10.0	10.0–50.0	10.0–30.0	27,090	7.1	422.5	14.1
5	High	3.0–15.0	40.0–50.0	30.0–50.0	3810	1.0	131.8	4.4
6	Very high	–	>70.0	>50.0	11,830	3.1	1421	47.3
	Total				379,450	100	3005.1	100

sands, while the very low resources of groundwater occur in metamorphic and igneous rocks (Jadambaa & Tserenjav, 2012).

18.2.6 Surface Water

Surface water in this region plays a crucial role in the North Arctic Ocean drainage basin because this region contains major headwater rivers feeding Selenge River in northern Mongolia and Lake Baikal in southeastern Siberia. River networks in this region are dense with 0.4–0.5 km/km^2 on average, ranging from 0.8–1.0 km/km^2 in the high Khangai and 0.1–0.5 km/km^2 in medium-high mountains within the

Table 18.5 Morphometric components of large rivers in the Khangai region

#	Rivers	Tributary to terminus	Drainage basin (km^2)	Length (km)	Width (km)		Depth (m)	
					Upstream	Downstream	Deep	Shallow
1	Selenge	Lake Baikal	282,000	992	60–70	100–250	4.5–5.0	4.0–1.0
2	Ider	Selenge R.	24,500	452	40–60	80–100	3.0–3.5	0.7
3	Chuluut	Ider R.	10,700	415	20–30	70–80	2.5	0.5
4	Khanui	Selenge R.	14,600	421	20–30	50–80	2.5–3.0	0.4–0.5
5	Orkhon	Selenge R.	132,835	1124	10–60	120–150	4.0	0.5–1.5
6	Tamir	Orkhon R.	13,100	185	–	40–70	2.5–3.0	0.4

Selenge River drainage basin to 0.08–0.1 km/km^2 in the intermontane valley between Khangai and Khentii mountain ranges (Munkhuu, 1979).

Table 18.3 shows surface water resource in some large river drainage basins of central Mongolia. Orkhon River (Fig. 18.2a, b), a major tributary river of Selenge River, is fed by large Tamir, Tuul, Kharaa, and Yeruu rivers (Fig. 18.1; Table 18.5) and small rivers, and it constitutes the largest drainage basin holding the maximum water resources in central Mongolia (Table 18.3; Orkhonselenge, 2021a, 2021b). Selenge River (see Chap. 15), the second largest drainage basin, is fed by Ider, Chuluut (see Chap. 19), Khanui, Khunui, Tamir, and Orkhon rivers, and in turn it feeds Lake Baikal (Table 18.5). Selenge River and its tributaries are fed by groundwater for about 40% of their inventories (Tserensodnom, 1971). Figure 18.4 shows a long-term trend of discharges of Orkhon, Selenge, and Tuul rivers and precipitation and evaporation in their river drainage basins.

Lakes in central Mongolia (see Sect. 18.3) are few in number and small in area (see Chaps. 1 and 5). There are 617 lakes in total each covering an area over 0.1 km^2, and they cover 461.8 km^2 in total (see Table 1.2). The lakes in this region constitute 20.2% of the total number and 2.9% of the total area of lakes in Mongolia (see Chaps. 1 and 5). Most of the lakes in the region have origins such as tectonic, volcanic, fluvial, glacial, and thermokarst (see Chap. 4), and they are freshwater lakes (see Chap. 5).

18.3 Lakes

Lakes in central Mongolia (Fig. 1.1) are medium in number but the smallest in area (Table 1.2; see Chap. 5). Lakes in the northern Khangai Mountain Range and western Selenge River drainage basin receive high precipitations of 250–400 mm/year. Lakes in the northern Khangai Mountain Range in central Mongolia, especially lakes in river valleys, are tectonic, glacial, karst, and volcanic lakes (see Chap. 4), whereas lakes on the plateau of the range have no outlets and are saline. Here, the

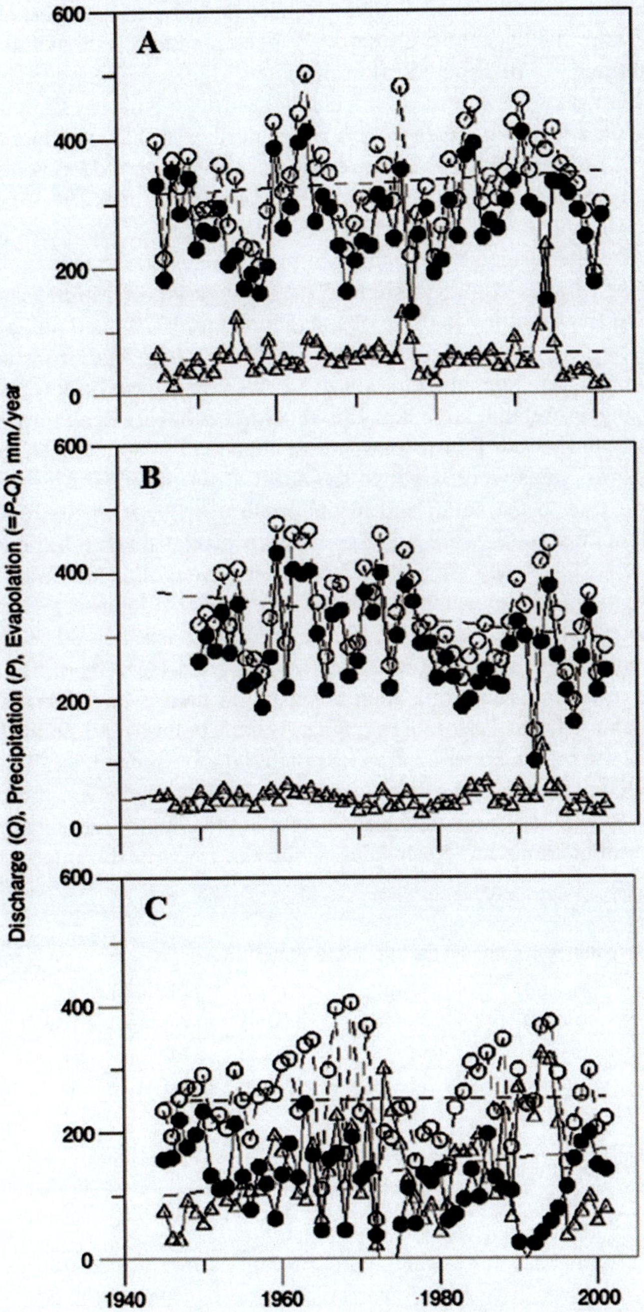

Fig. 18.4 Discharges of (**a**) Orkhon River in Orkhon soum, Bulgan aimag, (**b**) Selenge River in Khutag Undur soum, Bulgan aimag, and (**c**) Tuul River in Ulaanbaatar (Fig. 18.1) and precipitation and evaporation in the river drainage basins. Modified from Davaa et al. (2007)

largest lakes are Lakes Telmen, Sangiin Dalai, Ulaagchnii Khar, Khunguin Khar, Oigon, and Terkhiin Tsagaan (see Chap. 19). Tectonic lakes such as Lakes Sangiin Dalai and Telmen are distributed in the Khangai Mountain Range (Table 1.2; see Chap. 4). There are several tens of volcanic lakes in the Khanui, Khunui, Orkhon, Tamir, Chuluut, and Terkh River valleys in the northern Khangai Mountain Range (see Chap. 4). Some of these lakes have outlets, and they are deep and freshwater (see Chap. 5), for instance, Lake Terkhiin Tsagaan (see Chap. 19). Glacial Lakes Binderya, Khar, and Khukh are distributed in the Khangai Mountain Range (see Chap. 5). Lakes are formed by damming of river valleys due to tectonic and volcanic causes (e.g., Lakes Khukh, Shireet, Tsagaan Azarga, Terkhiin Tsagaan, etc.), permafrost melting resulting in thermokarst (Munkhuu, 1979), and fluvial processes forming oxbows (e.g., Lake Ugii; see Chap. 20) (see Chap. 4). For instance, in the valleys of Khanui, Selenge, and Orkhon rivers, there are many oxbow lakes.

In the Khangai foreland, large lakes are regarded as having been formed contemporaneously with the late Pleistocene intermontane valley glaciers (Klimek et al., 1980). Their lake areas were larger to the north of the Khangai Mountain Range, compared to those to the south and in the eastern lowlands in the region. In the Khentii Mountain Range, there are also modern glacial lakes (Orkhonselenge & Uuganzaya, 2018), but they are included in eastern Mongolia (see Chap. 6).

In this region, lakes elevated at 1001–1500 m a.s.l. (Table 4.1) are abundant among the total 617 lakes, and the 563 lakes of them constitute ~39.5% of the total 1426 lakes with the same elevation range in Mongolia (see Chap. 5). Table 18.6 shows lakes in central Mongolia each covering an area over 5.0 km^2. There are many small lakes such as Lakes Oigon, Khar, Bust, Kholboo, and Zuun. The majority of lakes in the region cover an area less than 5.0 km^2. Indeed, 95.4% of lakes in the region occupy an area less than 5.0 km^2 (Table 4.1); however, they constitute 20.7% of the total lakes covering the same area in the country (see Chap. 5). Lakes in the high mountainous are freshwater, while the lakes in the intermontane and

Table 18.6 Morphometric components of lakes in central Mongolia

#	Lakes	Origin/ genesis	Elevation (m a.s.l.)	Area (km^2)	Depth (m)		Description
					Average	Maximum	
1	Ikh Tukhum	Saline	1261	9.8	1.0	2.5	–
2	Ulaan	Saline	1375	5.0	0.4	1.0	–
3	Duruu Tsagaan	Tectonic	1712	7.8	3.7	7.8	–
4	Ikh	Freshwater	1505	7.9	2.5	4.4	–
5	Khuduu	Freshwater	2061	10.9	1.8	3.5	–
6	Khukh	Freshwater	2668	6.7	–		–
7	Terkhiin Tsagaan	Volcanic	2060	62.0	6.0	20.0	Chapter 19
8	Ugii	Oxbow	1332	23.8	6.6	15.3	Chapter 20
9	Telmen	Tectonic	1789	198.0	13.0	27.0	Section 18.4.5

river valleys are saline (see Chap. 5). Representative two large lakes in the region are discussed in detail at Chaps. 19 and 20.

18.4 Lake Evolution

Lakes in central Mongolia have rapidly shrunk and dropped their water levels for the past two decades. These phenomena have been related to the rising air temperature and decreasing precipitation during the last 60 years, strengthening aridity and human impacts such as intensive mining operations (e.g., gold) and irrigation. For instance, Lake Tsaidam has been dropping its water level, whereas Lakes Lun, Khar, and Tsagaan (Fig. 18.5) have entirely dried out and shifted into playa lakes. The latter lakes are temporally filled with water during the heavy rainfall. The warming-induced evolution in these lake basins is recorded in annually laminated varve lake sediments, sensitively responding to climate changes. These lakes decreasing their water levels and shrinking show how arid central Mongolia has become due to the strength of the westerly winds.

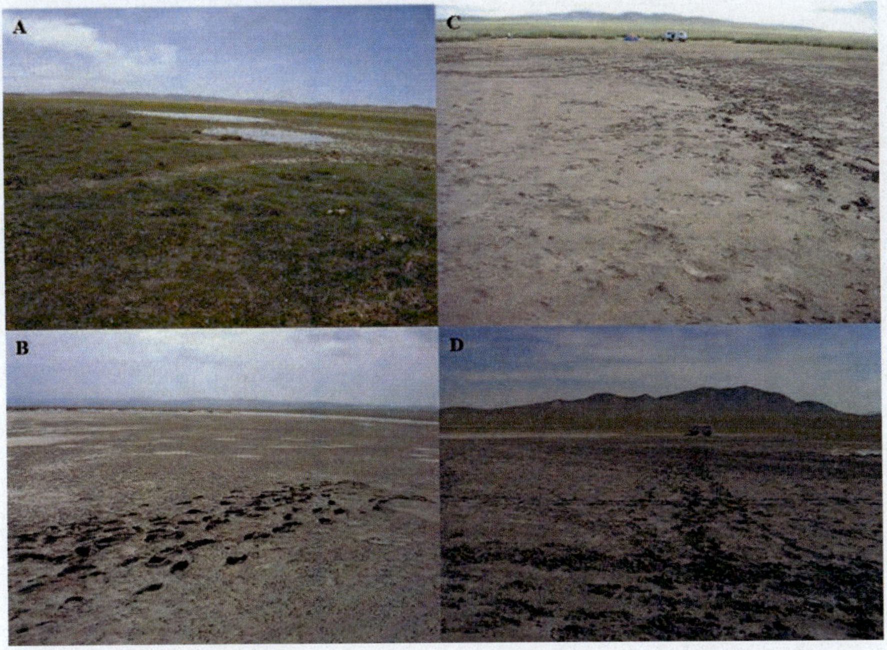

Fig. 18.5 (**a**) Lake Gun viewing toward the east. (**b**) Lake Tsaidam viewing toward the northeast. (**c**) Lake Khar viewing toward the southwest. (**d**). Lake Shariin Tsagaan viewing toward the southwest. Photos by A. Orkhonselenge on August 3–6, 2009

18.4.1 Lake Lun

Lake Lun (Fig. 18.5a) is located at the elevation of 1018 m a.s.l. in the Tuul River drainage basin within the intermontane valley between the Khangai and Khentii mountain ranges. The lake surface is entirely covered by grasses and rushes such as hummocks and tussocks, and it has gradually shifted to become a wetland where peat is formed. The lithological stratigraphy of a 340-cm-long sediment core L09-1 collected from the lake shows brown gray silts and muds at 0–25 cm, yellow silts and sands at 25–50 cm, muds with sands at 50–125 cm, silts in 125–150 cm, and yellow muds at 150–340 cm (Orkhonselenge et al., unpublished).

18.4.2 Lake Tsaidam

Lake Tsaidam (Fig. 18.5b) is located at the elevation of 1050 m a.s.l. in the Ugtaal Tsaidam valley within the intermontane valley between Khangai and Khentii mountain ranges. The lake is filled during seasonal heavy rainfall. Its littoral parts are covered by reeds and grasses. The lithological stratigraphy of a 215-cm-long sediment core TS09-5 consists mostly of muds down to the 190 cm depth, and brown sands occur at 190–215 cm. Radiocarbon dating of the lake sediments indicates a sedimentation rate of 2.7 cm/year (Orkhonselenge et al., unpublished). In the lake sediments deposited during the late Holocene, carbonates are more concentrated than biogenic silica (Fig. 18.6).

18.4.3 Lake Khar

Lake Khar (Fig. 18.5c) is located at the elevation of 973 m a.s.l. in the Tseel valley within the intermontane valley between the Khangai and Khentii mountain ranges. The lake is an intermittent lake covered by consolidated muds, and it is temporally filled with water during heavy rainfall. The lithological stratigraphy of a 265-cm-long sediment core KH09-1 collected from the lake consists of brown muds at 0–30 cm, yellow sands at 30–70 cm, light-yellow muds with silts at 70–95 cm, brown coarse sands at 95–120 cm where groundwater level was found, muds at 120–140 cm, brownish yellow muds with gravels at 140–150 cm, light-yellow muds at 150–170 cm, brownish yellow muds at 170–200 cm, reddish yellow muds at 200–210 cm, light yellowish gray muds at 210–230 cm, reddish brown muds at 230–255, and red muds at 255–265 cm (Orkhonselenge et al., unpublished).

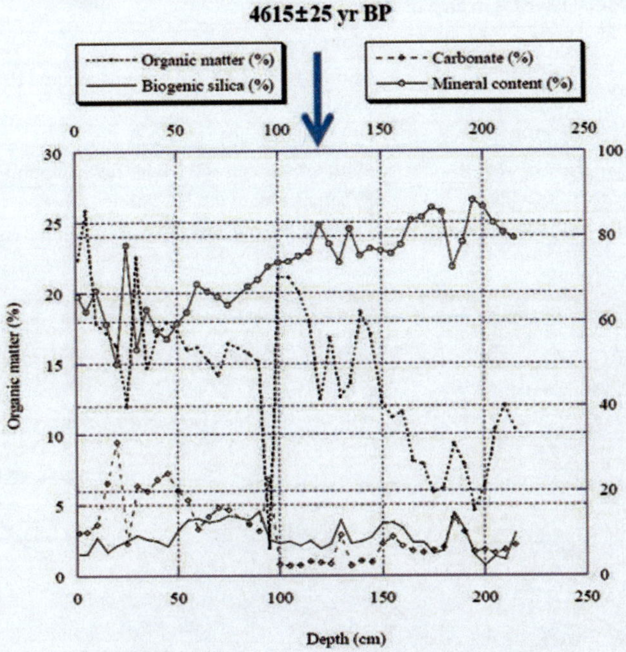

Fig. 18.6 Variations in physical and chemical components of Lake Tsaidam sediments (Fig. 18.5b). Modified from Orkhonselenge et al. (unpublished)

18.4.4 Lake Shariin Tsagaan

Lake Shariin Tsagaan (Fig. 18.5d) is located at the elevation of 691 m a.s.l. in the Shar River drainage basin in the northwestern Khentii Mountain Range. The lake has remained as a swampy environment, and it fills seasonally. The lithological stratigraphy of a 245-cm-long sediment core SH09-1 taken from a central part of the lake consists mainly of light gray muds at 0–30 cm, light gray muds with fine silts at 30–65 cm, gray sands at 65–80 cm, brown sands at 80–90 cm, muds with fine silts at 90–100, medium sands at 100–110 cm, yellow fine sands at 110–155 cm, muds at 155–170 m, and fine sands at 170–245 cm (Orkhonselenge et al., unpublished).

18.4.5 Lake Telmen

Lake Telmen is located at the elevation of 1789 m a.s.l. in a large intermontane depression in the Khangai Mountain Range (Fig. 1.1). Lake Telmen extends for 27.0 km in length, 16.0 km in width, and 93.4 km along the shoreline length (Tserensodnom, 1971). The lake occupies an area of 194.0 or 198.0 km^2 (Table 18.7). The lake's depth is about 10.0–20.0 m in the western and 27.0 m in the southeastern

Table 18.7 Temporal changes in area of Lake Telmen

#	Year	Area (km^2)	Authors
1	1929	284.0	Kondratiev (1929) in Tserensodnom (1971)
2	1948	194.0	Murzaev (1952)
3	1969	198.6	Tsegmid (1969)
4	1970	197.83	Orkhonselenge and Uuganzaya, unpublished
5	1971	194.0	Tserensodnom (1971, 2000)
6	2014	198.0	Orkhonselenge and Uuganzaya, unpublished

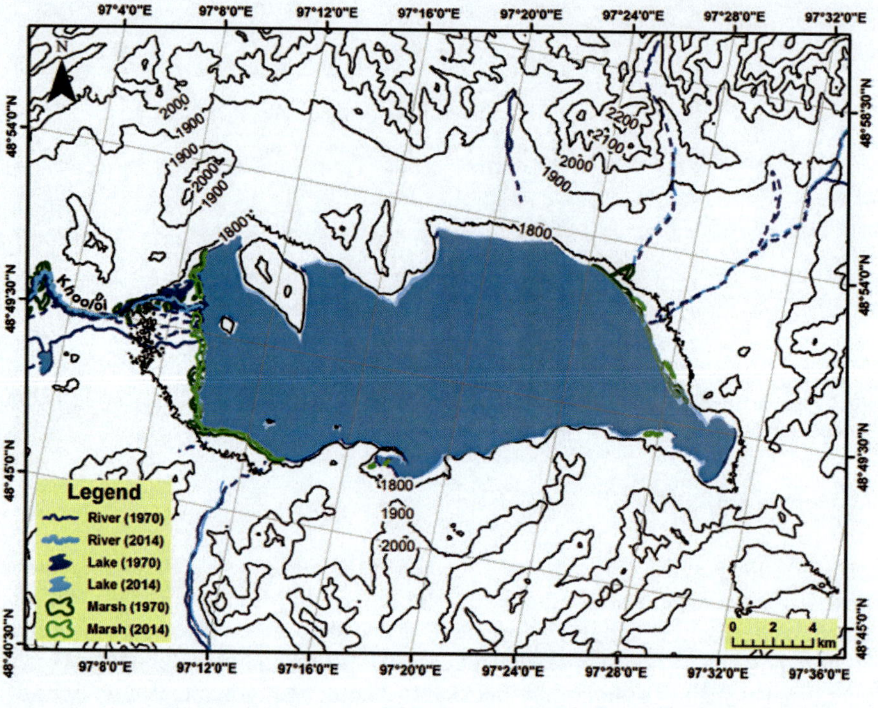

Fig. 18.7 Spatial distributions of areas and marshes of Lake Telmen in 1970 and 2014. Modified from Orkhonselenge and Uuganzaya (unpublished)

parts of the lake (Tserensodnom, 1971; Table 18.6). The lake level increased after 6.3 cal ka BP and from about 2.7 to 1.3 cal ka BP due to the enhanced effective moisture balance (Peck et al., 2002). The major water resources of Lake Telmen are rainfall, runoff, and groundwater. The lake drains from an area of 3940 km^2 in the Tarvagatai Mountain Range of the Khangai (Tsegmid, 1969). The lake is fed by Khooloi River, the largest perennial inflow, in the west (Fig. 18.7). Lake Telmen is the most salinized lake in the Khangai Mountain Range (Tsegmid, 1969). The mineralization values of Lake Telmen are 6.8 g/L at the surface, 6.9 g/L at 21.0 m depth (Tserensodnom, 2000), 8.2 g/L in the east (Tserensodnom, 1971), and 7.64 g/L dominated with Cl^-, SO_4^{2-}, HCO_3^-, and Na^+ (Ariunbileg et al., 2020). The salinity level of the lake was approximately 20 g/L between 7.1 and 6.3 cal ka BP, but it has been dropped to 4 g/L today (Peck et al., 2002).

Lake Telmen responds to the climate trend with a slow rate. During the last century (Table 18.7), the lake faced a decreasing trend of its area. The lake occupied a large area during the early 1930s, and since then the lake lost an area by almost 90.0 km^2 over the nearly 20 years between 1929 and 1948 (Table 18.7). Since ~1950s, the lake was relatively stabilized with small fluctuations. During the last half century, the lake area showed an increasing trend (Fig. 18.7). The lake gained an area of 4.6 km^2 over the 20 years since the 1950s (Table 18.7). The lake gained an area by ~0.2 km^2 during the last 44 years (Orkhonselenge and Uuganzaya, unpublished; Table 18.7; Fig. 18.7), and this increase of the area may have been contributed by the gradual increase of the annual precipitation since 1993 recorded at the Numrug meteorological station located 12.0 km northwest of the lake. Changes in the lake area show its sensitivity to variations in annual and seasonal precipitations within the lake basin. However, the local climate data recorded between 1993 and 2019 at the station are too scarce to analyze linear regression and correlation coefficients regarding the relationships between the lake area and the annual average air temperature and precipitation.

18.5 Summary

Central Mongolia plays an important role in climatological and hydrological systems and landform evolution in Mongolia and Central Asia. It exhibits particular landscapes with a combination of the high Khangai mountains in the west and a wide intermontane valley between eastern Khangai and western Khentii mountain ranges in the east. The landscapes in the region are formulated by large rivers in the north and dense river networks constituting a large part of the surface water resource of Mongolia and Eurasia (e.g., Selenge River, a major headwater of Lake Baikal belonging to the North Arctic Ocean drainage basin). In central Mongolia, volcanic, tectonic, fluvial, and glacial lakes are abundant, and lakes in the region occupy the smallest area in the country (2.9%). Lakes in this region are relatively few and small; however, most of the lakes have been shifting into intermittent or playa lakes under the strengthening aridity caused by the global warming. Geomorphological, geochemical, and geochronological data from the lakes in central Mongolia help reconstructing the paleoenvironmental histories of the Selenge River and Orkhon River drainage basins in the Khangai and the adjacent Tarvagatai, Bulnai, Buren, and Khentii mountain ranges and the southeastern Siberian region.

References

Ariunbileg, S., Isupov, V. P., Vladimirov, A. G., Kolpakova, M. N., & Kuibida, L. V. (2020). *Chemical evolution and geochemistry of mineralized lakes of Mongolia*. Admon Printing House, 132 p. [In Mongolian].

Badarch, G., Cunningham, W. D., & Windley, B. F. (2002). A new terrane subdivision for Mongolia: Implications for the Phanerozoic crustal growth of Central Asia. *Journal of Asian Earth Sciences, 21*, 87–110.

Dashkhuu, D., Kim, J. P., Chun, J. A., & Lee, W. S. (2015). Long-term trends in daily temperature extremes over Mongolia. *Weather and Climate Extremes, 8*, 26–33.

Davaa, G., Oyunbaatar, D., & Sugita, M. (2007). *Surface Water of Mongolia*. pp/ 55–68.

Davaagatan, T., Orkhonselenge, A., Fukushi, K., & Yabe, M. (2015). Sedimentary records in intermontane and prairie lakes in central Mongolia: Preliminary results from Lake Terkhiin Tsagaan and Lake Ugii. *Journal of Geographical Issues, 14*(408), 19–26.

Davaakhuu, T., & Orkhonselenge, A. (2015). Landscape development of Mt. Ovoot in the northeast of Khangai Mountain Range. Journal of Geographical. *Issues, 14*(408), 27–32.

Hara, H., Kurihara, T., Tsukada, K., Kon, Y., Uchino, T., Suzuki, T., Takeuchi, M., Nakane, Y., Nuramkhaan, M., & Chuluun, M. (2013). Provenance and origins of a Late Paleozoic accretionary complex within the Khangai–Khentei belt in the Central Asian Orogenic Belt, central Mongolia. *Journal of Asian Earth Sciences, 75*, 141–157.

Jadambaa, N., & Tserenjav, G. (2012). Resource distribution of groundwater in economic regions. In J. Byamba (Ed.), *Geology and mineral resources of Mongolia* (Hydrogeology) (Vol. VIII, pp. 253–263). [In Mongolian].

Jambaljav, Y., Gansukh, Y., Temuujin, K., Tsogt-Erdene, G., Undrakhtsetseg, T., Saruulzaya, A., Amarbayasgalan, Y., Dashtseren, A., & Narangerel, S. (2016). *Report of new map of permafrost distribution in Mongolia: Scale 1:1000000*. Institute of Geography and Geoecology, Mongolian Academy of Sciences, 90 p.

Jigj, S. (1975). *Primary feature of Mongolian landforms*. Institute of Geography & Permafrost. Mongolian Academy of Sciences (MAS), 126 p. [In Mongolian].

Jigj, S. (1976). *A brief description of paleoglaciations and paleoglaciers*. 51 p. [In Mongolian].

Khandsuren, P., Seong, Y. B., Oh, J. S., Rhee, H. H., Sandag, K., & Yu, B. Y. (2019). Late Quaternary glacial history of Khentey Mountains, Central Mongolia. *Boreas, 48*(3), 779–799.

Klimek, K. (1980). Major physico-geographical features of the southern slope of the Khangai Mountains. In K. Klimek & L. Starkel (Eds.), *Vertical zonality in the southern Khangai Mountains (Mongolia). Geographical studies* (Vol. 136, pp. 9–13). Ossolineum.

Klimek, K., Lomborenchin, R., & Starkel, L. (1980). Historical review of physico-geographical investigations in the Khangai Mountains. In K. Klimek & L. Starkel (Eds.), *Vertical zonality in the southern Khangai Mountains (Mongolia). Geographical studies* (Vol. 136, pp. 14–18). Ossolineum.

Klimek, K., & Starkel, L. (1980). *Vertical zonality in the southern Khangai Mountains (Mongolia). Geographical studies* (vol. 136). Polish Academy of Sciences and Institute of Geography and Spatial Organisation. 107 p.

Kröner, A., & Rojas-Agramonte, Y. (2014). The Altaids as seen by Eduard Suess, and Present Thinking on the Late Mesoproterozoic to Palaeozoic Evolution of Central Asia. *Austrian Journal of Earth Sciences, 107*(1), 156–168.

Ministry of Environment and Green Development. (2013). *Integrated water management plan of Mongolia.* 339 p.

Munkhuu, Z. (1979). *Characteristics of structural geomorphology of Dornod Khangai*. Sukhbaatar Publishing, 65 p. [In Mongolian].

Murzaev, E. M. (1952). *The description of physical geography of Mongolia* (2nd ed.). [In Russian].

Narantsetseg, T., & Khishigsuren, S. (2017). Classification and description of Quaternary deposits in Mongolia. *Khaiguulchin, 58*, 65–70.

Orkhonselenge, A. (2021a). Orkhon and Tuul Rivers valleys in Mongolian Plateau: Geomorphological, historical, and cultural landscapes. In *LATR2021 Virtual Conference: Life Along the River: Interactions between Human Societies and Valley Environments in the Convergence Zone of the Inner Asian Highlands, 1600–1950s. Presentation*. Columbia University. July 22–23, 2021. New York.

Orkhonselenge, A. (2021b, December 13–17). *Climate-driven channel change of Orkhon River, Mongolia*. AGU 2021 Fall Meeting. Presentation #1193. New Orleans, LA. https://doi.org/10.1002/essoar.10508953.1.

Orkhonselenge, A., & Harbor, J. M. (2018). Impacts of modern glacier changes on surface water resources in western and northern Mongolia. *Journal of Water Resource and Protection, 10*(6), 559–576.

Orkhonselenge, A., & Uuganzaya, M. (2018). Glacial geomorphology of Mt. Asralt Khairkhan and Mt. Baga Khentii Saridag in Khentii Mountain Range, Northeastern Mongolia. *International Journal of Geosciences, 9*(6), 308–319.

Orshikh, N., Enkhbayar, G., Garig, G., Enkhbat, R., & Chantuu, B. (2014). *Atlas of Mongolia with a scale of 1:800,000*. Monsudar Press, 117 p. [In Mongolian].

Peck, J. A., Khosbayar, P., Fowell, S. J., Pearce, R. B., Ariunbileg, S., Hansen, B. C. S., & Soninkhishig, N. (2002). Mid to Late Holocene climate change in north central Mongolia as recorded in the sediments of Lake Telmen. *Palaeogeography, Palaeoclimatology, Palaeoecology, 183*, 135–153.

Rother, H., Lehmkuhl, F., Fink, D., & Nottebaum, V. (2014). Surface exposure dating reveals MIS-3 glacial maximum in the Khangai Mountains of Mongolia. *Quaternary Research, 82*, 297–308.

Schwanghart, W., Schütt, B., & Walther, M. (2008). Holocene climate evolution of the Ugii Nuur Basin, Mongolia. *Advances in Atmospheric Sciences, 25*(6), 986–998.

Tarasov, P. E., Guinot, J., Cheddadi, R., Andreev, A. A., Bezusko, L. G., Blyakharchuk, T. A., Dorofeyuk, N. I., Filimonova, L. V., Volkova, V. S., & Zernitskaya, V. P. (1999). Climate in northern Eurasia 6000 years ago reconstructed from pollen data. *Earth and Planetary Science Letters, 171*, 635–645.

Tian, F., Herzschuh, U., Dallmeyer, A., Xu, Q., Mischke, S., & Biskaborn, B. K. (2013). Environmental variability in the monsoonewesterlies transition zone during the last 1200 years: Lake sediment analyses from central Mongolia and supraeregional synthesis. *Quaternary Science Reviews, 73*, 31–47.

Tsegmid, S. (1969). *Physical geography of Mongolia*. 759 p. [In Mongolian].

Tserensodnom, J. (1971). *Lakes of Mongolia*. State Publishing, 202 p. [In Mongolian].

Tserensodnom, J. (2000). *A catalog of lakes in Mongolia*. Shuvuun Saaral Publishing, 141 p. [In Mongolian].

Tumurbaatar, D. (2004). *Seasonal and permanent permafrost of Mongolia*. Mongolian Academy of Sciences, 147 p. [In Mongolian].

UNESCO. (2020). *World Heritage List*. Retrieved from https://whc.unesco.org/en/list/&order=country#alphaM.

Wang, W., Feng, Z., Lee, X., Zhang, H., Ma, Y., An, C., & Guo, L. (2004). Holocene abrupt climate shifts recorded in Gun Nuur lake core, northern Mongolia. *Chinese Science Bulletin, 49*(5), 520–526.

Wang, W., Ma, Y., Feng, F., Narantsetseg, T., Liu, K., & Zhai, X. (2011). A prolonged dry mid-Holocene climate revealed by pollen and diatom records from Lake Ugii Nuur in central Mongolia. *Quaternary International, 229*, 74–83.

Zhang, C., Zhang, W., Feng, Z., Mischke, S., Gao, X., Gao, D., & Sun, F. (2012). Holocene hydrological and climatic change on the northern Mongolian Plateau based on multi-proxy records from Lake Gun Nuur. *Palaeogeography, Palaeoclimatology, Palaeoecology, 323–325*, 75–86.

Zorigt, M., Myagmar, K., Orkhonselenge, A., van Beek, E., Kwadijk, J., Tsogtbayar, J., Yamkhin, J., & Dechinlkhundev, D. (2020). Modelling permafrost distribution over the river basins of Mongolia using remote sensing and analytical approaches. *Environmental Earth Sciences, 79*(308), 1–11.

Chapter 19
Lake Terkhiin Tsagaan

Abstract Lake Terkhiin Tsagaan, a volcanic freshwater lake, occurs in the Terkh River valley of the Khangai Mountain Range in central Mongolia. The lake plays an important role in the hydrological system of the North Arctic Ocean drainage basin through the activities of Terkh, Suman, Chuluut, Ider, and Selenge Rivers and Lake Baikal the lake constitutes an extraordinary scenery in the Khangai region, and Terkh River in the west, Suman River in the east, and the basaltic landscape of Mt. Khorgo in the northeast are notable. Lake Terkhiin Tsagaan responds to the climate change with a slow rate as the lake gained an area for only 0.51 km^2 during the last 45 years. The short-term changes of the lake area are directly related to the fluctuations of annual and winter precipitations. Meanwhile, the long-term changes of the lake area are related to the rises in annual, summer, autumn, and spring average air temperatures, leading to thawing of snow cover and permafrost in the Khangai Mountain Range to discharge into the lake. The linear regression analysis shows that the lake area between 1970 and 2015 was moderately related to the annual precipitation ($R^2 = 0.47$), but not clearly related to the annual average air temperature ($R^2 = 0.0186$). The correlation analysis shows that the lake area during 1970–2015 had a moderate positive correlation with the precipitation ($r = 0.685$) and a weak negative correlation with the air temperature ($r = -0.136$). Physical and chemical properties of stratigraphic sequences of sediments in Lake Terkhiin Tsagaan indicate this lake to be a montane type with high contents of organic matter and biogenic silica and fine sediments, implying warm and humid climate conditions in the Khangai region within central Mongolia at that time of their deposition.

Keywords Lake Terkhiin Tsagaan · Terkh River · Mt. Khorgo · Khangai Mountain Range · Climate change · Central Mongolia

A. Orkhonselenge et al., *Lakes of Mongolia*, Syntheses in Limnogeology,
https://doi.org/10.1007/978-3-030-99120-3_19

19.1 Introduction

Lake Terkhiin Tsagaan is classified as a volcanic lake (see Chap. 4), and it is located at the highest elevation among the lakes in the Khangai Mountain[1] Range of central Mongolia (Fig. 1.1). The lake archives valuable information on understanding large-scale dynamic changes of the lake basin evolution in the Khangai region and also regional paleoclimatic and paleoenvironmental histories in central Mongolia. Lake Terkhiin Tsagaan is a freshwater lake, and it plays an important role in the hydrological system of central Mongolia and southern Siberia through the activities of Terkh, Suman, Chuluut, Ider, and Selenge[2] Rivers and Lake Baikal[3] hydrological elements belonging to the North Arctic Ocean drainage basin.

Lake Terkhiin Tsagaan has attracted limited scientific attention to date. Lake Terkhiin Tsagaan was initially inferred as a volcanic lake by M.V. Pevtsov in 1878–1879 (Tserensodnom, 1971). The lake formed at 6.9 ± 0.1 ka BP when the Terkh River valley was dammed by a basaltic lava (Tserensodnom, 1971; Sevastyanov et al., 1989; Sanjmyatav, 2009) erupted at ca. 7.0 ka BP (Sevastyanov et al., 1989) from Mt. Khorgo (Myagmarjav & Davaa, 1999; Fig. 18.3). Recent studies have focused mostly on the lake in terms of geomorphological and sedimentological properties (Davaagatan, 2014; Davaagatan et al., 2015; Orkhonselenge & Davaagatan, 2016) and hydrochemical characteristics (Fukushi et al., 2015). However, geochronology to reconstruct the evolution of Lake Terkhiin Tsagaan and geomorphological processes in the lake basins in the wider context of central Mongolia has not been comprehensive.

Sedimentological study in lake basins, based on lithological stratigraphy of lake sediments, is important to reveal the geomorphological processes and dynamic evolution of lacustrine systems. In particular, studying sedimentary features of lake sediments is a key to deciphering local climate-induced environmental conditions of the lake basins. Because climate change is the major driving force behind environmental changes in lake basins (Andrén et al., 2015), interpreting sequences of lake sediments is essential for understanding the environmental and geomorphological evolution histories of the lake basins caused by climate change.

Lake Terkhiin Tsagaan represents fluvial and lacustrine systems in mountainous region because it occurs within an intermontane valley in the high Khangai Mountain Range (Fig. 1.1). Fluctuations of parameters sensitive to climate change (e.g., organic matter, biogenic silica, carbonates) suggest that lacustrine sedimentation is controlled by the major hydrological regime of the lake basin (Davaagatan et al., 2015). In Lake Terkhiin Tsagaan, sedimentation rates were recorded as the rates of 1.1 mm/year at 4.8–6.8 ka BP and 0.9 mm/year during the late Holocene (Khosbayar, 2005).

---

[1] *Khangai* has been misspelled as *Khangay* (or *Hangay*) and *Hangai* in many publications. Khangai is the right English transliteration from Mongolian *Хангай*.

[2] *Selenge* has been often incorrectly written as *Selenga* internationally based on the Russian pronunciation. Selenge is the correct English transliteration from Mongolian *Сэлэнгэ*.

[3] *Baigali* meaning *nature* has been misspelled as *Baikal* by the Russians. Baigali is the right English transliteration from Mongolian *Байгаль*.

In this chapter, the previous results from Lake Terkhiin Tsagaan are synthesized, and changes in area of the lake during the last 45 years and sedimentological characteristics of the lake are presented in detail. Paleoclimatic implications from Lake Terkhiin Tsagaan in central Mongolia are given in detail in Chap. 21.

19.2 Physiographic Condition

Lake Terkhiin Tsagaan is located at the elevation of 2060 m a.s.l. within the Terkh River valley (Fig. 19.1) in the southern Tarvagatai Mountain Range belonging to the Khangai Mountain Range of central Mongolia (Fig. 1.1).

Morphometric components of Lake Terkhiin Tsagaan are shown in Tables 1.3 and 18.6. The lake (Figs. 19.1 and 19.2) extends for 16.0 km in length, 4.0 km on average width, 6.0 km at the maximum width, and 66.0 km along the shoreline length (Table 1.3). The lake's water volume is 0.368 km^3 (Tserensodnom, 1971). Lake Terkhiin Tsagaan is the second largest lake in area within central Mongolia (Table 18.6). The lake covers an area of 61.1 km^2 (Tsegmid, 1969) or about 62.0 km^2 (Orkhonselenge and Uuganzaya, unpublished; see Sect. 19.6).

The lake sediments consist of clay, silts, sands, basalts, and gravels (Tserensodnom, 1971), and it ranges from very fine silts to clays (Davaagatan, 2014; see Sect. 19.7).

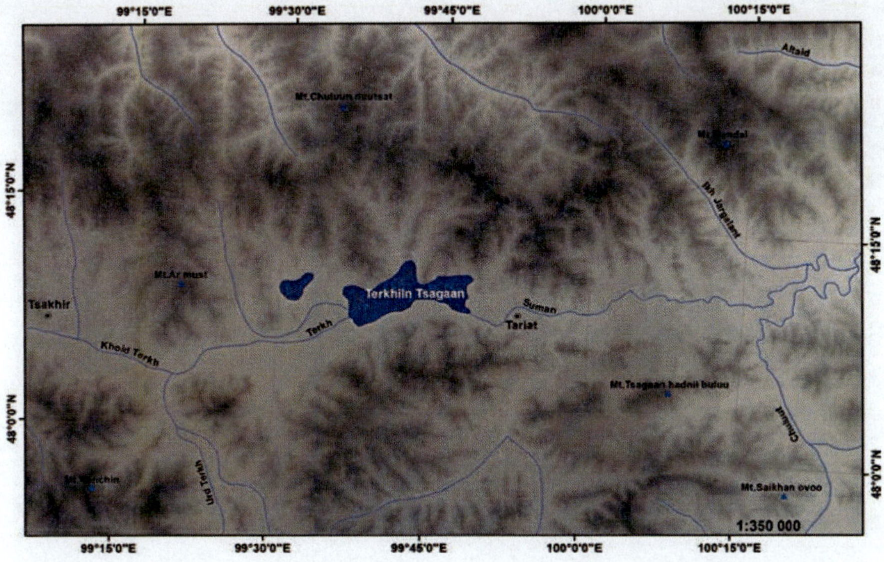

Fig. 19.1 Physiography of Lake Terkhiin Tsagaan basin. After Orkhonselenge and Davaagatan (2016)

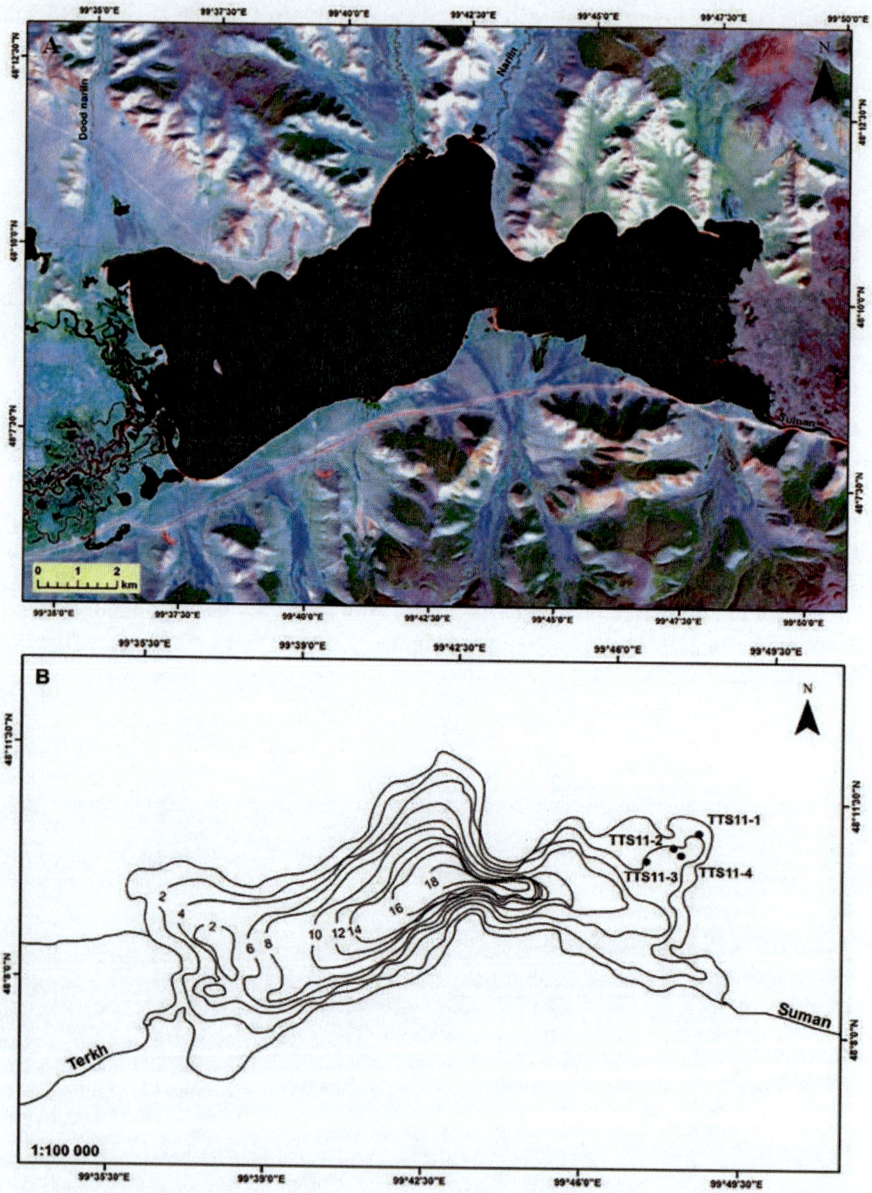

Fig. 19.2 (**a**) Landsat 8 image of Lake Terkhiin Tsagaan on June 16, 2015 (path of 134 and row of 29). Modified from Orkhonselenge and Uuganzaya (unpublished). (**b**) Bathymetry of Lake Terkhiin Tsagaan with locations of sedimentary cores (see Sect. 19.7). Modified from Orkhonselenge and Davaagatan (2016)

19.3 Geological and Geomorphological Settings

Lake Terkhiin Tsagaan is tectonically surrounded by the Devonian geosynclinal region in the north, Cenozoic basaltic plain in the east, Permian Orogenic Complex in the southeast, and Carboniferous to Permian geosynclinal region in the southwest and west (Academy of Sciences of Mongolian and Academy of Sciences of USSR, 1990). According to Badarch et al. (2002), the lake lies on the Tarvagatai cratonal block consisting of the Quaternary nonmarine clastic sedimentary rocks and Jurassic intermediate volcanic and volcaniclastic rocks. In the lake basin, various geological formations occur including the Neogene-Quaternary volcanic basalts in the northeast; Cenozoic plutonic tonalites, plagiogranites, granosyenites, granodiorites, and granites in the south, northwest, and north; and a continental platform consisting of the Cenozoic red and gray terrigenic sands, gravels, and pebbles in the southwest (Academy of Sciences of Mongolian and Academy of Sciences of USSR, 1990). The bedrock in the lake basin is composed of sandstones, limestone, the Devonian granites, and the Quaternary volcanic rocks (Myagmarjav, 2010).

Lake Terkhiin Tsagaan is an elliptically shaped volcanic lake extending from the west to the east (Figs. 19.1 and 19.2), and it is surrounded by numerous inactive volcanoes such as Mt. Khorgo (Fig. 18.3), Mt. Gichgene and Mt. Uran Dush, and steep basaltic cliffs (Tserensodnom, 2000). These mountains extend into the lake, forming numerous estuaries and bays in both northern and southern margins of the lake (Fig. 19.3). In the lake basin, there are high mountains such as Mt. Deed Bayan (2393 m a.s.l.) in the northwest, Mt. Ikh Khavchuu (2813 m a.s.l.) in the north, Mt. Tseenet Buluu (2908 m a.s.l.) and Mt. Khuush (2391 m a.s.l.) in the northeast, and

Fig. 19.3 Lake Terkhiin Tsagaan viewing toward the (**a**) northeast from the Mt. Khorgo (Fig. 18.3), (**b**) northeast, (**c**) north, and (**d**) east. Photos by A. Orkhonselenge on August 15, 2011

Mt. Uran Mandal (2800 m a.s.l.) in the south (Davaagatan, 2014). There are some small islands in the west of the lake (Fig. 19.2a), and the largest one is called *Tolgoi*, meaning *Head*, standing 30 m above the lake level. In the southwest, west, and northwest of the lake, there are numerous small freshwater lakes such as Lakes Khunt and Noot in the southwest; Lakes Salt, Shivert, Undur, and Tsokhio in the west; and Lake Khuduu in the northwest (Fig. 19.1; Table 18.6).

19.4 Climate Condition

In the Lake Terkhiin Tsagaan basin (Fig. 19.1), the annual average air temperature and precipitation are approximately −3.5 °C and 160 mm (Davaagatan, 2014; Davaagatan et al., 2015). The local climate data recorded at the Tariat meteorological station located 3.2 km east of the lake (Fig. 19.1) show the annual average air temperature of −2.4 °C and annual precipitation of 240 mm during the last 50 years (Fig. 19.4).

The monthly average air temperatures are −22.3 °C in January and 13.0 °C in July. The annual, summer, autumn, and spring average air temperatures are gradually rising between 1969 and 2020 (Figs. 19.4a, c). The summer precipitation is 179.1 mm with 47.8 mm in June, 72.3 mm in July, and 58.9 mm in August (Fig. 19.4d). The winter precipitation is slightly rising with a rate slower than other seasonal precipitations during 1969–2020 (Fig. 19.4d).

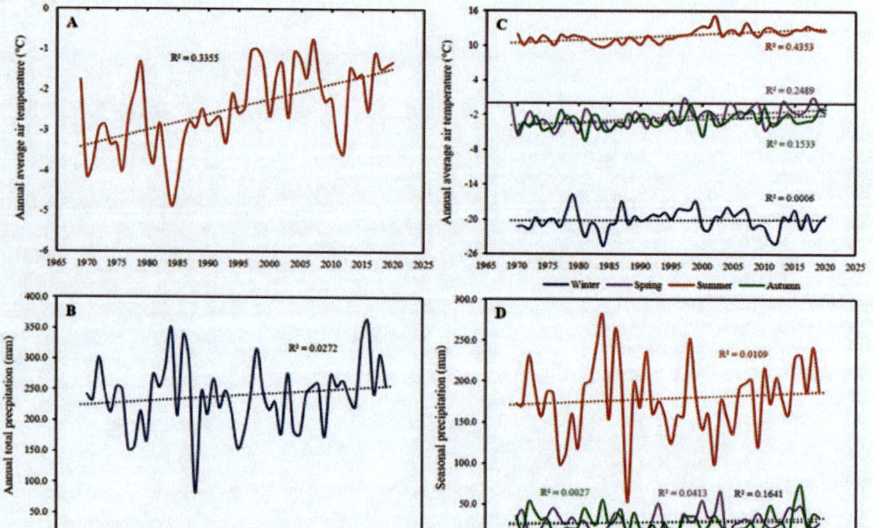

Fig. 19.4 Annual average (**a**) air temperature and (**b**) precipitation and seasonal average (**c**) air temperature and (**d**) precipitation recorded at the Tariat meteorological station (Fig. 19.1) during 1969–2020

19.5 Hydrological Condition

Lake Terkhiin Tsagaan is a freshwater lake (see Chap. 5), and it is fed by more than ten rivers (e.g., Terkh River in the west in Figs. 19.1 and 19.2) draining from an area of 7880 km^2 (Tsegmid, 1969) in the Khangai Mountain Range. The lake feeds Suman River in the east (Figs. 19.1 and 19.2), a major tributary of Chuluut River (Figs. 19.1 and 19.5; see Chap. 18). The Chuluut River is in turn a major tributary of Ider River, one of major headwaters of Selenge River (see Chaps. 15 and 18). Lake Terkhiin Tsagaan's depth reaches 6.0 m on average and 20.0 m at its maximum (Table 18.6), and it is particularly deep in the center (Fig. 19.2b). The lake's water level fluctuates for approximately 1.5 m/year (Tserensodnom, 1971).

The lake water temperature is 15–19 °C at the surface during summer and 4.0–4.3 °C at the bottom in winter (Table 19.1). The lake is frozen from the end of October to the end of May when it is covered by a 1.5–1.8-m-thick ice (Tserensodnom, 1971). Hydrochemically, the lake water is the Ca^{2+}-HCO_3-type (Fukushi et al., 2015). The mineralization value of Lake Terkhiin Tsagaan ranges at 0.16–0.85 g/L, and HCO_3^- is the dominant ion in the hydrochemical composition of the lake (Table 19.2).

19.6 Changes in Area

Lake Terkhiin Tsagaan has slightly expanded its area during the past 45 years although there have been significant fluctuations (Table 19.3). The lake gained an area of ~0.9 km^2 between 1969 and 2015 and 0.51 km^2 from 1970 to 2015 (Orkhonselenge and Uuganzaya, unpublished; Table 19.3, Fig. 19.6). Similar to other mountainous lakes such as Lakes Khoton (see Chap. 14), Targan (see Chap. 15), and Telmen (see Chap. 18), Lake Terkhiin Tsagaan has experienced an expansion of its area for less than ~1.0 km^2 in total.

The short-term decrease of lake area by 0.48 km^2 from 1970 to 1971 (Table 19.3) coincided with the drop in the annual precipitation from 242.1 mm in 1970 to 232.8 mm in 1971 (Fig. 19.4b). However, the long-term decrease in the lake area by 0.8 km^2 from 1971 to 2010 was consistent with the increased annual average air temperature in 1979 and 1997–2008 ranging between −1.6 °C and −0.7 °C except for 2000 and 2003 with the air temperatures of −2.6 °C (Fig. 19.4a). The short-term decrease in the lake area of 1.8 km^2 from 2010 to 2013 (Table 19.3) coincided with the drop in the annual precipitation from 273.5 mm in 2010 to 238.2 mm in 2013 (Fig. 19.4b). The short-term increase in the lake area of 1.99 km^2 (~2.0 km^2) during 2013–2015 may have been related to the rise in the annual precipitation from 238.2 mm in 2013 to 334.8 mm in 2015 (Fig. 19.4b) and the increased air temperature from −3.5 °C to −1.4 °C since 2013 (Fig. 19.4a) which largely contributed to the permafrost and snow melting in the surrounding mountains and their discharges into the lake.

Fig. 19.5 Chuluut River viewing toward the (**a**) north and (**b**) southwest. Photos by A. Orkhonselenge on August 14, 2011

The linear regression analysis (Fig. 19.7) shows that the lake area between 1970 and 2015 was moderately related to the annual precipitation ($R^2 = 0.47$), but not clearly related to the annual average air temperature ($R^2 = 0.0186$). The correlation analysis shows that the lake area during 1970–2015 had a moderate positive correlation with the precipitation ($r = 0.685$) and a weak negative correlation with the air temperature ($r = -0.136$).

Table 19.1 Water temperature of Lake Terkhiin Tsagaan

#	Depth (m)	Summer (°C)	Winter (°C)
1	0.0	15.0–19.0	–
2	2.0	–	1.2
3	4.0–6.0	–	2.0–2.3
4	8.0–10.0	–	3.0
5	Bottom	8.0–12.0	4.0–4.3

Table 19.2 Hydrochemical composition of Lake Terkhiin Tsagaan

Ions (mg/L)	Na^{+}+K^{+}	Mg^{2+}	Ca^{2+}	Cl^{-}	SO_4^{2-}	HCO_3^{-}	Total	Authors
July, 1968	20–30	10.0	10–20	10–20	20–30	100–120	170.0	Tserensodnom (1971)
High level	15	5.0	22.0	15.0	1.5	102.0	160.0	Davaa (2015)
Medium	17	5.0	23.0	20.0	1.5	104.0	171.0	Davaa (2015)
Low level	20	6.0	24.0	25.0	1.5	109.0	185.0	Davaa (2015)

Table 19.3 Temporal changes in area of Lake Terkhiin Tsagaan

#	Year	Area (km^2)	Volume (km^3)	Authors
1	1969	61.1	–	Tsegmid (1969)
2	1970	61.48	–	Orkhonselenge and Uuganzaya, unpublished
3	1971	61.0	0.368	Tserensodnom (1971)
4	2010	61.8	–	Davaa (2015)
5	2013	60.0	0.369	Davaa (2015)
6	2015	61.99	–	Orkhonselenge and Uuganzaya, unpublished

The increase in area of the lake during the last 45 years (Fig. 19.6) may have been related to the gradually rising annual and seasonal precipitations, especially winter precipitation (Fig. 19.4b, d), and the rapid rises in annual, summer, autumn, and spring average air temperatures (Fig. 19.4a, c) contributing to thawing of snow cover and permafrost to discharge into the lake.

19.7 Sedimentological Review

The physical and chemical properties of sediments in short cores TTS11-1 to TTS11-4 (Fig. 19.2b) obtained from Lake Terkhiin Tsagaan indicate the sedimentary environment as the montane-type. Water content of the core sediments is 15% on average and the value increases equally with the depth (Davaagatan, 2014).

The organic matters of the core sediments range from 23 to 35% on average in the core sediments (Davaagatan, 2014; Fig. 19.8). The organic matter increases with depth in the marginal parts of the lake, but it decreases with depth toward the center of the lake (Fig. 19.8). The high content of organic matter in the marginal parts may be correlated with direct accumulation of soil and plant residues that are eroded and

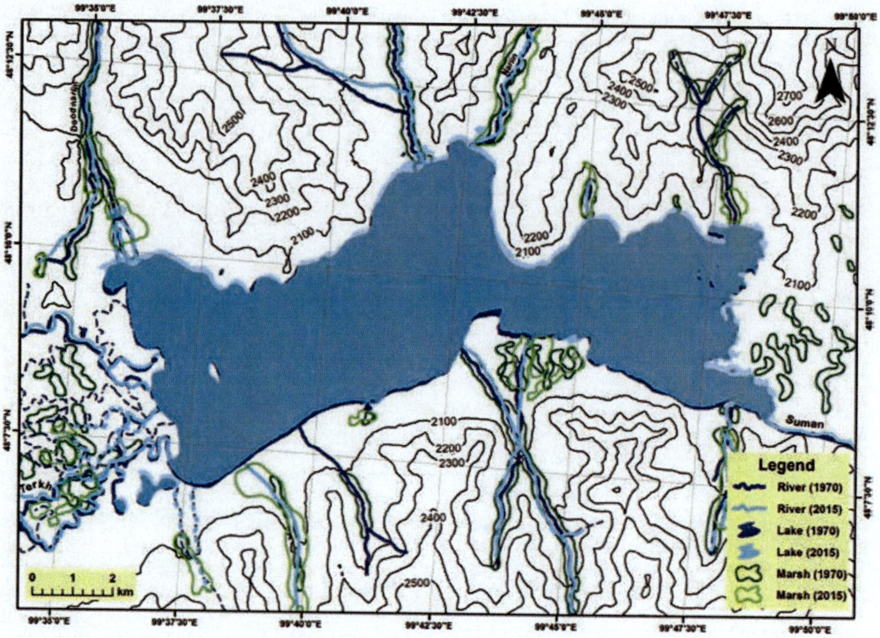

Fig. 19.6 Spatial distributions of areas and marshes of Lake Terkhiin Tsagaan in 1970 and 2015. Modified from Orkhonselenge and Uuganzaya (unpublished)

transported from the exposed surfaces of lake beaches during the low lake levels (Davaagatan et al., 2015; Orkhonselenge & Davaagatan, 2016).

The carbonates of the core sediments vary between 1.8 and 2.5% on average (Davaagatan, 2014; Fig. 19.8). In general, the carbonates increase in the direction from the shallower part toward the deeper part of the lake (Fig. 19.8). The spatial pattern of the vertical fluctuation of the carbonates indicates an increase in solubility in the direction from the marginal shallow part toward the central deep part of the lake, and the solubility may correspond to the increase of water content with the depth (Davaagatan et al., 2015; Orkhonselenge & Davaagatan, 2016; Fig. 19.8). This trend is consistent with the accumulation of monohydrate carbonates in the lake (Fukushi et al., 2015).

The biogenic silica of the core sediments ranges at 25–45% on average (Davaagatan, 2014; Fig. 19.8). The general trend of its spatial distribution shows the increase in the transection zone from the marginal shallower part toward the central deeper part of the lake. The vertical distribution of the biogenic silica shows an increase with the depth in the center of the lake (Davaagatan et al., 2015; Orkhonselenge & Davaagatan, 2016). This pattern is consistent with the increased biogenic silica (5–45%) in Lake Khuvsgul during the warm/wet interglacial periods (Murakami et al., 2010) and the high dissolved SiO_2 concentrations in the lake water, which provide favorable conditions for the formation of biogenic silica (Fukushi et al., 2015).

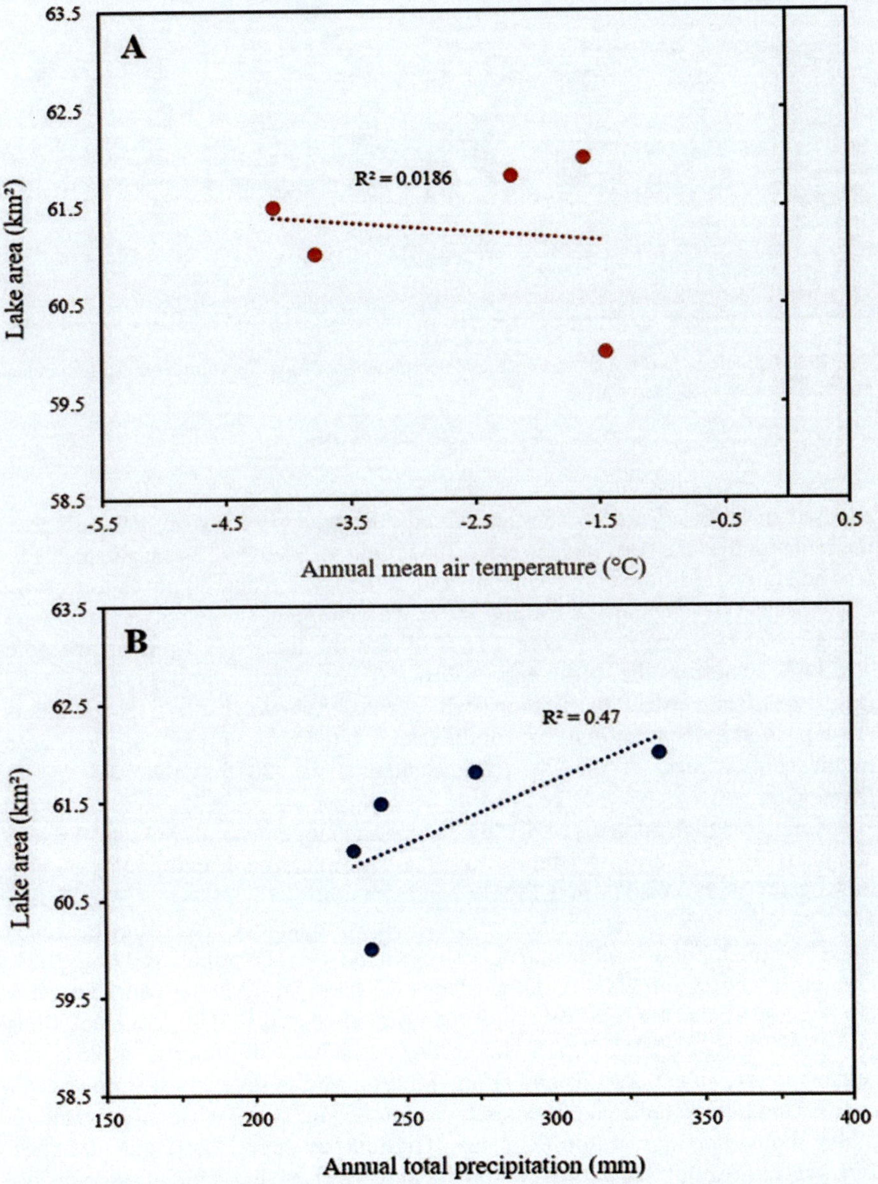

Fig. 19.7 Relationships between area of Lake Terkhiin Tsagaan and annual average (**a**) air temperature and (**b**) precipitation during 1970–2015

The mineral fractions of the core sediments are 25–40% on average (Davaagatan, 2014). The marginal parts of the lake contain larger minerals than those in the central part of the lake (Fig. 19.8). It implies that more sediments transported from the

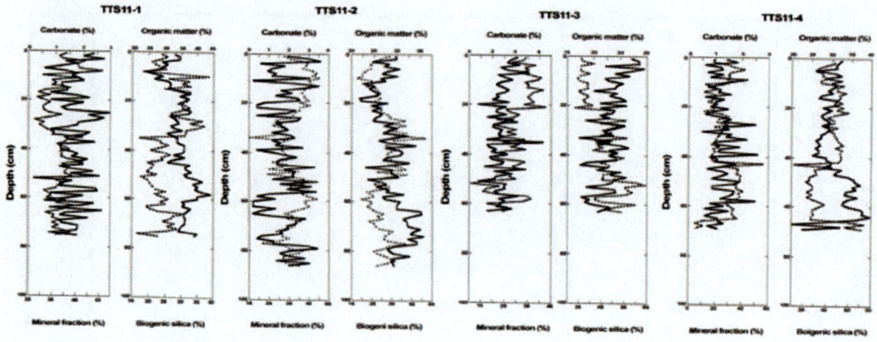

Fig. 19.8 Variations in physical and chemical components of Lake Terkhiin Tsagaan sediments. Modified from Davaagatan (2014)

exposed surfaces of lake beaches are directly deposited in the marginal parts than the center of the lake (Davaagatan et al., 2015; Orkhonselenge & Davaagatan, 2016).

The grain size of the core sediments ranges between 6 and 20 μm in total, with a few peaks (Davaagatan, 2014; Fig. 19.9). The sediments above 50 cm depth are the coarsest (~7 μm) in the core TTS11–4 and the finest (~13 μm) in the core TTS11-3, whereas the sediments below 50 cm depth show gradual shifts (Fig. 19.9). The vertical distribution of the grain size shows that coarser sediments are deposited in the marginal parts of the lake, while finer sediments are deposited in the central parts of the lake (Davaagatan et al., 2015; Orkhonselenge & Davaagatan, 2016).

Overall, the high contents of organic matter and biogenic silica in Lake Terkhiin Tsagaan (Fig. 19.8) are consistent with silica-diatomic organic carbon (C) (5–40% in LT98C3 and 20–80% in LT98C9) and organic C (8–14% in LT98C3 and 16–22% in LT98C9) rather than allochthonous organic C (Khosbayar, 2005) caused by the input of continental deposits into the lake (Lehmkuhl & Lang, 2001; Grunert & Lehmkuhl, 2004; Orkhonselenge & Davaagatan, 2016). Moreover, it is largely contributed by relatively high total organic carbon (TOC) concentrations prevalent during high lake levels, suggesting an additional influence of terrestrial organic matter (plant detritus and humus) transported to the lake by groundwater or overland flow (Xiao et al., 2006; Schwanghart et al., 2008) during high rainfalls under the warm and humid climates (Orkhonselenge & Davaagatan, 2016). Intervals of the high contents of carbonates and minerals (Fig. 19.8) are suggestive of low lake levels under cold and dry environments, and this implies the timing of transgressions of the lake, when abundant sediments are transported from the exposed beaches into the lake (Davaagatan et al., 2015; Orkhonselenge & Davaagatan, 2016).

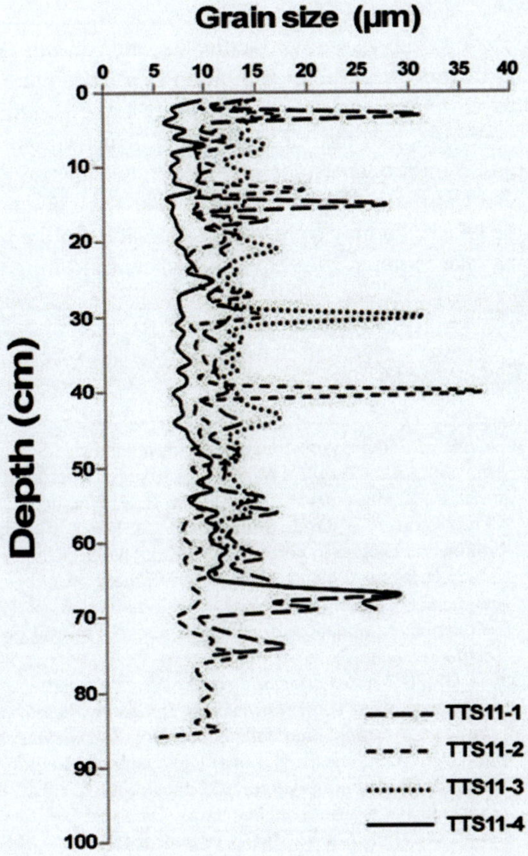

Fig. 19.9 Grain size distribution of Lake Terkhiin Tsagaan sediments. Modified from Davaagatan (2014)

19.8 Summary

Lake Terkhiin Tsagaan, a volcanic freshwater lake, reflects the climatological and hydrological dynamics in the Khangai region of central Mongolia. Lake Terkhiin Tsagaan has experienced climate changes since its initial formation due to damming of an erupted basaltic lava at ~7.0 ka BP. In the lake basin, a warming climate is observed with the rapid rises in annual and seasonal air temperatures over the past 50 years. The lake has expanded its area for 0.51 km^2 during the last 45 years. The short-term changes of the lake area during 1970–1971 and 2010–2013 were directly related to the fluctuations of annual and winter precipitations, whereas the long-term changes of the lake area during 1971–2010 were related to the rises in annual, summer, autumn, and spring average air temperatures which may have been causing thawing of snow cover and permafrost in the Khangai Mountain Range, which discharge the lake. The linear regression analysis shows that the lake area between 1970 and 2015 was moderately related to the annual precipitation ($R^2 = 0.47$), but it was not clearly related to the annual average air temperature ($R^2 = 0.0186$). The correlation analysis shows that the lake area during 1970–2015 had a moderate positive correlation with the precipitation ($r = 0.685$) and a weak negative correlation with the air temperature ($r = -0.136$). In Lake Terkhiin Tsagaan, intervals of increased organic matter and biogenic silica, synchronous with intervals of fine sediments, point to a montane-type lake which receives significant precipitation and suggests also warm and humid environments. The climate change and landscape feature in the lake basin may have been responsible for the sedimentological characteristics of the lake sediments in this semiarid central Mongolia.

References

Academy of Sciences of Mongolian and Academy of Sciences of USSR. (1990). *National Atlas of the Mongolian People's Republic.* 156 p. [In Russian].

Andrén, E., Klimaschewski, A., Angela, E. S., Natalie, S. A., Andrei, A. A., Keith, D. B., Daniel, J. C., Thomas, W. D. E., Solovieva, N., & Hammarlund, D. (2015). Holocene climate and environmental change in north-eastern Kamchatka (Russian Far East), inferred from a multi-proxy study of lake sediments. *Global and Planetary Change, 134*, 41–54.

Badarch, G., Cunningham, W. D., & Windley, B. F. (2002). A new terrane subdivision for Mongolia: Implications for the Phanerozoic crustal growth of Central Asia. *Journal of Asian Earth Sciences, 21*, 87–110.

Davaa, G. (2015). *Resource and regime of surface water of Mongolia.* 408 p. [In Mongolian].

Davaagatan, T. (2014). *Reconstructing Holocene climate change in the steppe region of Mongolia based on transportation and deposition of lake sediments: The cases of Lakes Terkhiin Tsagaan and Ugii.* MSc Thesis, National University of Mongolia, Ulaanbaatar. [In Mongolian].

Davaagatan, T., Orkhonselenge, A., Fukushi, K., & Yabe, M. (2015). Sedimentary records in intermontane and prairie lakes in central Mongolia: Preliminary results from Lake Terkhiin Tsagaan and Lake Ugii. *Journal of Geographical Issues, 14*(408), 19–26.

Fukushi, K., Katsuta, N., Jenkins, R. G., Matsubara, K., Takayama, B., Tanaka, Y., Davaasuren, D., Batkhishig, O., Hasebe, N., & Kashiwaya, K. (2015). Centennial-scale environmental changes in Terhiin Tsagaan Lake, Mongolia Inferred from Lacustrine Sediment: Preliminary Results. In K. Kashiwaya, J. Shen, & J. Y. Kim (Eds.), *Earth surface processes and environmental changes in East Asia: Records from Lake-catchment Systems* (pp. 25–44). Springer.

Grunert, J., & Lehmkuhl, F. (2004). Aeolian sedimentation in arid and semi-arid environments of Western Mongolia. In W. Smykatz-Kloss & P. Felix-Henningsen (Eds.), *Paleoecology of quaternary drylands* (pp. 195–218). Springer.

Khosbayar, P. (2005). *Mezozoic and Cenozoic Paleogeography and Paleoclimate of Mongolia. Transaction 15.* Institute of Geology and Mineral Resources, Mongolian Academy of Sciences, Ulaanbaatar. [In Mongolian].

Lehmkuhl, F., & Lang, A. (2001). Geomorphological investigations and luminescence dating in the southern part of the Khangay and the Valley of the Gobi Lakes (Central Mongolia). *Journal of Quaternary Science, 16*, 69–87.

Murakami, T., Katsuta, N., Yamamoto, K., Takamatsu, N., Takano, M., Oda, T., Matsumoto, G. I., Horiuchi, K., & Kawai, T. (2010). A 27-kyr record of environmental change in central Asia inferred from the sediment record of Lake Hovsgol, northwest Mongolia. *Journal of Paleolimnology, 43*, 369–383.

Myagmarjav, B. (2010). *Handbook of Lake Terkhiin Tsagaan.* Toonot Print Publishing. [In Mongolian].

Myagmarjav, B., & Davaa, G. (1999). *Surface water of Mongolia.* Inter Press. [In Mongolian].

Orkhonselenge, A., & Davaagatan, T. (2016). *Sedimentary feature in Lake Terkhiin Tsagaan and Lake Ugii, Central Mongolia*. Geophysical Research Abstracts, vol. 18, EGU2016-5666-2, EGU General Assembly 2016.

Sanjmyatav, Z. (2009). *Physical geography of Mongolia*. SHME Publishing. [In Mongolian].

Schwanghart, W., Schütt, B., & Walther, M. (2008). Holocene climate evolution of the Ugii Nuur basin, Mongolia. *Advances in Atmospheric Sciences, 25*(6), 986–998.

Sevastyanov, D. V., Dorofeyuk, N. I., & Liiva, A. A. (1989). The origin and evolution of the volcanic Terkhiin-Tsagan-Nur Lake in Central Hangai MPR. *Izvestiâ Vsesoûznogo Geografičeskogo Obŝestva, 121*, 137–223.

Tsegmid, S. (1969). *Physical geography of Mongolia*. 759 p. [In Mongolian].

Tserensodnom, J. (1971). *Lakes of Mongolia*. State Publishing, 202 p. [In Mongolian].

Tserensodnom, J. (2000). *Catalogy of Lakes in Mongolia*. Shuvuun Saaral Publishing, 141 p. [In Mongolian].

Xiao, J., Wu, J., Si, B., Liang, W., Nakamura, T., Liu, B., & Inouchi, Y. (2006). Holocene climate changes in the monsoon/arid transition reflected by carbon concentration in Dahai Lake of Inner Mongolia. *The Holocene, 16*, 551–560.

Chapter 20
Lake Ugii

Abstract Lake Ugii, an oxbow lake, resides in the Orkhon River valley between Khangai and Khentii Mountain Ranges in central Mongolia. The lake plays an important role in the hydrological system of the North Arctic Ocean drainage basin through the fluvial activities of Khugshin Orkhon, Orkhon, and Selenge rivers and Lake Baikal. Lake Ugii lost an area of ~2.0 km^2 between 1970 and 2014. This decrease of the lake area shows that the lake may have directly responded to the rising summer air temperature rather than gradually increasing summer precipitation over the last 35 years. The linear regression analysis shows that the lake area between 2000 and 2014 was weakly related to the annual average air temperature ($R^2 = 0.3458$) and precipitation ($R^2 = 0.2095$). The correlation analysis shows that the lake area during 2000–2014 had a moderately negative correlation with the air temperature ($r = -0.588$) and a weak negative correlation with the precipitation ($r = -0.457$). However, the limited available data are insufficient for further inference. Sedimentological characteristics of Lake Ugii show that the lake is a prairie-type lake with high contents of carbonates and minerals and coarse sediments. The stratigraphic sequence of the lake sediments suggests impact of climate changes on lake sedimentations. The large amounts of carbonates in the lake indicate a cold and arid climate leading to the regression of the lake and transportation of aeolian sediments, while the high contents of mineral fractions and coarse sediments imply precipitation-induced fluvial processes in the Orkhon River drainage basin in central Mongolia.

Keywords Lake Ugii · Orkhon River · Khugshin Orkhon · Khangai Mountain Range · Climate change · Central Mongolia

A. Orkhonselenge et al., *Lakes of Mongolia*, Syntheses in Limnogeology,
https://doi.org/10.1007/978-3-030-99120-3_20

20.1 Introduction

Lake sediments provide us much needed information on geomorphological processes in lake basins and impact of climate change on lake evolution (Orkhonselenge & Davaagatan, 2016; Orkhonselenge et al., 2019). The sediments that accumulate in a lake have both autochthonous and allochthonous origins (Peck et al., 2002), and these sediments show records of the lake basin responses to climate and environmental changes (Battarbee, 2000). This is because that the climate and environmental changes influence the type of sediment that accumulates in the lake (Peck et al., 2002).

Data on local and regional climate changes help us in understanding responses of various sensitive landscapes, especially lake basins in Mongolia, to present and future climate changes (Davaagatan et al., 2015; Orkhonselenge et al., 2019). Central Mongolia is located under the influences of the East Asian winter monsoon (EAWM), Asian low-pressure system (Yatagai & Yasunari, 1995), and westerly winds (Hilbig, 1995; see Chap. 18). At the present time, lakes in Mongolia respond to the climate warming by shrinkage (e.g., Kang & Hong, 2016; Szumińska, 2016; Orkhonselenge et al., 2018; Orkhonselenge & Bulgan, 2021) or with lake level changes (e.g., Kang et al., 2015). Because the landscapes of Mongolia are highly vulnerable to climate change (Gunin et al., 1999; Tarasov et al., 1999), understanding the evolution of lake basins is crucial. However, it is still not completely understood in central Mongolia.

To date, lake studies in Mongolia (see Chap. 2) have rarely focused on geomorphology and sedimentology of lake basins. Some geomorphological and sedimentological investigations in lake basins of central Mongolia have been conducted to understand large-scale regional climate changes (e.g., Peck et al., 2002; Fowell et al., 2003; Schwanghart et al., 2008, 2009; Wang et al., 2009, 2011; Tian et al., 2013; Davaagatan et al., 2015; Orkhonselenge & Davaagatan, 2016). However, sedimentological characterization of lakes to infer lake evolution and geomorphological processes leading to lake sedimentation in those lake basins has not been comprehensively performed in central Mongolia.

Lake Ugii, an oxbow lake of Orkhon River (Orkhonselenge, 2021; Fig. 20.1), represents the lake evolution in the wide intermontane valley between Khangai[1] and Khentii[2] Mountain Ranges (see Chap. 18) under the ongoing climate change of central Mongolia (Fig. 1.1). The lake was initially studied by M.V. Pevtsov in 1883 (Tserensodnom, 1971). Lake Ugii's depth was measured by the Russian physical geographer Murzaev (1952). In 1964, Mongolian scientists J. Tserensodnom and A. Dashdorj conducted hydrological and hydrobiological investigations in Lake Ugii with measurements of water regime, bottom deposits, and bathymetry (Tuvshinjargal, 2001). In 1981, national physical geographers J. Tserensodnom, Ts.

---

[1] *Khangai* has been misspelled as *Khangay* (or *Hangay*) and Hangai in many publications. Khangai is the right English transliteration from Mongolian *Хангай*.

[2] *Khentii* has been misspelled as *Khentey* (or *Hentey*) and *Khentei* (or *Hentei*) in many publications. Khentii is the right English transliteration from Mongolian *Хэнтий*.

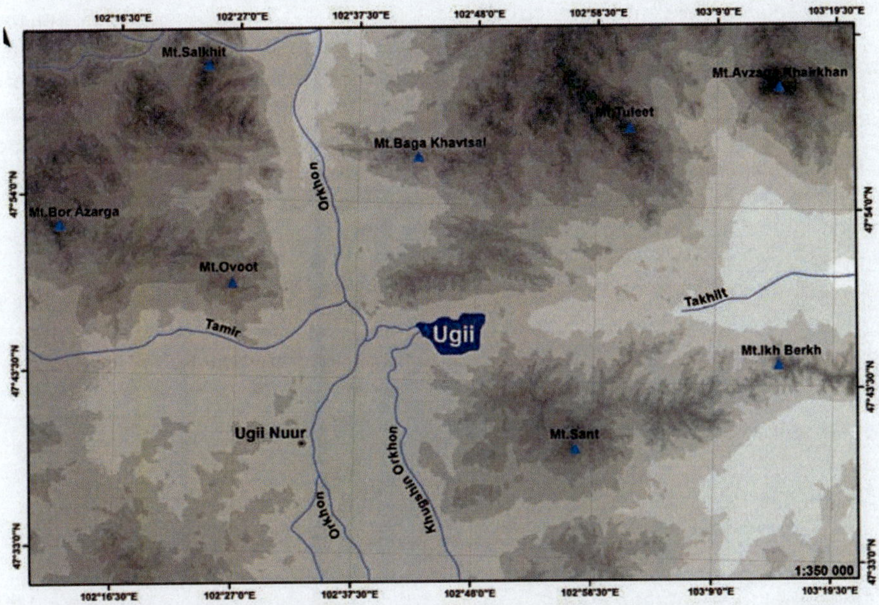

Fig. 20.1 Physiography of Lake Ugii basin in an intermontane valley between Khangai and Khentii Mountain Ranges. Modified from Orkhonselenge and Davaagatan (2016)

Sugar, Z. Sanjmyatav, and B. Avirmed studied hydrology and climatology in the Lake Ugii basin (Tuvshinjargal, 2001). Moreover, in 1987–1992, geoscientists from the Division of Physical Geography, Institute of Geography and Permafrost, Mongolian Academy of Sciences, studied water regime, balance of solar insolation and water level, and hydrothermal fluctuation in Lake Ugii (Tuvshinjargal, 2001).

In this chapter, previous results from Lake Ugii are synthesized, and changes in area of the lake during the last 45 years and sedimentological characteristics of the lake sediments are described in detail. Paleoclimatic implications from Lake Ugii in central Mongolia are given in Chap. 21.

20.2 Physiographic Condition

Lake Ugii is located at the elevation of 1337 m a.s.l. at the middle stream of Orkhon River (Fig. 20.1) in the northeast of Khangai Mountain Range in central Mongolia (Fig. 1.1), and the river runs into the Selenge[3] River and eventually feeds Lake Baikal[4].

[3] *Selenge* has been often incorrectly written as *Selenga* internationally based on the Russian pronunciation. Selenge is the correct English transliteration from Mongolian *Сэлэнгэ*.

[4] *Baigali* meaning ***nature*** has been misspelled as *Baikal* by the Russians. Baigali is the right English transliteration from Mongolian *Байгаль*.

The modern lake morphometrically extends for 7.4 km in length, 5.3 km in width, and 24.7 km along the shoreline length (Tserensodnom, 1971) as shown in Table 1.3. The lake occupies an area of 25.0 km^2 (Tsegmid, 1969) or ~23.8 km^2 (Orkhonselenge and Uuganzaya, unpublished; Table 18.6; see Sect. 20.6). Lake Ugii sediments are H_2S-rich clays, sands, and gravels distributed along the beaches (Tserensodnom, 2000), but bottom sediments are coarse silts to clays (Davaagatan, 2014; see Sect. 20.7). The mineral fraction of the lake sediments comprises of allogenic minerals dominated by quartz and feldspars (mainly the alkali polymorphs orthoclase, sanidine, and microcline) and authigenic minerals (e.g., TIC, Ca, Sr and S) (Schwanghart et al., 2008).

The Lake Ugii basin is situated within the area of dry steppes (Hilbig, 1995) dominated by grasses (*Cleistogenes*, *Stipa*), forbs (*Allium*), and shrubs (*Artemisia*, *Caragana*).

20.3 Geological and Geomorphological Settings

The Lake Ugii basin is located in the east of a perpendicular junction between the Tamir-Tuul basin and the Orkhon fault line (Khosbayar, 2005; Fig. 20.1). The bedrock in the hills and mountains around the lake consists of the late Paleozoic, Mesozoic, and Cenozoic rocks, but the bedrock in the lake basin includes the Miocene and Pliocene basalts, aeolian silts, and fine sands, which cover the Carboniferous schists and shales found along ridges and steep slopes, as well as the late Cretaceous and Paleogene conglomerates in cliffs exposed along the southern and eastern shores of the lake (Khosbayar, 2005; Schwanghart et al., 2008). The cliff lines along the southern and eastern shores are partly composed of conglomerates of probably the late Cretaceous and Neogene periods (Sladen & Traynor, 2000).

The modern topography of the Lake Ugii basin formed as circled-crescent shape (Figs. 20.1 and 20.2) due to the tectonic activity during the period between the Oligocene and Neogene (see Chap. 3) and lacustrine and fluvial networks were established in the Miocene (Jigj, 1975; Tuvshinjargal, 2001). During the Holocene, the lacustrine environment changed between open and closed lake conditions (Schwanghart et al., 2008). During the modification of the paleochannel of Orkhon River or Khugshin Orkhon[5] (Fig. 20.1) due to tectonic activity, Lake Ugii may have been originally formed as an oxbow lake in the middle stream of Orkhon River (Orkhonselenge, 2021; Fig. 20.1). Sedimentation in Lake Ugii started at 10.6 ka BP after a period of fluvial deposition by the Khugshin Orkhon (Walther & Gegeensuvd, 2005; Schwanghart et al., 2008).

The lake (Fig. 20.3) is surrounded by hills and low mountains in the wide intermontane valley between Khangai and Khentii mountain ranges (see Chap. 18). The mountains are Mt. Ikh Berkh (1646 m a.s.l.) in the east, Mt. Sant (1698 m a.s.l.) in the southeast, Mt. Ovoot (1622 m a.s.l.) in the northwest, and Mt. Ikh Zamt (1607 m a.s.l.) and Mt. Tuleet (1940 m a.s.l.) in the northeast (Davaagatan, 2014; Fig. 20.1).

[5] *Khugshin Orkhon* means *paleo-Orkhon* or a paleochannel of Orkhon River and it is transliterated from Mongolian *Хөгшин Орхон into English.*

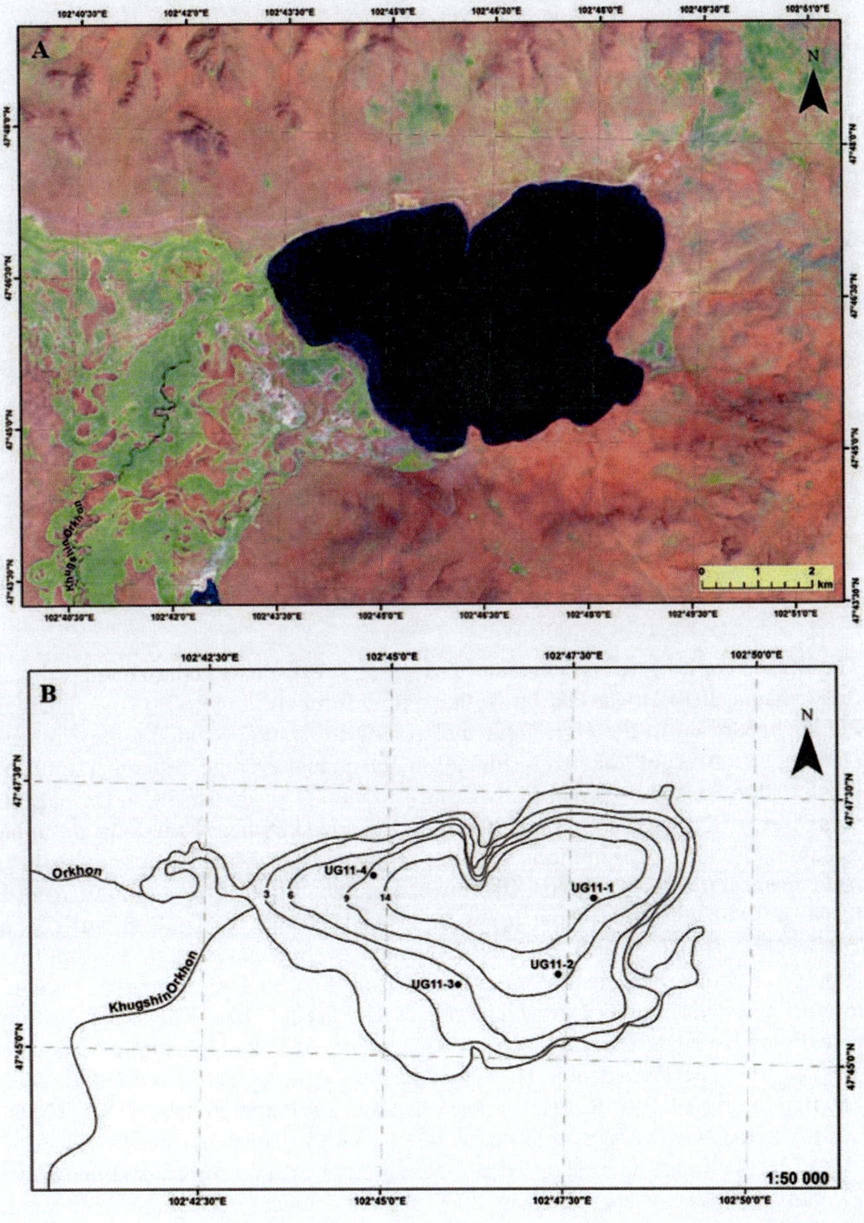

Fig. 20.2 (**a**) Landsat 8 image of Lake Ugii on August 16, 2014 (path of 134 and row of 27). Modified from Orkhonselenge and Uuganzaya (unpublished). (**b**) Bathymetry of Lake Ugii with locations of sedimentary cores (see Sect. 20.7). Modified from Orkhonselenge and Davaagatan (2016)

Fig. 20.3 Lake Ugii viewing toward the (**a**) north, (**b**) southeast, (**c**) east, and (**d**) southwest. Photos by A. Orkhonselenge on August 17, 2011

20.4 Climate Condition

The Lake Ugii basin is characterized by an extremely continental climate (Davaagatan, 2014). In the lake basin, the semiarid climate dominates (Tuvshinjargal, 2001). According to the Ugii Nuur meteorological station located approximately 15.0 km southwest of Lake Ugii (Fig. 20.1), the annual average air temperature and precipitation are approximately 1.5 °C and 130 mm (Davaagatan, 2014; Davaagatan et al., 2015). The highest precipitation rates generated by heavy rainfalls are found in summer, while precipitation in winter (November to March) occurs mostly in solid form and rarely exceeds 10 mm per month. The average rainfall records between 1940 and 1996 show drops in 1940–1960 and 1980–1996 and rise in 1960–1980 (Tuvshinjargal, 2001). The annual average precipitation is approximately 126.5 mm, and most of them occurs from May to August ranging from 4.7 to 94.9 mm per month (Tuvshinjargal, 2001; Table 20.1). The Khar Khorin[6] meteorological station located about 65.0 km south of Lake Ugii shows the annual average air temperature of about 0 °C and the annual precipitation of approximately 290 mm during 1962–1983. The January average air temperature is −18.2 °C contrasting to the July average air temperature of 16.8 °C (Tuvshinjargal, 2001).

The local climate data recorded at the Ugii Nuur meteorological station located 17.6 km southwest of the lake (Fig. 20.1) show the annual average air temperature of −1.0 °C and annual precipitation of 148.1 mm between 1985 and 2019 (Fig. 20.4a, b). The monthly average air temperature is −23.4 °C in January and 18.3 °C in July,

[6] *Khar Khorin* has been often misspelled as *Kara Korum* (or *Khara Khorum*) and *Har Horin* in many publications, Wikipedia, etc. *Khar Khorin* is the right English transliteration from Mongolian *Хар Хорин*.

Table 20.1 Precipitation in the Lake Ugii basin between 1987 and 1991

#	Precipitation (mm)					
	Period	May	June	July	August	Total
1	1987	–	9.9	37.3	94.9	142.1
2	1988	32.6	40.6	31.9	34.2	139.3
3	1990	8.7	81.2	4.7	–	94.6
4	1991	9.0	46.7	60.4	14.0	130.1

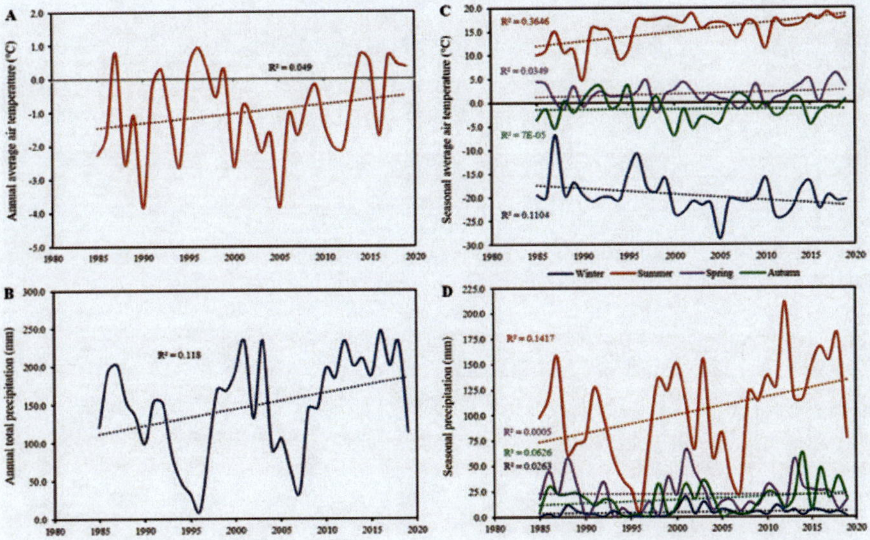

Fig. 20.4 Annual average (**a**) air temperature and (**b**) precipitation and seasonal average (**c**) air temperature and (**d**) precipitation recorded at the Ugii Nuur meteorological station (Fig. 20.1) during 1985–2019

and the monthly precipitation is 49.4 mm in July (Fig. 20.4c, d). The summer average air temperature (Fig. 20.4c) and precipitation (Fig. 20.4d) are rising during the last 35 years.

20.5 Hydrological Condition

Lake Ugii, a freshwater lake (see Chap. 5), is fed by Khugshin Orkhon River in the west (Figs. 20.1 and 20.2). According to Tserensodnom (2000), the lake is fed by Tsagaan River, a tributary of the Khugshin Orkhon River, draining from the southeastern Khangai Mountain Range. However, the lake feeds only Orkhon River (Figs. 20.1, 20.2, and 20.5b) through a channel or an outlet river in the west (Figs. 20.1, 20.2, and 20.5a), but the lake is predominantly fed by groundwater

Fig. 20.5 (a) Middle stream of the channel or an outlet river of Lake Ugii (Fig. 20.1) viewing toward the north. (b) Orkhon River viewing toward the north and flowing under the bridge located 14.0 km north of the Ugii Nuur meteorological station (Fig. 20.1) in Ugii Nuur soum, Arkhangai aimag. Photos by A. Orkhonselenge on August 16, 2011

(Schwanghart et al., 2008; Davaagatan et al., 2015). The lake basin opens toward the 12–20-km-wide Orkhon River valley in the west.

A paleochannel of Orkhon River called Khugshin Orkhon may have formed the lake in the past as an oxbow (Orkhonselenge, 2021; Fig. 20.1). The Khugshin Orkhon River constitutes an inlet in the western part of the lake (Figs. 20.1 and 20.2). The perennial Khugshin Orkhon and Orkhon rivers drain from an area of 4655 km^2 (Tserensodnom, 1971; Davaagatan, 2014) in the Mt. Suvarga Khairkhan in the southeastern Khangai Mountain Range (Orkhonselenge, 2021). The outlet river joins the modern Orkhon River after running over a distance of 7.0 km in the northwest of the lake (Figs. 20.1, 20.2, and 20.5a). This water course is active only

Table 20.2 Precipitation recorded at Lake Ugii and water level of the lake

#	Date	Precipitation (mm)	Water level (mm)
1	05.08.1987	10.4	7.5
2	06.22.1988	10.5	4.5
3	07.16.1991	19.5	5.3

Table 20.3 Bathymetric data with corresponding area and volume of Lake Ugii

#	Depth (m)	Area (km^2)	Volume (km^3)	Area (%)	Volume (%)
1	0–2	2.5	0.0329	9.7	19.3
2	2–4	4.75	0.0369	18.5	21.6
3	4–6	3.15	0.0306	12.2	17.9
4	6–8	2.95	0.0247	11.5	14.5
5	8–10	2.60	0.0195	10.1	11.4
6	10–12	2.65	0.0142	10.3	8.3
7	12–14	2.60	0.0090	10.1	5.3
8	14–15	1.95	0.0025	7.6	1.5
9	15–15.3	2.55	0.0003	10.0	0.2

when overflow from Lake Ugii occurs during or after heavy rainfalls (Davaagatan et al., 2015).

Lake Ugii's water level fluctuates depending on precipitation (Table 20.2), and consequently, area and volume of the lake vary with the depth (Tuvshinjargal, 2001; Table 20.3). The lake depth (Fig. 20.2b) reaches 6.6 m on average and 15.3 m at its maximum (Tserensodnom, 2000; Tuvshinjargal, 2001; Tables 18.6 and 20.3) or 17.0 m (Wang et al., 2011). The lake volume is 0.147 km^3 (Tserensodnom, 2000) or 0.171 km^3 (Davaa, 2015).

Lake Ugii's water temperature is 19.9–21.8 °C at its surface in summer and 3.7 °C at the bottom in winter (Tserensodnom, 2000). Lake Ugii is described as a thermally stratified, being oligo- to mesotrophic and dimictic with a transparency of 7.3 m, and the metalimnion is found between 10 m and 12 m above an unstable hypolimnion (Schwanghart et al., 2008).

20.6 Changes in Area

Lake Ugii fluctuated its area with a slow rate during the past 45 years (Table 20.4). The lake lost an area of 1.2 km^2 from 1969 to 2014 and of ~2.0 km^2 from 1970 to 2014 (Orkhonselenge and Uuganzaya, Unpublished; Table 20.4; Fig. 20.6). The decreases in area of the lake coincided with the records of the annual rainfall drop during 1980–1996 (Tuvshinjargal, 2001) and the summer air temperature rise within the lake basin since 1985 (Fig. 20.4c).

Figure 20.6 shows that the southeastern, southern, southwestern, and western shores of the lake have been sensitive to the rising air temperature (Fig. 20.4a, c)

Table 20.4 Temporal changes in area of Lake Ugii

#	Year	Elevation (m a.s.l.)	Area (km^2)	Authors
1	1969	1337	25.0	Tsegmid (1969)
2	1970	–	25.78	Orkhonselenge and Uuganzaya, unpublished
3	1971	1337	25.0	Tserensodnom (1971)
4	2000	1337	25.0	Tserensodnom (2000)
5	2010	–	23.4	Davaa (2015)
6	2013	1332	23.0	Davaa (2015)
7	2014	–	23.79	Orkhonselenge and Uuganzaya, unpublished

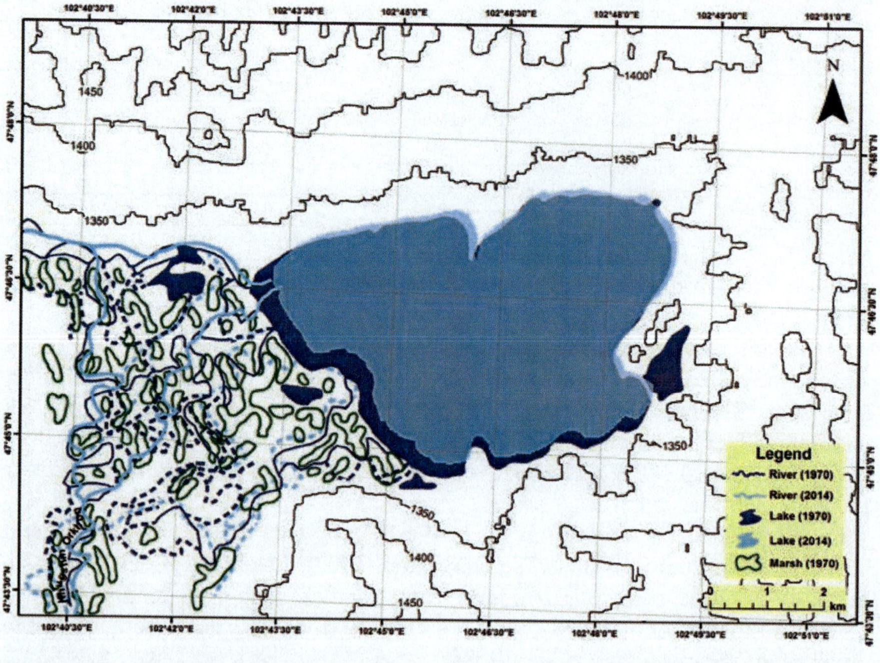

Fig. 20.6 Spatial distributions of areas and marshes of Lake Ugii in 1970 and 2014. Modified from Orkhonselenge and Uuganzaya (unpublished)

despite that summer precipitation is also rising since 1985 (Fig. 20.4d). The isolated small lakes in the east, southwest, and west of the lake disappeared by 2014 (Fig. 20.6). It is shown that Lake Ugii rapidly responds to the air temperature rise rather than precipitation. This trend is also observed for Lake Yakhi in the Dornod plain of eastern Mongolia (Orkhonselenge & Bulgan, 2021; see Chap. 6) and Lake Buun Tsagaan in the Govi[7] region of southern Mongolia (see Chap. 10).

[7] *Govi* has been often spelled as *Gobi* in the international literature. Govi is the correct English transliteration from Mongolian *Говь*.

The linear regression analysis (Fig. 20.7) shows that the lake area between 2000 and 2014 was weakly related to the annual average air temperature ($R^2 = 0.3458$) and precipitation ($R^2 = 0.2095$). The correlation analysis shows that the lake area during 2000–2014 had a moderate negative correlation with the air temperature ($r = -0.588$) and a weak negative correlation with the precipitation ($r = -0.457$). However, the periodically limited data prevent further trend analysis and inference for the lake.

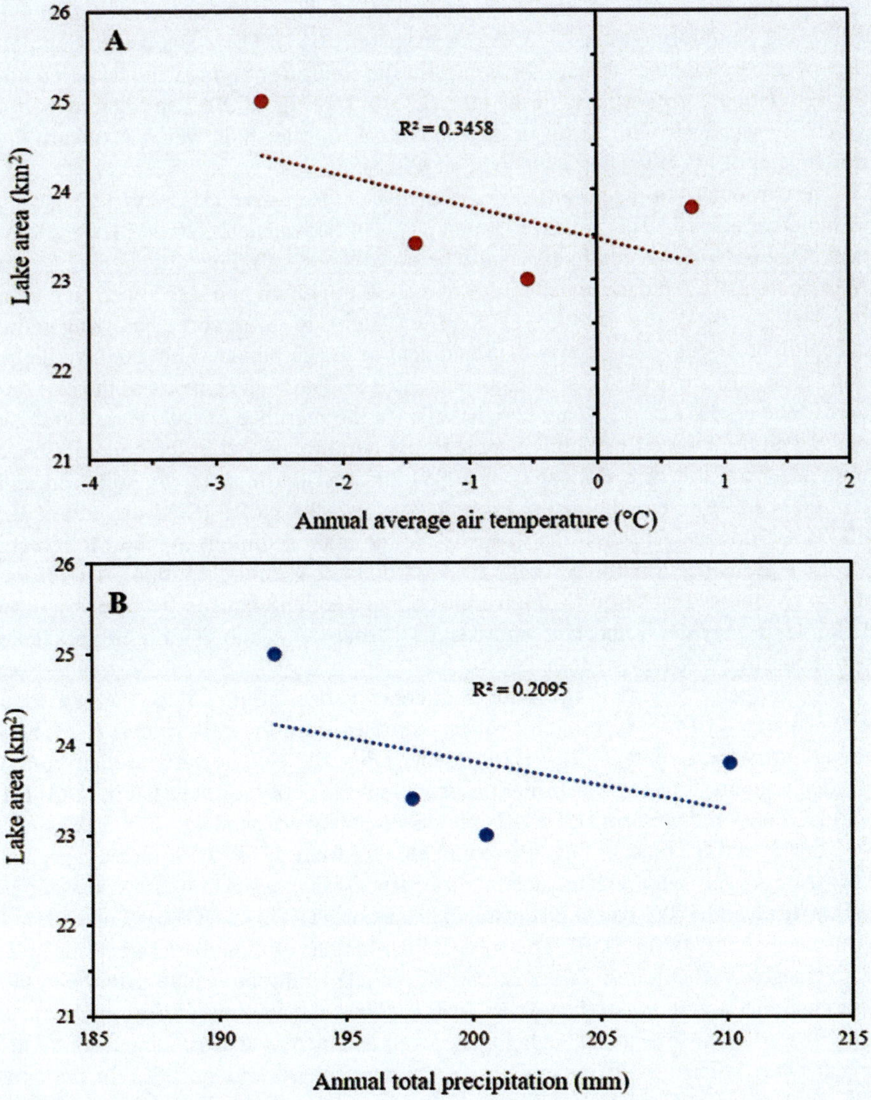

Fig. 20.7 Relationships between area of Lake Ugii and annual average (**a**) air temperature and (**b**) precipitation during 2000–2014

20.7 Sedimentological Review

The physical and chemical properties of sediments in the cores UG11-1 to UG11-4 (Fig. 20.2b) obtained from Lake Ugii indicate the sedimentary environments of the lake as the prairie type (Davaagatan, 2014; Davaagatan et al., 2015).

The water content of the core sediments is 15% on average. It increases gradually from 2.5 to 20% above 25 cm depth, and a continuous stable phase appears at approximately 20% below 25 cm depth (Davaagatan, 2014).

The organic matter of the core sediments changes rapidly with the depth (Fig. 20.8), and it ranges at 12–25% (Davaagatan, 2014). There is a decreasing pattern of the organic matter with the depth for the shallower parts of the lake in comparison with the deeper parts of the lake (UG11-1 in Fig. 20.8). The organic matter generally increases with depth in the direction from the shallower part toward the deeper part of the lake (Davaagatan et al., 2015).

The carbonates of the core sediments are 5% in the core UG11-1 and 12.5% each in the other cores UG11-2, UG11-3, and UG11-4 (Davaagatan, 2014). The carbonates decrease with the depth, especially below 20 cm depth. The latter cores contain high contents of the carbonates >12% above 20 cm depth and <12% below 20 cm depth (Fig. 20.8). The core UG11-1 differs from the other cores, showing more abrupt fluctuations, particularly in the content of the carbonates above 30 cm depth and below ~70 cm depth (Davaagatan et al., 2015). The high contents of the carbonates in the cores (Fig. 20.8) are consistent with the presence of calcite-rich grayish clays with the mineral gypsum above 110 cm depth observed in the core UN99C1 obtained from the lake (Khosbayar, 2005) and constant calcite precipitation and dominance of pyrite and gypsum in the lake at 4.2–2.8 ka BP (Schwanghart et al., 2008). Carbonaceous loess-like deposits are the main sediments of the Mongolian steppe regions (Lehmkuhl & Lang, 2001; Grunert & Lehmkuhl, 2004). Therefore, direct sediment deposition through transport of aeolian deposits by overflow into Lake Ugii may be important sources of carbonate accumulation in this lake (Orkhonselenge & Davaagatan, 2016).

The biogenic silica of the core sediments varies from 7.5 to 12% in total (Davaagatan, 2014). The biogenic silica increases from the eastern part (UG11-1) toward the western part (UG11-4) of the lake (Fig. 20.2b). The vertical distribution of the biogenic silica shows similar patterns with the depth for the eastern (UG11-1 and UG11-2) and western (UG11-3 and UG11-4) parts of the lake (Fig. 20.8).

The mineral fractions of the core sediments vary from 65 to 72.5% on average, and they increase generally with the depth. For instance, mineral fractions are <70% above 20 cm depth and >70% below 20 cm depth in the cores UG11-2, UG11-3, and UG11-4 (Davaagatan, 2014; Fig. 20.8). The vertical distributions of the mineral fractions indicate unstable patterns in the deepest part of the lake and more stable patterns in the marginal shallower parts of the lake (Fig. 20.8). The large amounts of the mineral fractions in the lake sediments reflect high lake levels and increased detrital input contributing for the abundances of minerals illite, chlorite, calcite, and smectite for the core UN99C1 at 4450 ^{14}C year BP (Khosbayar, 2005) and the current state of the thermally stratified water body (Schwanghart et al., 2008) consisting stratigraphically of fluvial sands at the bottom and clay to silty and organic poor loam in the overlying sediments.

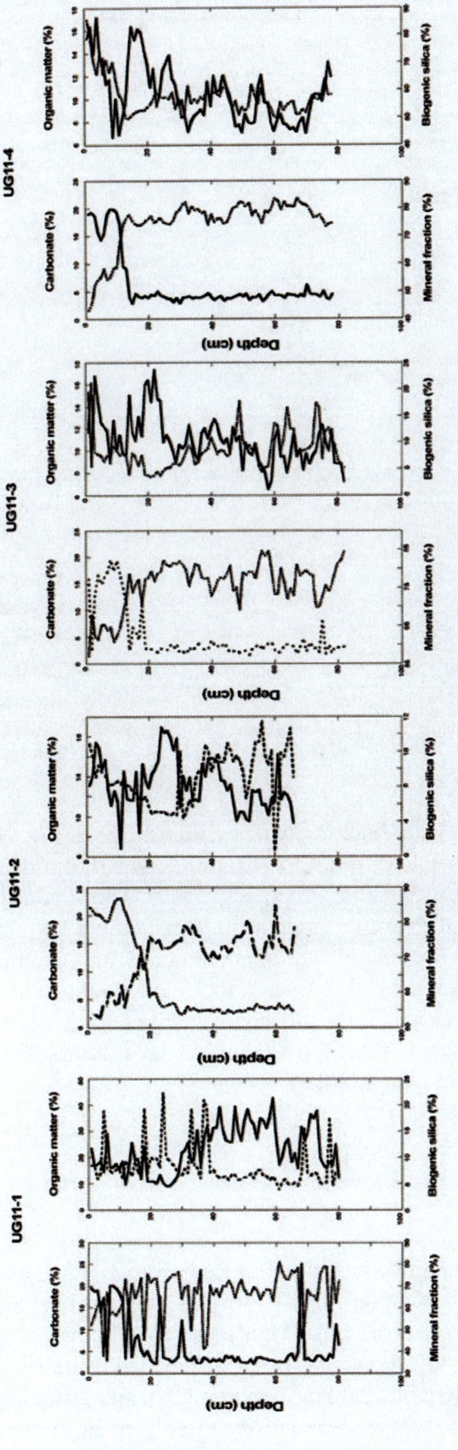

Fig. 20.8 Variations in physical and chemical components of Lake Ugii sediments. Modified from Davaagatan (2014)

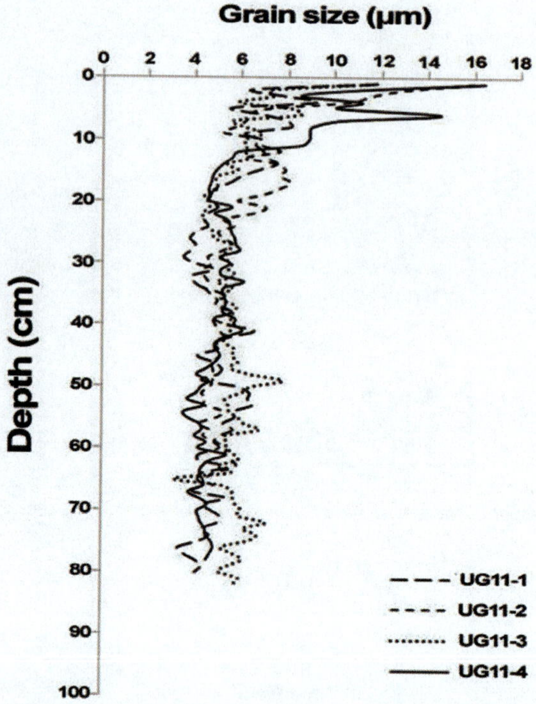

Fig. 20.9 Grain size distribution of Lake Ugii sediments. Modified from Davaagatan (2014)

The sediments shift from finer to coarser with the depth (Fig. 20.9). Grain size of the lake sediments ranges at ~6 µm to ~16 µm above 10 cm depth and at 4–8 µm below 10 cm depth (Davaagatan, 2014; Fig. 20.9). The spatial distribution of the grain sizes shows that the lake sediments are finer in the west than those in the east of the lake. The coarser sediments are in agreement with a gray clayey silt layer in the lake with median sizes ranging from 5 to 15 µm (Wang et al., 2011). The large amounts of the mineral fractions and coarse sediments imply the dominance of ephemeral or intermittent short-term, rapid, heavy rainfalls in the lake basin (Orkhonselenge & Davaagatan, 2016).

20.8 Summary

Lake Ugii in central Mongolia is a hydrologically important lake in the North Arctic Ocean drainage basin through the fluvial activities of Khugshin Orkhon, Orkhon, and Selenge rivers and Lake Baikal. Lake Ugii represents the hydrological dynamics of central Mongolia today by showing how rapid the lake responds to climate change. Lake Ugii has lost an area of ~2.0 km^2 between 1970 and 2014. The drop of the lake area shows that the lake was sensitive to the rapidly rising summer air temperature rather than the gradually rising precipitation during the last 35 years. The linear regression analysis shows that the lake area between 2000 and 2014 was weakly related to the annual average air temperature (R^2 = 0.3458) and precipitation (R^2 = 0.2095). The correlation analysis shows that the lake area during 2000–2014 had a moderately negative correlation with the air temperature (r = −0.588) and a weak negative correlation with the precipitation (r = −0.457). However, the short-term data are insufficient for further inference. Lake Ugii sediments indicate that the lake is a prairie-type within an arid environment characterized by high contents of carbonates and minerals and coarse sediments. The physical and chemical properties of the Lake Ugii sediments indicate the impact of the arid climate on the lake sedimentation. The stratigraphic records from Lake Ugii suggest that the lake has experienced the climate and environmental changes in semiarid central Mongolia in the past.

References

Battarbee, R. W. (2000). Palaeolimnological approaches to climate change, with special regard to the biological record. *Quaternary Science Review, 19*(1–5), 107–124.

Davaa, G. (2015). *Resource and regime of surface water of Mongolia*. 408 p. [In Mongolian].

Davaagatan, T. (2014). *Reconstructing Holocene climate change in the steppe region of Mongolia based on transportation and deposition of lake sediments: The cases of Lakes Terkhiin Tsagaan and Ugii*. MSc Thesis, National University of Mongolia, Ulaanbaatar. [In Mongolian].

Davaagatan, T., Orkhonselenge, A., Fukushi, K., & Yabe, M. (2015). Sedimentary records in intermontane and prairie lakes in central Mongolia: Preliminary results from Lake Terkhiin Tsagaan and Lake Ugii. *Journal of Geographical Issues, 14*(408), 19–26.

Fowell, S. J., Hansen, B. C., Peck, J. A., Khosbayar, P., & Ganbold, E. (2003). Mid to late Holocene climate evolution of the Lake Telmen basin, north central Mongolia, based on palynological data. *Quaternary Research, 59*(3), 353–363.

Grunert, J., & Lehmkuhl, F. (2004). Aeolian sedimentation in arid and semi-arid environments of Western Mongolia. In W. Smykatz-Kloss & P. Felix-Henningsen (Eds.), *Paleoecology of quaternary drylands* (pp. 195–218). Springer.

Gunin, P. D., Vostokova, E. A., Dorofeyuk, N. I., Tarasov, P. E., & Black, C. C. (1999). *Vegetation dynamics of Mongolia*. Kluwer.

Hilbig, W. (1995). *The vegetation of Mongolia*. SPB Academic Publishing.

Jigj, S. (1975). *Primary feature of Mongolian landforms*. Institute of Geography & Permafrost, Mongolian Academy of Sciences (MAS), 126 p. [In Mongolian].

Kang, S., & Hong, S. Y. (2016). Assessing Seasonal and Inter-Annual Variations of Lake Surface Areas in Mongolia during 2000–2011 Using Minimum Composite MODIS NDVI. *PLoS One, 11*(3), e0151395.

Kang, S., Lee, G., Togtock, C., & Jang, K. (2015). Characterizing regional precipitation-driven lake area change in Mongolia. *Journal of Arid Land, 7*(2), 146–158.

Khosbayar, P. (2005). *Mezozoic and Cenozoic paleogeography and paleoclimate of Mongolia. Transaction 15*. Institute of Geology and Mineral Resources, Mongolian Academy of Sciences, Ulaanbaatar. [In Mongolian].

Lehmkuhl, F., & Lang, A. (2001). Geomorphological investigations and luminescence dating in the southern part of the Khangay and the Valley of the Gobi Lakes (Central Mongolia). *Journal of Quaternary Science, 16*, 69–87.

Murzaev, E. M. (1952). *The description of physical geography of Mongolia* (2nd ed.). [In Russian].

Orkhonselenge, A. (2021, December 13–17). *Climate-driven channel change of Orkhon River, Mongolia*. AGU 2021 Fall Meeting. Presentation #1193. New Orleans, LA. https://doi.org/10.1002/essoar.10508953.1.

Orkhonselenge, A., & Bulgan, O. (2021). Geochemical studies and lacustrine geomorphology of Lake Yakhi basin in eastern Mongolia. *Géomorphologie: Relief, Processes, Environnement, 27*(3), 231–242. https://doi.org/10.4000/geomorphologie.15873

Orkhonselenge, A., & Davaagatan, T. (2016). *Sedimentary feature in Lake Terkhiin Tsagaan and Lake Ugii, Central Mongolia.* Geophysical Research Abstracts Vol. 18, EGU2016-5666-2, EGU General Assembly 2016.

Orkhonselenge, A., Komatsu, G., Davaagatan, T., & Uuganzaya, M. (2019). Sedimentary records from Lakes Reconstructing Late Quaternary Paleoclimate Changes in Mongolia. In *XX INQUA Congress. Presentation #2419*. Dublin, Ireland.

Orkhonselenge, A., Komatsu, G., & Uuganzaya, M. (2018). Climate-driven changes in lake areas for the last half century in the Valley of Lakes, Govi region, southern Mongolia. *Natural Science, 10*(7), 263–277.

Peck, J. A., Khosbayar, P., Fowell, S. J., Pearce, R. B., Ariunbileg, S., Hansen, B. C. S., & Soninkhishig, N. (2002). Mid to Late Holocene climate change in north central Mongolia as recorded in the sediments of Lake Telmen. *Palaeogeography, Palaeoclimatology, Palaeoecology, 183*, 135–153.

Schwanghart, W., Schütt, B., & Walther, M. (2008). Holocene climate evolution of the Ugii Nuur basin, Mongolia. *Advances in Atmospheric Sciences, 25*(6), 986–998.

Schwanghart, W., Frechen, M., Kuhn, N. J., & Schütt, B. (2009). Holocene environmental changes in the Ugii Nuur basin, Mongolia. *Palaeogeography, Palaeoclimatology, Palaeoecology, 279*, 160–171.

Sladen, C., & Traynor, J. J. (2000). Lakes during the evolution of Mongolia. In E. H. Gierlowski-Kordesch & K. R. Kelts (Eds.), *Lake basins through space and time: Studies in geology* (Vol. 46, pp. 35–57). American Association of Petroleum Geologists.

Szumińska, D. (2016). Changes in surface area of the Böön tsagaan and Orog lakes (Mongolia, Valley of the lakes, 1974–2013) compared to climate and permafrost changes. *Sedimentary Geology*, 1–37.

Tarasov, P. E., Guinot, J., Cheddadi, R., Andreev, A. A., Bezusko, L. G., Blyakharchuk, T. A., Dorofeyuk, N. I., Filimonova, L. V., Volkova, V. S., & Zernitskaya, V. P. (1999). Climate in northern Eurasia 6000 years ago reconstructed from pollen data. *Earth and Planetary Science Letters, 171*, 635–645.

Tian, F., Herzschuh, U., Dallmeyer, A., Xu, Q., Mischke, S., & Biskaborn, B. K. (2013). Environmental variability in the monsoonewesterlies transition zone during the last 1200 years: Lake sediment analyses from central Mongolia and supraeregional synthesis. *Quaternary Science Reviews, 73*, 31–47.

Tsegmid, S. (1969). *Physical geography of Mongolia.* 759 p. [In Mongolian].

Tserensodnom, J. (1971). *Lakes of Mongolia.* State Publishing, 202 p. [In Mongolian].

Tserensodnom, J. (2000). *Catalogy of lakes in Mongolia.* Shuvuun Saaral Publishing. [In Mongolian].

Tuvshinjargal, D. (2001). *Thermal regime of Lake Ugii.* Doctorate Thesis in Geography: Hydrology. Institute of Geography, Mongolian Academy of Sciences, Ulaanbaatar. 93 p. [In Mongolian].

Walther, M., & Gegeensuvd, T. (2005). *Ugii Nuur (Central Mongolia) Paleo Environmental Studies of Lake Level Fluctuations and Holocene Climate Change. Geographica-Geokologica,* 2. MOLARE Research Centre.

Wang, W., Ma, Y., Feng, Z., Meng, H., Sang, Y., & Zhai, X. (2009). Vegetation and climate changes during the last 8660 cal. a BP in central Mongolia, based on a high-resolution pollen record from Lake Ugii Nuur. *Chinese Science Bulletin, 54*(9), 1579–1589.

Wang, W., Ma, Y., Feng, Z., Narantsetseg, T., Liu, K. B., & Zhai, X. (2011). A prolonged dry mid-Holocene climate revealed by pollen and diatom records from Lake Ugii Nuur in central Mongolia. *Quaternary International, 229*, 74–83.

Yatagai, A., & Yasunari, T. (1995). Interannual variations of summer precipitation in the arid/semi-arid regions in China and Mongolia: Their regionality and relation to the Asian summer monsoon. *Journal of Meteorology Society, 73*, 909–923.

Chapter 21
Paleoclimatic Patterns Recorded in the Lakes of Mongolia

Abstract The lakes of Mongolia offer a reconstructive archive of paleoclimate changes extending back to the late Pleistocene with excellent natural conditions preserved due to the limited human impact in the country. The paleoclimatic archives such as sedimentary stratigraphic sequences stored in the lakes provide information needed for reconstructing spatial and temporal patterns of late Quaternary paleoclimate changes in Mongolia, Eurasia, and Central Asia during the late Pleistocene and Holocene. In this context, comparison with other records from adjacent regions in Eurasia and Central Asia is important. The spatial and temporal patterns of the paleoclimate changes show that in the late Pleistocene, the cold and dry climate dominated in western Mongolia, whereas the cold and wet climate dominated in northern Mongolia. In the Holocene, paleoclimate changes of Mongolia differed from region to region where lake basins were driven by local atmospheric circulations. In the early Holocene, the warm and wet climate prevailed in southern, western, and central Mongolia. Meanwhile, in northern Mongolia, the cold and wet climate persisted in the Darkhad and Khuvsgul basins, and the cold and dry climate prevailed in the downstream of the Selenge River basin and the intermontane valley between the Khuvsgul and Khentii Mountain Ranges. In the middle Holocene, the warm and dry climate prevailed in northern (Khuvsgul basin), western (Mongolian Altai Mountain Range), central, and southern Mongolia, whereas the warm and wet climate dominated in western Mongolia (Great Lakes Depression) and northern Mongolia (Darkhad basin and downstream of the Selenge River basin and the intermontane valley between the Khuvsgul and Khentii Mountain Ranges). In the late Holocene, the warm and wet climates prevailed in southern, western, and central Mongolia. Moreover, in northern Mongolia, the cool and dry climates prevailed in the Darkhad basin, downstream of the Selenge River basin and the intermontane valley between the Khuvsgul and Khentii Mountain Ranges, whereas the cool and wet climate dominated in the Khuvsgul basin. Overall, the warm and arid climate has been prevailing throughout Mongolia in the Anthropocene.

Keywords Lakes · Paleoclimate · Climate change · Pleistocene · Holocene · Anthropocene · Mongolia

A. Orkhonselenge et al., *Lakes of Mongolia*, Syntheses in Limnogeology,
https://doi.org/10.1007/978-3-030-99120-3_21

21.1 Introduction

Understanding the formation and evolution of lakes in Mongolia (see Chap. 3) is one of the essential acts necessary for reconstructing paleoclimate changes in Mongolia, Eurasia, and Central Asia. Many modern lakes in Mongolia respond to climate changes in dramatic ways (see Chaps. 6–20). For instance, some lakes in western (Mongolian Altai Mountain Range), northern (Khuvsgul[1]) Mountain Range), and central (Khangai[2] Mountain Range) Mongolia have expanded in area, whereas some lakes in eastern, southern, and western Mongolia (Govi[3] regions, especially the Valley of the Lakes and Great Lakes Depression) have faced serious decreases in area due to climate changes occurring during the last half century (Orkhonselenge et al., 2019; Orkhonselenge & Bulgan, 2021).

Sedimentary stratigraphic archives from lakes provide abundant information on how the lakes responded to the paleoclimate changes over geological timescales. Lake deposits are important resources because they are reckoned to preserve detailed records of paleoenvironmental conditions, especially the paleoclimate in and around the lake (Bridge & Demicco, 2008). Lakes play important roles in the reconstruction of paleoclimate changes because they are stable storages for sediments while also participating in the fluvial system with inflows and outflows. Lake sediments provide valuable proxy data imprinting impacts of local and regional climate changes on geomorphic evolution and hydrological dynamics in lake basins (Orkhonselenge et al., 2015, 2019).

Mongolia is considered to play an important role in the climate system over Eurasia and Central Asia because it occurs in the intercontinental position of Eurasia at the junction of three major large-scale climatic systems, the Siberian-Mongolian high-pressure anticyclone or East Asian Winter Monsoon (EAWM), Asian low-pressure cyclone, and westerly atmospheric circulations (Orkhonselenge et al., 2015, 2019). Mongolia is characterized by an extremely dry extra-continental climate developing under the interaction of the Siberian high- and Asian low-pressure cells and the westerlies (Hilbig, 1995; Gong & Ho, 2002; Panagiotopoulos et al., 2005), which is modulated by the North Atlantic Oscillations (NAO) (Visbeck, 2002). Arid and semiarid Asia, including semiarid Mongolia, represents one of the areas that are most sensitive to the large-scale climate changes (e.g., Feng et al., 1993; D'Arrigo et al., 2000; Jacoby et al., 2000).

---

[1] *Khuvsgul* has been misspelled as *Hovsgol* and *Khubsugul* in publications. *Khuvsgul* is the right English transliteration from Mongolian *Хөвсгөл*.

[2] *Khangai* has been misspelled as *Khangay* (or *Hangay*) and *Hangai* in many publications. Khangai is the right English transliteration from Mongolian *Хангай*.

[3] *Govi* has been often spelled as *Gobi* in the international literature. Govi is the correct English transliteration from Mongolian *Говь*.

Mongolia also raises particular interest as one place holding the valuable record of paleoclimate changes in Eurasia and Central Asia due to its highly elevated intercontinental location on the Central Asian Orogenic Belt (CAOB) occupying a transitional area between the Siberian taiga forest in the north and the Govi sandy desert in the south, where the Mongolian high-pressure system (MHPS) develops over the region (An et al., 2008). The MHPS has a strong influence on the winter climate of Central East Asia (Gong et al., 2001) which is in close relation to the NAO and North Pacific Oscillations (NPO) (Kerr, 1999; Gong et al., 2001; Hoerling et al., 2001). Paleoclimate change in Mongolia was a function of complicated regional responses to atmospheric conditions (Wang & Feng, 2013).

Numerous studies on reconstruction of paleoclimate and paleoenvironmental changes in Mongolia have been conducted (see Chap. 2), and these reconstructions were dominantly based on lake sediments (e.g., Dorofeyuk & Tarasov, 1998; Prokopenko et al., 2007; Mischke et al., 2020) to comparing with paleomagnetic records (e.g., Fedotov et al., 2004; Khosbayar, 2005), aeolian sediments (e.g., Feng et al., 2007; Felauer et al., 2012), fluvial sediments, paleosols (e.g., Lehmkuhl et al., 2012), glacier extents (e.g., Grosswald & Rudoy, 1996; Blomdin et al., 2018), permafrost, peats (e.g., Fukumoto et al., 2012; Orkhonselenge et al., 2021), tree ring records (e.g., Davi et al., 2013), and archeological traces (e.g., Rosen et al., 2019).

Reconstructing the past climate changes in Mongolia is a major contribution for predicting future climate changes in Eurasia and Central Asia (Orkhonselenge et al., 2015, 2019). To date, there has been only limited synthesized analytic picture presented in regard to the local climate changes in Mongolia reconstructed from high-resolution multi-proxy records of lake sediments. The lakes of Mongolia enable expanding the reconstructive records, possibly all the way up to the Mesozoic Era because of prevalence or stabilization of the lakes since the Triassic through the Mesozoic Era (see Chap. 3). However, the present high-resolution proxy data archived in lakes of Mongolia have been extended only to the Quaternary, especially the late Pleistocene, due to (1) the delayed development of limnology in Mongolia (see Chap. 2), (2) still insufficient paleoclimate data obtained from lakes, and (3) a lack of linkage of lake study with other fields in Earth science (e.g., geochronology, sedimentology, paleoclimatology, etc.) for interdisciplinary advances (see Chap. 22).

Relatively few proxy series of acceptable quality reflecting paleoclimate changes have been obtained from Mongolian lakes today (e.g., Grunert et al., 2000; Peck et al., 2002; Prokopenko et al., 2007; Wang et al., 2011; Narantsetseg et al., 2013; Lehmkuhl et al., 2018a; Barna et al., 2021). A general picture of the Holocene climate changes in Mongolia is still open to debate although some general climate changes recorded in lakes are reaching consensus, for instance, wet early Holocene, dry middle Holocene (e.g., Krivonogov et al., 2003; Prokopenko et al., 2005; An et al., 2008; Orkhonselenge et al., 2013), and humid (e.g., An et al., 2008) or arid (e.g., Orkhonselenge et al., 2018) late Holocene.

Deciphering climate systems prevailed in each region of Mongolia (Fig. 1.1) over geological timescales has not been advanced sufficiently to date (Orkhonselenge et al., 2015). Nevertheless, some studies have tried to understand the general patterns of paleo- and modern climate systems in Mongolia (Orkhonselenge et al., 2015, 2019). For instance, climate change differs within the country during the Holocene: the humid early Holocene characterized by the increasing temperature and humidity, the dry middle Holocene identified by the enhanced aridity, and the humid late Holocene marked by the dropped temperature and decreased evaporation (An et al., 2008). Moreover, western Mongolia has been under the influences of the East Asian Summer Monsoon (EASM) and the NAO climates during the Holocene (Sun et al., 2013), whereas eastern Mongolia has been influenced by the EASM at the transition time from the late Pleistocene to the early Holocene (see Chap. 6). The early works can be significantly improved by deepening the understanding of local and regional climatic variations in Mongolia.

Hence, in this chapter, the local spatial and temporal patterns of paleoclimate changes recorded in lakes of Mongolia are synthesized and reviewed (Tables 21.1, 21.2, 21.3, 21.4, and 21.5), and these records are compared with those from adjacent regions in Eurasia and Central Asia (Fig. 21.1; Table 21.6). In addition, the most

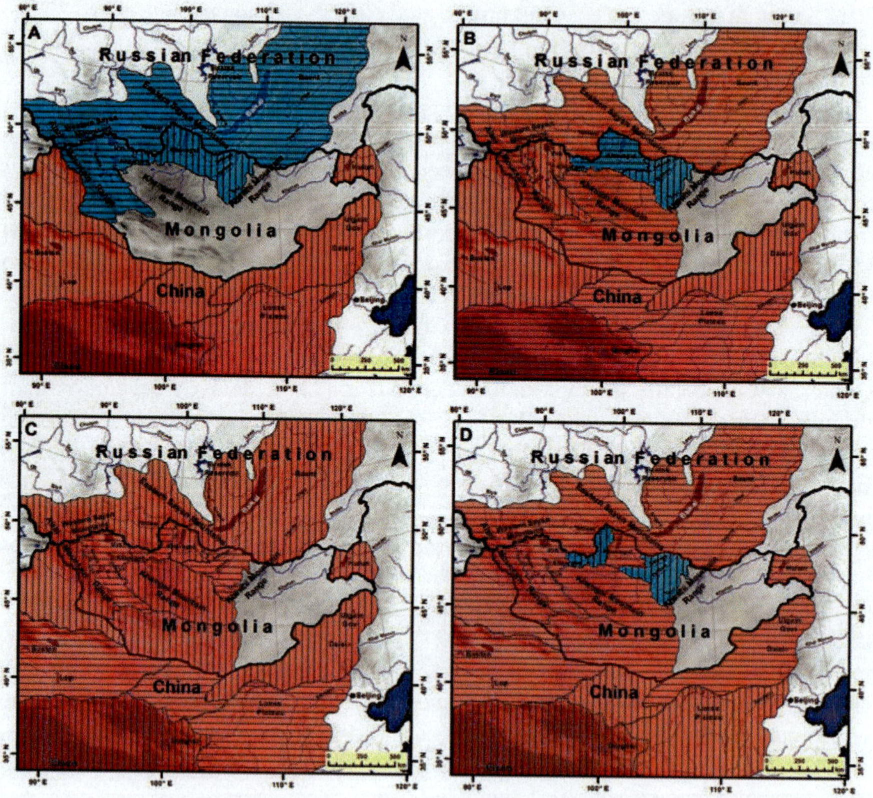

Fig. 21.1 Synthesized spatial patterns of paleoclimate changes recorded in lakes of Mongolia and adjacent areas of southern Russia and northern China during the **(a)** late Pleistocene, **(b)** early Holocene, **(c)** middle Holocene, and **(d)** late Holocene: cold (blue) and warm (red) climates with arid (vertical) and humid (horizontal) periods. These paleoclimate conditions are with respect to the present climate. After Orkhonselenge et al. (2019)

Table 21.1 Summary of paleoclimate changes reconstructed from Mongolian lakes during the Quaternary

#	Regions	Lakes	Description	Late Pleistocene (>11.7 ka BP)	Early Holocene (11.7–10.5 ka BP)	Middle Holocene (7.5–3.5 ka BP)	Late Holocene (<3.5 ka BP)
1	Eastern Mongolia	Buir	Chapter 7	–	–	–	Warm and humid
		Khagiin Khar	Chapter 6	Cold to mild	–	–	Humid
		Khukh	Chapter 8	–	–	–	Warm and arid
		Yakhi	Chapter 6	–	–	–	Warm and arid
2	Southern Mongolia	Adgiin Tsagaan	Chapter 9	Humid to arid	Humid	–	Humid
		Bayan Tukhum	Chapter 9	Humid to arid	Warm and humid	Humid	Arid
		Buun Tsagaan	Chapter 10	Humid to arid	Humid	Humid	Humid
		Orog	Chapter 9	Humid to arid	Humid	Humid	Humid
		Ulaan	Chapter 11	Arid to humid	Warm and humid	Humid	Arid
3	Western Mongolia	Achit	Chapter 12	Mild and arid	Mild and arid	Warm and humid	Warm and humid
		Bayan	Aeolian	Cold and arid to cold and humid	Humid	Warm and humid to cold and arid	Warm and humid
		Dayan	Glacial	–	–	–	Humid to arid
		Duruu	Freshwater	–	–	Arid	Warm and arid to cold and arid
		Ekhen	Tectonic	Humid	–	–	–
		Ereen	Aeolian	Humid	Humid	Humid to arid	Arid
		Khar	Glacial	Cold	–	–	Humid
		Khoton	Chapter 14	Cold and arid	Warm and humid	Warm and humid to warm and arid	Cold and semiarid
		Khurgan	Chapter 14	Humid	–	–	–

(continued)

Table 21.1 (continued)

#	Regions	Lakes	Description	Late Pleistocene (>11.7 ka BP)	Early Holocene (11.7–10.5 ka BP)	Middle Holocene (7.5–3.5 ka BP)	Late Holocene (<3.5 ka BP)
		Tsetseg	Chapter 12	Humid to arid	Humid	Arid	Humid to arid
		Uvs	Chapter 13	Cold and arid to cold and humid	Humid	Warm and humid to cold and arid	Warm and humid
		Zagas	Freshwater	–	–	Arid	Warm and arid to cold and arid
4	Northern Mongolia	Darkhad	Chapter 15	Cold and arid	Cold and humid to arid	Humid	Warm and arid
		Dood	Chapter 15	Cold and arid	Cold & humid	Warm and humid to cold and arid	Cold and arid
		Erkhel	Saline	Cold	Warm and humid	Arid	Arid
		Khargal	Chapter 17	–	–	–	Warm and arid
		Khuvsgul	Chapter 16	Cold and arid	Cold and humid	Warm and arid	Humid to arid
		Targan	Chapter 15	–	–	–	Warm and humid
5	Central Mongolia	Gun	Tectonic	Arid	Cold and arid	Warm and arid	Warm and humid
		Khuis	Glacial	–	–	–	Arid to humid
		Telmen	Chapter 18	–	–	Arid	Humid
		Terkhiin Tsagaan	Chapter 19	–	–	Cold and arid	Warm and humid
		Ugii	Chapter 20	Cold and arid	Warm and humid	Warm and arid	Warm and humid

Table 21.2 Paleoclimate changes reconstructed from Mongolian lakes during the late Pleistocene

MIS 6	MIS 5	MIS 4	MIS 3	MIS 2	BA	YD			
191–130 ka BP	130–71 ka BP	71–57 ka BP	57–29 ka BP	29.0–14.7 ka BP	14.7–12.9 ka BP	12.9–11.7 ka BP	Lakes	Regions	Details
–	–	–	–	Cold to mild	Mild	–	Khagiin Khar	Eastern Mongolia	Oh et al. (2019)
–	–	–	Humid	Cold and arid	–	–	Adgiin Tsagaan	Southern Mongolia	Komatsu et al. (2001)
–	–	–	–	–	Humid	Arid	Bayan Tukhum		Felauer et al. (2012)
Arid	Humid	Humid	Humid	Arid	Arid	Arid	Buun Tsagaan		Lehmkuhl et al. (2018a)
–	–	–	Humid	Cold and arid	–	–			Komatsu et al. (2001)
Cold and arid	Humid	Humid	Humid	Humid to arid	Arid	Arid	Orog		Lehmkuhl and Lang (2001); Lehmkuhl et al. (2018a); Yu et al. (2019)
–	–	–	Humid	Cold and arid	Cold and humid	–			Komatsu et al. (2001)
–	Humid	–	Humid	–	–	–			Nottebaum et al. (2021)
Humid	–	–	–	Warm and humid	Cold and arid	Cold and arid	Ulaan		Lee et al. (2013)
–	–	–	–	Cold and arid	Cold and arid	Humid			Mischke et al. (2020)

(continued)

Table 21.2 (continued)

MIS 6	MIS 5	MIS 4	MIS 3	MIS 2	BA	YD			
191–130 ka BP	130–71 ka BP	71–57 ka BP	57–29 ka BP	29.0–14.7 ka BP	14.7–12.9 ka BP	12.9–11.7 ka BP	Lakes	Regions	Details
–	–	–	–	–	Humid	–	Achit	Western Mongolia	Lehmkuhl and Haselein (2000)
–	–	–	–	Cold and humid	Cold and humid to mild and arid	Mild and arid			Sun et al. (2013)
–	–	Warm and humid	Warm and humid	Cold and arid	Cold and arid	Cold and humid	Bayan		Grunert et al. (2000); Lehmkuhl and Haselein (2000)
–	–	–	–	Humid	Humid	Humid	Ekhen		Klinge et al. (2021)
–	–	–	Humid	–	–	–	Ereen		Yang et al. (2004); Grunert et al. (2009)
Humid	Humid	–	–	–	–	–			Stolz et al. (2012)
–	–	–	–	Cold	–	–	Khar		Walther et al. (2017)
–	–	–	–	–	–	Arid	Khoton		Rudaya et al. (2009); Rudaya and Li (2013)
–	–	–	Humid	Humid	–	–			Klinge et al. (2021)
–	Humid	Humid	–	–	–	–	Khurgan		Klinge et al. (2021)
–	–	–	Arid	Arid	Humid	Arid	Tsetseg		Klinge and Lehmkuhl (2013)
Humid	Humid	Warm and humid	Warm and humid	Cold and arid	Cold and arid	Cold and humid	Uvs		Grunert et al. (2000); Lehmkuhl and Haselein (2000)

Table 21.2 (continued)

MIS 6	MIS 5	MIS 4	MIS 3	MIS 2	BA	YD			
191–130 ka BP	130–71 ka BP	71–57 ka BP	57–29 ka BP	29.0–14.7 ka BP	14.7–12.9 ka BP	12.9–11.7 ka BP	Lakes	Regions	Details
–	Cold	–	Cold	Cold	–	Cold	Darkhad	Northern Mongolia	Grosswald & Rudoy, 1996; Krivonogov et al. (2005); Komatsu et al. (2009)
–	–	–	–	–	–	Warm and arid			Krivonogov et al. (2012a)
–	–	–	Cold	Cold	–	–			Gillespie et al. (2008), Komatsu et al. (2009)
–	–	–	–	–	–	Cold and arid	Dood		Narantsetseg et al. (2013)
Humid	Humid	Humid	–	–	–	–			Lehmkuhl and Haselein (2000)
–	–	–	–	Cold	Cold	Cold	Erkhel		Watanabe et al. (2007)
–	–	–	–	–	Arid	Arid	Khuvsgul		Fedotov et al. (2004)
–	–	–	–	–	–	Warm and humid			Karabanov et al. (2004)
–	Cold and humid	–	–	–	–	–			Kashiwaya et al. (2010)
Cold	Warm	Cold	Warm	Cold and arid	–	–			Fedotov et al. (2015)
–	–	–	–	Arid	Humid	–			Prokopenko et al. (2005)
–	–	–	–	–	–	Humid			Orkhonselenge et al. (2013)
Arid	Arid	Arid	Arid	Arid	Arid	Arid			Prokopenko et al. (2009)
–	–	–	–	–	–	Arid	Gun	Central Mongolia	Zhang et al. (2012)
–	–	–	–	–	–	Cold and arid	Ugii		Schwanghart et al. (2009)

Table 21.3 Paleoclimate changes reconstructed from Mongolian lakes during the early Holocene

11.7–10.5 ka BP	10.5–9.5 ka BP	9.5–8.5 ka BP	8.5–7.5 ka BP	Lakes	Regions	Details
Warm and humid	Warm and humid	Warm and humid	Cold and arid	Bayan Tukhum	Southern Mongolia	Felauer et al. (2012)
–	–	–	Humid	Adgiin Tsagaan		Lehmkuhl and Lang (2001)
Warm and humid	Warm and humid	Warm and humid	Warm and humid			Komatsu et al. (2001)
Warm and humid	Warm and humid	Warm and humid	Warm and humid	Buun Tsagaan		Komatsu et al. (2001)
Humid	Humid	Humid	Humid	Orog		Yu et al. (2019)
Warm and humid	Warm and humid	Warm and humid	Warm and humid			Komatsu et al. (2001)
Humid	Humid	Humid	Humid			Nottebaum et al., 2021
Humid	Humid	Humid	Humid	Ulaan		Lee et al. (2011, 2013)
Humid	Humid	Arid	Arid			Mischke et al. (2020)
Mild and arid	Mild and arid	Mild and arid	Mild and arid	Achit	Western Mongolia	Sun et al. (2013)
Humid	Humid			Bayan		Grunert et al. (2000)
Humid	Humid	Humid	Humid	Ereen		Stolz et al. (2012)
Humid	–	–	–	Khoton		Lehmkuhl and Haselein (2000)
Cold and arid	Warm and humid	Warm and humid	Warm and humid			Rudaya et al. (2009)
Cold and arid to cold and humid	Warm and humid	Warm and humid	Warm and humid			Rudaya and Li (2013)
Humid	Humid	Humid	Humid	Tsetseg		Klinge and Lehmkuhl (2013)
Humid	Humid	–	–	Uvs		Grunert et al. (2000); Yang et al. (2004)
Warm and arid	Warm and arid	Humid	Humid	Darkhad	Northern Mongolia	Krivonogov et al. (2012a)
Cold and arid	–	–	–			Gillespie et al. (2008)
Cold and arid	Cold and arid	Cold and humid	Cold and humid	Dood		Narantsetseg et al. (2013)
Warm and humid	Warm and humid	Warm and humid	Warm and humid	Erkhel		Watanabe et al. (2007)

(continued)

Table 21.3 (continued)

11.7–10.5 ka BP	10.5–9.5 ka BP	9.5–8.5 ka BP	8.5–7.5 ka BP	Lakes	Regions	Details
–	–	–	Warm and humid (7.8–7.6)	Khargal		Barna et al. (2021)
Humid	Humid	Humid	Humid	Khuvsgul		Fedotov et al. (2004); Prokopenko et al. (2007); Orkhonselenge et al. (2013)
Warm and arid	Warm and arid	Warm and arid	Warm and arid to warm and humid (7.7)	Gun	Central Mongolia	Wang et al. (2004)
Cold and arid	Cold and arid	Cold and arid	Cold and arid			Feng et al. (2005)
Arid	Humid	Humid	Humid			Zhang et al. (2012)
Cold and arid	Cold and arid	Cold and arid	Cold and arid to warm and humid			Feng et al. (2013)
–	Humid	Humid	Humid			Wang and Feng (2013)
–	Arid	Arid	Arid to humid (7.9)	Ugii		Schwanghart et al. (2008)
Humid	Humid	Humid	Humid			Schwanghart et al. (2009)
Mild and semi-humid	Mild and semi-humid	Mild and semi-humid	Mild and semi-humid (cold and humid at 8.4–8.3)			Wang et al. (2009)

Table 21.4 Paleoclimate changes reconstructed from Mongolian lakes during the middle Holocene

7.5–6.5 ka BP	6.5–5.5 ka BP	5.5–4.5 ka BP	4.5–3.5 ka BP	Lakes	Regions	Details
Humid (<7.0)	Humid	Humid (with cold and arid at 5.0–4.5 ka BP)	Humid (>4.0)	Bayan Tukhum	Southern Mongolia	Felauer et al. (2012)
Humid (<7.0)	Humid	Humid	Humid (>4.0)			Grunert et al. (2009)
Humid	Humid	Humid	Humid	Buun Tsagaan		Lehmkuhl et al. (2018a, 2018b)
Humid	Humid	Humid	Humid	Orog		Lehmkuhl et al. (2018a, 2018b)
Humid to arid (7.0)	Arid to humid (6.5 and 5.5)	Humid to arid (>5.0) and arid to humid (<5.0)	Humid			Nottebaum et al., 2021

(continued)

Table 21.4 (continued)

7.5–6.5 ka BP	6.5–5.5 ka BP	5.5–4.5 ka BP	4.5–3.5 ka BP	Lakes	Regions	Details
Humid	Humid	Humid	Humid	Ulaan		Lee et al. (2013)
–	Humid	Humid	Humid			Orkhonselenge et al. (2018)
Arid to humid (6.8)	Humid	Humid	Humid			Mischke et al. (2020)
Mild and arid	Warm and humid	Warm and humid	Warm and humid	Achit	Western Mongolia	Sun et al. (2013)
Warm and humid	Warm and humid	Cold and arid	Cold and arid	Bayan		Grunert et al. (2000)
–	–	–	Humid (4.3)	Dayan		Unkelbach et al. (2019)
Humid	Humid	Arid	Arid	Ereen		Grunert et al. (2009)
Humid	Humid	–	–			Stolz et al. (2012)
Warm and humid	Warm and humid	Arid (<5.0)	Arid	Khoton		Rudaya et al. (2009)
Cold and humid	Cold and humid to cold and arid (6.0)	Cold and arid	Cold and arid			Rudaya and Li (2013)
Arid	Arid	Arid	Arid	Tsetseg		Klinge and Lehmkuhl (2013)
Warm and humid	Warm and humid	Cold and arid (<5.0)	Cold and arid	Uvs		Grunert et al. (2000)
Arid	Humid	Humid	Arid	Darkhad	Northern Mongolia	Krivonogov et al. (2012a)
Warm and humid	Warm and humid to cold and arid (5.8)	Cold and arid	Cold and arid	Dood		Narantsetseg et al. (2013)
Warm	Warm	Warm	Warm	Erkhel		Watanabe et al. (2007)
Warm and humid (7.4–7.3) to arid (7.0)	Arid to warm and humid (6.2–6.1)	–	Warm and humid (4.4–4.0)	Khargal		Barna et al. (2021)
Humid	Humid	Arid	–	Khuvsgul		Fedotov et al. (2004)
Humid to arid (6.6)	–	–	–			Prokopenko et al. (2005)
Cold and humid	Cold and humid to warm and arid (6.0)	Warm and arid	Warm and arid			Prokopenko et al. (2007)
Arid	Arid	Arid	Arid			Orkhonselenge et al. (2013)

(continued)

Table 21.4 (continued)

7.5–6.5 ka BP	6.5–5.5 ka BP	5.5–4.5 ka BP	4.5–3.5 ka BP	Lakes	Regions	Details
Cold and humid	Cold and humid (5.2)	Cold and humid	Cold and humid	Gun	Central Mongolia	Wang et al. (2004)
Warm and humid	Warm and humid	Warm and humid	Warm and humid			Feng et al. (2005, 2013)
Humid to arid (7.0)	Arid to humid (5.7)	Humid	Humid to arid (4.1)			Zhang et al. (2012)
Humid to warm and arid	Warm and arid	Warm and arid	Warm and arid to humid			Wang and Feng (2013)
Arid	Arid	Arid	Humid	Telmen		Peck et al. (2002), Fowell et al. (2003)
Cold and arid	Cold and arid	Cold and arid	Cold and arid	Terkhiin Tsagaan		Davaagatan et al. (2015)
Humid	Humid	Humid	Humid to arid (4.2)	Ugii		Schwanghart et al. (2008)
Warm and humid	Warm and humid	Warm and humid	Humid to arid			Schwanghart et al. (2009)
Warm and arid	Warm and arid	Warm and arid	Warm and arid			Wang et al. (2009)
Arid	Arid	Arid	Arid			Davaagatan et al. (2015)

Table 21.5 Paleoclimate changes reconstructed from Mongolian lakes during the late Holocene and Anthropocene

3.5–2.5 ka BP	2.5–1.5 ka BP	1.5–0.5 ka BP	0.5–0.05 ka BP	<0.05 ka BP[a]	Lakes	Regions	Details
–	–	–	–	Warm and humid	Buir	Eastern Mongolia	Chapter 7
–	–	–	–	Warm and arid	Khukh		Chapter 8
–	–	–	Humid	–	Khagiin Khar		Oh et al. (2019)
–	–	–	–	Warm and arid	Yakhi		Orkhonselenge and Bulgan (2021), Chapter 6
Arid	Arid	Arid	Arid	–	Bayan Tukhum	Southern Mongolia	Felauer et al. (2012)
–	–	Humid	Humid	Warm and arid	Adgiin Tsagaan		Lehmkuhl and Lang (2001), Komatsu et al. (2001), Chapter 9

(continued)

Table 21.5 (continued)

3.5–2.5 ka BP	2.5–1.5 ka BP	1.5–0.5 ka BP	0.5–0.05 ka BP	<0.05 ka BP[a]	Lakes	Regions	Details
–	–	Humid	Humid	Warm and arid	Buun Tsagaan		Lehmkuhl and Lang (2001), Komatsu et al. (2001), Chapter 10
–	–	Humid	Humid	Warm and arid	Orog		Lehmkuhl and Lang (2001), Komatsu et al. (2001), Chapter 9
Arid	Arid	Arid	Arid	Warm and arid	Ulaan		Lee et al. (2013), Orkhonselenge et al. (2018), Mischke et al. (2020), Chapter 11
Warm and humid	Warm and humid	Warm and arid	Warm and arid	Warm and arid	Achit	Western Mongolia	Sun et al. (2013), Chapter 12
Humid	Humid	Humid	Humid	–	Bayan		Grunert et al. (2000)
Humid	Humid	Humid to arid	Arid	–	Dayan		Unkelbach et al. (2019)
–	–	Cold and humid	Cold and humid to warm and arid (0.1)	–	Duruu		Shinneman et al. (2010)
–	–	–	Arid	–	Ereen		Stolz et al. (2012)
–	–	–	Warm	–	Khar		Walther et al. (2017)
Arid	Arid	Arid	Arid	Warm and humid	Khoton		Rudaya et al. (2009), Chapter 14
Cold and semiarid	Cold and semiarid	Cold and semiarid	Cold and semiarid				Rudaya and Li (2013), Chapter 14
Humid	Humid	Arid	Arid	–	Tsetseg		Klinge and Lehmkuhl (2013)
Humid	Humid	Humid	Humid	Warm and arid	Uvs		Grunert et al. (2000), Chapter 13
–	–	Cold and humid	Cold and humid to warm and arid (0.1)	–	Zagas		Shinneman et al. (2010)

(continued)

Table 21.5 (continued)

3.5–2.5 ka BP	2.5–1.5 ka BP	1.5–0.5 ka BP	0.5–0.05 ka BP	<0.05 ka BP[a]	Lakes	Regions	Details
Warm	Warm	Warm	Warm	Warm and humid	Darkhad	Northern Mongolia	Krivonogov et al. (2012a), Chapter 15
Cold and arid	Cold and arid	Cold and arid	Cold and arid	Warm and humid	Dood		Narantsetseg et al. (2013)
–	Humid	Humid	–	Warm and arid	Khargal		Barna et al. (2021), Chapter 17
Humid	Humid	Humid to arid (1.1)	Arid	Warm and arid	Khuvsgul		Prokopenko et al. (2007), Chapter 16
Humid	Humid to arid (2.0)	Arid	Arid to humid				Orkhonselenge et al. (2013), Chapter 16
–	–	–	–	Warm and humid	Targan		Chapter 15
Cold	–	–	Cold	–	Gun	Central Mongolia	Wang et al. (2004)
Warm and humid	Warm and humid	–	–				Feng et al. (2005)
Humid to arid (3.0)	Humid	Humid	Humid				Zhang et al. (2012)
Warm and humid	Warm and humid	Warm and humid	Warm and humid				Feng et al. (2013)
Humid	Humid	Humid	Humid				Wang and Feng (2013)
–	–	Arid (1.2–1.0) to humid (1.0–0.6)	Cold and humid (>0.1)	–	Khuis		Tian et al. (2013)
Arid to humid (2.7)	Humid	Humid (arid at 1.6–1.2)	Humid	Warm and arid	Telmen		Peck et al. (2002), Fowell et al. (2003), Chapter 18
–	–	–	Humid (0.2) to arid (0.1)	Warm and humid	Terkhiin Tsagaan		Fukushi et al. (2015), Chapter 19
Warm	–	–	–				Khosbayar (2005), Chapter 19

(continued)

Table 21.5 (continued)

3.5–2.5 ka BP	2.5–1.5 ka BP	1.5–0.5 ka BP	0.5–0.05 ka BP	<0.05 ka BP[a]	Lakes	Regions	Details
Warm and humid	Warm and humid	Warm and humid	Warm and humid	Warm and humid	Terkhiin Tsagaan		Davaagatan et al. (2015), Orkhonselenge and Davaagatan (2016), Chapter 19
Arid to humid (2.8)	Humid	Humid	Humid	Warm and arid	Ugii		Khosbayar (2005); Schwanghart et al. (2008, 2009), Chapter 20
Warm and humid	Warm and humid	Warm and humid	Warm and humid				Davaagatan et al. (2015); Orkhonselenge and Davaagatan (2016), Chapter 20
Arid	Humid	Humid to arid	Arid				Wang et al. (2009, 2011), Chapter 20

[a]Anthropocene (<0.05 ka BP): Results are inferred from this work (see Chaps. 6–20)

Table 21.6 Paleoclimate changes recorded in lakes of southern Russia and northern China

Regions	Late Pleistocene (>11.7 ka BP)	Early Holocene (11.7–7.5 ka BP)	Middle Holocene (7.5–3.5 ka BP)	Late Holocene (<3.5 ka BP)	Lakes	Details
Southern Russia	Cold and arid	Warm and humid (cold at 9.5–8.5 and arid at 8.0–7.0)	Warm and arid (warm and humid at 6.5–5.7 and cold at 5.5–4.5)	Warm and arid (cold at 3.0–2.5)	Baikal	Demske et al. (2005)
	Humid to arid	Humid to arid (9.2–9.0 and <8.6)	Arid (>6.8) to humid (cold at 6.0–5.0)	Humid (cold at 3.0–2.5)		Bezrukova et al. (2005)
	–	Humid	Warm (centered at 6.0)	–		Prokopenko et al. (2006)
	–	Humid	Warm and arid (at 7.5–6.5) to warm and humid	Warm and arid		Prokopenko et al. (2007)

(continued)

Table 21.6 (continued)

Regions	Late Pleistocene (>11.7 ka BP)	Early Holocene (11.7–7.5 ka BP)	Middle Holocene (7.5–3.5 ka BP)	Late Holocene (<3.5 ka BP)	Lakes	Details
	Warm and humid	Warm and humid (centered at 9.0–7.0)	–	–		Tarasov et al. (2007a); Prokopenko et al. (2010)
	Humid	–	–	Humid		Mackay et al. (2011)
	–	–	–	Cold (3.4–2.3) and arid (2.8–1.7) to warm and humid (1.3–0.2) and warm and arid (<0.2)	Beloye	Krivonogov et al. (2012b)
	Cold and arid	Warm and humid	Warm and humid to cold and arid (6.0)	Cold and arid	Grusha, Akkol	Blyakharchuk et al. (2007)
	–	–	Warm and humid (6.8–6.4) to cold (6.4–4.1)	Warm and humid	Khara Nur	Alexander and Bezrukova (2019)
	Cold and arid	Warm and humid (centered at 10.8)	Warm and humid (>7.3) and cold and arid (centered at 7.0–6.5)	Cold and arid	Kotokel	Tarasov et al. (2009)
	Humid	Humid	Warm and arid	Warm and arid		Bezrukova et al. (2011)
	Humid	Humid to arid (10.5)	Humid (7.3–6.2) to arid (6.2–3.8)	Humid (>2.0) to arid (<2.0)	Tere-Khol	Panin et al. (2012)
	–	–	Arid (centered at 6.5)	Arid	Tsagan-Tyrm	Sklyarov et al. (2010)
	Warm and arid	Warm and humid	Cold and arid	Cold and arid	Uzungol, Kendegelukol, Tashkol	Blyakharchuk et al. (2004)

(continued)

Table 21.6 (continued)

Regions	Late Pleistocene (>11.7 ka BP)	Early Holocene (11.7–7.5 ka BP)	Middle Holocene (7.5–3.5 ka BP)	Late Holocene (<3.5 ka BP)	Lakes	Details
Northern China	–	–	Arid (7.4–6.4)	–	Anguli-nuur	Wang et al. (2010)
	–	–	Humid to warm and arid (6.7–6.2) and cold and arid (5.4–4.7)	–	Baahar	Guo et al. (2007)
	–	–	Arid (7.7–6.8)	–	Bai Nuur	Wang et al. (2012)
	–	Cold (8.0)	Arid (6.5)	Arid	Bayanchagan	Jiang et al. (2006)
	–	–	Cold and arid (5.0–4.5)	–	Bosten	Wünnemann et al. (2003)
	–	Cold and arid (9.2–7.8) to warm and arid (<7.8)	Warm and arid (>7.1)	–	Diaojiao	Shi and Song (2003)
	–	–	Cold and arid (6.4–4.4)	–	Hulun (or Dalai in Chaps. 6 and 7)	Wen et al. (2010)
	–	Arid	Arid to humid (5.4) and humid to arid (3.9)	Arid	Juyan	Herzschuh et al. (2004)
	–	Arid	Cold and arid (7.4–6.0) to warm and humid (6.0–4.0)	Cold and arid (>0.8) to warm and humid (<0.8)	Qigai Nuur	Sun and Feng (2013)
	–	–	Arid (7.2–6.1)	–	Ulan Nuur	Wang et al. (2012)
	–	Arid	Arid to humid (6.7)	Humid (<0.5)	Wulungu	Liu et al. (2008)

[a]*Dalai* meaning *ocean* has been often spelled as *Hulun* in many publications. Dalai is the right English transliteration from Mongolian *Далай*

recent climate trends during the last half century are primarily analyzed from our data (Table 21.5). The studied lakes are presented in the alphabetical order in each region of Mongolia (Tables 21.1, 21.2, 21.3, 21.4, and 21.5) and the adjacent areas (Table 21.6) in the tables, where results are ordered in publication years.

21.2 Local Spatial Pattern of Paleoclimate

Paleoclimatic records obtained from Mongolian lakes are synthesized for each region (Table 21.1). In addition, the lakes of Mongolia providing major conclusions for understanding the paleoclimate changes are listed and described in detail (Tables 21.2, 21.3, 21.4, and 21.5). For instance, Lake Khuvsgul's sedimentary record is highly informative for northern Mongolia, giving direct evidence for changes of atmospheric precipitation during the late Pleistocene and Holocene (Fedotov et al., 2004). During the early Holocene (see Sect. 21.3.2), in northern Mongolia (see Sect. 21.2.4), a warm and humid climate prevailed in the Lakes Khuvsgul and Darkhad basins, and in the northern part of central Mongolia (see Sect. 21.2.5), a warm and humid climate was recorded in the Lake Gun basin. Moreover, in western Mongolia (see Sect. 21.2.3), a cold and wet climate dominated in the Lake Khoton basin, while a warm and dry climate dominated in the Lake Achit basin (Orkhonselenge et al., 2019).

The revealed spatial pattern of paleoclimate changes from lakes in Mongolia depends on each regional landscape, each lake type, and local climatic regime. For instance, there are similar patterns of paleoclimate changes (1) between the Mongolian Altai Mountain Range of western Mongolia and the Khangai Mountain Range of central Mongolia and (2) between the Great Lakes Depression in western Mongolia and the intermontane valley between the Khangai[4] and Khentii[5] mountain ranges in central Mongolia.

21.2.1 Eastern Mongolia

In eastern Mongolia (see Chap. 6), paleoclimate has not been reconstructed frequently from lakes in comparison with those in other regions of Mongolia (Orkhonselenge et al., 2015, 2019; Tables 21.1, 21.2, 21.3, 21.4, and 21.5). In the Lake Khagiin Khar basin within the central Khentii Mountain Range, cold and warm climates prior to and after the Last Glacial Maximum (LGM) are noted (Table 21.2). The high lakeshores in Lake Buir basin during the middle Pleistocene (Lehmkuhl & Haselein, 2000) may imply a humid climate in eastern Mongolia at that time.

[4] *Khangai* has been misspelled as *Khangay* (or *Hangay*) and *Hangai* in many publications. Khangai is the right English transliteration from Mongolian *Хангай*.

[5] *Khentii* has been misspelled as *Khentey* (or *Hentey*) and *Khentei* (or *Hentei*) in many publications. Khentii is the right English transliteration from Mongolian *Хэнтий*.

Whether eastern Mongolia is still under the influence of the EASM today is much arguable among the scientists who study the climate system of Mongolia. Only the Khalkh and Numrug River drainage basins and the Khyangan[6] Mountain Range in the easternmost tip of Mongolia are clearly recognized to be affected by the EASM today (Tsegmid, 1969).

Today, in the northern and northwestern parts of the region, around the eastern Khentii Mountain Range, a relatively cool climate is dominant under the EAWM in winter and the westerlies in summer, whereas in eastern and southeastern parts of the region, around the branch mountains of the Khyangan Mountain Range and Degee and Khalkh River drainage basins, climate is warm under the EASM (Badarch et al., 1965; Chen et al., 2008). However, at the present time, eastern Mongolia is considered to be located in the outer zone of the EASM (e.g., Herzschuh, 2006; Wang et al., 2010; Li & Morrill, 2010).

Eastern Mongolia is drying since the late 1900s accentuated by a change from a particularly wet period (Davi et al., 2013). A significant trend in this large-scale drying is found across eastern Mongolia during 1951–2005 (Li et al., 2009). The warming eastern Mongolia will be strengthened during the summer, and the annual average air temperature will rise 1.4 °C by 2020, 2.8 °C by 2050, and 4.9 °C by 2080 (Gomboluudev, 2006). The local climate trends in Lakes Buir and Khukh basins show a trend of rising air temperature (Tables 21.1 and 21.5; see Chaps. 6–8), and it is also observed in Lake Yakhi basin during the last half century (Orkhonselenge & Bulgan, 2021; Table 21.1).

21.2.2 Southern Mongolia

In southern Mongolia (see Chap. 9), paleoclimatic study based on lakes has rarely been conducted except for some recent few works since 2000s (Tables 21.1, 21.2, 21.3, 21.4, and 21.5).

An arid climate was hypothesized to have been present because of the disappearance of lakes during the marine oxygen isotope stage (MIS) 6, while a humid environment existed contributing to large paleolakes between the MIS 5e and the beginning of the LGM (Lehmkuhl et al., 2018a). A relatively humid environment in the Lake Orog basin was reconstructed during the MIS 3 and the early Holocene, and the harsh climatic periods at ~43.0 ka, ~36.0 ka, during the LGM and the Younger Dryas (YD) resulting in playa phases (Yu et al., 2019; Table 21.1). The expansions of Lakes Buun Tsagaan, Adgiin Tsagaan, and Orog in the Valley of the Lakes were likely contributed by the humid climate during a period before the full glacial period corresponding to MIS 3 (Komatsu et al., 2001; Table 21.1). A cold and arid climate may have prevailed during the MIS 2, possibly leading to low levels of Lakes Orog and Adgiin Tsagaan (Komatsu et al., 2001; Table 21.2), whereas humid climates at 8.5 ka and 1.5 ka (Tables 21.1, 21.3, and 21.5) were reconstructed

[6] *Khyangan* has been misspelled as *Khingan* (or *Hingan*) and *Hyangan* in many publications. Khyangan is the right English transliteration from Mongolian *Хянган*.

with the high lake levels of Lake Adgiin Tsagaan (Lehmkuhl & Lang, 2001), with a last semipermanent lake probably existing there about several hundred years ago (Komatsu et al., 2001).

The history of paleoclimatic and paleoenvironmental conditions in the Lake Bayan Tukhum basin during the late Quaternary was reconstructed as follows: a change from wet to dry conditions during the late glacial, warm, and wet climate at the onset of the Holocene, the humid period of the Holocene lasting until about 4.0 cal ka BP, and afterward increased aridity during the late Holocene (Felauer et al., 2012; Table 21.2). Paleoclimate changes in the Lake Ulaan basin during the last 17.0 ka BP were reconstructed as follows: relatively warm and humid at 16.5–15.0 ka BP, cold and dry at 15.0–11.3 ka BP, the most humid in the early Holocene (11.3–8.8 ka BP), humid in the middle Holocene (8.6–4.7 ka BP), and dry in the late Holocene (after 3.0 ka BP) (Lee et al., 2013; Table 21.1). Holocene dominated by a distinct high productivity alkaline environment with felsic and alkaline input relative to the late Pleistocene, which is attributed to an increased hydrodynamic strength of riverine inflow and/or intensified erosion and weathering of felsic source rocks in the Lake Orog basin (Yu et al., 2017), implies arid to humid climates (Table 21.1). High lake levels in the Valley of the Lakes from the early Holocene to the beginning of middle Holocene (Klinge & Sauer, 2019) suggest a humid climate. A climatic shift from humid in the middle Holocene to arid in the late Holocene was reconstructed in the Lake Ulaan basin (Orkhonselenge et al., 2018) (Tables 21.1, 21.4, and 21.5). Nottebaum et al. (2021) found large extensions of Lake Orog occurring during MIS 5 and 3 and the Holocene.

21.2.3 Western Mongolia

In western Mongolia (see Chap. 12), the paleoclimate changes have been reconstructed mostly from a few lakes (Tables 21.1, 21.2, 21.3, 21.4, and 21.5).

In the Lake Khoton basin (see Chap. 14) within the Mongolian Altai Mountain Range, a humid period during the YD to the early Holocene (12.2–10.1 ka BP) was suggested by the evidence for a higher level of the lake (Lehmkuhl & Haselein, 2000; Table 21.1). The climate in the Lake Khoton basin was cold and arid during the late Pleistocene and early Holocene (11.5–10.7 ka), warm and humid in the middle Holocene (9.3–6.5 ka), and warm and dry after 6.5 ka and over the last 3.0 ka BP (Rudaya et al., 2008). In addition, the pollen-based quantitative reconstruction of precipitation suggests that the regional climate in the Lake Khoton basin (Tables 21.3, 21.4, and 21.5) was relatively humid before 11.5 ka BP and arid at 11.5–10.7 cal ka BP and became substantially humid and warm at 10.7–5.0 cal ka BP and dry climate since 5.0 ka BP (Rudaya et al., 2009). The diverse data obtained from Lake Khoton suggest its climates as warm and wet at 11.5–6.0 ka, dry at 6.0–4.0 ka, and cold and wet after 4.0 ka BP (Rudaya & Li, 2013).

In the Lake Tsetseg basin, climate conditions during the late Pleistocene and Holocene were recorded with high lake levels during the late Pleistocene, a dry-out period with prevailing aeolian processes in the middle Holocene, and a wet phase in the late Holocene when the lake developed again (Klinge & Lehmkuhl, 2013; Tablc 21.1).

The pollen-based biomization and precipitation-average annual air temperature quantitative reconstruction from Lake Achit showed a cold and wet climate at ~22.6–13.2 cal ka BP, mild and dry conditions at ~13.2–6.4 cal ka BP, cooling and humid conditions at ~6.4–1.6 cal ka BP, and dry periods at the last ~1.6 cal ka BP (Sun et al., 2013; Tables 21.1 and 21.2). In addition, lacustrine sediments and diatom assemblages at Lake Achit provided evidence for a high level at 13.3 ka BP (Lehmkuhl & Haselein, 2000) implying a humid climate (Table 21.2).

The climate in the Great Lakes Depression was reconstructed based on geomorphological evidence and fluctuations of lake levels at Lakes Uvs and Bayan: generally humid before the LGM with one arid phase, dry and cold climate during the second half of the LGM, and wet and cold during the late glacial (Grunert et al., 2000). High lake levels in the Great Lakes Depression in the late glacial (Klinge & Sauer, 2019) suggest a humid climate (Table 21.1 and 21.2). Klinge et al. (2021) showed the late Pleistocene developments of Lakes Khurgan, Khoton, and Ekhen in the Mongolian Altai Mountain Range.

21.2.4 Northern Mongolia

In northern Mongolia (see Chap. 15), Lake Khuvsgul and remnant lakes of the paleolake Darkhad have been widely studied to extract information on the local and regional paleoclimate changes (Tables 21.1, 21.2, 21.3, 21.4, and 21.5).

In the Lake Khuvsgul basin (see Chap. 16), the arid climate-induced dramatic low-standing lake in the middle to late Pleistocene was noted together with the complicated sediment accumulation (Prokopenko et al., 2009; Table 21.1). Evidence for a warm and wet Bölling/Allërod (BA) period at 15.4–13.7 cal ka BP and a cold and dry YD at 13.7–11.7 cal ka BP is recorded in the lake sediments (e.g., Karabanov et al., 2004; Prokopenko et al., 2005; Narantsetseg et al., 2006; Orkhonselenge et al., 2013; Table 21.2). The wet early Holocene from ~9.5 to ~6.0 cal ka BP, the dry and warm middle Holocene from ~6.0 to ~3.5 cal ka BP, and the relatively wet late Holocene after ~3.5 cal ka BP were reconstructed by the pollen-based forest-steppe index (Prokopenko et al., 2007; Tables 21.3, 21.4, and 21.5). The climate humidification between 11.0 and 9.0 ka BP with a significant peak of magnetic susceptibility record reflected inputs of clastic sediment during the increase of lake level, and a short-term drought at 5.5 ka BP resulting in the decrease in lake level of ca. 2.5 m was found (Fedotov et al., 2004; Tables 21.1 and 21.5).

In the Lake Erkhel basin, cold periods at ~24.0–11.0 ka BP and warm periods at ~11.0–3.0 ka BP were reflected in the sedimentation rates of 5.2 cm/ka and 31.0 cm/ka (Watanabe et al., 2007, 2009; Tables 21.2, 21.3, and 21.4). In the Lake Khargal basin (see Chap. 17), palynological, macrofossil, and geochemical analyses of cores allow reconstruction of vegetation and climate changes in northern Mongolia throughout the Holocene (Barna et al., 2021; Tables 21.3, 21.4, and 21.5).

The paleolake Darkhad (see Chap. 15) contains valuable paleoclimatic information within a long sedimentary record, and stages of the lacustrine development are recognizable in landforms and subsurface sediments (Krivonogov et al., 2005). For instance, in the paleolake basin, paleoclimate changes were recovered from Lake Dood as a cold and dry climate in the late glacial, a relatively cold and wet climate from ~9.5 cal ka BP to ~7.8 cal ka BP, onset of the Holocene climatic optimum (the most humid and warm stage) between 7.8 and 5.8 cal ka BP, and cooler and drier climate after 5.8 cal ka BP (Narantsetseg et al., 2013; Tables 21.2, 21.3, and 21.4). The high lakeshore of Lake Dood during the middle Pleistocene (Lehmkuhl & Haselein, 2000) may imply a humid climate in northern Mongolia at that time (Tables 21.2). In addition, mainly cold climates with glacier advances resulting in ice-damming of the paleolake Darkhad are envisaged for ~53.0–35.0 ka (Gillespie et al., 2008; Komatsu et al., 2009), ~19.0–17.0 ka (Gillespie et al., 2008), or ~26.0–18.0 ka (Komatsu et al., 2009) and later at ~11.0–10.0 ka (Gillespie et al., 2008; Tables 21.2 and 21.3).

21.2.5 Central Mongolia

In central Mongolia (see Chap. 18), paleoclimatic records from lakes are used to understand the local climate dynamics of the past (Tables 21.1, 21.2, 21.3, 21.4, and 21.5).

In the Lake Telmen basin (see Chap. 18), climate changes for the past 7.1 cal ka BP were reconstructed based on sedimentological and geomorphological evidence as hyperarid at 7.1–6.3 cal ka BP, still arid conditions with the increased effective moisture balance at 6.3–4.4 cal ka BP, generally more humid conditions since 4.4 cal ka BP, the present-day effective moisture balance at 2.7–1.3 cal ka BP, and an increase in effective moisture during the past approximately 0.7 cal ka BP (Peck et al., 2002; Table 21.1). Palynological and sedimentological data reconstructed the climate as relatively arid at 7.5–4.5 cal ka BP, the maximum humid at ~4.5–1.6 cal ka B.P, an aridification centered at 1.4 cal ka BP, and humid periods at 0.9–0.6 and 0.4–0.05 cal ka BP (Fowell et al., 2003; Tables 21.4 and 21.5).

In the Lake Terkhiin Tsagaan basin (see Chap. 19), dry and cold environments were found by the presence of abundant carbonates and minerals during the middle Holocene, while the warm and humid environments were suggested because of the increased organic matter and biogenic silica in the late Holocene (Davaagatan et al., 2015; Orkhonselenge & Davaagatan, 2016; Table 21.1). The shift from cold to warm since 3.6 ^{14}C ka BP was reflected in the curves of S ratio for magnetic susceptibility in the lake sediments (Khosbayar, 2005).

In the Lake Ugii basin (see Chap. 20), arid climates were suggested in the middle Holocene because of large amounts of carbonates and minerals found in the lake sediments (Davaagatan et al., 2015; Orkhonselenge & Davaagatan, 2016), coinciding with the abundant minerals of illite-chlorite-calcite-smectite at 4.4 ^{14}C ka BP (Khosbayar, 2005; Table 21.1). Warm and wet environments recorded in the lake basin were suggested with the increases of organic matter and biogenic silica in the lake sediments (Davaagatan et al., 2015), which were contributed by the presence of stable high lake levels from 2.8 ka BP to the present (Schwanghart et al., 2008) and by the abundant organics and minerals of kaolinite-montmorillonite since 2.8 ^{14}C ka BP (Khosbayar, 2005) (Tables 21.4 and 21.5). Moreover, a warm and arid climate is suggested by the low lake levels occurring from 10.6 to 7.9 ka BP, arid conditions prevailed at 4.2–2.8 ka BP, and a stable humid phase followed until today (Schwanghart et al., 2008). A humid climate with slightly high moisture occurred at 10.0–8.0 cal ka BP, more humidity at 8.0–4.0 cal ka BP, increased aridity at 4.0–2.8 cal ka BP, and an increased moisture level during the last 2.0 ka (Schwanghart et al., 2009; Table 21.1). The multi-proxy data from Lake Ugii reconstructed a mild climate with moderate moisture conditions at ~9.1–6.7 cal ka BP, a dry phase at ~6.7–3.3 cal ka BP, a wet period at 2.6–1.0 cal ka BP, and a relatively dry period at ~1.0 ka BP (Wang et al., 2009, 2011; Tables 21.3, 21.4, and 21.5).

In the Lake Gun basin (see Chap. 18) in the downstream of Selenge[7] River basin, cold and dry climates at 10.7–7.7 and 2.2–1.5 cal ka BP and wet and warm climates at 7.7–2.2 cal ka BP were reconstructed based on lake level fluctuations (Feng et al., 2005; Table 21.1). Furthermore, a wetter and warmer climate at 9.3–7.7 ka BP and a wetter and cooler climate at 5.6–4.1 ka BP were recorded with magnetic susceptibility in the lake sediments (Wang et al., 2004). A high lake level was noted in Lake Gun during the past 8.0 ka (Dorofeyuk & Tarasov, 1998). Moreover, the Holocene hydrology and climate were reconstructed as dry conditions during the earliest Holocene (>10.8–10.3 cal ka BP), wetter conditions at 10.3–7.0 cal ka BP, a dry phase from 7.0 to 2.5 cal ka BP (with two short wet periods at 5.7–4.1 and 3.6–3.0 cal ka BP), and wetter conditions since 2.5 cal ka BP (Zhang et al., 2012; Tables 21.2, 21.3, 21.4, and 21.5). The Holocene climates were recorded from pollen as low temperature and moisture during the early Holocene (from ~10.0 to ~8.0 cal ka BP), their gradual rise from ~8.0 to ~6.0 cal ka BP, and warm and wet with high-amplitude fluctuations during the last ~6.0 cal ka BP (Feng et al., 2013). The regional average moisture index curve indicates a relative wet period from ~10.0 to 7.0 cal ka BP and a middle Holocene dry phase lasting from ~7.0 to ~3.8 cal ka BP, followed by a considerably wet period since ~3.8 cal ka BP (Wang & Feng, 2013; Tables 21.2, 21.3, 21.4, and 21.5).

21.3 Local Temporal Pattern of Paleoclimate

21.3.1 Late Pleistocene

Paleoclimate reconstructions from Mongolian lakes during the late Pleistocene are synthesized in Table 21.2 and Fig. 21.1a.

In eastern Mongolia (see Sect. 21.2.1), a climatic shift from cold to mild was demonstrated by the advance at 21.0–19.0 ka BP and the retreat at 15.9 ± 0.6 ka BP

[7] *Selenge* has been often incorrectly written as *Selenga* internationally based on the Russian pronunciation. Selenge is the correct English transliteration from Mongolian *Сэлэнгэ*.

of glaciers in the Lake Khagiin Khar basin (Oh et al., 2019) of the Khentii Mountain Range (Table 21.2).

In southern Mongolia (see Sect. 21.2.2), arid conditions at MIS 6 were suggested by the disappearance of large lakes (Table 21.2). Humid conditions at MIS 5e lasting probably until the beginning of the last ice age were recorded with large paleolakes caused by a strong EASM (Lehmkuhl et al., 2018a). The MIS 6 aridity coincided with the cold stage (Lehmkuhl et al., 2018b). A humid climate at MIS 3 was recorded with the high lake level of Lake Orog, and evidence for an arid MIS 2 was found in the Valley of the Lakes (Lehmkuhl et al., 2018a). The humid environment lasting to the early stage of the last glaciation around 70 ka was recorded with a huge paleolake in the Lake Orog basin (Lehmkuhl & Lang, 2001). An increased precipitation was recorded in the Lake Orog basin with higher lake levels during the MIS 3 and early MIS 2 (~35–~24 ka BP), and arid conditions during the late Pleistocene were found with the lower area of the lake basin dominated by the *Artemisia* steppe (Yu et al., 2019). Komatsu et al. (2001) hypothesized that wetter periods corresponding to MIS 3 prior to the LGM explained the reconstructed high lake stands in large stable Lakes Buun Tsagaan, Adgiin Tsagaan, and Orog covering extensive areas of the Valley of the Lakes. Additionally, they suggested a cold or wet climate during the terminal late glacial period after LGM as another possible period for the lake high stands. The climate during the late glacial changed from wet conditions at 15.0–12.0 cal ka BP toward a short phase of drier conditions at 12.0–11.0 cal ka BP in the Lake Bayan Tukhum basin (Felauer et al., 2012). The most humid time in southern Mongolia at the Pleistocene and Holocene border was observed in the Lake Ulaan basin (Lee et al., 2011) where the relatively warm and humid conditions at 16.5–15.0 ka BP and the cold and dry climates at 15.0–11.3 ka BP were recorded (Lee et al., 2013). Others state that the arid conditions in the Lake Ulaan basin shifted into the humid conditions since 12.3 ka BP causing the rise in lake level and the decrease in salinity (Mischke et al., 2020; Table 21.2).

In western Mongolia (see Sect. 21.2.3), the humid climate in the Lake Uvs basin during the middle Pleistocene or the first phase of the late Pleistocene before the LGM coincided with a high lake level of Lake Bayan at ~80 m above the present level, and then the climate shifted to become dry during the LGM before 13.2 ka when levels of Lakes Uvs and Bayan dropped and the lakes disappeared altogether (Table 21.2). And it turned to become wet during the late glacial since 13.2 ka when the lake basins had filled again, reaching a further high stand of 48 m above the present lake level of Lake Bayan (Grunert et al., 2000). The humid conditions were also supported by the presence of a deep Lake Ereen in the eastern Mongol Els sand dunefield at the transition from MIS 6 to MIS 5 (Stolz et al., 2012), MIS 3 (Yang et al., 2004) and at 39.0 ka BP (Grunert et al., 2009). Dry conditions at 33.0 ka BP contributed to the aeolian sand accumulation coinciding with the interstadial MIS 3 of the last glacial cycle, and the wet conditions contributing to the highest lake level (12.0 m above the modern lake) between the LGM and late glacial were recorded in the Lake Tsetseg basin (Klinge & Lehmkuhl, 2013). Cold and wet conditions at ~22.6 to 13.2 cal ka BP and mild and dry conditions after 13.2 cal ka BP were reconstructed with pollen data in the Lake Achit basin (Sun et al., 2013). The cold period at MIS 2 was reconstructed with a glacier advance in the western Kharganat River valley and blocked it, resulting in the formation of the endorheic Lake Khar basin (Walther et al., 2017). The dry conditions were suggested with the existence of a relatively deep and oligotrophic/mesotrophic Lake Khoton prior to 11.5 cal ka BP (Rudaya et al., 2009; Rudaya & Li, 2013). Damming of the terminal moraines at Lake Khoton occurred at 16.2 ± 2.0 and 18.3 ± 2.2 ka (Klinge et al., 2021) implies the humid climates at that time. Klinge et al. (2021) show the development of Lake Ekhen in the Mongolian Altai Mountain Range since ~17.9 cal ka BP indicating evidence of a late glacial paleolake with the high lake level from the late glacial period to the early Holocene. Moreover, the development of a proglacial lake during brief retreats and readvances of moraine at Lake Khurgan was found at 88.5 ± 14.5 and 85.5 ± 10.4 ka between the late MIS 5 and MIS 4 in an early state of the last glacial period (Klinge et al., 2021; Table 21.2).

In northern Mongolia (see Sect. 21.2.4), warm MIS 3, MIS 5, MIS 7, and MIS 9 with the increase in $\delta^{18}O$ values, a high content of Br, a low amount of clay particles, and high bio-productivity were found in Lake Khuvsgul (Fedotov et al., 2015; Table 21.2). There, positive $\delta^{13}C$ values linked with high inflowing dissolved inorganic carbonates and forest-steppe vegetation were evidenced during cold MIS 8, MIS 6, MIS 4, and MIS 2, but glaciers during MIS 2 were the smallest in sizes due to the high regional aridity (Fedotov et al., 2015). The MIS 5d was a severely cold stage, one of the lowest water levels (severe aridity period) in the Lake Khuvsgul basin (Kashiwaya et al., 2010). The arid climate was reflected in the enriched carbonate CO_2 (up to 10%) and water-extractable SO_4^{2-} (up to 300 times greater than that in the Holocene) due to the low (ca. 110 mm) regional precipitation at the end of the Pleistocene, when Lake Khuvsgul was only 100 m deep, and its volume was ca. ten times less than today (Fedotov et al., 2004). The dominance of the warm and wet climate from ~13.0 to 12.0 ^{14}C ka BP was suggested by the redeveloped planktonic communities and the increased primary production in Lake Khuvsgul (Karabanov et al., 2004). The dry climate before 15.4 cal ka BP during the LGM was recorded with the level of Lake Khuvsgul ~100 m lower than the present level, and then the lake level rose after 15.4 ka BP (Prokopenko et al., 2005). The high lake level at 13.0–12.0 cal ka BP of Lake Khuvsgul showed humid conditions at that time (Orkhonselenge et al., 2013).

Cold periods at ~24.0–11.0 ka BP were reflected in the sedimentation rate of 5.2 cm/ka in the Lake Erkhel basin (Watanabe et al., 2007). The cold climates promoted the maximum glacier advances at MIS 5 coinciding with the earlier paleolake Darkhad (Krivonogov et al., 2005) and also ~53.0–35.0 ka (Gillespie et al., 2008; Komatsu et al., 2009) and ~19.0–17.0 ka BP (Gillespie et al., 2008) or ~26.0–18.0 (Komatsu et al., 2009) for the later lakes. The cold climates contributed to the late Pleistocene and Holocene formations of paleolakes in the Darkhad basin because of glacier advances in the Sayan mountains blocking the drainage (Grosswald & Rudoy, 1996; Krivonogov et al., 2005; Komatsu et al., 2009), but the sedimentary damming mechanism may not be excluded for smaller lakes. Nevertheless, warm conditions in the Darkhad basin were suggested by the minimal or disappearance of the lake at 12.0–9.6 cal ka BP (Krivonogov et al., 2012a). A cold and dry climate of the late glacial recorded in Lake Dood of the Darkhad basin changed after 9.5 cal ka BP (Narantsetseg et al., 2013).

In central Mongolia (see Sect. 21.2.5), a dry and cold pre-Holocene period was found in the Lake Ugii basin (Schwanghart et al., 2009; Table 21.2). Dry conditions were recorded also during the earliest Holocene (>10.8 cal ka BP) with low lake levels and prevailing aeolian sediment deposition in Lake Gun (Zhang et al., 2012).

21.3.2 *Early Holocene*

Paleoclimate changes recorded in Mongolian lakes during the early Holocene are presented in Table 21.3 and Fig. 21.1b.

In southern Mongolia (see Sect. 21.2.2), rising air temperature and precipitation at 11.0 cal ka BP, interrupted by a short dry and cold reversal at 8.5–7.5 ka cal BP, was reconstructed in the Lake Bayan Tukhum basin (Felauer et al., 2012; Table 21.3). Humid climate periods at 11.1–9.6 ka BP (Lee et al., 2011) and 11.3–8.8 and 8.6–7.6 ka BP (Lee et al., 2013) are indicated in the Lake Ulaan basin. The wettest climate in the Lake Ulaan basin causing the highest inflows and the highest productivity was recorded in the early Holocene; however, the arid climate caused declining of lake level and increase in salinity at 9.4–6.8 ka BP (Mischke et al., 2020). The relatively humid environment in the early Holocene was characterized also in the Lake Orog basin (Yu et al., 2019) with larger extents of lakes in the Valley of the Lakes, especially high lake levels of Lake Adgiin Tsagaan, at about 8.5 ka (Lehmkuhl & Lang, 2001). The wet period during the warmest early Holocene was considered to be responsible for some of the high lake stands in Lakes Buun Tsagaan, Adgiin Tsagaan, and Orog (Komatsu et al., 2001). The expansion of Lake Orog during the early Holocene was caused by an increase in precipitation and meltwater inflow from the surrounding mountains (Nottebaum et al., 2021).

In western Mongolia (see Sect. 21.2.3), high lake levels and humid conditions during the onset of the Holocene were found in the Lake Uvs basin (Grunert et al., 2000; Yang et al., 2004; Table 21.3). The high stands of 40 m above the present level for Lake Uvs during the transgressive phase and the high stands of over 10 m above the present level of Lake Uvs and 4.4 m for adjacent Lake Bayan during the regressive phase in the late glacial and early Holocene between 11.2 ± 60 and 9.7 ka BP (Grunert et al., 2000) imply that the climate was humid at that time. The generally short-term arid climate at 11.5–10.7 cal ka BP, with abrupt climate changes at ~11.5 and ~7.5 ka BP related to glacial activities (Rudaya & Li, 2013), is implied with the presence of scarce and dry vegetation, while a noticeable increase in precipitation from 200–250 mm/year to 450–550 mm/year at about 10.0 cal ka BP was recorded with a quantitative reconstruction of vegetation dynamics in the Lake Khoton basin (Rudaya et al., 2009). The mild and dry climate at 13.2–6.4 cal ka BP was noted in the Lake Achit basin (Sun et al., 2013). The humid climate in the early Holocene is recognizable from the high lake levels in the Great Lakes Depression (Lehmkuhl et al., 2018a). The humid conditions in the early to middle Holocene are supported by the formation of Lake Ereen in the eastern Mongol Els sand dunefield (Stolz et al., 2012; Table 21.3).

In northern Mongolia (see Sect. 21.2.4), the pollen-based forest-steppe index from Lake Khuvsgul showed that a higher abundance of conifers occurred at 9.0–7.5 cal ka BP suggesting higher precipitation, and increases of diatom abundance and taxonomic diversity at 11.0–7.0 ka BP were coincident with the early Holocene humidity maximum (Prokopenko et al., 2007). ~6 m higher at >12.0–10.6 ka BP and ~2 m higher at 10.2–9.0 ka BP lake levels of Lake Khuvsgul than the present lake level (Orkhonselenge et al., 2013) also contributed to the humid conditions in the basin at that time (Table 21.3). The humid climate at the transition to the Holocene was reflected in increases of total diatoms from 0 to 10^8 g^{-1}, of BiSi from 1% to 20%, and of organic matter from <1% to >6% in Lake Khuvsgul (Fedotov et al., 2004). Warm periods at ~11.0–3.0 ka BP were recorded with the sedimentation rates of 9.1–36.6 cm/ka in the Lake Erkhel basin, and especially the humidification with warming at about 11.0 ka BP may have caused the abrupt increase of sedimentation rate from 5.2 to 31.0 cm/ka (Watanabe et al., 2007). The minimal lake level at 12.0–9.6 cal ka BP suggesting warm and arid conditions and a relatively deep lake at 9.6–7.1 cal ka BP suggesting humid conditions were found in the Lake Darkhad basin (Krivonogov et al., 2012a). However, a relatively cold and dry climate that occurred prior to 9.5 cal ka BP in the Lake Dood basin changed to a relatively cold and wet climate prevailing up until 7.8 cal ka BP (Narantsetseg et al., 2013). A brief cold climate at ~11.0–10.0 ka may have prompted a glacier readvance, and a shallow paleolake Darkhad was formed possibly because of the glacier outwash damming (Gillespie et al., 2008). In Lake Khargal, relatively high values of TOC at ~7.8–7.6 cal ka BP are interpreted as interval of increased temperature or precipitation in the lake basin (Barna et al., 2021; Table 21.3).

In central Mongolia (see Sect. 21.2.5), arid conditions are suggested by low lake levels at 10.6–7.9 ka BP in the Lake Ugii basin (Schwanghart et al., 2008; Table 21.3). Slightly more humid conditions in the early Holocene still not sufficient to generate a dense vegetation cover to cease aeolian dynamics of sand mobilization were found in the Lake Ugii basin (Schwanghart et al., 2009). In the lake basin, mild and semi-humid conditions before 7.8 cal ka BP with a noticeable cool and humid interval at 8.4–8.3 cal ka BP are shown in a high-resolution pollen record (Wang et al., 2009). In the Lake Gun basin, the warm and dry conditions with the presence of dry lake before 9.3 ka BP were reversed into a warm and humid climate at 7.7 cal ka BP (Wang et al., 2004). There, dry conditions were recorded during the earliest Holocene (>10.8–10.3 cal ka BP) with low lake levels and the prevailing aeolian sediment deposition, and then the lake expanded under significantly wetter conditions between 10.3 and 7.0 cal ka BP (Zhang et al., 2012). The interval of warmer and dryer climate during the cold and arid early Holocene was marked with the carbonate-rich laminated layer in Lake Gun (Feng et al., 2005).

The humidity throughout Mongolia at the last period of the early Holocene (Table 21.3) coincided with the simulated high precipitation and snow accumulation in Central Asia with a peak at 8.0–7.5 ka BP (Bush, 2005).

21.3.3 Middle Holocene

Paleoclimate changes recorded in Mongolian lakes during the middle Holocene are synthesized in Table 21.4 and Fig. 21.1c.

In southern Mongolia (see Sect. 21.2.2), humid climates at 8.6–3.1 ka BP (Lee et al., 2013) and 6.0–3.4 cal ka BP with higher sedimentation rates (Orkhonselenge et al., 2018) were recorded in the Lake Ulaan basin (Table 21.4). Moreover, the arid climate in the Lake Ulaan basin causing the decline in the lake level and the increase of salinity ended at 6.8 ka BP (Mischke et al., 2020). Lake Bayan Tukhum represents moderate pluvial conditions in the basin due to higher humidity between 7.0 and 4.0 ka BP (Grunert et al., 2009). The humid period of the Holocene lasted until about 4.0 cal ka BP, interrupted by a shorter dry and cold reversal at 5.0–4.5 cal ka BP, in the Lake Bayan Tukhum basin (Felauer et al., 2012). A humid climate in the middle Holocene was recorded with the high lake levels in Valley of the Lakes (Lehmkuhl et al., 2018a, 2018b). The water level of Lake Orog declined at 7.0–5.0 ka BP, but the middle Holocene highstands (*c.* 6.5 ka and 5.5 ka) fed by higher precipitation in the Govi Altai Mountain Range were reconstructed by alpine vegetation development (Nottebaum et al., 2021; Table 21.4). The humidity in southern Mongolia in the middle Holocene coincided with the simulated high precipitation and snow accumulation in Central Asia with a peak at 4.5 ka BP (Bush, 2005).

In western Mongolia (see Sect. 21.2.3), during a generally regressive phase between 7.3 ± 0.09 and 3.2 ± 70 ka BP, either a lengthy standstill or a slight short-term rise in the generally falling lake level in Lake Uvs pointed to lower temperatures and less precipitation from ca. 5.0 ka BP onward (Grunert et al., 2000; Table 21.4). The aridity was also found with a shallow Lake Ereen in the eastern Mongol Els sand dunefield at 5.5 ka BP (Grunert et al., 2009). A dry-out period in the middle Holocene contributing to the gradual lake shrinking and activity of aeolian processes was recorded in the Lake Tsetseg basin, and since the middle Holocene, the lake dried-out completely and lake deposits were blown out as sandy concretions (Klinge & Lehmkuhl, 2013). A relatively dry and cool episode during 6.0–4.0 ka BP occurred (Rudaya & Li, 2013), and a decline of the forest vegetation and a return to predominance of open vegetation types in the Lake Khoton basin occurred after 5.0 cal ka BP when precipitation decreased to 250–300 mm/year (Rudaya et al., 2009). A mild and dry phase until 6.4 cal ka BP and a relatively warm and wet phase after 6.4 cal ka BP were recorded with pollen data in the Lake Achit basin (Sun et al., 2013). The humid conditions were suggested by the formation of Lake Ereen in the eastern Mongol Els sand dunefield in the early to middle Holocene (Stolz et al., 2012; Table 21.4).

In northern Mongolia (see Sect. 21.2.4), organic C in diatomaceous silt deposits in Lake Khuvsgul decreased abruptly at ca. 5.5 ka BP suggesting a short-term drought (Fedotov et al., 2004; Table 21.4). Low lake levels between 8.0 and 4.0 cal ka BP with short events of higher-rate sedimentation at 7.4–7.1 and 4.8–4.5 cal ka BP (Orkhonselenge et al., 2013) reflect arid climates. Aridity was suggested also by the declined diatom productivity at 6.6 cal ka BP in Lake Khuvsgul

(Prokopenko et al., 2005). A drier and warmer stage was documented to have the pollen-based forest-steppe index at ca. 6.0–3.5 cal ka BP, and it was associated with the summer temperature maximum as a result of higher soil temperatures and lower moisture availability (Prokopenko et al., 2007). However, the pollen records showed a cool and humid stage between 7.5 and 5.5 cal ka BP (Prokopenko et al., 2007). In the Darkhad basin, a relatively deep lake at 6.4–4.5 cal ka BP suggesting humid conditions and a shallow lake level at 7.1–6.4 cal ka BP and the disappearance of the lake after 4.5 cal ka BP suggesting the arid conditions were generally found in the Lake Darkhad basin (Krivonogov et al., 2012a). The generally warm and humid period prevailed in the Lake Dood basin from ~7.8 to ~5.8 cal ka BP, and the cool and dry stage prevailed since 5.8 cal ka BP (Narantsetseg et al., 2013; Table 21.4). Moreover, warm periods prior to 3.0 ka BP were recorded with the sedimentation rates of 9.1–36.6 cm/ka in the Lake Erkhel basin (Watanabe et al., 2007). Relatively high values of TOC at ~7.4–7.3, 6.2–6.1, and 4.2–4.1 cal ka BP in Lake Khargal sediments indicate increased temperature or precipitation in the lake basin, whereas mean grain size of the sediments with the sandy horizons between fine silt and clay to fine sand indicates increased runoff at ~4.4–~4.0 cal ka BP, and the presence of birch woodland with herbaceous taxa at ~7.0 cal ka BP (Barna et al., 2021) implies the climate change in the middle Holocene (Table 21.4).

In central Mongolia (see Sect. 21.2.5), pollen records from Lake Telmen show that the high aridity index indicating the prevailing climate was more arid than the present, and the lake was at a relative low stand between 7.1 and 4.4 cal ka BP (Peck et al., 2002). A relatively arid phase from 7.5 to 4.5 cal ka BP was recorded in the Lake Telmen basin (Fowell et al., 2003; Table 21.4). A warm and arid condition in the Lake Ugii basin at 7.8–3.2 cal ka BP was found by the presence of semidesert steppe plants (Wang et al., 2009). The dry phase was also likely a warm phase (Wang & Feng, 2013). However, high moisture supplies and high lake levels were recorded in the Lake Ugii basin between 7.9 and 4.2 ka BP (Schwanghart et al., 2008). The middle Holocene points to favorable moisture conditions and warming associated with the change from sand mobilization to loess accumulation (Schwanghart et al., 2009). A relatively high abundance of benthic diatoms at Lake Gun implies that the climate was relatively dry at 6.5–5.4 ka BP (Feng et al., 2005), but a cold event was found at 5.2 ka BP (Wang et al., 2004). Unstable climate conditions with intermediate lake levels existed at 7.0–2.5 cal ka BP, and three periods of low lake levels and significantly drier conditions were recorded at 7.0–5.7, 4.1–3.6, and 3.0–2.5 cal ka BP in the Lake Gun basin (Zhang et al., 2012).

21.3.4 Late Holocene

Paleoclimate changes recorded in Mongolian lakes during the late Holocene are synthesized in Table 21.5 and Fig. 21.1d.

In eastern Mongolia (see Sect. 21.2.1), paleoclimate data derived from lakes are relatively limited due to insufficient studies. Lake Khagiin Khar's level since

0.4 cal ka BP (Oh et al., 2019) suggests a comparatively humid condition in the lake basin (Table 21.5). However, recent climate change records from lake basins in this region during the last five decades (or the Anthropocene in Mongolia) differ among each other (see Chaps. 6–8). For instance, the Lake Buir basin is humidifying, while the Lakes Khukh and Yakhi basins are drying, although all the lake basins are warming during the last three to six decades or the Anthropocene (Orkhonselenge & Bulgan, 2021; Table 21.5).

In southern Mongolia (see Sect. 21.2.2), an arid condition after 4.0 cal ka BP accompanied by the strengthening aeolian processes, dune remobilization, and lake desiccation was recorded in the Lake Bayan Tukhum basin (Felauer et al., 2012; Table 21.5). The arid climate caused declining of effective moisture after 3.0 ka BP (Lee et al., 2013), lower sedimentation rate after 3.2–2.7 cal ka BP (Orkhonselenge et al., 2018), and shrinkage of the lakes after 3.2 ka BP (Mischke et al., 2020) in the Lake Ulaan basin. However, evidence for a relatively humid period around 1.5 ka existed in the Valley of the Lakes (Lehmkuhl & Lang, 2001). This wet period during the late Holocene may be accompanied with higher lake stands in Lakes Buun Tsagaan, Adgiin Tsagaan, and Orog in the Valley of the Lakes (Komatsu et al., 2001). The arid climate in the late Holocene was reflected in generally low lake levels and complete disappearance of most lakes in the last 50 years due to the strong human usage of water resources (Lehmkuhl et al., 2018a, 2018b). A significant rise in temperature and a gradual increase in evaporation caused decreasing of areas for Lakes Buun Tsagaan and Orog (Szumińska, 2016). The recent climate changes recorded in the lake basins in this region during the last three to four decades within the Anthropocene show that the lake basins are warming and drying (Table 21.5; see Chaps. 9–11).

In western Mongolia (see Sect. 21.2.3) in Lake Tsetseg basin, a wet phase in the late Holocene contributed to the lake level rising ~1 m above the modern lake, and the recent dry phase has made the lake shrink to the contemporary size (Klinge & Lehmkuhl, 2013; Table 21.5). In the Lake Khoton basin, stable cool and semiarid conditions with moderate moisture during the past 4.0 ka were reconstructed (Rudaya & Li, 2013). The most recent evidence of Lakes Uvs and Bayan standing ~1 m higher than the present level represents a humid climate within the Lake Uvs basin starting at around 3.0 ka BP or 2.0 ka BP (Grunert et al., 2000). Relatively warm and wet periods until 1.6 cal ka BP and a dry period after 1.6 cal ka BP were reconstructed with pollen data in the Lake Achit basin (Sun et al., 2013). A humid phase in the Lake Dayan basin was suggested by the presence of an alpine meadow since 4.3 cal ka BP and by a substantially greater and/or denser forest occurrence starting around 2.4 cal ka BP, while a dry phase was recorded with a marked forest decline after 1.0 cal ka BP (Unkelbach et al., 2019). Extended periods of above-average temperatures at 1.4–1.0 and 0.6–0.4 ka and warming from 0.1 ka up to today were recorded with the increased salinity, while cooler intervals at 1.0–0.6 ka and 0.4–0.1 ka were generally the times of lower salinity in Lakes Zagas and Duruu (Shinneman et al., 2010). The recent dryness linked to the shrinking of Lake Ereen for many years is consistent with the lowering of the groundwater level underneath the alluvial plain of the lake basin (Stolz et al., 2012). A relatively stable Lake Khar remaining in recent years (Walther et al., 2017) suggests a warm condition in the lake basin contributing to glacier melting to feed the lake. However, the recent climate changes recorded in the lake basins in this region during the last three to eight decades or the Anthropocene (Table 21.5) differed among each other (see Chaps. 12–14). For instance, the Lake Khoton basin is humidifying, while the Lakes Achit and Uvs basins are drying, although all these basins are warming (Table 21.5).

In northern Mongolia (see Sect. 21.2.4), an increase of regional moisture in the Lake Khuvsgul basin contributed to the forest expansion after 3.5 to 1.1 cal ka BP (Prokopenko et al., 2007; Table 21.5). Lake Khuvsgul's increased levels at 4.0–2.0 cal ka BP and also in very recent years reflect humid climate, while the dropped level after 2.0 cal ka BP (Orkhonselenge et al., 2013) suggests dry climate. A warm condition after 4.5 cal ka BP was suggested based on the disappearance of large lakes in the Lake Darkhad basin (Krivonogov et al., 2012a). The climate in the Lake Dood basin became cooler and drier after 5.8 cal ka BP (Narantsetseg et al., 2013). However, the recent climate changes recorded in the lake basins in this region during the last five decades or the Anthropocene (Table 21.5) differed among each other (see Chapters 15–17). For instance, the Lake Darkhad basin is humidifying, while the Lakes Khuvsgul and Khargal basins are drying, although all these basins are warming (Table 21.5). In Lake Khargal basin, abundant conifer needles record growth of larch near the lake by ~1.5 cal ka BP, possibly due to increased moisture availability and southward expansion of the taiga during the late Holocene (Barna et al., 2021; Table 21.5).

In central Mongolia (see Sect. 21.2.5), a relatively humid condition at 2.7–1.3 cal ka BP in Lake Telmen basin was signaled by the low aridity index or an increase of effective moisture since 1.3 ka BP (Peck et al., 2002). This humid condition was in response to the maximum humidity at 4.5–1.6 cal ka BP during the Subboreal and Subatlantic periods with a short dry period at 1.6–1.2 cal ka BP (Fowell et al., 2003; Table 21.5). Arid conditions at 4.0–2.8 ka BP and a stable humid phase since 2.8 ka BP prevailed in the Lake Ugii basin (Schwanghart et al., 2008, 2009) where the humid climate was considered to be consistent with the retreat of semidesert steppe and expansion of forest steppe (Wang et al., 2009, 2011). In the Lake Khuis basin, a series of climate changes since 1.2 ka BP were recorded by the vegetation as dry steppe dominated at ca. 1.2–1.0 ka BP, forest steppe at ca. 1.0–0.6 ka BP, meadow dominated cold-wet period at ca. 0.6–0.1 ka BP, and forest steppe after 0.1 ka BP (Tian et al., 2013). The humid condition is suggested by the high biogenic response in Lake Terkhiin Tsagaan at ~0.2 ka BP, while an arid condition is indicated by the low biogenic response at ~0.1 ka BP (Fukushi et al., 2015). The most humid conditions seem to have occurred from 4.5 to 2.5 (possibly up to 1.6) ^{14}C ka BP in the Lake Gun basin (Feng et al., 2005). Another study states that in the Lake Gun basin, low lake levels and a significantly drier condition were recorded at 3.0–2.5 cal ka BP, wet conditions were established around 2.5 cal ka BP and the lake level rose again due to the higher effective moisture, and the relatively wet conditions of the present were achieved ca. 1.6 cal ka BP (Zhang et al., 2012). However, the recent climate changes recorded in the lake basins in this region during the last five decades or the Anthropocene (Table 21.5) differed among each other (see Chaps. 18–20). For instance, the Lake Terkhiin Tsagaan basin is humidifying, while the Lakes Telmen and Ugii basins are drying, although all these basins are warming (Table 21.5).

21.4 Regional Pattern of Paleoclimate

The late Quaternary paleoclimate changes revealed from lakes in adjacent regions of Eurasia and Central Asia are synthesized in Table 21.6. The northern hemisphere summer insolation reached the maximum during the early Holocene (Wang, 2011). This led to the strengthening of the EASM and westerlies as well as the higher effective moisture in most parts of Eurasia (e.g., Morril et al., 2003; Herzschuh, 2006; Tarasov et al., 2007b). In a large regional synthesis, Lehmkuhl and Haselein (2000) attributed the increased lake levels at the neoglacial period (4.0–3.0 ka) to about 3.0–2.0 ka with climate conditions wetter and cooler than today in Central Asia. Here, we compare the adjacent lake records from southern Russia and northern China (Table 21.6, Fig. 21.1) with the records from Mongolian lakes (Tables 21.1, 21.2, 21.3, 21.4, and 21.5).

21.4.1 Southern Russia

The late Pleistocene and Holocene climates in Mongolia are approximately synchronous with the records from southern Russia (Table 21.6, Fig. 21.1), in particular with the climate in southern Siberia (including the Lake Baikal[8], Sayan, and Altai regions).

21.4.1.1 Baikal Region

In the Lake Baikal region, high-resolution pollen data showed climate variability as warm and humid prior to 16.0 cal ka BP, cool and dry at BA and YD, warm and humid at 11.0–7.5 ka BP with cooling at 9.5–8.5 ka BP and drying at 8.0–7.0 ka BP, the Holocene thermal optimum (6.5–5.7 ka BP), and cooling at 5.5–4.5 and 3.0–2.5 ka BP (Demske et al., 2005; Table 21.6). The distribution of Siberian spruce (*Picea obovata*) and *Picea* pollen and forest-steppe vegetation index review the past climatic conditions in the Lake Baikal region since the late glaciation as expansion of Siberian spruce in BA time, steppe landscape prior to 9.2–9.0 cal ka, declining of Siberian spruce from YD to 8.6–6.8 cal ka, and decreasing temperatures and higher

[8] *Baigali* meaning *nature* has been misspelled as *Baikal* based on the Russians' pronunciation. Baigali is the right English transliteration from Mongolian *Байгаль*.

moisture availability with higher *Picea* abundance and higher forest-steppe index at 6–5 cal ka and 3–2.5 cal ka and abundance of *P. obovata* in the late Holocene (Bezrukova et al., 2005). The biogenic silica (BioSi) signal in Lake Baikal sediments during the present interglacial is antiphased with the early Holocene regional humidity indices, and the BioSi follows a significant rise of the regional annual temperature predicted by global climate modeling (GCM) simulations around ca. 6.0 ka (Prokopenko et al., 2006). The diatom and BioSi proxies in Lake Baikal respond to the annual air temperature increases after 6.0 ka when the TOC maximized with a rapid increase in the BioSi production and diatom abundance at ca. 4.0–2.5 ka (Prokopenko et al., 2007). The high moisture availability in the lake basin during the early Holocene declined rapidly between 7.5 and 6.5 ka (Prokopenko et al., 2007). Pollen records from Lake Baikal shows dominances of tundra and steppe biomes at 15.0–13.3 cal ka BP, dominance of taiga after 13.3 cal ka BP, and onset of relatively warm and wet conditions after 10.0 cal ka BP with the warmest and wettest climate at 9.0–7.0 cal ka BP (Tarasov et al., 2007a; Table 21.6). The dominance of the boreal conifer (taiga) forest vegetation in the Baikal region shows that the warm conifer phase during the MIS 11 lasted to 15.0 cal ka BP, and the conifer optimum dominated/centered at 420–405 cal ka with abundant *Abies sibirica*, indicative of climate significantly warmer and less continental than today, was found at 9.0–7.0 cal ka (Prokopenko et al., 2010). The moist conditions prevailed in the Lake Baikal region during the latter stages of the YD, and fluvial input at 3.3–2.0 cal ka BP coincided with high precipitation in the region (Mackay et al., 2011).

In the saline Lake Tsagan-Tyrm basin within the Ol'khon area of the Lake Baikal region, various characteristics of essentially carbonate sediments show that the Ol'khon regional climate has become more arid in recent 6.5 ka (Sklyarov et al., 2010). In the Lake Kotokel basin in the eastern Lake Baikal, pollen-based climate was reconstructed as lower temperature (~ −38 °C in January, ~12 °C in July) and annual precipitation (~270–300 mm) prior to 13.5 cal ka BP and between 12.8 and 11.6 ka BP with open steppe and tundra communities, a noticeable increase in precipitation and temperatures with boreal woodland between 11.5 and 10.5 ka BP, a warmer and wetter climate at 10.8–7.3 cal ka BP with the boreal forest (taiga) communities, a gradual decrease in precipitation and temperature with a decrease in woods at about 7.0–6.5 ka BP and since that time gradual decrease in precipitation and average January temperature towards their present day (Tarasov et al., 2009). The pollen record from Lake Kotokel agrees with the general climatic trend in the less continental climate of the early Holocene (~11.0 ka) to ~7.0 ka and, later on, more continental climate, which led to a significant decrease in the annual average precipitation and winter air temperature and an increase in the average summer air temperature (Bezrukova et al., 2011; Table 21.6).

21.4.1.2 Sayan Region

In the Lake Tere-Khol basin at the southwestern edge of the Baikal Rift Zone (BRZ) within the Sayan-Tuva Highland, past environments were reconstructed (Panin et al., 2012) as wet environment before 10.5 ka BP with wetland plants, dry climate at 10.5–7.3 ka BP with decreased plants, wet climate at 7.3–6.2 ka BP with deep lakes and increased plants, drying tendency at 6.2–3.8 ka BP, wet climate at 3.8–2.0 ka BP with diatom growth, and dry climate since 2.0 ka BP contributing shallow lakes (Table 21.6).

In Lake Khara Nur within the east Sayan mountains, layers of clayish sand rich in organic matter and peat formed at 6.8–6.4 ka showing the highest level of summer insolation and the development of woody vegetation and the reduced abundance of the arboreal pollen, especially dark coniferous pollen, which attributes to the post-optimal cooling. But the expansion of the dark coniferous associations at 4.1–0.6 ka, caused by the atmospheric humidity increase, and the increased forest vegetation around 0.6 ka lasting until the present day are also found (Alexander & Bezrukova, 2019).

21.4.1.3 Altai Region

In the Lakes Uzungol, Kendegelukol, and Tashkol basins of the Ulagan high mountains in the Russian Altai, the vegetation history showed the warm and dry conditions prevailing in the late glacial and Holocene, warmer and somewhat moister climatic conditions at 12.0–9.5 cal ka BP, strong influence of moist air after 9.5 cal ka BP, and cooler and more continental climatic conditions from 7.5 cal ka BP lasting to the present (Blyakharchuk et al., 2004; Table 21.6). In glacial Lakes Grusha and Akkol of the Russian Altai, pollen records reconstructed the climate changes as a cool and dry climate with tundra-steppe and meadow-steppe and low organic content during the YD period, a warm and humid climate with increased organic content and expanded forests (e.g., *Pinus sibirica*, *P. sylvestris*, *Picea obovata*, and *Abies sibirica*) at 11.0 cal ka BP, a cooler and drier climate recorded with declined forests after 6.0 cal ka BP, and a cooling and increased aridity recorded with increased alpine herbs after 6.0 cal ka BP (Blyakharchuk et al., 2007).

In western Siberia, climate-driven lake level stands of Lake Beloye showed a cold stage at 3.4–2.3 cal ka BP, a dry stage at 2.8–1.7 cal ka BP, a warm and wet stage at 1.3–0.2 cal ka BP, and a dry stage after 0.2 cal ka BP (Krivonogov et al., 2012b; Table 21.6). In northeastern Siberia, major paleoclimate trends recorded from a small shallow lake in the downstream of Lena River infer a cold period at ~12.7 to 11.6 cal ka BP with the occurrence of pyrite, oligotrophic system with alkaline and cold conditions during the earliest Holocene, moderately warmer climates at ~9.1 to 5.7 cal ka BP with a warmer hydrological regime at ~5.7 cal ka BP during the Holocene thermal maximum, declined diatom valve concentrations starting ~2.8 cal ka BP as the result of neoglacial cooling, and rising diatom again since 0.6 cal ka BP which is interpreted to be related to the subsequent recent climate warming (Biskaborn et al., 2012).

21.4.2 Northern China

In this review of the regional pattern, the late Pleistocene and Holocene climatic trends in Mongolia recorded from lakes are consistent with data from northern, northeastern, and northwestern China (Table 21.6, Fig. 21.1).

The paleomonsoon records from China demonstrate an abrupt strengthening of the EASM at 12.0 ka BP and associated increases of precipitation and lake levels between 11.0 and 8.0 ka BP, followed by the stepwise attenuation of the monsoon circulation and climate aridization between ~8.0 and 4.0 cal ka BP toward the modern level (Chen et al., 2003).

21.4.2.1 Northern Region

Several reconstructions carried out in arid and semiarid China demonstrate the middle Holocene as a generally dry period (e.g., Guo et al., 2000; Chen et al., 2003). The aridity is expected at 7.5–5.0 ka BP with the Holocene optimum, and the EASM maximum, a response to the changing seasonality related to orbital forcing of the climate, indicates the progressive increases of summer solar radiation anomaly returned after 8.0–7.0 ka BP in northern China (An et al., 2000). The Community Climate Model 0 (CCM0) shows that a minimum of 19% deviation of precipitation at 6.0 ka BP in northern east-central China (An et al., 2000) reflects clearly the Holocene optimum (effective moisture maximum) intensifying at 9.0 ka B.P in northern China and shifting southward and weakening by 6.0 ka BP, but the drying began after 8.0–7.0 ka BP.

The climatic optimum was the peak of the megathermal climate which occurred between ~8.5 and ~ 3.0 ka BP (Shi & Kong, 1992) in the entire northern China from ~7.2 to ~6.0 ka BP (Guo et al., 2007).

21.4.2.2 Northeastern Region

A lowered lake level recorded in saline Lake Baahar located on the Ordos Plateau documented a warm and relatively dry episode from ~6.7 to ~6.2 ka BP and a generally drier and cooler climate from ~5.4 to ~4.7 ka BP (Guo et al., 2007; Table 21.6). The transition from a humid to an arid climate after ~5.9 ^{14}C ka BP in the woodland-steppe ecotone of the southern Inner Mongolia in northeastern China coincided with the enhanced aeolian activity and deciduous woodlands, and the transitions led to a warm and dry period from 5.9 to 2.9 ^{14}C ka BP, and the present desertification is due to the weakening of the EASM and the continuing increase of winter air temperature (Liu et al., 2002).

A high-resolution pollen analysis of a 2.4-m-long core obtained from Lake Diaojiao in the central part of Inner Mongolia in the northeastern China reconstructs the Holocene climates as arid-cold at 9.2–7.8 ka BP and arid-warm at 7.8–7.1 ka BP (Shi & Song, 2003). A multi-proxy study of the aeolian deposits in central Inner Mongolia reconstructed an arid climate after ~5.6 ka (Sun et al., 2006). A high-resolution pollen record from Lake Bayanchagan, southern Inner Mongolia, indicates that the Holocene cooling started at approximately 8.0 cal ka BP, 1.5 cal ka BP earlier than the start of the drying phase resulting from surface temperature responding more directly than precipitation to the decrease of solar output (Jiang et al., 2006; Table 21.6).

The magnetic data of core sediments in combination with particle size, TOC, and C/N and pollen analyses performed on Lake Anguli-nuur sediments in southeastern Inner Mongolia indicated that the environment began to deteriorate from 7.4 to 6.9 cal ka BP when the lake water level was apparently dropping, which corresponds to an enhanced drought, and wind activities apparently intensified around the lake during 6.9–6.4 cal ka BP (Wang et al., 2010). Pollen-assemblage data from Lake Hulun in northeastern Inner Mongolia suggested that the climate became cooler and drier from 6.4 to 4.4 cal ka BP as indicated by the decreased *Artemisia/Chenopodiaceae* ratios, declined broad-leaved forests, and expanded coniferous forests (Wen et al., 2010).

The sediment sequences recovered from Lakes Bai Nuur and Ulaan Nuur situated in the transition zone between semi-humid and semiarid climate regimes in southeastern Inner Mongolia showed an environmental deterioration toward aridity commenced at 7.7 or 7.6 cal ka BP and further enhanced at 6.9 or 6.8 cal ka BP around Bai Nuur, and the environments began to deteriorate at 7.2 or 7.1 cal ka BP and remarkably strengthened at 6.2 or 6.1 cal ka BP around Ulan Nuur (Wang et al., 2012; Table 21.6).

21.4.2.3 Northwestern Region

Fossil pollen data recorded in the eastern Juyan paleolake showed a rather dry climate ca. 10.7–5.4 cal ka BP, moisture conditions at 5.4–3.9 cal ka BP, and dry conditions at about 3.9 cal ka BP (Herzschuh et al., 2004; Table 21.6). Several lake records from northwestern China indicate moist and warm conditions between 7.0 and 5.0 cal ka BP (Wünnemann et al., 2003).

The dry and cold conditions during 5.0–4.5 cal ka BP are in agreement with the Lake Boston dated to be 5.5 cal ka BP indicating a weakening trend of the westerlies (Wünnemann et al., 2003; Table 21.6). Lake Wulungu in northern Xinjiang experienced a shallow lake phase or wetland environment during the early Holocene (~10.0–6.7 cal ka BP) as indicated by high pollen-based wetland biome scores (Xiao et al., 2006; Liu et al., 2008). A general wetting trend since 6.7 cal ka BP is suggested by the increasing of *Artemisia/Chenopodiaceae* ratio and the decreasing of pollen-based desert biome scores, suggesting that the Holocene moisture optimum occurred between 4.2 and 0.5 cal ka BP (Liu et al., 2008).

21.5 Summary

The synthetic records inferred from lakes of Mongolia with high-resolution multiproxy data provide a picture of the late Quaternary climate changes in Mongolia and in the adjacent regions of Eurasia and Central Asia. This reconstruction shows that the spatial and temporal patterns of the paleoclimate changes varying from region to region within Mongolia and the adjacent regions are dependent on driving impacts of atmospheric circulations derived from the East Asian Winter Monsoon, East Asian Summer Monsoon, and westerlies of the northern hemisphere during the late Quaternary. There are similar spatial trends regarding paleoclimate changes among southern, western, and central Mongolia. The spatial patterns of the paleoclimate changes of Mongolia are synthesized as temporal patterns as the cold and humid late Pleistocene, the warm and humid early Holocene, the warm and arid middle Holocene, the cool and humid late Holocene, and the warm and arid Anthropocene. Sufficient geochronological controls on sedimentary sequences or archives from more lakes in Mongolia, especially eastern Mongolia, are necessary for a further complete understanding of the late Pleistocene and Holocene climate changes in Mongolia and to expand our knowledge of the overall paleoclimatic history in Mongolia, Eurasia, and Central Asia.

References

Alexander, S., & Bezrukova, E. V. (2019). Lakes of the Jom-Bolok Volcanoes Valley in the East Sayan Mts., Baikal region: Morphogenesis and potential for regional paleoenvironmental studies. *Journal of Geographical Science, 29*, 1–18.

An, Z., Stephen, C. P., John, E. K., Wu, X., Wang, S., Liu, X., Li, X., & Zhou, W. (2000). Asynchronous Holocene optimum of the East Asian monsoon. *Quaternary Science Reviews, 19*, 743–762.

An, C. B., Chen, F. H., & Barton, L. (2008). Holocene environmental changes in Mongolia: A review. *Global and Planetary Change, 63*, 283–289.

Badarch, N., Tsegmid, S., & Tserensodnom, J. (1965). *Landscapes and natural zones of eastern Mongolia.* Press of Mongolian Academy of Sciences, 117 p. [In Mongolian].

Barna, J., Bigelow, N., Orkhonselenge, A., Fowell, S., Gardner, W. R. M., Burentogtokh, J., & Honeychurch, W. (2021). Holocene paleoenvironmental changes in the forest-steppe ecotone of Northern Mongolia based on palynological and sedimentological analyses of cores from Lake Khargal. In *GSA Annual Meeting—The Geological Society of America. Presentation #27–12. Portland, Oregon.* October 10–13, 2021. https://doi.org/10.1130/abs/2021AM-370452.

Bezrukova, E. V., Abzaeva, A. A., Letunova, P. P., Kulagina, N. V., Vershinin, K. E., Belov, A. V., Orlova, L. A., Danko, L. V., & Krapivina, S. M. (2005). Post-glacial history of Siberian spruce (*Picea obovata*) in the Lake Baikal area and the significance of this species as a paleo-environmental indicator. *Quaternary International, 136*(1), 47–57.

Bezrukova, E. V., Tarasov, P. E., Kulagina, N. V., Abzaeva, A. A., Letunova, P. P., & Kostrova, S. S. (2011). Palynological study of Lake Kotokel' bottom sediments (*Lake Baikal region*). *Russian Geology and Geophysics, 52*(4), 458–465.

Biskaborn, K. B., Herzschuh, U., Bolshiyanov, D., Savelieva, L., & Diekmann, B. (2012). Environmental variability in northeastern Siberia during the last ~13,300 yr inferred from lake diatoms and sediment–geochemical parameters. *Palaeogeography, Palaeoclimatology, Palaeoecology, 329–330*, 22–36.

Blomdin, R., Stroeven, A. P., Harbor, J. M., Gribenski, N., Caffee, M. W., Heyman, J., Rogozhina, I., Ivanov, M. N., Petrakov, D. A., Walther, M., Rudoy, A. N., Zhang, W., Orkhonselenge, A., Hättestrand, C., Lifton, N. A., & Jansson, K. N. (2018). Timing and dynamics of glaciation in the Ikh Turgen mountains, Altai Region, High Asia. *Quaternary Geochronology, 47*, 54–71.

Blyakharchuk, T. A., Wright, H. E., Borodavko, P. S., Van Der Knaap, W. O., & Ammann, B. (2004). Late Glacial and Holocene vegetational changes on the Ulagan high-mountain plateau, Altai Mountains, southern Siberia. *Palaeogeography, Palaeoclimatology, Palaeoecology, 209*, 259–279.

Blyakharchuk, T. A., Wright, H. E., Borodavko, P. S., Van Der Knaap, W. O., & Ammann, B. (2007). Late Glacial and Holocene vegetational history of the Altai Mountains (southwestern Tuva Republic, Siberia). *Palaeogeography, Palaeoclimatology, Palaeoecology, 245*, 518–534.

Bridge, J., & Demicco, R. (2008). *Earth surface processes, landforms and sediment deposits*. Cambridge University Press, 815 p. ISBN: 978-0-521-85780-2.

Bush, A. B. G. (2005). CO_2/H_2O and orbitally driven climate variability over central Asia through the Holocene. *Quaternary International, 136*(1), 15–23.

Chen, T., Chen, A., Lan, H., Lou, J., & Chen, Y. (2003). The Dry Holocene Megathermal in Inner Mongolia. *Palaeogeography, Palaeoclimatology, Palaeoecology, 193*, 181–200.

Chen, F., Yu, Z., Yang, M., Ito, E., Wang, S., Madsen, D. B., Huang, X., Zhao, Y., Sato, T., Birks, H. J. B., Boomer, I., Chen, J., An, C., & Wunnemann, B. (2008). Holocene moisture evolution in arid central Asia and its out-of-phase relationship with Asian monsoon history. *Quaternary Science Reviews, 27*, 351–364.

D'Arrigo, R., Jacoby, G., Pederson, N., Frank, D., Buckley, B., Nachin, B., Mijiddorj, R., & Dugarjav, C. (2000). Mongolian tree-rings, temperature sensitivity and reconstructions of Northern Hemisphere temperature. *The Holocene, 10*(6), 669–672.

Davaagatan, T., Orkhonselenge, A., Fukushi, K., & Yabe, M. (2015). Sedimentary records in intermontane and prairie lakes in central Mongolia: Preliminary results from Lake Terkhiin Tsagaan and Lake Ugii. *Journal of Geographical Issues, 14*(408), 19–26.

Davi, N. K., Pederson, N., Leland, C., Nachin, B., Suran, B., & Jacoby, G. C. (2013). Is eastern Mongolia drying? A long-term perspective of a multidecadal trend. *Water Resources Research, 49*, 151–158.

Demske, D., Heumann, G., Granoszewski, W., Nita, M., Mamakowa, K., Tarasov, P. E., & Oberhänsli, H. (2005). Late glacial and Holocene vegetation and regional climate variability evidenced in high-resolution pollen records from Lake Baikal. *Global and Planetary Change, 46*, 255–279.

Dorofeyuk, N. I., & Tarasov, P. E. (1998). Vegetation and lake levels of northern Mongolia since 12,500 yr B.P. based on the pollen and diatom records. *Stratigraphy and Geology Correlation, 6*(1), 70–83.

Fedotov, A. P., Chebykin, E. P., Semenov, M. Y., Vorobyova, S. S., Osipov, E. Y., Golobokova, L. P., Pogodaeva, T. V., Zheleznyakova, T. O., Grachev, M. A., Tumurhuu, D., Oyunchimeg, T., Narantsetseg, T., Tumurtogoo, O., Dolgikh, P. T., Arsenyuk, M. I., & Batist, M. D. (2004). Changes in the volume and salinity of Lake Khubsugul (Mongolia) in response to global climate changes in the upper Pleistocene and the Holocene. *Paleogeography, Paleoclimatology, Paleoecology, 209*, 245–257.

Fedotov, A. P., Ignat'ev, A. V., & Velivetskaya, T. A. (2015). Reconstruction of deglaciation of Northern Mongolia for the last 330 ka BP, inferred from ostracod stable isotope records from Lake Khubsugul. *Environmental Earth Science, 74*, 2041–2054.

Felauer, T., Schlütz, F., Murad, W., Mischke, S., & Lehmkuhl, F. (2012). Late Quaternary climate and landscape evolution in arid Central Asia: A multiproxy study of lake archive Bayan Tohomin Nuur, Gobi desert, southern Mongolia. *Journal of Asian Earth Sciences, 48*, 125–135.

Feng, Z. D., Ma, Y. Z., Zhang, H. C., Narantsetseg, T., & Zhang, X. S. (2013). Holocene climate variations retrieved from Gun Nuur lake-sediment core in the northern Mongolian Plateau. *The Holocene, 23*(12), 1721–1730.

Feng, Z. D., Thompson, L. G., Mosley-Thompson, E., & Yao, T. D. (1993). Temporal variations of climate in China during the past 10,000 years. *The Holocene, 3*, 174–180.

Feng, Z. D., Wang, W. G., Guo, L. L., Khosbayar, P., Narantsetseg, T., Jull, A. J. T., An, C. B., Li, X. Q., Zhang, H. C., & Ma, Y. Z. (2005). Lacustrine and eolian records of Holoceneclimate changes in the Mongolian Plateau: Preliminary results. *Quaternary International, 136*, 25–32.

Feng, Z. D., Zhai, X. W., Ma, Y. Z., Huang, C. Q., Wang, W. G., Zhang, H. C., Khosbayar, P., Narantsetseg, T., Liu, K. B., & Rutter, N. W. (2007). Eolian environmental changes in the Northern Mongolian Plateau during the past ~ 35,000 yr. *Palaeogeography, Palaeoclimatology, Palaeoecology, 245*(3–4), 505–517.

Fowell, S. J., Hansen, B. C., Peck, J. A., Khosbayar, P., & Ganbold, E. (2003). Mid to late Holocene climate evolution of the Lake Telmen basin, north central Mongolia, based on palynological data. *Quaternary Research, 59*(3), 353–363.

Fukumoto, Y., Kashima, K., Orkhonselenge, A., & Ganzorig, U. (2012). Holocene environmental changes in northern Mongolia inferred from diatom and pollen records of peat sediment. *Quaternary International, 254*, 83–91.

Fukushi, K., Katsuta, N., Jenkins, R. G., Matsubara, K., Takayama, B., Tanaka, Y., Davaasuren, D., Batkhishig, O., Hasebe, N., & Kashiwaya, K. (2015). Centennial-scale environmental changes in Terhiin Tsagaan Lake, Mongolia Inferred from lacustrine sediment: Preliminary results. In K. Kashiwaya, J. Shen, & J. Y. Kim (Eds.), *Earth surface processes and environmental changes in East Asia—Records from lake-catchment systems* (pp. 25–44). Springer.

Gillespie, A. R., Burke, R. M., Komatsu, G., & Bayasgalan, A. (2008). Late Pleistocene glaciers in Darhad Basin, northern Mongolia. *Quaternary Research, 69*(2), 169–187.

Gomboluudev, P. (2006). Future climate change of Mongolia under Special Report Emission Scenarios (SRES). In *Proceedings of Fifth Mongolia-Korea Joint Seminars on Environmental Changes of Northeast Asia. October 10–14. Ulaanbaatar, Mongolia.*

Gong, D. Y., & Ho, C. H. (2002). The Siberian High and climate change over middle to high latitude Asia. *Theoretical and Applied Climatology, 72*, 1–9.

Gong, D. Y., Wang, S. W., & Zhu, J. H. (2001). East Asian winter monsoon and arctic oscillation. *Geophysical Research Letters, 28*, 2073–2076.

Grosswald, M. G., & Rudoy, A. N. (1996). Quaternary glacier-dammed lakes in the mountains of Siberia. *Polar Geography, 20*(3), 180–198.

Grunert, J., Lehmkuhl, F., & Walther, M. (2000). Paleoclimatic evolution of the Uvs Nuur basin and adjacent areas (Western Mongolia). *Quaternary International, 65–66*, 171–192.

Grunert, J., Stolz, C., Hempelmann, N., Hilgers, A., Hulle, D., Lehmkul, F., Felauer, T., & Dasch, D. (2009). The evolution of small lake basin in the Gobi desert in Mongolia. *Quaternary Sciences, 29*(4), 678–686.

Guo, Z. T., Petit-Maire, N., & Kropelin, S. (2000). Holocene non-orbital climatic events in present-day arid areas of northern Africa and China. *Global Planetary Change, 26*, 97–103.

Guo, L. L., Feng, Z. D., Li, X. Q., Liu, L. Y., & Wang, L. X. (2007). *Holocene climatic and environmental changes recorded in Baahar Nuur Lake core in the Ordos Plateau, Inner Mongolia of China*. Chinese Science Bulletin. Science in China Press Springer-Verlag.

Herzschuh, U. (2006). Palaeo-moisture evolution in monsoonal Central Asia during the last 50,000 years. *Quaternary Science Reviews, 25*(1–2), 163–178.

Herzschuh, U., Tarasov, P., Wünnemann, B., & Hartmann, K. (2004). Holocene vegetation and climate of the Alashan Plateau, NW China, reconstructed from pollen data. *Palaeogeography, Palaeoclimatology, Palaeoecology, 211*(1–2), 1–17.

Hilbig, W. (1995). *The vegetation of Mongolia*. SPB Academic Publishing, 260 p.

Hoerling, M. P., Hurrell, J. W., & Xu, T. (2001). Tropical origins for recent North Atlantic climate change. *Science, 292*, 90–92.

Jacoby, G., Lovelius, N., Shumilov, O., Raspopov, O., Kurbinov, J., & Frank, D. (2000). Long-term temperature trends and tree growth in the Taymir region of northern Siberia. *Quaternary Research, 53*, 312–318.

Jiang, W., Guo, Z., Sun, X., Wu, H., Chu, G., Yuan, B., Hatte, C., & Guiot, J. (2006). Reconstruction of climate and vegetation changes of Lake Bayanchagan (Inner Mongolia): Holocene variability of the East Asian monsoon. *Quaternary Research, 65*, 411–420.

Karabanov, E., Williams, D., Kuzmin, M., Sideleva, V., Khursevich, G., Prokopenko, A., Solotchina, E., Tkachenko, L., Fedenya, S., Kerber, E., Gvozdkov, A., Khlustov, O., Bezrukova, E., Letunova, P., & Krapivina, S. (2004). Ecological collapse of Lake Baikal and Lake Hovsgol ecosystems during the Last Glacial and consequences for aquatic species diversity. *Palaeogeography, Palaeoclimatology, Palaeoecology, 209*(1–4), 227–243.

Kashiwaya, K., Ochiai, S., Sumino, G., Tsukamoto, T., Szyniszewska, A., Yamamoto, M., Sakaguchi, A., Hasebe, N., Sakai, H., Watanabe, T., & Kawai, T. (2010). Climato-hydrological fluctuations printed in long lacustrine records in Lake Hövsgöl, Mongolia. *Quaternary International, 219*, 178–187.

Kerr, R. A. (1999). A new force in high-latitude climate. *Science, 284*, 241–242.

Khosbayar, P. (2005). *Mezozoic and Cenozoic Paleogeography and Paleoclimate of Mongolia.* Institute of Geology and Mineral Resources, Mongolian Academy of Sciences, Transaction 15, 184 p. [In Mongolian].

Klinge, M., & Lehmkuhl, F. (2013). Geomorphology of the Tsetseg Nuur basin, Mongolian Altai—Lake development, fluvial sedimentation and aeolian transport in a semi-arid environment. *Journal of Maps, 9*(3), 361–366.

Klinge, M., Schlütz, F., Zander, A., Hülle, D., Batkhishig, O., & Lehmkuhl, F. (2021). Late Pleistocene lake level, glaciation and climate change in the Mongolian Altai deduced from sedimentological and palynological archives. *Quaternary Research, 99*, 168–189. https://doi.org/10.1017/qua.2020.67

Klinge, M., & Sauer, D. (2019). Spatial pattern of Late Glacial and Holocene climatic and environmental development in Western Mongolia—A critical review and synthesis. *Quaternary Science Reviews, 210*, 26–50.

Komatsu, G., Brantingham, P. J., Olsen, J. W., & Baker, V. R. (2001). Paleoshoreline geomorphology of Boon Tsagaan Nuur, Tsagaan Nuur and Orog Nuur: The Valley of Lakes, Mongolia. *Geomorphology, 39*, 83–98.

Komatsu, G., Arzhannikov, S. A., Gillespie, A. R., Burke, R. M., Miyamoto, H., & Baker, V. R. (2009). Quaternary paleolake formation and cataclysmic flooding along the upper Yenisei River. *Geomorphology, 104*, 143–164.

Krivonogov, S. K., Bezrukova, E. V., Takahara, H., & Riedel, F. (2003). The Hovsgol Lake in the Late Pleistocene and Holocene: On-land geologic evidences for the change of its level. In K. Kashiwaya (Ed.), *Long continental records from Lake Baikal* (pp. 187–206). Springer-Verlag.

Krivonogov, S. K., Sheinkman, V. S., & Mistruykov, A. A. (2005). Stages in the development of the Darhad dammed lake (Northern Mongolia) during the Late Pleistocene and Holocene. *Quaternary International, 136*(1), 83–94.

Krivonogov, S. K., Takahara, H., Yamamuro, M., Pries, Y. I., Khazina, I. V., Khazin, L. B., Kuzmin, Y. V., Safonova, I. Y., & Ignatova, N. V. (2012a). Regional to local environmental changes in southern Western Siberia: Evidence from biotic records of mid to late Holocene sediments of Lake Beloye. *Palaeogeography, Palaeoclimatology, Palaeoecology, 331–332*, 177–193.

Krivonogov, S. K., Yi, S., Kashiwaya, K., Kim, J. C., Narantsetseg, T., Oyunchimeg, T., Safonova, I. Y., Kazansky, A. Y., Sitnikova, T., Kim, J. Y., & Hasebe, N. (2012b). Solved and unsolved problems of sedimentation, glaciation and paleolakes of the Darhad Basin, Northern Mongolia. *Quaternary Science Reviews, 56*(21), 142–163.

Lee, M. K., Lee, Y. I., Lim, H. S., Lee, J. I., Choi, J. H., & Yoon, H. I. (2011). Comparison of radiocarbon and OSL dating methods for a Late Quaternary sediment core from Lake Ulaan, Mongolia. *Journal of Paleolimnology, 45*, 127–135.

Lee, M. K., Lee, Y. I., Lim, H. S., Lee, J. I., & Yoon, H. I. (2013). Late Pleistocene–Holocene records from Lake Ulaan, southern Mongolia: Implications for east Asian palaeomonsoonal climate changes. *Journal of Quaternary Science, 28*(4), 370–378.

Lehmkuhl, F., Hilgers, A., Fries, S., Hülle, D., Schlütz, F., Shumilovskikh, L., Felauer, T., & Protze, J. (2012). Holocene geomorphological processes and soil development as indicator for environmental change around Karakorum, Upper Orkhon Valley (Central Mongolia). *Catena, 87*(1), 31–44.

Lehmkuhl, F., Grunert, J., Hülle, D., Batkhishig, O., & Stauch, G. (2018a). Paleolakes in the Gobi region of southern Mongolia. *Quaternary Science Reviews, 179*, 1–23.

Lehmkuhl, F., & Haselein, F. (2000). Quaternary paleoenvironmental change on the Tibetan Plateau and adjacent areas (Western China and Western Mongolia). *Quaternary International, 65–66*, 121–145.

Lehmkuhl, F., & Lang, A. (2001). Geomorphological investigations and luminescence dating in the southern part of the Khangay and the Valley of the Gobi Lakes (Central Mongolia). *Journal of Quaternary Science, 16*, 69–87.

Lehmkuhl, F., Nottebaum, V., & Hülle, D. (2018b). Aspects of late Quaternary geomorphological development in the Khangai Mountains and the Gobi Altai Mountains (Mongolia). *Geomorphology, 312*, 24–39.

Li, J., Cook, E. R., D'Arrigo, R., Chen, F., & Gou, X. (2009). Moisture variability across China and Mongolia: 1951–2005. *Climate Dynamics, 32*, 1173–1186.

Li, Y., & Morrill, C. (2010). Multiple factors causing Holocene lake-level change in monsoonal and arid central Asia as identified by model experiments. *Climate Dynamics, 35*, 1119–1132.

Liu, H., Xu, L., & Cui, H. (2002). Holocene history of desertification along the Woodland-Steppe Border in Northern China. *Quaternary Research, 57*, 259–270.

Liu, X., Dong, H., Rech, J. A., Ryo, M., Yang, B., & Wang, Y. (2008). Evolution of Chaka Salt Lake in NW China in response to climatic change during the Latest Pleistocene–Holocene. *Quaternary Science Reviews, 27*, 867–879.

Mackay, A. W., Swann, G. E. A., Brewer, T. S., Leng, M. J., Morley, D. W., Piotrowska, N., Rioqual, P., & White, D. (2011). A reassessment of late glacial—Holocene diatom oxygen isotope record from Lake Baikal using a geochemical mass-balance approach. *Journal of Quaternary Science, 26*(6), 627–634.

Mischke, S., Lee, M. K., & Lee, Y. I. (2020). Climate history of Southern Mongolia since 17 ka: The ostracod, gastropod and charophyte record from Lake Ulaan. *Frontiers in Earth Science, 8*(221), 1–15.

Morril, C., Overpeck, J. T., & Cole, J. E. (2003). A synthesis of abrupt changes in the Asian summer monsoon since the last deglaciation. *The Holocene, 13*(4), 465–476.

Narantsetseg, Ts., Tomurhuu, D., & Oyunchimeg, T. (2006). Lithology and biogeochemistry of sediment of the Lake Hovsgol: Implication for the paleoclimate changes. In: *Proceedings of the Institute of Hydrology and Meteorology, Ulaanbaatar, Mongolia*, pp. 139–148. [In Mongolian].

Narantsetseg, T., Krivonogov, S. K., Oyunchimeg, T., Uugantsetseg, B., Burr, G. S., Tomurhuu, D., & Dolgorsuren, K. (2013). Late Glacial to Middle Holocene climate and environmental changes as recorded in Lake Dood sediments, Darhad Basin, northern Mongolia. *Quaternary International, 311*, 12–24.

Nottebaum, V., Stauch, G., van der Wal, J. L. N., Zander, A., Schlütz, F., Shumilovskikh, L., Reicherter, K., Batkhishig, O., & Lehmkuhl, F. (2021). Late Quaternary landscape evolution and paleoenvironmental implications from multiple geomorphic dryland systems, Orog Nuur Basin, Mongolia. *Earth Surface Processes and Landforms.*, 1–23. https://doi.org/10.1002/esp.5247

Oh, J. S., Seoung, Y. B., Hong, S. C., & Yu, B. Y. (2019). Paleo-shoreline changes in moraine dammed lake Khagiin Khar, Khentey Mountains, Central Mongolia. *Journal of Mountain Science, 16*(6), 1215–1230.

Orkhonselenge, A., & Bulgan, O. (2021). Geochemical studies and lacustrine geomorphology of Lake Yakhi basin in eastern Mongolia. *Géomorphologie: Relief, Processes, Environnement, 27*(3), 231–242. https://doi.org/10.4000/geomorphologie.15873

Orkhonselenge, A., & Davaagatan, T. (2016). *Sedimentary feature in Lake Terkhiin Tsagaan and Lake Ugii, Central Mongolia*. Geophysical Research Abstracts Vol. 18, EGU2016-5666-2, EGU General Assembly 2016.

Orkhonselenge, A., Komatsu, G., Davaagatan, T., & Uuganzaya, M. (2019). Sedimentary Records from Lakes Reconstructing Late Quaternary Paleoclimate Changes in Mongolia. In: *XX INQUA Congress. Presentation #2419*. Dublin, Ireland.

Orkhonselenge, A., Komatsu, G., & Uuganzaya, M. (2018). Middle to late Holocene sedimentation dynamics and paleoclimatic conditions in the Lake Ulaan basin, southern Mongolia. *Géomorphologie: Relief, Processus, Environnement, 24*(4), 351–363.

Orkhonselenge, A., Krivonogov, S. K., Kashiwaya, K., Nakamura, T., & Davaagatan, T. (2015). Paleoclimate changes in Mongolia: Lake Records. In: *Extended Abstracts of International Association of Geomorphologists (IAG) Regional Conference on Gradualism vs Catastrophism in Landscape Evolution. Barnaul, Russia*. p. 175.

Orkhonselenge, A., Krivonogov, S. K., Mino, K., Kashiwaya, K., Safonova, I. Y., Yamamoto, M., Kashima, K., Nakamura, T., & Kim, J. Y. (2013). Holocene sedimentary records from Lake

Borsog, eastern shore of Lake Khuvsgul, Mongolia, and their paleoenvironmental implications. *Quaternary International, 290-291*, 95–109.

Orkhonselenge, A., Uuganzaya, M., Davaagatan, T., & Enkhbayar, G. (2021). Late Holocene Peatland Evolution in Terelj and Tuul Rivers Drainage Basins in the Khentii Mountain Range of Northeastern Mongolia. *Water, 13*, 562. https://doi.org/10.3390/w13040562

Panagiotopoulos, F., Shahgedanova, M., Hannachi, A., & Stephenson, D. B. (2005). Observed trends and teleconnections of the Siberian high: A recently declining center of action. *Journal of Climate, 18*, 1411–1422.

Panin, A. V., Bronnikova, M. A., Uspenskaya, O. N., Arzhantseva, I. A., Konstantinov, E. A., Koshurnikov, E. A., Selezneva, V. V., Fuzeina, Yu, N., & Sheremetskaya, E. D. (2012). Evolution of Tere-Khol Lake and the Holocene Dynamics of the Environment in the Southeastern Part of the Sayan-Tuva Highland. *Doklady Earth Sciences, 446*(2), 1204–1210.

Peck, J. A., Khosbayar, P., Fowell, S. J., Pearce, R. B., Ariunbileg, S., Hansen, B. C. S., & Soninkhishig, N. (2002). Mid to Late Holocene climate change in north central Mongolia as recorded in the sediments of Lake Telmen. *Palaeogeography, Palaeoclimatology, Palaeoecology, 183*, 135–153.

Prokopenko, A. A., Bezrukova, E. V., Khursevich, G. K., Solotchina, E. P., Kuzmin, M. I., & Tarasov, P. E. (2010). Climate in continental interior Asia during the longest interglacial of the past 500,000 years: The new MIS 11 records from Lake Baikal, SE Siberia. *Climate of the Past, 6*, 31–48.

Prokopenko, A. A., Hinnov, L. A., Williams, D. F., & Kuzmin, M. I. (2006). Orbital forcing of continental climate during the Pleistocene: A complete astronomically tuned climatic record from Lake Baikal, SE Siberia. *Quaternary Science Reviews, 25*, 3431–3457.

Prokopenko, A. A., Khursevich, G. K., Bezrukova, E. V., Kuzmin, M. I., Boes, X., Williams, D. F., Fedenya, S. A., Kulagina, N. V., Letunova, P. P., & Abzaeva, A. A. (2007). Paleoenvironmental proxy records from Lake Hovsgol, Mongolia, and a synthesis of Holocene climate change in the Lake Baikal watershed. *Quaternary Research, 68*, 2–17.

Prokopenko, A. A., Kuzmin, M. I., Li, H., Woo, K., & Catto, N. R. (2009). Lake Hovsgol basin as a new study site for long continental paleoclimate records in continental interior Asia: General context and current status. *Quaternary International, 205*, 1–11.

Prokopenko, A. A., Kuzmin, M. I., Williams, D. F., Gelety, V. E., Kalmychkov, G. V., Gvozdkov, A. N., & Solotchin, P. A. (2005). Basin-wide sedimentation changes and deglacial lake-level rise in the Hovsgol basin, NW Mongolia. *Quaternary International, 136*, 59–69.

Rosen, A. M., Hart, T. C., Farquhar, J., Schneider, J. S., & Yadmaa, T. (2019). Holocene vegetation cycles, land-use, and human adaptations to desertification in the Gobi Desert of Mongolia. *Vegetation History and Archaeobotany, 28*, 295–309.

Rudaya, N. A., Tarasov, P. E., Dorofeyuk, N. I., Kalugin, I. A., Andreev, A. A., Diekmann, B., & Daryin, A. V. (2008). Paleenvironment. The stone age. *Archaeology Ethnology and Anthropology of Eurasia, 36*(4), 2–14.

Rudaya, N., Tarasov, P., Dorofeyuk, N., Solovieva, N., Kalugin, I., Andreev, A., Daryin, A., Diekmann, B., Riedel, F., Narantsetseg, T., & Wagner, M. (2009). Holocene environments and climate in the Mongolian Altai reconstructed from the Hoton-Nur pollen and diatom records: A step towards better understanding climate dynamics in Central Asia. *Quaternary Science Reviews, 28*, 540–554.

Rudaya, N., & Li, H. (2013). A new approach for reconstruction of the Holocene climate in the Mongolian Altai: The high-resolution ^{13}C records of TOC and pollen complexes in Hoton-Nur Lake sediments. *Journal of Asian Earth Sciences, 69*, 185–195.

Schwanghart, W., Schütt, B., & Walther, M. (2008). Holocene climate evolution of the Ugii Nuur Basin, Mongolia. *Advances in Atmospheric Sciences, 25*(6), 986–998.

Schwanghart, W., Frechen, M., Kuhn, N. J., & Schütt, B. (2009). Holocene environmental changes in the Ugii Nuur basin, Mongolia. *Palaeogeography, Palaeoclimatology, Palaeoecology, 279*, 160–171.

Sklyarov, E. V., Solotchina, E. P., Vologina, E. G., Ignatova, N. V., Izokh, O. P., Kulagina, N. V., Sklyarova, O. A., Solotchin, P. A., Stolpovskaya, V. N., Ukhova, N. N., Fedorovskii, V. S., & Khlystov, O. M. (2010). Detailed Holocene climate record from the carbonate section of saline Lake Tsagan-Tyrm (*West Baikal area*). *Russian Geology and Geophysics, 51*(3), 237–258.

Shi, Y. F., & Kong, Z. C. (1992). *Mid-Holocene Megathermal Environments in China* (pp. 1–18). China Ocean Press. [In Chinese].

Shi, P., & Song, C. (2003). Palynological records of environmental changes in the middle part of Inner Mongolia, China. *Chinese Science Bulletin, 48*(14), 1433–1438.

Shinneman, A. L. C., Umbanhowar, C. E., Edlund, M. B., & Soninkhishig, N. (2010). Late-Holocene moisture balance inferred from diatom and lake sediment records in western Mongolia. *The Holocene, 20*(1), 123–138.

Stolz, C., Hülle, D., Hilgers, A., Grunert, J., Lehmkuhl, F., & Dash, D. (2012). Reconstructing fluvial, lacustrine and aeolian process dynamics in Western Mongolia. *Zeitschrift für Geomorphologie, 56*(3), 267–300.

Sun, J., Li, S., Han, P., & Chen, Y. (2006). Holocene environmental changes in the central Inner Mongolia, based on single-aliquot-quartz optical dating and multi-proxy study of dune sands. *Palaeogeography, Palaeoclimatology, Palaeoecology, 233*, 51–62.

Sun, A., & Feng, Z. (2013). Holocene climatic reconstructions from the fossil pollen record at Qigai Nuur in the southern Mongolian Plateau. *The Holocene, 23*(10), 1391–1402.

Sun, A., Feng, Z., Ran, M., & Zhang, C. (2013). Pollen-recorded bioclimatic variations of the last ~22,600 years retrieved from Achit Nuur core in the western Mongolian Plateau. *Quaternary International, 311*, 36–43.

Szumińska, D. (2016). Changes in surface area of the Böön tsagaan and Orog lakes (Mongolia, Valley of the lakes, 1974–2013) compared to climate and permafrost changes. *Sedimentary Geology, 340*, 62–73.

Tarasov, P., Bezrukova, E., Karabanov, E., Nakagawa, T., Wagner, M., Kulagina, N., Letunova, P., Abzaeva, A., Granoszewski, W., & Riedel, F. (2007a). Vegetation and climate dynamics during the Holocene and Eemian interglacials derived from Lake Baikal pollen records. *Palaeogeography, Palaeoclimatology, Palaeoecology, 252*, 440–457.

Tarasov, P., Bezrukova, E., Karabanov, E., Nakagawa, T., Wagner, M., Kulagina, N., Letunova, P., Abzaeva, A., Granoszewski, W., & Riedel, F. (2007b). Satellite- and pollen-based quantitative woody cover reconstructions for northern Asia: Verification and application to late-Quaternary pollen data. *Earth and Planetary Science Letters, 264*(1–2), 284–298.

Tarasov, P. E., Bezrukova, E. V., & Krivonogov, S. K. (2009). Late Glacial and Holocene changes in vegetation cover and climate in southern Siberia derived from a 15 kyr long pollen record from Lake Kotokel. *Climate of the Past Discussions, 5*, 127–151.

Tian, F., Herzschuh, U., Dallmeyer, A., Xu, Q., Mischke, S., & Biskaborn, B. K. (2013). Environmental variability in the monsoonewesterlies transition zone during the last 1200 years: Lake sediment analyses from central Mongolia and supraregional synthesis. *Quaternary Science Reviews, 73*, 31–47.

Tsegmid, S. (1969). *Physical geography of Mongolia*. State Press, 405 p. [In Mongolian].

Unkelbach, J., Kashima, K., Enters, D., Dulamsuren, C., Punsalpaamuu, G., & Behling, H. (2019). Late Holocene (Meghalayan) palaeoenvironmental evolution inferred from multi-proxy-studies of lacustrine sediments from the Dayan Nuur region of Mongolia. *Palaeogeography, Palaeoclimatology, Palaeoecology, 530*(15), 1–14.

Visbeck, M. (2002). The ocean's role in Atlantic climate variability. *Science, 297*, 2223–2224.

Walther, M., Dashtseren, A., Kamp, U., Temuujin, K., Meixner, F., Pan, C. G., & Gansukh, Y. (2017). Glaciers, Permafrost and Lake Levels at the Tsengel Khairkhan Massif, Mongolian Altai, During the Late Pleistocene and Holocene. *Geosciences, 7*(73), 1–20.

Wang, Y. (2011). *Late Glacial to Holocene climate and vegetation changes on the Tibetan Plateau inferred from fossil pollen records in lacustrine sediments*. Dissertation zur Erlangung des akademischen Grades "doctor rerum naturalium" (Dr. rer. nat.) in der Wissenschaftsdisziplin. "Palaeoecology and Paleolimnology". Eingereicht an der Mathematisch-Naturwissenschaftlichen Fakultät der Universität Potsdam. Potsdam. 182 p.

Wang, W., & Feng, Z. (2013). Holocene moisture evolution across the Mongolian Plateau and its surrounding areas: A synthesis of climate records. *Earth-Science Reviews, 122*, 38–57.

Wang, W., Feng, Z., Lee, X., Zhang, H., Ma, Y., An, C., & Guo, L. (2004). Holocene abrupt climate shifts recorded in Gun lake core, northern Mongolia. *Chinese Science Bulletin, 49*(5), 520–526.

Wang, H., Liu, H., Zhu, J., & Yin, Y. (2010). Holocene environmental changes as recorded by mineral magnetism of sediments from Anguli-nuur Lake, southeastern Inner Mongolia Plateau, China. *Palaeogeography, Palaeoclimatology, Palaeoecology, 285*, 30–49.

Wang, H., Liu, H., Zhao, F., Yin, Y., Zhu, J., & Snowball, I. (2012). Early- and mid-Holocene palaeoenvironments as revealed by mineral magnetic, geochemical and palynological data of sediments from Bai Nuur and Ulan Nuur, southeastern inner Mongolia Plateau, China. *Quaternary International, 250*, 100–118.

Wang, W., Ma, Y., Feng, Z., Meng, H., Sang, Y., & Zhai, X. (2009). Vegetation and climate changes during the last 8660 cal. a BP in central Mongolia, based on a high-resolution pollen record from Lake Ugii Nuur. *Chinese Science Bulletin, 54*(9), 1579–1589.

Wang, W., Ma, Y. Z., Feng, Z. D., Narantsetseg, T., Liu, K. B., & Zhai, X. W. (2011). A prolonged dry mid-Holocene climate revealed by pollen and diatom records from lake Ugii Nuur in central Mongolia. *Quaternary International, 229*(1–2), 74–83.

Watanabe, T., Nakamura, T., & Kawai, T. (2007). Radiocarbon dating of sediments from large continental lakes (Lakes Baikal, Hovsgol and Erhel). *Nuclear Instruments and Methods in Physics Research B, 259*, 565–570.

Watanabe, T., Nakamura, T., Nara, F. W., Kakegawa, T., Horiuchi, K., Senda, R., Oda, T., Nishimura, M., Matsumoto, G. I., & Kawai, T. (2009). High-time resolution AMS 14C data sets for Lake Baikal and Lake Hovsgol sediment cores: Changes in radiocarbon age and sedimentation rates during the transition from the last glacial to the Holocene. *Quaternary International, 205*, 12–20.

Wen, R., Xiao, J., Chang, Z., Zhai, D., Xu, Q., Li, Y., Itoh, S., & Lomtatidze, Z. (2010). Holocene climate changes in the mid-high-latitude-monsoon margin reflected by the pollen record from Hulun Lake, northeastern Inner Mongolia. *Quaternary Research, 73*, 293–303.

Wünnemann, B., Chen, F., Riedel, F., Zhang, C., Mischke, S., Chen, G., Demske, D., & Ming, J. (2003). Holocene lake deposits of Bosten Lake, southern Xinjiang, China. *Chinese Science Bulletin, 48*, 1429–1432.

Xiao, J., Wu, J., Si, B., Liang, W., Nakamura, T., Liu, B., & Inouchi, Y. (2006). Holocene climate changes in the monsoon/arid transition reflected by carbon concentration in Dahai Lake of Inner Mongolia. *The Holocene, 16*, 551–560.

Yang, X., Rost, K. T., Lehmkuhl, F., Zhenda, Z., & Dodson, J. (2004). The evolution of dry lands in northern China and in the Republic of Mongolia since the Last Glacial Maximum. *Quaternary International, 118–119*, 69–85.

Yu, K., Lehmkuhl, F., Diekmann, B., Zeeden, C., Nottebaum, V., & Stauch, G. (2017). Geochemical imprints of coupled paleoenvironmental and provenance change in the lacustrine sequence of Orog Nuur, Gobi Desert of Mongolia. *Journal of Paleolimnology, 58*, 511–532.

Yu, K., Lehmkuhl, F., Schlütz, F., Diekmann, B., Mischke, S., Grunert, J., Murad, W., Nottebaum, V., Stauch, G., & Zeeden, C. (2019). Late Quaternary environments in the Gobi Desert of Mongolia: Vegetation, hydrological, and palaeoclimate evolution. *Palaeogeography, Palaeoclimatology, Palaeoecology, 514*, 77–91.

Zhang, C., Zhang, W., Feng, Z., Mischke, S., Gao, X., Gao, D., & Sun, F. (2012). Holocene hydrological and climatic change on the northern Mongolian Plateau based on multi-proxy records from Lake Gun Nuur. *Palaeogeography, Palaeoclimatology, Palaeoecology, 323–325*, 75–86.

Part III
Scientific, Economic and Touristic Significances of Lakes in Mongolia

Chapter 22
Future Directions of Lake Study in Mongolia

Abstract Lake study in Mongolia plays a major role not only in Earth science but also in the country's society and economy. Because lake study represents an Earth science field, it is closely related to geology, geomorphology, sedimentology, geochemistry, geochronology, and paleoclimatology. However, their interdisciplinary field approaches (e.g., limnogeology, hydrogeology, hydrogeomorphology, hydrogeochemistry) are still far from a well-developed stage in the country. The policy for higher education in Mongolia should be more geared toward developing basic sciences, especially Earth science and its fields (e.g., hydrology, geomorphology, geochemistry), supporting to educate a new generation of specialists qualified at the international level. In Mongolia, its liberal arts system remains under planning or in its embryonic state today although the Ministry of Education has initiated the basic preparation for the research-based university development since 2014. Today, higher education in Mongolia is still practiced with the traditional way of reading, receiving, memorizing, and writing theories with roles of lecturers and learners clearly separated during the class followed by the conventional Russian education style since the nineteenth century. It is important to improve and renew the present situation (1) with rapid changes in our society even though it is still a developing country and (2) by young and bright Mongolian generations who are educated at top-level universities in developed countries.

Keywords Lake study · Limnology · Earth science · Higher education · Mongolia

22.1 Introduction

The study of inland waters is very important in the development of Earth science and for solving social and economic issues, and it includes many different scientific disciplines. Limnology covers all inland waters, which may be lakes, rivers, streams, and creeks but also reservoirs, ponds, groundwater, and wetlands. These are often freshwater systems, but limnology also includes inland salt and brackish or slightly salty waters. The limnology is the science for lake study that can provide improved

A. Orkhonselenge et al., *Lakes of Mongolia*, Syntheses in Limnogeology,
https://doi.org/10.1007/978-3-030-99120-3_22

understanding of lake ecosystem dynamics and information that can lead to better management policies (Garn et al., 2003). For instance, the limnology incorporates multiple scientific disciplines into one, including physics, chemistry, and biology.

Lakes can be taken as an indicator of climate change, and they experience expansion or shrinkage in response to the changing climate (e.g., Brown, 1995). Because lakes are container, accumulator, and repository of sediments (Davaagatan et al., 2015), it can be a major climatic and environmental archive and one of main clues to show local and regional landscape evolutionary changes spanning over various geologic timescales (Orkhonselenge et al., 2015).

As more studies are conducted on a variety of lake systems, accumulated information leads to the development of general concepts about how lakes respond to environmental changes (Garn et al., 2003) and ongoing climate changes at the present (see Chap. 21). With constantly expanding databases and increased knowledge, limnologists and hydrologists can develop comprehensive models that can be used to predict how lakes might evolve in the future. In this context, development of lake study is important to provide a full understanding or predict conditions of any particular lake and its basin. A limnologist or hydrologist cannot accurately assess a lake system or suggest a management strategy without data and analysis from the particular lake and its environment (Garn et al., 2003). Each lake system is unique, and its dynamics can be understood only to a limited degree based on information obtained from other lakes. The most important basic factors that give unique characters to each lake ecosystem are atmospheric input, physiography, land use, and morphometry of lakes (Garn et al., 2003). Moreover, the lake study dominantly requires knowledge of Earth science fields of climatology, geology, geomorphology, glaciology, cryology, hydrology, sedimentology, geochemistry, and geochronology (see Part II).

Lake study, which contributes to general and specific knowledge of lake basins, requires commitment of considerable effort, but this commitment leads to major improvement of the likelihood of developing reasonable and effective lake management strategies (Garn et al., 2003). For instance, water-related socioeconomic analysis based on a modeling of Lake Tonle Sap in Cambodia can significantly contribute to water modeling and impact analysis for the Integrated Water Resources Management (IWRM) (Keskinen, 2006). Moreover, settlements on Neolithic lakeshores offered excellent preservation conditions for organic finds (e.g., wood, botanical remains, animal bones), which provide exact dendrochronological dating and detailed information about nutrition of the people and their environment (Schibler & Jacomet, 2013).

In Mongolia, IWRM has not been applied in any lake basin to date, except for the International Geosphere-Biosphere Program in the Lake Uvs basin for monitoring global warming (see Chaps. 13 and 24). IWRM has been considered only in river drainage basins by the Ministry of Environment and Green Development (2013) with case studies for the Selenge[1] River (Karthe et al., 2014; Karthe, 2018), a major

[1] *Selenge* has been often incorrectly written as *Selenga* internationally based on the Russian pronunciation. Selenge is the correct English transliteration from Mongolian *Сэлэнгэ*.

headwater of Lake Baikal[2], and Kharaa River (Karthe et al., 2015) drainage basins in northern Mongolia. One plan for the IWRM, aiming to secure a long-term water supply for socioeconomic development, was set to be completed in river drainage basins of Mongolia between 2015 and 2021 in the context of water use, demand, resources, and balance (Ministry of Environment and Green Development, 2013). In this IWRM plan, some lakes in the river drainage basins were considered together, with examples of the Lake Uvs-Tes River, Lake Buun Tsagaan-Baidrag River, and Lake Khar-Khovd River drainage basins, where to provide (1) safe drinking water, (2) agriculture water, (3) industry and mining water, (4) a balance in water resources, and (5) implementation of water management. Moreover, a sustainable water resource management plan in an integrated glacier-based catchment with numerous rivers and lakes basins in Mongolia was considered by Orkhonselenge and Harbor (2018). This attempt was made through a linkage between glaciological and hydrological studies contributing to the scientific base knowledge of changing precipitation-glacier-inflow-runoff-outflow-storage relationships that are necessary components of IWRM. The lack of IWRM in Mongolia is directly related to the poor development of basic science and higher education today.

In this chapter, further scientific significance of lake study in Mongolia, the present condition of lake study, and feasible study types with multi-proxy criteria are presented.

22.2 Scientific Significance of Lake Study

At the present day, lake study in Mongolia is still developing to utilize advanced techniques (see Chap. 2); however, it is going on at a slow rate than in other countries. To date, lake study in Mongolia is focused mostly on general description of the lakes (see Part I) and limited reconstructions of climatic and environmental dynamics in the lake basins (see Part II). The important examples of Mongolian lake studies published during the last two decades are listed in Table 22.1.

In Mongolia, lake study should be conducted to meet specific purposes such as reconstructing formation and evolution of lakes and their basins, understanding response of lakes and their basins to the present climate change, water management planning for lake basins, or achieving combinations of purposes such as development and implementation of lake management plans and protection of lake water quality, which support decision-making by local administrative organizations or governmental organizations such as the Ministry of Environment. In Mongolia, there are some nongovernmental organizations (NGO), associations, and local groups acting to protect river drainage basins in recent years, for example, *Ongi River Protection Movement* and *Orkhon River Protection Movement* (Fig. 22.1),

[2] *Baigali* meaning *nature* has been misspelled as *Baikal* by the Russians. Baigali is the right English transliteration from Mongolian *Байгаль*.

Table 22.1 Major lake studies conducted in Mongolia during the last two decades

#	Lakes	Studies	Authors
1	L. Uvs	Paleoclimate change, paleoenvironmental evolution	Grunert et al. (2000)
2	L. Ugii	Water balance, thermal regime, water resource	Tuvshinjargal (2001)
3	L. Buun Tsagaan, L. Orog, L. Adgiin Tsagaan	Paleolake landforms	Komatsu et al. (2001)
4	The Valley of the Lakes	Paleoshoreline reconstruction	Lehmkuhl and Lang (2001), Komatsu et al. (2001)
5	L. Dood	Lake-level fluctuations, paleoclimate changes	Peck et al. (2001), Narantsetseg et al. (2013)
6	L. Telmen	Lake basin evolution, paleoclimate change	Peck et al. (2002), Fowel et al. (2003)
7	L. Khuvsgul	Lake-level fluctuations	Krivonogov et al. (2003)
8	L. Gun	Holocene hydrology, climate changes	Wang et al. (2004), Zhang et al. (2012)
9	Darkhad basin	Paleolake reconstruction	Krivonogov et al. (2005, 2012), Gillespie et al. (2008), Komatsu et al. (2009), Pechersky et al. (2013)
10	L. Khuvsgul	Holocene climate change	Prokopenko et al. (2007)
11	L. Ugii	Lake-level fluctuations, Holocene climate changes	Schwanghart et al. (2008, 2009), Wang et al. (2009, 2011)
12	Tukhum, Mongol, and Khongor sand dunefield depressions	Pluvial lake reconstruction, paleoclimate changes	Grunert et al. (2009), Stolz et al. (2012)
13	L. Khoton	Lake reconstruction, Holocene climate dynamics	Rudaya et al. (2009)
14	6 lakes in the Great Lakes Depression	Changes in the precipitation-to-evaporation balance	Shinneman et al. (2010)
15	L. Khuvsgul	Sedimentary records	Kashiwaya et al. (2010)
16	All lakes	Changes in lake area	Davaa (2010)
17	12 lakes	Physicochemical properties and characterization of organic matters of peloids	Tserenpil et al. (2010)
18	L. Ulaan	Pleistocene and Holocene sediment-transport mechanism, paleoclimate changes	Lee et al. (2011, 2013)
19	L. Orog	Lake evolution	Fassnacht et al. (2011), Krylov et al. (2011)
20	L. Khuvsgul	Lake-level fluctuations, paleoclimate changes	Orkhonselenge et al. (2011, 2013, 2014)
21	Hypersaline soda closed basin lakes	^{238}U concentration	Isupov et al. (2011)
22	Hyper-alkaline soda lakes	Sr, U concentrations	Linhoff et al. (2011)

(continued)

Table 22.1 (continued)

#	Lakes	Studies	Authors
23	L. Bayan Tukhum	Lake evolution history, late Quaternary climate, and paleoenvironmental conditions	Felauer et al. (2012)
24	L. Khuis	Late Holocene climate changes	Tian et al. (2013)
25	L. Achit	Pleistocene and Holocene vegetation variations and paleoclimate changes	Sun et al. (2013)
26	L. Khoton	High-resolution ^{13}C records of TOC and pollen complexes	Rudaya and Li (2013)
27	L. Shaazgai	^{238}U, As concentrations	Isupov et al. (2013, 2015), Gaskova et al. (2015)
28	L. Terkhiin Tsagaan, L. Ugii	Sedimentary features, paleoclimate changes	Davaagatan (2014), Davaagatan et al. (2015), Orkhonselenge and Davaagatan (2016)
29	165 lakes	Changes in lake area	Kang et al. (2015)
30	185 lakes	Water balance, water level, and area	Davaa (2015)
31	L. Ulaan	Water resource reconstruction, paleohydrological system	Sternberg and Paillou (2015), Holguín and Sternberg (2016)
32	All lakes	Paleoclimate changes	Orkhonselenge et al. (2015, 2019b)
33	73 lakes	Lake surface areas	Kang and Hong (2016)
34	L. Buun Tsagaan, L. Orog	Changes in surface area	Szumińska (2016)
35	L. Buun Tsagaan, L. Orog	Hydrochemistry	Szopińska et al. (2016a, 2016b)
36	L. Ganga	Water-level fluctuation, water balance	Oyunbaatar et al. (2017)
37	L. Khar	Lake-level fluctuations	Walther et al. (2017)
38	Thermokarst lakes	Lake dynamics in permafrost regions	Saruulzaya (2017)
39	L. Orog	Hydrological variability	Yu et al. (2017, 2019)
40	L. Buun Tsagaan, L. Orog	Hydrochemistry	Chida et al. (2018)
41	6 lakes in the Valley of the Lakes	Changes in lake area	Orkhonselenge et al. (2018a)
42	L. Ulaan	Holocene sedimentary dynamics	Orkhonselenge et al. (2018b)
43	Lakes in southern and western Mongolia	Paleolake reconstructions	Lehmkuhl et al. (2018)
44	Lake Khagiin Khar	Paleoshoreline changes	Oh et al. (2019)
45	L. Toson, L. Burd	Changes in lake area	Gerelsaikhan and Orkhonselenge et al. (2019a)
46	L. Khargal	Changes in lake area	Orkhonselenge et al. (2019a), Gerelsaikhan and Orkhonselenge et al. (2019b)

(continued)

Table 22.1 (continued)

#	Lakes	Studies	Authors
47	L. Buun Tsagaan basin	Shoreline features	Matsuyama et al. (2019)
48	L. Achit	Catastrophic flood-induced drainage	Agatova et al. (2020)
49	L. Ulaan	Paleoclimate changes	Mischke et al. (2020)
50	Lakes in eastern, southern, and western Mongolia	Geochemistry of mineralized lakes	Ariunbileg et al. (2020)
51	L. Yakhi	Changes in lake area, geochemical review	Orkhonselenge et al. (2020), Orkhonselenge and Bulgan (2021)
52	L. Khargal	Holocene vegetation and climate changes and geochemical dynamics	Barna et al. (2021), Orkhonselenge et al. (2021)
53	L. Orog	Late Pleistocene lake-level changes	Nottebaum et al. (2021)
54	L. Orog, L. Adgiin Tsagaan, L. Shargiin Tsagaan, L. Khyargas	OSL dating applications	Wang et al. (2021)
55	L. Khoton, L. Khurgan, L. Ekhen	Late Pleistocene lake-level changes	Klinge et al. (2021)

Note: Lakes are listed in order of publication year. The references are given in Chap. 2 and Part II

Fig. 22.1 "Orkhon River Protection Movement" NGO, a member of the Sustainable Development Solutions Network (SDSN), initiated in 2011

counteracting illegal activities (e.g., mining, fishing, car washing, uses of toxic chemicals, and waste water diluting to explore gold) for the purpose of reducing pollution of river water, protecting river ecosystem, and educating local citizens on water management. However, governmental and nongovernmental organizations and/or associations are not particularly active for protecting and managing playas (see Chap. 4), shrinking lakes (see Chap. 5), and deteriorating lake ecosystems caused by human activities. Today, local lake associations and private organizations in addition to the state organizations to transfer factual voice about how lakes of Mongolia face challenges for surviving under the rapidly strengthening climate warming are missing. For example, Lake Ulaan (Fig. 22.2) frequently shrinks during the last half century (Orkhonselenge et al., 2018a, 2018b) and finally shifts to a playa environment (see Chap. 11) due to that Ongi River, a major source of surface water in the Govi[3] region of southern Mongolia, cannot supply water or feed the lake because mining companies in its basin overuse its water for their operation.

This chapter commends advanced research performed with high-resolution multi-proxy data for certain lakes (see Part II) in each region of Mongolia (Fig. 1.1), which use modern methods and sophisticated techniques. There is a need to clarify types of lake studies ranging from simple to complex at different levels. For example, the types of lake study should be distinct from region to region throughout Mongolia because each region is differently affected by major controlling factors such as geology, climate (see Chaps. 6, 9, 12, 15, and 18), and land use. Here, we provide proper types of lake studies for each region of Mongolia based on our analytic and logistic objectives and investigations (Table 22.2). Our mission is to (1) provide geomorphological, hydrological, and other scientific information to describe the current characteristics and conditions of lakes in Mongolia (see Chaps. 6–20), (2) identify and understand evolution of lake basins under the present climate condition (see Chap. 21), and (3) determine effective management actions to protect or restore the lakes. Detailed or precise information about the lakes of Mongolia is found in Parts I and II. Lake study in Mongolia is not only of scientific importance to develop interdisciplinary fields within Earth science but also to be followed by the economic significance (see Chap. 23) and touristic aspects (see Chap. 24) for the country.

As indicated in Table 22.2, the types of lake study to be possibly performed in Mongolia (Table 22.3) are (1) monitoring to assess current conditions and evaluate further trends, (2) diagnosing to identify causes of ecological problems, (3) assessing feasibility for possible solutions and effectiveness of management practices, and (4) investigating scientifically valuable lakes with high-resolution multi-proxy data. The monitoring, diagnosing, and assessing feasibility in lake studies are today qualitatively and quantitatively supported by the Ministry of Environment and Tourism that is a government organization and also by state agencies such as the Institute of Meteorology, Hydrology, and Environment. However, the investigation of scientifically valuable lakes should be performed by research institutes, research teams, and

[3] *Govi* has been often spelled as *Gobi* in the international literature. Govi is the correct English transliteration from Mongolian *Говь*.

Fig. 22.2 Exposed floors of Lake Ulaan in the Govi region of southern Mongolia. Photos by A. Orkhonselenge on October 24, 2015 for (**a**) and June 18, 2018 for (**b**)

individual skilled researchers. The proposed specific programs in Mongolia for lake studies (Table 22.2) must be financially supported by the state and local administration grants. It would also be helpful if lake basin administrations, lake associations, and other qualified local organizations and agencies for lake planning and management activities will be established in Mongolia.

Table 22.2 Proposed types of lake study

#	Types	Criteria	Details of criteria
1	Monitoring	Simple[1] and complex[1]	Simple[1]: physiography (origin, altitude, geology, climate, etc.) and morphometry (shape, depth, area, etc.)
2	Diagnostic	Simple[2] and complex[1]	Simple[2]: ecological and biological indicators (nutrients, micro- and macrophytes, algae, pollen, etc.)
3	Feasibility	Simple[1,2] and complex[1]	Complex[1]: physical (transparency, density, temperature, etc.) and chemical (pH, salinity, biogenic silica, dissolved oxygen, etc.) traces
4	Scientific	Simple[1,2] and complex[2]	Complex[2]: advanced geotraces (major and trace elements, minerals, dating, magnetic susceptibility, etc.)

Table 22.3 Proposed types of lake study for each region of Mongolia

#	Regions	Details and descriptions	Types of lake studies
1	Eastern Mongolia	Chapter 6	Monitoring, diagnostic, feasibility, and scientific
2	Southern Mongolia	Chapter 9	Monitoring, diagnostic, and scientific
3	Western Mongolia	Chapter 12	Monitoring and scientific
4	Northern Mongolia	Chapter 15	Monitoring and scientific
5	Central Mongolia	Chapter 18	Monitoring, diagnostic, feasibility, and scientific

22.3 Environment of Lake Study

Today, there is not a nationally established research institution nor laboratory specialized for lake study in Mongolia. There was an active lake study group in the country between 1960 and 2000. For instance, a national research team (e.g., N. Badarch, Sh. Tsegmid, J. Tserensodnom, Ts. Sugar, N. Batnasan, D. Tuvshinjargal) from the Institute of Geography and Geoecology, Mongolian Academy of Sciences, led lake studies in Mongolia (see Chap. 2). However, their works were limited to provide a general description of each lake, and they did not conduct research in geomorphology, geochemistry, sedimentology, geochronology, and paleoclimatology in lake basins. There is still no expert, except for some physical geographers, in the field of hydrology conducting lake study in Mongolia.

The Institute of Chemistry and Biology, Mongolian Academy of Sciences, has investigated the hydrochemistry and hydrobiology of lakes in Mongolia since the 1960s (see Chap. 2). Moreover, the Institute of Meteorology, Hydrology, and Environment began collecting meteorological and hydrological data from meteorological stations near lakes since the 1960s when measurement instruments were initially installed. However, the works at these institutes had never advanced to the contexts of geomorphology, geochemistry, and paleoclimatology of the lakes. The research team (e.g., D. Tumurkhuu, Ts. Narantsetseg, Ts. Oyunchimeg) from the

Institute of Geology and Paleontology, Mongolian Academy of Sciences, investigated geology and sedimentology for a few lakes during the last decade within the frameworks of international projects such as the Khuvsgul[4] Drilling Project (see Chaps. 15 and 16) and the Darkhad Drilling Project (see Chap. 15). For example, they presented the results from paleolake Darkhad basin in Narantsetseg et al. (2006, 2010, 2013). The joint research team including geochemist S. Ariunbileg from the Institute of Geology and Paleontology, Mongolian Academy of Sciences, and researchers (e.g., V.P. Isupov, A.G. Vladimirov, M.N. Kolpakova) from the Russian Academy of Sciences investigated hydrochemistry and geochemistry of mineralized lakes in 2009 (see Chap. 23), and their results are presented in Ariunbileg et al. (2020).

Today, the two Ministries of Environment and Tourism and of Education call for research projects in the nation aimed for national scientists and binational research teams (e.g., Mongolia and Russia, Mongolia and China, Mongolia and Taiwan), and the Scientific Technology Foundation funds the projects. However, these research projects are concerned mostly about general environmental issues related to deterioration of biodiversity, deforestation, and other ecological issues due to mining, overgrazing, desertification, etc. instead of detailed specific topics.

Nevertheless, the situation is slowly changing. For example, a research team from the Laboratory of Geomorphology and Geochemistry, National University of Mongolia, has initiated research project of lake study in Mongolia in the context of geomorphological, geochemical, sedimentological, and paleoclimatological implications since the laboratory's establishment in 2015, and they have been publishing their research results (e.g., Davaagatan et al., 2015; Orkhonselenge et al., 2018a, 2018b, 2021; Gerelsaikhan & Orkhonselenge, 2019a, 2019b; Orkhonselenge & Bulgan, 2021). Furthermore, international research teams from the USA (e.g., Peck et al., 2002; Fowell et al., 2003; Barna et al., 2021), Germany (e.g., Grunert et al., 2009; Yu et al., 2017; Lehmkuhl et al., 2018), Russia (e.g., Prokopenko et al., 2005, 2007; Krivonogov et al., 2003, 2012), Poland (e.g., Szumińska, 2016; Szopińska et al., 2016a, 2016b), Japan (e.g., Kashiwaya et al., 2010; Sekine et al., 2020), and others (e.g., Lee et al., 2013; Sternberg & Paillou, 2015; Mischke et al., 2020) have been studying lakes in Mongolia intensely in recent years (Table 22.1; see Part II).

22.4 Challenges Facing Lake Study

In Mongolia, study of lakes has been gradually expanding in recent years due to the presence of useful multi-proxy data and the opportunity to obtain valuable information about the lake basin evolution and the local and regional climate changes (see Part II). However, these multi-proxies and their valuable information have been

[4] *Khuvsgul* has been misspelled as *Hovsgoll* and *Khubsugul* in publications. Khuvsgul is the right English transliteration from Mongolian *Хөвсгөл*.

primarily provided by international research project teams who collect samples from Mongolian lakes and analyze them at their research laboratories with advanced techniques and instruments. Among the lake studies in Mongolia during the last two decades (Table 22.1), foreign researchers accounted for 73.3% and Mongolian researchers only 26.7% in their roles played for all the lake studies. In addition, most of the Mongolian researchers have been linked with the foreign researchers' projects.

The Mongolian government and the Ministry of Education tried to reform the higher education system in 2014 through shifting the system from the Soviet Union's traditional style, which is primarily based on reading, receiving, memorizing, and writing theories, into the western-style liberal arts phase, which is based on empirical experiments, research-based results, and science and technology-based innovation. Today, the reform in higher education has not been achieved fully throughout universities of Mongolia, except for some minor changes in number and name of schools and curriculums. In the liberal arts system, the faculty should be educated at the advanced level in which they should learn new sophisticated teaching methods and research techniques at the higher education qualified at the international levels. Then, the teaching staff should be ready to teach at research-based universities and to reform new curriculums suited for the higher education. The nation's mindset to understand and evaluate worthiness or important role of the Earth science in the society, especially lake study involving geomorphology, geochemistry, sedimentology, and climate change, would be hopefully changed in the near future or within the next decade. Lake study in Mongolia is moving forward slowly with a rate of turtle today.

22.5 Summary

Lake study in Mongolia is very important for the development of Earth science and its interdisciplinary fields. It is also valuable for the protection and management of lake basin systems through advanced research on lakes and response of the lake basins to the present climate change. High-resolution measurements in lake study are largely lacking in Mongolia today except for those obtained by a limited number of foreign researchers' investigations. Such measurements should be performed with the fields of geomorphology, geochemistry, sedimentology, and geochronology. These efforts would be the base not only for appropriate protection management plans but also for detailed data archiving utilized in further advanced investigations in each lake basin. The present phase of lake study in Mongolia misses the presence of not only of skilled national scientists and researchers qualified at the international advanced level but also of research results with high-resolution multi-proxy data. For a proper future development of lake study in Mongolia, national plans and policies on the higher education must be led by well-trained professionals and specialists. A big gap existing between national plans and policies and their actual implementation on the ground must be filled with sophisticated and wise ways. There is much work left for the young Mongolian generations who can lead on the development of basic science and higher education in this country.

References

Ariunbileg, S., Isupov, V. P., Vladimirov, A. G., Kolpakova, M. N., & Kuibida, L. V. (2020). *Chemical evolution and geochemistry of mineralized lakes of Mongolia*. Admon Printing House, 132 p. [In Mongolian].

Barna, J., Bigelow, N., Orkhonselenge, A., Fowell, S., Gardner, W. R. M., Burentogtokh, J., & Honeychurch, W. (2021). Holocene paleoenvironmental changes in the forest-steppe ecotone of Northern Mongolia based on palynological and sedimentological analyses of cores from Lake Khargal. In *GSA Annual Meeting—The Geological Society of America. Presentation #27–12. Portland, Oregon. October 10–13, 2021*. https://doi.org/10.1130/abs/2021AM-370452.

Brown, A. G. (1995). Vegetation and lake-level change. In K. J. Gregory, L. Starkel, & V. R. Baker (Eds.), *Global continental palaeohydrology* (pp. 131–150). Wiley.

Davaagatan, T., Orkhonselenge, A., Fukushi, K., & Yabe, M. (2015). Sedimentary records in intermontane and prairie lakes in central Mongolia: Preliminary results from Lake Terkhiin Tsagaan and Lake Ugii. *Journal of Geographical Issues, 14*(408), 19–26.

Fowell, S. J., Hansen, B. C., Peck, J. A., Khosbayar, P., & Ganbold, E. (2003). Mid to late Holocene climate evolution of the Lake Telmen basin, north central Mongolia, based on palynological data. *Quaternary Research, 59*(3), 353–363.

Garn, H. S., Elder, J. F., & Robertson, D. M. (2003). *Why study lakes? An overview of USGS Lake Studies in Wisconsin*. U.S. Department of the Interior U.S. Geological Survey. Fact Sheet 063–03. 8 p.

Gerelsaikhan, D., & Orkhonselenge, A. (2019a). Water pollution and area changes of Lakes Toson and Burd. In *Proceedings of Conference on The Environment-2019*, p. 48. Ulaanbaatar. [In Mongolian].

Gerelsaikhan, D., & Orkhonselenge, A. (2019b). Recent lake area changes of Lake Khargal in northern Mongolia. *The Environment, 1*(1), 15–20. [In Mongolian with English abstract].

Grunert, J., Stolz, C., Hempelmann, N., Hilgers, A., Hulle, D., Lehmkul, F., Felauer, T., & Dasch, D. (2009). The evolution of small lake basin in the Gobi desert in Mongolia. *Quaternary Sciences, 29*, 678–689.

Karthe, D. (2018). Environmental changes in central and east Asian drylands and their effects on major river-lake systems. *Quaternary International, 475*, 91–100.

Karthe, D., Chalov, S., & Borchardt, D. (2014). *Water resources and their management in central Asia in the early twenty first century: Status, challenges and future prospects*. Environmental Earth Science.

Karthe, D., Heldt, S., Houdret, A., & Borchardt, D. (2015). IWRM in a country under rapid transition: Lessons learnt from the Kharaa River Basin, Mongolia. *Environmental Earth Science, 73*, 681–695.

Kashiwaya, K., Ochiai, S., Sumino, G., Tsukamoto, T., Szyniszewska, A., Yamamoto, M., Sakaguchi, A., Hasebe, N., Sakai, H., Watanabe, T., & Kawai, T. (2010). Climato-hydrological fluctuations printed in long lacustrine records in Lake Hövsgöl. *Mongolia Quaternary International, 219*(1–2), 178–187.

Keskinen, M. (2006). The lake with floating villages: Socio-economic analysis of the Tonle Sap Lake. *International Journal of Water Resources Development, 22*(3), 463–480.

Krivonogov, S. K., Bezrukova, E. V., Takahara, H., & Riedel, F. (2003). The Hovsgol Lake in the Late Pleistocene and Holocene: On-land geologic evidences for the change of its level. In K. Kashivaya (Ed.), *Long continental records from Lake Baikal* (pp. 187–206). Springer-Verlag.

Krivonogov, S. K., Yi, S., Kashiwaya, K., Kim, J. C., Narantsetseg, T., Oyunchimeg, T., Safonova, I. Y., Kazansky, A. Y., Sitnikova, T., Kim, J. Y., & Hasebe, N. (2012). Solved and unsolved problems of sedimentation, glaciation and paleolakes of the Darhad Basin, Northern Mongolia. *Quaternary Science Reviews, 56*, 142–163.

Lee, M. K., Lee, Y. I., Lim, H. S., Lee, J. I., & Yoon, H. I. (2013). Late Pleistocene–Holocene records from Lake Ulaan, southern Mongolia: Implications for east Asian palaeomonsoonal climate changes. *Journal of Quaternary Science, 28*(4), 370–378.

Lehmkuhl, F., Grunert, J., Hülle, D., Batkhishig, O., & Stauch, G. (2018). Paleolakes in the Gobi region of southern Mongolia. *Quaternary Science Reviews, 179*, 1–23.

Ministry of Environment and Green Development. (2013). *Integrated water management plan of Mongolia*. 339 p.

Mischke, S., Lee, M. K., & Lee, Y. I. (2020). Climate history of Southern Mongolia since 17 ka: The Ostracod, Gastropod and Charophyte record from Lake Ulaan. *Frontiers in Earth Science, 8*(221), 1–15.

Narantsetseg, T., Krivonogov, S. K., Oyunchimeg, T., Uugantsetseg, B., Burr, G. S., Tomurhuu, D., & Dolgorsuren, K. (2013). Late Glacial to Middle Holocene climate and environmental changes as recorded in Lake Dood sediments, Darhad Basin, northern Mongolia. *Quaternary International, 311*, 12–24.

Narantsetseg, Ts., Tomurhuu, D., & Oyunchimeg, T. (2006). Lithology and biogeochemistry of sediment of the Lake Hovsgol: Implication for the paleoclimate changes. In *Proceedings of the Institute of Hydrology and Meteorology, Ulaanbaatar, Mongolia*, pp. 139–148. [in Mongolian].

Narantsetseg, T., Tomurhuu, D., Oyunchimeg, T, Tuvshingargal, B., Idermunkh, T., Krivonogov, S., Kuzmin, Y., & Burr, G. (2010). Late Pleistocene/Early Holocene climate conditions of northern Mongolia based on carbon and oxygen stable isotopes of Dood Lake sediments. In *Abstract of the 8th International Symposium on Environmental Processes of East Eurasia, Asian Monsoon Changes and Interplay of High and Low Latitude Climates, Kunming, China*, pp. 13–16.

Orkhonselenge, A., & Bulgan, O. (2021). Geochemical studies and lacustrine geomorphology of Lake Yakhi basin in eastern Mongolia. *Géomorphologie: Relief, Processes, Environnement, 27*(3), 231–242. https://doi.org/10.4000/geomorphologie.15873

Orkhonselenge, A., Bulgan, O., Gerelsaikhan, D., Davaagatan, T., & Altansukh, N. (2020). *Hydroclimatic fluctuation in Lake Yakhi, eastern Mongolia*. EGU General Assembly 2020. EGU2020-686 presentation. https://doi.org/10.5194/egusphere-egu2020-686.

Orkhonselenge, A., Fowel, S., Josh, B., Nancy, G., Camillie, T., & Gerelsaikhan, D. (2019a). *A report from fieldwork measurement and mapping of changes in lake area of Lake Khargal*. 15 p. [In Mongolian].

Orkhonselenge, A., Fowel, S., Nancy, G., Josh, B., Gerelsaikhan, D., & Camillie, T. (2021). Geochemical properties of Lake Khargal sediments in northern Mongolia. In *Goldschmidt2021 Virtual Conference Presentation. July 4–9, 2021, Lyon, France*. https://doi.org/10.7185/gold2021.4898.

Orkhonselenge, A., & Harbor, J. M. (2018). Impacts of modern glacier changes on surface water resources in western and northern Mongolia. *Journal of Water Resource and Protection, 10*(6), 559–576.

Orkhonselenge, A., Komatsu, G., Davaagatan, T., & Uuganzaya, M. (2019b). Sedimentary Records from Lakes Reconstructing Late Quaternary Paleoclimate Changes in Mongolia. In *XX INQUA Congress. Presentation #2419. Dublin, Ireland*.

Orkhonselenge, A., Komatsu, G., & Uuganzaya, M. (2018a). Climate-driven changes in lake areas for the last half century in the Valley of Lakes, Govi region, southern Mongolia. *Natural Science, 10*(7), 263–277.

Orkhonselenge, A., Komatsu, G., & Uuganzaya, M. (2018b). Middle to late Holocene sedimentation dynamics and paleoclimatic conditions in the Lake Ulaan basin, southern Mongolia. *Géomorphologie: Relief, Processus, Environnement, 24*(4), 351–363.

Orkhonselenge, A., Krivonogov, S. K., Kashiwaya, K., Nakamura, T., & Davaagatan, T. (2015). Paleoclimate changes in Mongolia: Lake Records. In *Extended Abstracts of International Association of Geomorphologists (IAG) Regional Conference on Gradualism vs Catastrophism in Landscape Evolution. Barnaul, Russia*. p. 175.

Peck, J. A., Khosbayar, P., Fowell, S. J., Pearce, R. B., Ariunbileg, S., Hansen, B. C. S., & Soninkhishig, N. (2002). Mid to Late Holocene climate change in north central Mongolia as recorded in the sediments of Lake Telmen. *Palaeogeography, Palaeoclimatology, Palaeoecology, 183*, 135–153.

Prokopenko, A. A., Khursevich, G. K., Bezrukova, E. V., Kuzmin, M. I., Boes, X., Williams, D. F., Fedenya, S. A., Kulagina, N. V., Letunova, P. P., & Abzaeva, A. A. (2007). Paleoenvironmental proxy records from Lake Hovsgol, Mongolia, and a synthesis of Holocene climate change in the Lake Baikal watershed. *Quaternary Research, 68*, 2–17.

Prokopenko, A. A., Kuzmin, M. I., Williams, D. F., Gelety, V. F., Kalmychkov, G. V., Gvozdkov, A. N., & Solotchin, P. A. (2005). Basin-wide sedimentation changes and deglacial lake-level rise in Lake Hovsgol basin, NW Mongolia. *Quaternary International, 136*, 59–69.

Schibler, J., & Jacomet, S. (2013). Short climatic fluctuations and their impact on human economies and societies: The potential of the Neolithic lake shore settlements in the Alpine foreland. *The Journal of Human Palaeoecology, 15*(2), 173–182.

Sekine, Y., Kitajima, T., Fukushi, K., Gankhurel, B., Tsetsgee, S., Davaasuren, D., Matsumiya, H., Chida, T., Nakamura, M., & Hasebe, N. (2020). Hydrogeochemical study on closed-basin lakes in cold and semi-arid climates of the Valley of the Gobi Lakes, Mongolia: Implications for hydrology and water chemistry of paleolakes on Mars. *Minerals, 10*, 792.

Sternberg, T., & Paillou, P. (2015). Mapping potential shallow groundwater in the Gobi Desert using remote sensing: Lake Ulaan Nuur. *Journal of Arid Environments, 118*, 21–27.

Szopińska, M., Szumińska, D., Polkowska, Ż., Machowiak, K., Lehmann, S., & Chmiel, S. (2016a). The chemistry of river–lake systems in the context of permafrost occurrence (Mongolia, Valley of the Lakes). Part I. Analysis of ion and trace metal concentrations. *Sedimentary Geology, 340*, 74–83.

Szopińska, M., Dymerski, T., Polkowska, Ż., Szumińska, D., & Wolska, L. (2016b). The chemistry of river–lake systems in the context of permafrost occurrence (Mongolia, Valley of the Lakes) Part II. Spatial trends and possible sources of organic composition. *Sedimentary Geology, 340*, 84–95.

Szumińska, D. (2016). Changes in surface area of the Böön tsagaan and Orog lakes (Mongolia, Valley of the lakes, 1974–2013) compared to climate and permafrost changes. *Sedimentary Geology*, 1–37.

Yu, K., Lehmkuhl, F., Diekmann, B., Zeeden, C., Nottebaum, V., & Stauch, G. (2017). Geochemical imprints of coupled paleoenvironmental and provenance change in the lacustrine sequence of Orog Nuur, Gobi Desert of Mongolia. *Journal of Paleolimnology, 58*, 511–532.

Chapter 23
Economic Values of Lake Study in Mongolia

Abstract The study of minerals in bottom sediments within the presently live and playa lakes is highly needed to discover sources of ore deposits which can support the economy of Mongolia. Moreover, brine resources in lakes (e.g., halite, thenardite) provide one of the economic strategic resources of the country. In particular, lakes in eastern, western, and central Mongolia have the great potential to develop brine industries in the nation. Lakes in eastern Mongolia provide 39.1% of the total thenardite deposits in Mongolia, whereas lakes in central Mongolia provide 10.9% of the same. Lake Gurvan Tes in southern Mongolia contains the most abundant food salt resource in Mongolia, while Lake Khar in central Mongolia provides the second most abundant. In western Mongolia, lakes in the Great Lakes Depression contain three types of brines and occupy 4.3% of the total thenardite deposits in Mongolia, and they have abundant uranium resources in their bottom sediments. In northern Mongolia, Lake Zeg alone provides 2.2% of the total thenardite deposits in Mongolia. Playa lakes in eastern, southern, and central Mongolia, increasing in number due to the rising air temperature and fluctuating precipitation during the last half century, display abundant halite, thenardite, and soda on their surfaces. Quantitative and qualitative sedimentological, mineralogical, and geochemical investigations from lakes of Mongolia could provide valuable information on minerals and associated rare earth elements in the lake sediments.

Keywords Lakes · Brines · Salts · Minerals · Economy · Mongolia

23.1 Introduction

Lakes are reservoirs of freshwater, sources of fish, shipping corridors, and sites of chemical sedimentation (Bridge & Demicco, 2008). For instance, the Great Lakes in the USA are a major and direct source of drinking water for tens of millions of residents, and they provide the foundation for billions of dollars in the region's economy (Donovan, 2012). Moreover, clean lakes maintain lakeshore property

A. Orkhonselenge et al., *Lakes of Mongolia*, Syntheses in Limnogeology,
https://doi.org/10.1007/978-3-030-99120-3_23

values, contribute to the economic status of the entire communities, and provide low-cost drinking water (University of Maine, 2019). Lake water is one of the most treasured natural resources. Some terrestrial, especially lake, sediments have economic significance because they contain important quantities of natural gas and petroleum, coal, oil shale, and uranium (Boggs Jr., 2012).

Paleolake deposits are important sources of sands and gravels for construction materials, terrestrial fossils, various evaporite minerals, and oil shale (Bridge & Demicco, 2008). Saline lakes can store rich unconventional resources of uranium (U) compounds due to the input of surface water and groundwater transporting U leached from bedrocks (Isupov et al., 2011). U compounds, Br, I, B, Li, and other elements in saline lakes are an important raw material base for the production of advanced materials (Isupov et al., 2015). Playa lakes also provide such primary raw materials and valuable rare earth elements (REE) for advanced industrial products (e.g., catalysts, ceramics, metallurgies, glasses). To date there have been many studies of U in lake sediments for nuclear industries in the world.

In Mongolia, the importance of lake study needs to be understood not only for academic advances in Earth science (see Chap. 22) but also for economic worth. Previous studies on Mongolian lakes since the 1950s (see Chap. 2) determined water, brine, and biological resources in the lakes as factors with economic values in brine and fish industries. For instance, Dashdorj (1956) noted the possibility to develop a fish industry based on the fish resource of Lake Buir in eastern Mongolia (see Chaps. 6 and 7). Avirmed (1959) determined the brine composition of Lake Devteer in western Mongolia. Luvsandorj (1959) revealed physical and chemical properties of brine ores and saline lakes in the northern Khyargas basin in western Mongolia. Davaasuren (1961) determined the chemical composition of brines in Lakes Uvs and Duruu in western Mongolia (see Chap. 13). Vlasov and Luvsandorj (1964) suggested how to utilize brines of Lake Gurvan Tes in southern Mongolia. Dashdorj and Tomilov (1965) highlighted the significance of fishing in Lake Khuvsgul[1] in northern Mongolia (see Chap. 16). Tsend revealed the hydrochemical composition of lakes within the Dornod region in eastern Mongolia (Tsend, 1966a), the Govi[2] region in southern Mongolia (Tsend, 1965), and the Darkhad basin in northern Mongolia (Tsend, 1966b). Luvsandorj (1967) investigated the hydrochemical composition of mineral lakes in Mongolia. Isupov et al. (2011) analyzed the major and trace element compositions in hypersaline soda lakes such as Lakes Uureg, Khyargas, Khar Us, and Shaazgai in the Great Lakes Depression (see Chap. 12), and they found the maximum resource of U amounting to 6000 tons present in Lake Khyargas of western Mongolia. Some lakes in the Choibalsan U province of

[1] *Khuvsgul* has been misspelled as *Hovsgol* and *Khubsugul* in publications. Khuvsgul is the right English transliteration from Mongolian *Хөвсгөл*.

[2] *Govi* has been often spelled as *Gobi* in the international literature. Govi is the correct English transliteration from Mongolian *Говь*.

eastern Mongolia contain high contents of U resources measuring around $n \times 10^{-4}\%$ (Isupov et al., 2011).

Impacts of geological, geomorphological, geochemical, and climatological factors on the accumulation of U in soda Lake Shaazgai located in the U ore region of western Mongolia are evaluated (Isupov et al., 2013). U is deposited not only in the lake water but also in bottom sediments of the lake as well (Isupov et al., 2013). Moreover, U and arsenic (As) speciation in solution and bottom sediments of Lake Shaazgai are represented by a mixture of U oxides-hydroxides, as As is adsorbed by iron hydroxides and compatible U-As phases (Gaskova et al., 2015). Soda-type lakes maintain unchanged ion composition, while chloride-type and sulfate-type lakes change their chemical compositions according to the sequence of mineral precipitation (Kolpakova & Gaskova, 2018). Ariunbileg et al. (2020) provided synthesized hydrochemical and geochemical data obtained from the mineral lakes in eastern (e.g., Lakes Burd, Tsaidam, Khotont), southern (e.g., Lakes Davsan, Kholbooljin, Biger), and western (e.g., Lakes Tonkhil, Ikhes, Oigon) Mongolia, and they showed concentrations of macroelements (e.g., Na, K, Mg, Ca) and microelements (e.g., Li, U, Sr, B, Br) in water and sediments of the lakes and also concentrations of brines (e.g., thenardites, halites, bledites) and minerals (e.g., quartz, feldspars, kaolinites) in the lake sediments. Recently, mineral compositions in lake sediments are indicated in Lakes Yakhi in eastern Mongolia (Orkhonselenge & Bulgan, 2021), Ulaan in southern Mongolia (Orkhonselenge et al., in press), and Khargal in northern Mongolia (Orkhonselenge et al., 2021).

Integrated study of a lake within a frame of interdisciplinary fields in Earth science (see Chap. 22) would contribute to revealing economic values from lake sediments because the lake is a storage and/or a container of natural resources. Saline and/or mineral lakes have been classified in each region of Mongolia (see Chap. 5), and those lakes may have stored economically valuable natural resources. In this chapter, economic significance of lake study in Mongolia is presented. This is the evaluation of economic values of lakes in each region (Table 23.1) and brine resources (Table 23.2) in lakes of each region of Mongolia (Table 23.3) based on mineralization values for lakes covering an area over 0.5 km^2 estimated by Tserensodnom (2000) (Table 23.1).

23.2 Economic Significance of Lake Study

Lake study contributes to the general and specific knowledge about lake basin systems (see Part II). However, discovering that how much resources each lake contains is economically important leads to major improvement in developing successful lake management strategies at the local level (see Chap. 22). As shown in Part II, assessing the present water level and surface area of lakes in Mongolia and integrating lake-river drainage basin systems throughout the country are vital to estimate

Table 23.1 Economic values in lakes of Mongolia

#	Regions	Lakes	Type	Mineralization (g/L)	Salt for forage	Lake bottom muds for traditional therapy
1	Eastern Mongolia	Duruu	F	0.236	–	–
		Angirt	S	4.12	–	–
		Ikh Dalai	M	148.83	+	–
		Davsan	S	290.58	–	–
		Khyart	S	28.04	–	–
		Khukh	–	0.87 (1964), 1.36 (1968), 1.6 (1989)	–	–
		Davst	S	148.83	–	–
		Jirmiin Tsagaan	S	231.29	–	–
		Khar	–	121.12	–	–
		Ukhert	–	135.57	+	–
		Sum	M	4.14	–	–
		Bayan Erkhet	–	7.75	–	–
		Dund Bayan	S	287.07	–	–
		Shar Burd	–	152.42	–	–
		Yakhi	M	5.0–5.17	–	–
		Shar Bukh	S	170.14	+	–
		Baruun Shavar	S	–	+	–
		Shorvog	–	170.14	–	–
		Bulan Shavar	S	11.88	–	–
		Sangiin Dalai	M	306.86	H	–
		Shar Shoroot	S	–	–	+
		Baruun Duruu	S	0.73	–	–
		Khotont	–	284.98	+	+
		Buir	F	298.7–365.9 (west) 200.5–294.1 (east)	–	–
		Burd	S	333.33	–	+
		Toson	S	72.63	+	H

(continued)

Table 23.1 (continued)

#	Regions	Lakes	Type	Mineralization (g/L)	Salt for forage	Lake bottom muds for traditional therapy
2	Southern Mongolia	Khukh	F	0.034–0.045	–	–
		Khukh Burd	S	0.093	–	–
		Adgiin Tsagaan	S	3.04–5.87	+	–
		Davsan	S	309.6	H	–
		Kholbooljin	S	4.78	–	–
		Buun Tsagaan	F	0.304–0.574	–	–
		Orog	S	1.75–5.01	–	–
		Borvoi	S	214.1	H	–
		Tonkhil	S	271.7	H	–
		Khulma	S	282.9	H	–
		Taigan	M	15.6	–	–
		Biger	S	173.9	H	H
		Ikhes	S	155.4	H	–
		Shargiin Tsagaan	M	–	+	–
		Sangiin Dalai	M	45.9	–	–
		Duruu	S	–	H	–
		Tsookhor	S	–	H	–
3	Western Mongolia	Duruu	S	2.27	–	–
		Khurgan	F	0.07–0.09	–	–
		Khoton	F	0.095	–	–
		Dayan	F	0.05	–	–
		Tolbo	F	0.66–0.67	–	–
		Tal	F	0.269	–	–
		Kholboo	S	–	+	–
		Oigon	–	27.32 (0.0 m), 20.25 (7.0 m)	–	–
		Kholboo	–	1.833	–	–
		Khunt	–	1.154	–	–
4	Northern Mongolia	Khargal	S	9.15–9.85	–	+
		Airkhan	S	12.8–16.46 (1969) 5.39–7.53 (1978)	H	–
		Borogchin Taagaan	S	13.51	+	–
		Chingiin Tsagaan	S	1.466	–	H
		Tsegeen	S	0.793	–	+
		Sharga	S	0.81–0.83	–	+
		Yamaat	–	0.43	–	–
		Davst	S	–	+	–

(continued)

Table 23.1 (continued)

#	Regions	Lakes	Type	Mineralization (g/L)	Salt for forage	Lake bottom muds for traditional therapy
5	Central Mongolia	Khashaat	S	–	+	–
		Kheltgii Tsagaan	S	–	+	–
		Duruu Tsagaan	S	10.1–11.5	H	–
		Airag	S	–	–	–
		Ikh	F	0.46–0.48	–	–
		Gun	–	0.52–0.58	–	–
		Khuduu	–	1.89	–	–
		Terkhiin Tsagaan	F	0.17–0.22	–	–
		Ugii	F	0.49–0.52	–	–
		Tsaidamiin Ikh	S	159.5	H	H

Note: *S* saline, *F* freshwater, *M* mineral, *H* high, *L* low, *(+)* exists, *(–)* no information

Table 23.2 Brine deposits in Mongolia

Total brine deposits	Solid salt	Soda salt	Food salt in modern lakes	Food salt and thenardite (or Glauber salt)
57	1	2	8	46
Places	Mt. Shuden	Lakes Buult and Shar Burd	Lakes Gurvan Tes, Khar, Sangiin Dalai, Nergui, Davst, Alag, Takhilt, and Yolkhon	Figure 23.1

and predict surface water resources for further Integrated Water Resources Management (IWRM). Furthermore, the IWRM leads to social and economic plans such as hydroelectric station management.

To date, the numerous studies regarding the Mongolian lakes since 1950–1960s intended to determine whether the lakes had sufficient resources to develop brine and/or fish industries and to detect salt level and hydrochemical composition of the lakes (see Sect. 23.3). Study of minerals in bottom sediments in the presently live and playa lakes in Mongolia is much advantageous for further exploitation of mineral resources and REE. Recently, a few researchers (e.g., V.P. Isupov, O.L. Gaskova, S. Ariunbileg, A. Orkhonselenge) emphasize the significance of raw materials, especially minerals (see Sect. 23.4), in some lakes of Mongolia. This is because these raw materials from the lakes are the sources of advanced materials to be used in chemical, construction, energy, and electronic industries. Table 23.1 shows the economic value of lakes in each region of Mongolia. Lakes in eastern, southern, and western Mongolia have higher resources of salts, brines, and minerals than those in northern and central Mongolia (Table 23.1).

Table 23.3 Types of brine deposits in lakes of Mongolia

#	Regions	Soda salt	Food salt	Food salt and thenardite (or Glauber salt)	Soda salt and thenardite
1	Eastern Mongolia	Lake Shar Burd	–	Lakes Sangiin Dalai and Tsaidam	Lakes Baga Dalai, Ikh Dalai, Jirmiin Tsagaan, Zuun Ulziit, Zuun Shorvog, Shorvog, Baruun Shorvog, Khukh, Tukhum, Shorvog Khonkhor, and Buun Tsagaan
2	Southern Mongolia	–	Lake Gurvan Tes	Lakes Leg, Tukhum, and Biger	–
3	Western Mongolia	Lake Buult	Lakes Sangiin Dalai, Nergui, Davst, Alag, Takhilt, and Yolkhon	Lakes Davsan, Suuj, Devteer, and Ikh San	Lakes Davst and Tsetseg
4	Northern Mongolia	–	–	–	Lake Zeg
5	Central Mongolia	–	Lake Khar	Lakes Ulziit and Borog	Lakes Ikh Tsaidam, Tseel and, Tsaidam

23.3 Study of Brines in Lakes

Brine is one of the strategic raw materials in the country (Ganbaatar, 2012), and it is abundantly found in lakes of Mongolia (Tables 23.1, 23.2, and 23.3, Fig. 23.1). Larger saline and mineral lakes (e.g., Lakes Uvs, Khyargas, Durgun) dominantly reside in the Great Lakes Depression in western and central (e.g., Lakes Sangiin Dalai, Telmen, Bust) Mongolia, whereas smaller saline and mineral lakes are commonly found in eastern (e.g., Lakes Sangiin Dalai, Ulziit Shavar, Bayan Burd) and southern (e.g., Lakes Buuriin Davst, Khutag, Khaya) Mongolia (see Chap. 5; Table 23.1). The types of brines and hydrochemical composition of saline lakes in Mongolia were presented by Luvsandorj (1967).

Brines in modern lakes are formed as a thin layer (as white and nonconsolidated thenardite) on the lake surfaces due to evaporation under the warm and arid climates (Orkhonselenge & Bulgan, 2021) or in lake sediments extending downward to 1.3 m depth (Ganbaatar, 2012). Halite, Glauber salts, and soda are dominant in Mongolia, and they contain various minerals (e.g., thenardite, mirabilite, epsomite). The brines in modern lakes are classified according to the brine composition as halite-dominated lakes (e.g., Lakes Gurvan Tes, Davsan, Baga, Tsavdan, Sangiin Dalai), mirabilite-dominated lakes (e.g., Lakes Ikh Tsaidam, Khar, Buun Tsagaan, Biger, Khyargas), and mirabilite-soda-dominated lakes (e.g., Lakes Buult, Shar Burd) (Ganbaatar, 2012). Some lakes contain abundant soda, thenardite, and Glauber salts (Table 23.3) which are used for forage, whereas mineral muds are used for traditional therapy (Table 23.1).

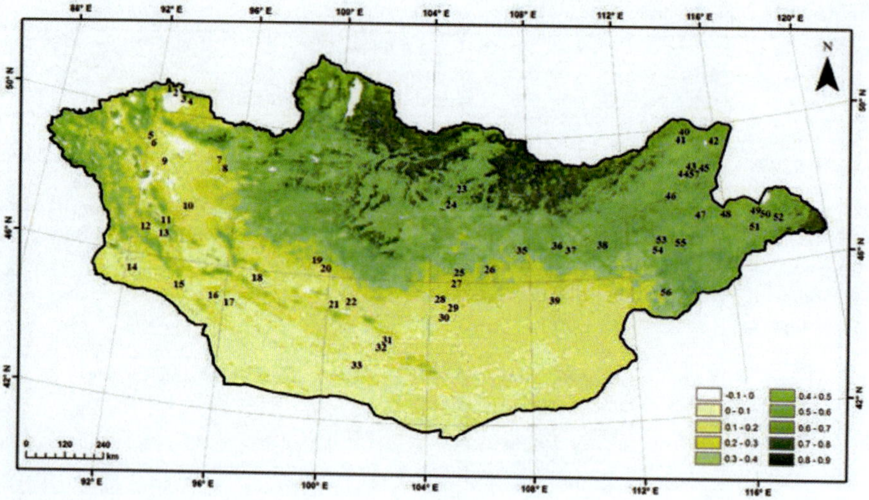

Fig. 23.1 Brine distribution in lakes of Mongolia with paleohalite ore (1), food salt (3, 5, 7, 9, 10, 12–15, 20, 21, 31–37, 39–51, 53, 55, 56), Glauber salt (2, 6, 8, 11, 16–19, 22, 25, 29, 30, 38, 52, 54), and soda (4, 57): 1-Mt. Shuden, 2-Davst, 3-Nergui, 4-Buult, 5-Lake Davsan, 6-Lake Suuj, 7-Lake Takhilt, 8-Lake Devteer, 9-Lake Sangiin Dalai, 10-Tsookhor, 11-Lake Ikh San, 12-Lake Davst, 13-Lake Tsetseg, 14-Lake Yolkhon, 15-Lake Alag, 16-Lake Leg Tukhum, 17-Lake Buuri, 18-Lake Biger, 19-Bayan Tsagaan, 20-Khoid uud, 21-Khunugt, 22-Kholboo, 23-Lake Tseel, 24-Lake Tsaidam, 25-Lake Ulziit, 26-Lake Khar, 27-Tsaidam I, 28-Davst, 29-Baruun Tsagaan Us, 30-Lake Borog, 31-Alag Tsav II, 32-Alag Tsav I, 33-Gurvan Tes, 34-Lake Zeg, 35-Lake Ikh Tsaidam, 36-Lake Shorvog Khonkhor, 37-Buun Tsagaan, 38-Lake Tsaidam, 39-Khujiriin Togoon, 40-Lake Baga Dalai, 41-Lake Ikh Dalai, 42-Lake Jirmiin Tsagaan, 43-Nergui I, 44-Nergui II, 45-Lake Zuun Ulziit, 46-Sumber, 47-Baruun Shavar, 48-Lake Zuun Shorvog, 49-Lake Shorvog, 50-Lake Baruun Shorvog, 51-Sangiin Dalai, 52-Lake Khukh, 53-Lake Tukhum, 54-Shuten, 55-Burden Davsan, 56-Shine Tsagaan Us, 57-Lake Shar Burd. Modified from Ganbaatar (2012)

In eastern Mongolia (see Chap. 6), there are ores of soda and Glauber salt in Lake Shar Burd (Fig. 23.1) and many thenardite deposits (Table 23.3). Lake Shar Burd is a circle-shaped lake covering an area of 1.5 km^2, and it contains 0.3–1.3-m-thick brines in the south and 3.2-m-thick brines in the east consisting of Glauber salt for 10.3–38.2%, soda for 0.3–20.7%, and food salts (Ganbaatar, 2012). Lakes Sangiin Dalai and Tsaidam also contain Glauber salts. The Lake Sangiin Dalai basin extends for 3.2 km in length and 2.2 km in width, and it contains strata of brines with a 0.2–1.5-cm-thick thenardite layer on top, 50.7% food salt of 0.3 m thick in the middle, and 41.7% Glauber salt of 11.0 m thick at the bottom (Ganbaatar, 2012). Lakes in this region provide 39.1% of the total thenardite deposits in Mongolia (Table 23.3). Research in playa and intermittent or ephemeral lakes (see Chap. 5) indicates a wide variety in the exploration of new minerals and even brine sources. For example, the light-gray and white spars (feldspar, calcite, barite) such as thenardite, soda, and brine are widespread on the exposed floor of Lake Yakhi at the present warming and drying climatic conditions (Orkhonselenge & Bulgan, 2021).

In southern Mongolia (see Chap. 9), Lake Gurvan Tes contains the most abundant food salt resource in the country (Tables 23.1 and 23.2; Fig. 23.1). The Lake Gurvan Tes basin extends for 2.1–2.3 km in length and 0.9–1.0 km in width, and it contains a 0.2–1.6-m-thick halite layer on top and a mirabilite layer of 0.5–2.3-m-thick at the bottom with the composition of 62.0–92.7% halite, 0.2–2.9% Ca, 0.2–2.7% Mg, and 0.5–6.7% SO_4 (Ganbaatar, 2012). The region provides also 6.5% of the total thenardite deposits existing in Mongolia, but they are not in lakes (Table 23.3). Lakes Leg, Tukhum, and Biger contain Glauber salts. Thenardite, quartz, bledite, and halite are dominant in brines of Lakes Orog, Kholbooljin, Davsan (Fig. 23.2), Biger, and Bor Ovoo (or Borvoi) in the Valley of the Lakes of the Govi region (Ariunbileg et al., 2017). For instance, Lake Davsan was observed with the numerous circle-shaped pale red brine deposit coverages on its surface (Fig. 23.2) in 2009 when Lake Biger was also covered by abundant soda (Ariunbileg et al., 2020). Brines of Lake Bor Ovoo (or Borvoi) have been widely used for food, and its water is enriched with lithium (Li) (Ariunbileg et al., 2020). In Lake Ulaan (see Chap. 11), the evaporated and precipitated cations Ca, Na, and K of the lake water since the 1950s, during which the lake has been frequently exposed aerially, contribute brines to be bleached at its surface (Orkhonselenge et al., in press).

In western Mongolia (see Chap. 12), lakes in the Great Lakes Depression contain three types of brines (Table 23.3). Lakes Sangiin Dalai, Nergui, Davst, Alag, Takhilt,

Fig. 23.2 Lake Davsan in the Valley of the Lakes of the Govi region, southern Mongolia. Photo by S. Ariunbileg on August 6, 2009

and Yolkhon have food salt resources (Table 23.3, Fig. 23.1). Only Lake Buult contains soda ores, and this lake located at 900–1000 m a.s.l. in the eastern Lake Uvs basin contains a soda layer on top and Glauber and food salts at the bottom with the composition of 63.1% Na_2CO_3, 18.9% Na_2SO_4, 2.5% $NaHCO_3$, 1.8% NaCl, and < 1.0% other brines (Ganbaatar, 2012). Lakes in this region provide 4.3% of the total thenardite deposits in Mongolia (Table 23.3).

In northern Mongolia (see Chap. 15), only Lake Zeg is the provider of thenardite with 2.2% of the total deposits in Mongolia (Table 23.3). In central Mongolia (see Chap. 18), Lake Khar has the second most abundant food salt resource in the country (Tables 23.2 and 23.3, Fig. 23.1). Lakes Ulziit and Borog contain Glauber salts. Lakes Ikh Tsaidam, Tseel, and Tsaidam in this region provide 10.9% of the total thenardite deposits in Mongolia (Table 23.3).

Brines have been widely used in food (e.g., butter, sausage, fish, pickled foods, etc.), tannery, chemical, and medical industries and also in veterinary and human hospitals in Mongolia since the 1940s. At the present time, brine is listed as the 16th share of the gross domestic product (GDP), and it is broadly used in relation to energy, agriculture, hospital, beauty, and daily life. In Mongolia, the domestic brine industrialization was cut by 18 times in 1995 comparing to that in 1985 (Ganbaatar, 2012) due to utilization of brines imported from Russia and China instead of the domestic brines, and this was caused by the governmental policy to shut down national tannery and feeding industries since the early 1990s.

23.4 Study of Minerals in Lakes

Sedimentological investigations of lakes on provenance signatures, tectonic settings, source area weathering, source rock composition, and mineral type could result in valuable information not only for various fields of Earth science (Orkhonselenge et al., 2021 see Chap. 22) but also for economy. For instance, sediments of highly mineralized lakes and/or playa lakes under highly evaporative arid climate contain exploitable microelements (e.g., Li, Sr, Br, B, U, etc.), REE (e.g., Y, Sc, La, Gd, Ey, etc.), raw materials (e.g., sands, gravels, organic matters, etc.), and diverse minerals (e.g., quartz, diamonds, calcite, etc.) for producing further advanced products. This kind of study of Mongolian lakes is still undervalued today.

In eastern Mongolia, Lake Yakhi sediments (see Chap. 6) are composed mostly of arkose and shale, while the composition of source rocks in the lake basin shows derivation of the lake sediments from granites to granodiorites and tonalites in the K-feldspar-plagioclase source terrains (Orkhonselenge & Bulgan, 2021; Fig. 23.3).

In southern Mongolia, Lake Orog (see Chap. 9) is rich with minerals such as mica, chlorite, quartz, plagioclase, K-feldspar, hornblende, and calcite in coarse intervals and authigenic productivity (Ca, Cl, and $CaCO_3$) in an alkaline environment (Yu et al., 2017). In Lake Ulaan (see Chap. 11), the mafic igneous and the quartzose sedimentary provenances have been detected (Lee et al., 2013;

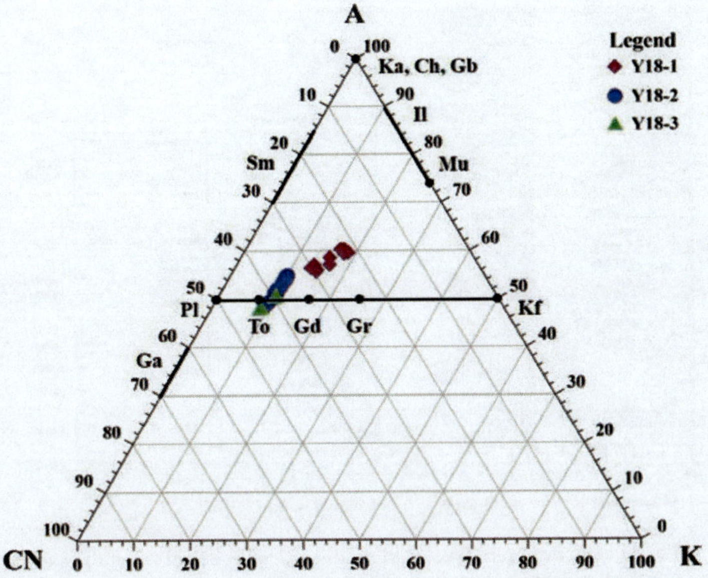

Fig. 23.3 A-CN-K (Al_2O_3-CaO + Na_2O-K_2O) ternary diagram showing minerals in Lake Yakhi sediments (Fig. 6.5 for locations of the core sediments): Ka (kaolinite), Ch (chlorite), Gb (gibbsite), Il (illite), Ms. (muscovite), Kf (K-feldspar), Sm (smectite), Pl (plagioclase), To (tonalite), Gd (granodiorite), Gr (granite), and Ga (gabbro). Modified from Orkhonselenge and Bulgan (2021)

Orkhonselenge et al., in press). Quartz, spar (feldspar, calcite, barite), Ca(Mg)CO_3, and halite are dominant in clays of Lakes Ulaan, Zeges, Burd, Taatsiin Tsagaan, Orog, Kholbooljin, Davsan (Fig. 23.2), Adgiin Tsagaan, Buun Tsagaan, Biger, and Bor Ovoo (or Borvoi) in the Valley of the Lakes (Ariunbileg et al., 2017). U and Li are increased in saline Lakes Burd, Zeges, Kholbooljin (0.1 mg/L ^{238}U), Davsan (0.13 mg/L ^{238}U), and Bor Ovoo (or Borvoi) (0.12 mg/L ^{238}U and 1.0 mg/L Li) (Ariunbileg et al., 2017). In water of these lakes, a cation ratio is exchanged as $Na^+ + K^+ > Mg^{2+} > Ca^{2+}$ and concentrations of brine and soda increased, while in water of Lake Biger, the cation ratio is exchanged as $Mg^{2+} > Ca^{2+} > Na^+$+K^+ and concentrations of S-As and Sr rose (Ariunbileg et al., 2020). Moreover, Lake Ulaan sediments are mainly derived from the tonalites, granites, and granodiorites, and the lake sediments primarily contain shale, graywacke, and litharenite (Orkhonselenge et al., in press).

In western Mongolia (see Chap. 12), Lake Shaazgai favors accumulation of U. It is a soda-type lake characterized mainly by the HCO_3, HCO_3-Cl-Na-Ca composition, and coprecipitation of a significant portion of U within clay minerals (e.g., kaolinite, montmorillonite) in the lake sediments during their formation (Isupov et al., 2013). Lake Khyargas, fed by Zavkhan River draining U-bearing rocks, stores the greatest amount of U (~6000 tons) in Mongolia. The lakes in the Great Lakes Depression are enriched with abundant U deposits showing the highly potential U province in Central Asia (Isupov et al., 2011).

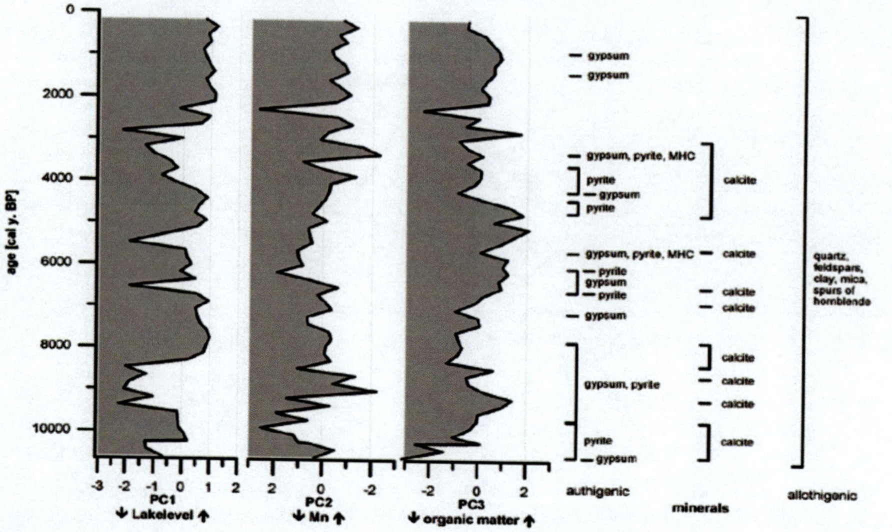

Fig. 23.4 Mineral composition in Lake Ugii sediments comprising the authigenic and allogenic mineral fractions. After Schwanghart et al. (2009)

In northern Mongolia (see Chap. 15), minerals in Lakes Dood and Erkhel sediments were studied by Khosbayar (2005). In Lake Khuvsgul, biogenic silica (BiSi) and carbonate ($CaCO_3$) precipitations (e.g., Prokopenko et al., 2007; Kashiwaya et al., 2010; Orkhonselenge et al., 2013) and carbonate-water $^{18}O/^{16}O$ fractionation factors (Fedotov et al., 2015) were detected.

In central Mongolia (see Chap. 18), Lake Telmen sediments of Ca-aggregate lamina with calcite crystals are abundantly found (Peck et al., 2002). Minerals in Lakes Telmen, Terkhiin Tsagaan, and Ugii sediments were identified by Khosbayar (2005). For instance, the large amounts of mineral fractions in Lake Ugii sediments reflect increased detrital input contributing for the abundances of minerals illite, chlorite, calcite, and smectite (Khosbayar, 2005). The Lake Ugii sediments (see Chap. 20) are dominated by allogenic minerals such as quartz, feldspars (orthoclase, sanidine, microcline, and albite), and silicates (chlorite, hornblende, kaolinite, and muscovite), but pyrite and gypsum are predominantly found in association with calcite in the lake sediments (Schwanghart et al., 2009; Fig. 23.4). Moreover, in Lake Gun sediments, calcite, aragonite, and pyrite are dominant (Zhang et al., 2012).

23.5 Summary

Comprehensive study of lakes in Mongolia provides considerable information not only valued in Earth science but also useful for industries leading to economic growth. Lakes of Mongolia are economically valuable sources because the lakes maintain various types of natural resources in the form of brines, minerals with exploitable microelements, rare earth elements, and other raw materials which can be broadly used for electricity, energy, technology, construction, agriculture, and other sectors. There is substantial possibility to develop the brine industry based on brine resources of Lakes Sangiin Dalai and Gurvan Tes in eastern and southern Mongolia and soda resources of Lakes Shar Burd and Buult in eastern and western Mongolia. Geochemical and sedimentological studies regarding the resources and compositions of brine and minerals in Mongolian lakes could lead to exploration of strategically valuable minerals extracted from lake sediments and to develop robust domestic brine industries. Filling in the serious research gap existing in sparse quantitative and qualitative studies of strategically valuable resources such as brines and minerals for the Mongolian lakes is scientifically and economically crucial today.

References

Ariunbileg, S., Isupov, V. P., Vladimirov, A. G., Kolpakova, M. N., & Kuibida, L. V. (2020). *Chemical evolution and geochemistry of mineralized lakes of Mongolia*. Admon Printing House, 132 p. [In Mongolian].

Ariunbileg, S., Isupov, V. P., Vladimirov, A. G., Orkhonselenge, A., & Shatskaya, S. S. (2017). *Impact of climate change on hydrogeochemical characteristics of lakes in the Valley of Lakes, southern Mongolia*. Goldschmidt2017: Presentation #2169. Paris.

Avirmed, A. (1959). Salt compositions of Lake Devteer. *Journal of Chemistry*, 3–4. [In Mongolian].

Boggs, S., Jr. (2012). *Principles of Sedimentology and Stratigraphy* (5th ed.). Pearson Prentice Hall, Pearson Education, 585 p.

Bridge, J., & Demicco, R. (2008). *Earth surface processes, landforms and sediment deposits*. Cambridge University Press, 815 p. ISBN: 978-0-521-85780-2.

Dashdorj, A. (1956). Fishing in Lake Buir. *Scientific Journal*, 1. [In Mongolian].

Dashdorj, A., & Tomilov, A. A. (1965). Significance of fishing in Lake Khuvsgul. *Journal of Institute of Limnology, 6*(26). [In Russian].

Davaasuren, D. (1961). Physical and chemical characteristics of Lakes Uvs and Duruu and chemical compositions of salts in their basins. *Journal of National University of Mongolia*. [In Mongolian].

Donovan, J. (2012). *New Study Evaluates Ecological, Economic Impacts of Water Use in Great Lakes Region*. Retrieved from https://www.mtu.edu/news/stories/2012/august/new-study-evaluates-ecological-economic-impacts-water-use-great-lakes-region.html.

Fedotov, A. P., Ignat'ev, A. V., & Velivetskaya, T. A. (2015). Reconstruction of deglaciation of Northern Mongolia for the last 330 ka BP, inferred from ostracod stable isotope records from Lake Khubsugul. *Environmental Earth Science, 74*, 2041–2054.

Ganbaatar, T. (2012). Salt. In *Geology and Mineral Resources of Mongolia, 7, Chapter 3, 43–49*. Ministry of Mining and Erdenet, Ltd.. [In Mongolian].

Gaskova, O. L., Isupov, V. P., Vladimirov, A. G., Shvartsev, S. L., & Kolpakova, M. N. (2015). Thermodynamic Modeling of the Behavior of Uranium and Arsenic in Mineralized Shaazgai_ Nuur Lake (Northwest Mongolia). *Doklady Earth Sciences, 465*(1), 1159–1163.

Isupov, V. P., Ariunbileg, S., Razvorotneva, L. I., Lyakhov, N. Z., Shvartsev, S. L., Vladimirov, A. G., Kolpakova, M. N., Shatskaya, S. S., Chupakhina, L. E., Moroz, E. N., & Kuibida, L. V. (2013). Geochemical Model of Uranium Accumulation in Shaazgai-Nuur Lake (Northwestern Mongolia). *Doklady Earth Sciences, 448*(1), 143–148.

Isupov, V. P., Vladimirov, A. G., Lyakhov, N. Z., Shvartsev, S. L., Ariunbileg, S., Kolpakova, M. N., Shatskaya, S. S., Chupakhina, L. E., Kuibida, L. V., & Moroz, E. N. (2011). Uranium in Saline Lakes of Northwestern Mongolia. *Doklady Earth Sciences, 437*(1), 352–356.

Isupov, V., Vladimirov, A., Sodov, A., Kolpakova, M., Shvartsev, S., & Volkova, N. (2015). Hydromineral resources of Saline Lakes of Mongolia and Russian Altai. *Advanced Materials Research, 1085*, 166–170.

Kashiwaya, K., Ochiai, S., Sumino, G., Tsukamoto, T., Szyniszewska, A., Yamamoto, M., Sakaguchi, A., Hasebe, N., Sakai, H., Watanabe, T., & Kawai, T. (2010). Climato-hydrological fluctuations printed in long lacustrine records in Lake Hövsgöl. *Mongolia Quaternary International, 219*, 178–187.

Khosbayar, P. (2005). *Mezozoic and Cenozoic Paleogeography and Paleoclimate of Mongolia. Transaction 15*. Institute of Geology and Mineral Resources, Mongolian Academy of Sciences. [In Mongolian].

Kolpakova, M. N., & Gaskova, O. L. (2018). Major ions behaviour during evaporation of different saline type water of Western Mongolian lakes (geochemical modelling). *Hydrology Research, 49*(1), 163–176.

Lee, M. K., Lee, Y. I., Lim, H. S., Lee, J. I., & Yoon, H. I. (2013). Late Pleistocene–Holocene records from Lake Ulaan, southern Mongolia: Implications for east Asian palaeomonsoonal climate changes. *Journal of Quaternary Science, 28*(4), 370–378.

Luvsandorj, S. (1959). Preliminary physical and chemical results from salt ores and water of saline lakes in northern Khyargas depression. *Journal of Chemistry*, 3–4. [In Mongolian].

Luvsandorj, S. (1967). *Mineral lakes of Mongolia*. Press of Mongolian Academy of Sciences. [In Mongolian].

Orkhonselenge, A., & Bulgan, O. (2021). Geochemical studies and lacustrine geomorphology of Lake Yakhi basin in eastern Mongolia. *Géomorphologie: Relief, Processes, Environnement, 27*(3), 231–242. https://doi.org/10.4000/geomorphologie.15873

Orkhonselenge, A., Fowel, S., Nancy, G., Josh, B., Gerelsaikhan, D., & Camillie, T. (2021). Geochemical properties of Lake Khargal sediments in northern Mongolia. In *Goldschmidt2021 Virtual Conference Presentation. July 4–9, 2021. Lyon, France*. https://doi.org/10.7185/gold2021.4898.

Orkhonselenge, A., Krivonogov, S. K., Mino, K., Kashiwaya, K., Safonova, I. Y., Yamamoto, M., Kashima, K., Nakamura, T., & Kim, J. Y. (2013). Holocene sedimentary records from Lake Borsog, eastern shore of Lake Khuvsgul, Mongolia, and their paleoenvironmental implications. *Quaternary International, 290–291*, 95–109.

Orkhonselenge, A., Uuganzaya, U., & Davaagatan, T. Late Holocene sedimentation dynamics in the Lake Ulaan basin, southern Mongolia: A history of a playa lake. *Environmental Earth Science*. In press.

Peck, J. A., Khosbayar, P., Fowell, S. J., Pearce, R. B., Ariunbileg, S., Hansen, B. C. S., & Soninkhishig, N. (2002). Mid to Late Holocene climate change in north central Mongolia as recorded in the sediments of Lake Telmen. *Palaeogeography, Palaeoclimatology, Palaeoecology, 183*, 135–153.

Prokopenko, A. A., Khursevich, G. K., Bezrukova, E. V., Kuzmin, M. I., Boes, X., Williams, D. F., Fedenya, S. A., Kulagina, N. V., Letunova, P. P., & Abzaeva, A. A. (2007). Paleoenvironmental proxy records from Lake Hovsgol, Mongolia, and a synthesis of Holocene climate change in the Lake Baikal watershed. *Quaternary Research, 68*, 2–17.

Schwanghart, W., Frechen, M., Kuhn, N. J., & Schütt, B. (2009). Holocene environmental changes in the Ugii Nuur basin, Mongolia. *Palaeogeography, Palaeoclimatology, Palaeoecology, 279*, 160–171.

Tsend, N. (1965). *Chemical composition of water in the Govi lakes. A report on Institute of Husbandry*. [In Mongolian].

Tsend, N. (1966a). Chemical composition of some lakes in the Dornod. *Journal of Institute of Husbandry*, 15. [In Mongolian].

Tsend, N. (1966b). *Chemical composition of water in the Darkhad lakes. A report on Institute of Husbandry*. [In Mongolian].

Tserensodnom, J. (2000). *Catalogy of Lakes in Mongolia*. Shuvuun Saaral Publishing, 141 p. [In Mongolian].

University of Maine. (2019). *The Economics of Lakes—Dollars and Sense*. Retrieved from https://www.maine.gov/dep/water/lakes/research.html.

Vlasov, N. A., & Luvsandorj, S. (1964). Physical and chemical characteristics of Lake Gurvan Tes, and methodology to use its salts. *Journal of Natural Sciences*, 1. [In Russian].

Yu, K., Lehmkuhl, F., Diekmann, B., Zeeden, C., Nottebaum, V., & Stauch, G. (2017). Geochemical imprints of coupled paleoenvironmental and provenance change in the lacustrine sequence of Orog Nuur, Gobi Desert of Mongolia. *Journal of Paleolimnology, 58*, 511–532.

Zhang, C., Zhang, W., Feng, Z., Mischke, S., Gao, X., Gao, D., & Sun, F. (2012). Holocene hydrological and climatic change on the northern Mongolian Plateau based on multi-proxy records from Lake Gun Nuur. *Palaeogeography, Palaeoclimatology, Palaeoecology, 323–325*, 75–86.

Chapter 24
Touristic Prospects of Lakes in Mongolia

Abstract Mongolia is a fascinating destination in Eurasia for scientists, visitors, and tourists because of its spacious territory, extraordinary landscape, untouched nature, spectacular scene, and nomadic life, occupying the southern Siberian taiga forest in the north, wide steppe plains in the east, northern Govi desert in the south, and modern glaciers in the west with spectacular lakes isolated with the high Khuvsgul, Khentii), Khangai, Mongolian Altai, and Govi Altai Mountain Ranges. Since the 1990s, tourism in Mongolia is still at its initial stage of development for attracting visitors to lakes. Most tourists visit Lake Khuvsgul in northern Mongolia and other lakes such as Lakes Uvs, Khyargas, and Khoton in western Mongolia; Lakes Buir, Khukh, and Khagiin Khar in eastern Mongolia; and Lakes Terkhiin Tsagaan, Ugii, and Khuis in central Mongolia for diverse activities. Lakes strongly attract international and domestic visitors recently because of their inspiring landscapes and great resources preserved in their basins. However, according to the Ministry of Environment and Tourism, only six events around lakes were included among the 93 touristic events planned to be held throughout Mongolia in 2020. Research-based information is essentially needed for fostering the tourism in lake basins of Mongolia. Moreover, the Ministry of Education and the Ministry of Environment and Tourism are required to pay more attention in the quality of higher education and the development of basic science and highly qualified experts and specialists today.

Keywords Lakes · Tourism · Nature · Landscape · Wildlife · Nomads · Mongolia

24.1 Introduction

Lakes are sites of recreational activity, and they provide us with fascinating scenery, bird and wildlife viewing, drinking water, fishing, refreshment, and leisure (Bridge & Demicco, 2008). Lake basins are one of the geoheritages enabling the establishment of geosites or geoparks which promote geodiversities. Geoheritage

A. Orkhonselenge et al., *Lakes of Mongolia*, Syntheses in Limnogeology,
https://doi.org/10.1007/978-3-030-99120-3_24

encompasses valuable and significant geological and geomorphological elements of the landscape worthy of conservation, and geosite is one of important components of natural heritage (Faccini et al., 2018).

Preservation of lakes for their beauty and recreational benefits, as well as for their value as habitats for fish and wildlife, is of paramount importance. Lacustrine landscape in many regions of Mongolia is also deeply connected with mountainous regions. For instance, the high Altai, Khangai[1], Khuvsgul[2], and Khentii[3] Mountain Ranges hold extraordinary freshwater and saline lakes (see Parts I, II). For these reasons, lakes have recently been one of privileged destinations for researchers and travelers attracted by their fascinating landscapes and spectacular scene in the country.

Lakes are highly valued not only for economic benefits (see Chap. 23) but also for their recreational, esthetic, and scenic qualities (Garn et al., 2003). Lakes constitute important habitats and food resources for a diverse array of bird, fish, aquatic life, and wildlife. For instance, in eastern Mongolia, Lake Yakhi serves for nomadic birds and steppe antelopes, while in western Mongolia Lake Uvs serves for birds (see Sect. 24.4). Lake ecosystems can undergo rapid environmental changes, often leading to significant declines in their esthetic, recreational, and aquatic ecosystem functions (Garn et al., 2003). The condition of a lake at a given time is the result of the interaction of many factors such as tectonic setting, geological background, geomorphological processes, climate change, human influence, and characteristics of the lake itself. As exposed to external effects from the atmosphere, their watersheds and groundwater, lakes are subject to evolve through time. However, a good lake management to control ecology of lake basins (see Chap. 22) is always required because human activity can further accelerate the rate of evolution in lake ecosystems. For example, in the State of Maine, USA, 6000 scenic lakes are as woven into the quality of life as they are parts of the landscape, and as many as 50% of Mainers list swimming as a primary activity and total recreational use exceeds 12 million user days, including both Maine residents and out of state users (University of Maine, 2019).

Mongolia has attracted foreign tourists with its untouched vast territory, spectacular landscapes, natural beauty, wildlife, and nomadic culture recently (see Sect. 24.3). Landscape diversity in each region of the country (Fig. 24.1) is deeply connected with long-term geological evolution (see Chap. 3) and climate changes over geological timescales (see Chap. 21) and with human imprints during the historical time. The natural, historical, cultural, and spiritual importance of the landscape features in Mongolia are reflected in countless artifacts and archeological sites and in the contemporary life, knowledge, arts, resource use, songs, and legends of local

[1] *Khangai* has been misspelled as *Khangay* (or *Hangay*) and *Hangai* in many publications. Khangai is the right English transliteration from Mongolian *Хангай*.

[2] *Khuvsgul* has been misspelled as *Hovsgol* and *Khubsugul* in publications. Khuvsgul is the right English transliteration from Mongolian *Хөвсгөл*.

[3] *Khentii* has been misspelled as *Khentey* (or *Hentey*) and *Khentei* (or *Hentei*) inin many publications. Khentii is the right English transliteration from Mongolian *Хэнтий*.

Fig. 24.1 (**a**) Kherlen River in eastern Mongolia (Figs. 6.3b and 18.1) viewing toward the east, (**b**) Bayan Zag Flaming Cliff in southern Mongolia viewing toward the west, (**c**) Mt. Tsambagarav in western Mongolia (Figs. 12.1 and 12.2) viewing toward the south, and (**d**) Lake Khuvsgul in northern Mongolia (Figs. 15.3 and 16.2) viewing toward the north. Photos by A. Orkhonselenge on July 12, 2020 for (**a**), October 25, 2015 for (**b**), August 10, 2013 for (**c**), and August 25, 2008 for (**d**)

and indigenous communities (UNESCO, 2020). For these reasons, Mongolia has been a privileged destination for scientists, travelers, and artists attracted by the fascinating landscapes. New York Times (2020) wrote that sparsely populated Mongolia was becoming spoiled with tourists. Its tourist numbers are on the rise, and in 2019 Mongolia attracted 530,000 tourists rising from 150,000 recorded in 2000.

Today, both international and domestic tourists excited with fishing, swimming, and recreating in lakes, lake shores, and lake basins of Mongolia (see Sect. 24.4) have been increasing year after year. International tourists are also strongly interested in national cultural festivals *Naadam* in summer and *Tsagaan Sar* in winter showing the nation's uniqueness or distinctiveness. *Naadam* is a celebration festival for the *National Revolution Day*, and it attracts tourists with the three most popular national games in Mongolia of wrestling, archery, and long-distance horse racing. *Tsagaan Sar* is a celebration festival for the *Lunar Calendar New Year* in Mongolia.

In recent years, the Government of Mongolia informs that mining and tourism are the most important economic sectors, and it plans to deal with appropriate policies. For instance, in a frame of the policies, firstly, the present political party changed the name of the Ministry of Environment and Green Development to the Ministry of Environment and Tourism (see Sect. 24.3) and the name of main square from Chinggis[4] to Sukhbaatar in 2016. Secondly, the government intends to open a newly established airport near Ulaanbaatar,[5] the capital of Mongolia, besides the Chinggis Khaan International Airport (formerly Buyant Ukhaa International Airport). This new airport opened in 2021 is expected to accommodate triple the number of tourists over the current one. Thirdly, the government has given tax breaks to travel agencies and tour operators for establishing touristic resorts and constructing hotels, and it has broadly abolished licenses for tourism businesses.

However, in addition to changing the names of places and organizations and enhancing tourist facilities and service areas (e.g., hotels, camps, restaurants), the government has not led any policy to upgrade the advanced level of education for experts and specialists in the tourism sector. Today, national and private universities offer tourism curriculums for 4 years, while some travel agencies and tour operators organize short-term tourism trainings with a certificate. Nevertheless, the Ministry of Education does not control their curriculums, teaching methods, training requirements, and quality of the services to be comparable to those at the international level. As a result, many graduates majored in tourism enter the tourism sector today without taking any course in physical geography, landscape, history, and culture of Mongolia. Moreover, universities, travel agencies, and tour operators open job positions for lecturers, instructors, managers, and guides focusing only on foreign language skills of the graduates instead of their essential education backgrounds and knowledge in the abovementioned fields. Therefore, the government and the Ministries of Education and of Environment and Tourism should pay more attention in qualification of higher education and highly qualified human resources trained at universities and professional organizations.

In this chapter, the data related to tourism activity (Table 24.1; see Sect. 24.3) for Mongolian lakes are synthesized (see Sect. 24.4). The tour events and assets in each region of Mongolia (Tourism News of Mongolia, 2020) are summarized for lakes covering an area over 0.5 km^2 estimated by Tserensodnom (2000) (Table 24.2).

24.2 Impact of Lake Study on Tourism

For tourism linked to lakes in Mongolia, the targeted scientific research, Integrated Water Resources Management (IWRM), and other advanced plans in lake basins (see Chap. 22) can represent very useful tools for further development of tourism, management, and promotion.

Lakes in Mongolia have substantially attracted scholars recently (see Chaps. 2, 21, and 22) because the lakes preserve archives in their sediments and basins useful

[4] *Chinggis Khaan* has been often misspelled as *Genghis Khan* in the international literature. Chinggis Khaan is the correct transliteration from Mongolian *Чингис Хаан* into English.

[5] *Ulaanbaatar* has been often misspelled as *Ulan Bator* or *Ulanbator* by the Russians and *Ulan Bataar* by others in the literature, publications, Wikipedia, etc. Ulaanbaatar is the right transliteration from Mongolian *Улаанбаатар* into English.

Table 24.1 Lakes included in protected areas of Mongolia

#	Natural reserves (32)	Natural monuments (24)	Strictly protected areas (32)	National parks (18)	Details	Regions
1	Lake Yakhi	–	–	–	Chapter 6	Eastern Mongolia
2	–	Lake Khangal	–	–	Chapter 6	
3	–	–	Lake Uvs	–	Chapter 13	Western Mongolia
4	–	–	–	Lake Khyargas	Chapter 12	
5	–	–	–	Lake Khar Us	Chapter 12	
6	–	–	–	Lake Ulaagchnii Khar	Chapter 12	
7	–	–	–	Lake Khuvsgul	Chapter 16	Northern Mongolia
8	–	–	–	Lake Terkhiin Tsagaan	Chapter 19	Central Mongolia
9	–	Lake Khuis	–	–	Chapter 18	

for reconstructing the histories of landscape evolution and climate change in Eurasia. The lakes in each region of Mongolia (see Chap. 1) are diverse with respect to their origins due to tectonic activities and geomorphological processes over geological timescales (see Chaps. 3 and 4). Among them, freshwater and saline lakes (see Chap. 5) have been privileged destinations for national and international scientists and travelers. For example, Lake Khuvsgul in northern Mongolia (Fig. 24.1d) has been intensely visited and studied by scholars since the mid-1990s for obtaining valuable high-proxy reconstructive data (see Chap. 16) and also by tourists since 2000s for taking recreation activities such as fishing, swimming, and ice festival (see Sect. 24.4).

Understanding the water regime of a lake and the landscape in the lake basin through detailed investigations on geology, geomorphology, sedimentology, geochemistry, cryology, hydrology, geochronology, and climatology is crucial for leading touristic activities based upon the research-based decision-making regarding human communicating with natural elements (e.g., lake, river, glacier, permafrost, swamp, rock, cave, forest, etc.) and for knowing how to plan constructions for accommodations with appropriate materials to be used for roads, camps, and other objects. For example, the knowledge of headwater and water table in the lake basins saves time and expenditure to manage water exploitation and consumption for touristic resorts. Moreover, geomorphology helps to organize suitable land use in the lake basins based on landform diversity, and it can also provide precaution about natural hazards such as flooding, creep, rockslide, collapse, snow avalanche, and others, which can occur in the lake basin. Earth science could give tourism much useful integrated information on proper complex solutions for further planning of exploration, consumption, and preservation in lake basins.

Geoheritage has aroused a strong interest within the scientific community due to its scientific, social, and economic significances (Faccini et al., 2018). Protected

Table 24.2 Touristic assets of lakes in Mongolia

#	Regions	Lakes	Types	Recreation	Fishing	Migrating and seabirds	Swimming	Traditional therapy
1	Eastern Mongolia	Bayan Burd	M	–	–	+	–	–
		Baga Dalai	–	–	–	+	–	–
		Khoigiin Tsagaan	S	–	–	+	–	–
		Ganga	M	–	–	H	–	–
		Galuut	M	–	–	+	–	–
		Bus	M	–	–	+	–	–
		Duruu	F	–	H	H	–	–
		Shavart	M	–	–	+	–	–
		Shukh	M	+	–	–	–	–
		Khukh	F	–	+	+	+	–
		Davst	S	–	+	+	–	–
		Zuun Tukhum	M	–	–	+	–	–
		Bulan Shavar	S	–	–	+	–	–
		Shar Shoroot	S	–	–	–	–	+
		Dug	M	–	–	+	–	–
		Buir	F	+	H	–	+	+
		Bayan	F	–	+	+	–	–
		Baga Bayan	F	–	+	+	–	–
2	Southern Mongolia	Khukh	F	+	H	H	–	–
		Sangiin Dalai	S	–	–	+	–	–
		Buun Tsagaan	F	+	H	H	–	–
		Orog	S	+	H	H	–	–
		Khukh Nuuriin Deed	F	+	+	+	+	–
		Sangiin Dalai	M	–	–	+	–	+
		Ereen	F	+	H	H	–	–
		Nogoon	F	–	–	+	–	–

(continued)

Table 24.2 (continued)

#	Regions	Lakes	Types	Recreation	Fishing	Migrating and seabirds	Swimming	Traditional therapy
3	Western Mongolia	Usan Khooloi	F	–	H	–	+	–
		Tolbiin Baga	–	–	H	H	–	–
		Khongor Ulun	F	–	H	–	–	–
		Duruu	S	–	–	+	+	–
		Angirt	F	–	H	+	+	–
		Tugrug	S	–	+	–	–	–
		Khukh Gol	F	–	–	+	+	–
		Ikh Khag	F	–	–	+	–	–
		Khil	F	–	+	–	–	–
		Tsagaan	–	–	+	–	+	–
		Shigertei 1	F	–	+–	–	+	–
		Sordog	–	–	H	H	–	–
		Sagsag	S	–	H	–	–	–
		Shigertei 2	F	–	+	+	+	–
		Saikhan Bulag	F	–	+	–	+	–
		Khurgan	F	+	H	H	+	–
		Khoton	F	+	H	H	+	–
		Dayan	F	+	H	H	+	–
		Tolbo	F	+	H	H	+	–
		Tal	F	–	H	H	+	–
		Ikh Tsagaan	F	–	+	–	+	–
		Nariin Khairtiin Deed	F	–	H	–	–	–
		Ikher	F	–	H	–	+	–
		Nariin Khairtiin Dood	F	–	H	–	–	–
		Zagast	F	–	H	–	+	–
		Turgenii Ikh	F	+	H	–	+	–
		Khukh	–	–	H	H	+	–
		Tsagaan Asgat	F	–	H	–	–	–
		Oigon	–	–	–	+	–	–
		Kholboo	–	+	–	+	–	–
		Tsetseg	–	+	–	+	–	–
		Khunt	–	–	–	+	+	+

(continued)

Table 24.2 (continued)

#	Regions	Lakes	Types	Recreation	Fishing	Migrating and seabirds	Swimming	Traditional therapy
4	Northern Mongolia	Khargal	S	+	+	+	+	+
		Olon	–	–	H	H	–	–
		Airkhan	S	–	–	+	+	–
		Urmiin Tsagaan	–	–	+	+	+	–
		Borogchingiin Taagaan	S	–	–	+	+	–
		Chingiin Tsagaan	S	–	–	H	+	+
		Tsegeen	S	–	+	+	+	+
		Sharga	S	+	–	+	+	+
		Yamaat	–	–	+	+	–	–
		Dund	–	–	+	+	–	–
		Zegest	–	+	+	–	–	–
		Tsungurug	–	–	+	+	+	–
5	Central Mongolia	Duruu Tsagaan	S	+	–	H	+	–
		Buurug Tsagaan	F	–	–	+	–	–
		Ikh	F	+	–	+	–	–
		Duut	F	+	H	H	–	–
		Gun	–	+	+	+	+	–
		Davaa	F	–	H	H	–	–
		Salt/Sult	F	–	H	H	–	–
		Shivirt	F	–	H	H	–	–
		Undur Tsokhio	F	–	H	H	–	–
		Noot	F	–	+	+	–	–
		Khuduu	–		H	+	+	–
		Khukh	–	+	–	–	–	–
		Terkhiin Tsagaan	F	+	H	H	+	–
		Ugii	F	+	H	H	+	–
		Tsaidamiin Ikh	S	–	–	–	–	+
		Khukhiin Khar	F	–	H	–	–	–
		Gichgenii Ikh	F	–	H	–	–	–
		Ovoot	F	–	H	–	–	–
		Suul Tolgoi	–	–	H	H	–	–

Note: *S* saline, *F* freshwater, *M* mineral, *H* abundant, *(+)* possible, *(–)* no information areas represent conservation of geoheritage through initiatives of public use (e.g., geotourism products, field excursions, educational trainings, and sightseeing). In Mongolia, terms of geoheritage, geosite, and geopark have not been grounded professionally and publicly, but their concerns are included within the protected areas. For example, four primary categories such as *strictly protected areas*, *national parks*, *natural reserves*, and *natural monuments* constitute the national protected areas (Protected Areas of Mongolia, 2017). Today, nine lake basins are included in these categories (Table 24.1) implying that they are geoheritages, geosites, and geoparks. For instance, Lake Uvs (see Chap. 13), one of the strictly protected areas, has attracted interest globally as a World Heritage (see Sect. 24.4). According to the Government of Mongolia (1994), academic, touristic, and therapeutic activities are not legally prohibited in the protected areas.

To date, there is no scientific literature about landscape and lakes of Mongolia on the subjects of geoheritage, geosite, geopark, and spectacular spots for international and national tourists in the context of tourism development. Therefore, considering the present situation of tourism in Mongolia developing at a slow rate and of the vision of the Ministry of Environment and Tourism to develop the tourism intensely, this book, particularly the Part II and Sects. 24.3 and 24.4, intends to provide an academic background, research results, and scientific informative data about some specific lakes in Mongolia attracting domestic and international tourists. Much academic work is still needed in this field of lake tourism in Mongolia.

24.3 Present Status of Tourism in Mongolia

Mongolia is the 47th among the 52 places to travel in 2020 (New York Times, 2020), and Mongolia is listed as the fifth place among the fascinating countries to travel and be excited with landscapes and nomadic lifestyle (Telegraph, 2020). Today, there are more than 400 travel companies, 300 hotels, and 600 tourist resorts and camps in Mongolia. According to the National Statistics Office of Mongolia (2020), 32,929,519 international and domestic tourists entered through the Mongolian customs between 1999 and 2018. Moreover, a total of 7,352,850 international tourists visited Mongolia between 2002 and 2019, and Chinese, Russian, Korean, Japanese, and American tourists were dominant among them. Tourists from the Asia-Pacific Ocean region were the maximum in numbers, i.e., they constituted 65.0% of the total international tourists between 2006 and 2018. The favorable seasons in which the international and domestic tourists (in total of 56,920,430 which visited Mongolia between 2010 and 2019) were summer reaching 38.2% and autumn with 34.1% among the total, i.e., 2,414,752 and 1,938,542 tourists on average landed in Mongolia (National Statistics Office of Mongolia, 2020) in these seasons per year during the last decade. Some of them visited Mongolia for fishing and swimming in lakes and recreation on lake shores. Among the planned events for domestic and international travelers in 2020 (Tourism News of Mongolia, 2020), ITM-2020 International Tourism Exhibition was organized by the Mongolian Tourism

Association and supported by the Ministry of Environment and Tourism on October 30 and 31, 2020 in Ulaanbaatar.

According to the government of Mongolia, tourism in Mongolia is emphasized as one of important economic sectors, and it should be developed extensively. In this framework, the government replaced the vision of the Ministry of Environment and Green Development with that of the Ministry of Environment and Tourism in 2016, and it increased the number of the international flights. Since the Ministry of Environment and Tourism has been established, it has focused on only multiplying tour operators, tourist camps, touristic resorts, service areas, and facilities. However, the Ministry has not yet consulted with academic backgrounds to provide human resources at the professional level and scientific information to the tourism sector. In reality, the Ministry is still not ready to supply research-based information about any touristic spots for domestic and international tourists. Today, international visitors intending to appreciate the fascinating and inspiring natural beauty of Mongolia are hungry with such research-based information and interpretation about the Mongolian extraordinary landscapes. Due to a lack of detailed information and uncertainties about the landscape of Mongolia, there are still many cases in which foreign tourists return home with unsatisfied feeling because they could not read the Mongolian language and consequently not understand everything. However, they can be excited with only having insight, making videos, and taking photos of the spectacular nature during scouting the countryside and remote areas in Mongolia. Recently, western travelers have been attempting backpacking, bicycle, and motorbike tours for experiencing the wilderness in Mongolia.

General information about Mongolia has been provided through various publications. For example, Baatar et al. (2000) introduced history, politics, geography, economy, culture, and population of Mongolia in the book *Mongolia: In brief* (Fig. 24.2a). Shagdar (2008) published the book *A Hundred Road to Journey Through Mongolia*, in the Mongolian language. Becker (2008) published the book *Mongolia: Travels in the Untamed Land* (Fig. 24.2b). Moreover, Gilmour (2017) published about history and culture of Mongolia in the book *Among the Mongols*. Robinson (2010), Kohn (2005, 2006, 2008, 2017), Kohn and Starnes (2011), Kohn et al. (2014), and Holden et al. (2018) have published books providing very valuable introductory information for international travelers visiting Mongolia. However, foreign tourists often misspell geographical place names of Mongolia because they search for information about Mongolia in Google and Wikipedia where basic information about the country has been written with incorrect spellings by the Russians based on their pronunciations of Mongolian vocabularies. For example, Ulaanbaatar is often misspelled as Ulan Bator (see Sect. 24.1), and Selenge[6] River (see Chaps. 15 and 18) is also incorrectly written as Selenga River following the Russians' pronunciations. Moreover, *ger*,[7] a round-shaped traditional dwelling of Mongolians, is

[6] *Selenge* has been often misspelled as *Selenga* internationally based on the Russian pronunciation. Selenge is the correct transliteration from Mongolian *Сэлэнгэ* into English.

[7] *Ger* is often written as *Yurt* as translated in the Russian. Ger is the correct transliteration from Mongolian *Гэр* into English.

Fig. 24.2 Publications about Mongolia: (**a**) *Mongolia: In brief* by Baatar et al. (2000), (**b**) *Mongolia: Travels in the Untamed Land* by Becker (2008), and (**c**) *Mongolia: Nomadic by Nature* by "Heritage of Mongols" NGO (2017)

often written as *yurt* when translated in the Russian language. Due to the recent increase of international tourists visiting Mongolia, some national teams and independent writers have begun to introduce Mongolia by themselves. For example, the *Heritage of Mongols* nongovernmental organization (NGO) (2017) provided basic information about geography, nature, population, economy, and culture of Mongolia through a documentary film in six languages (Fig. 24.2c).

Today, most of tour operators and tourist agencies in Mongolia hire guides who can communicate in foreign languages. As a result, many private universities established by economists, linguists, and even politicians participate in the promotion of education in tourism. However, the graduates of these institutes, as guides, do not have any knowledge about the natural features of the country, and they cannot explain formation, evolution, and distinctiveness of such marvelous landscapes to international visitors. Domestic tourists are also facing the same situation regarding the introductory information about the uniqueness of natural features around the touristic spots. For instance, they simply head to a place where they are excited about recreation and refreshment under the natural beauty and spectacular scene without understanding it. This situation shows that tourism in Mongolia is in general eager to earn money instead of paying attention to advanced education in tourism through curriculums and experts or specialists, of giving valuable guidance to tourists, and of providing satisfaction to fill their expectations. Local public museums in each administrative unit in Mongolia were left out after the 1990s due to the political shift from the socialism to the democracy because the state at that time had no policy to deal with or handle public properties such as the museums, libraries, industries, airports, and other organizations, except for the Mongolian Academy of

Sciences and five state universities. For example, as a recent case in 2019, some politicians shut down the National Museum of Nature.

In the context of the government's intention to develop tourism in Mongolia rapidly and to raise the economy of the country while earning money from its activity, an integrated policy including the following basic steps is urgently required today. The steps as indicated below are simple but also powerful and practical.

1. Lecturers and instructors in tourism should be educated at highly qualified and advanced levels.
2. Higher education curriculums in tourism should be reformed or renewed at the international level in order to send well-educated and skilled graduates as instructors or guides to professional organizations in tourism.
3. Research-based information in tourism at the professional level including guidebooks, handbooks, introductory material, supplementary materials, and other data sources should be provided publicly,
4. Names in Mongolia including geography, history and culture, and even also people should be written and translated by specialists with correct spellings and meanings for data source platforms in English (e.g., Google, Wikipedia, etc.),
5. Research-based policy to develop and enhance the national geoheritages, geosites, and geoparks is required because they provide valuable touristic spots.
6. The local museums abandoned since the 1990s should be reconstructed publicly and supported by the government finance.

24.4 Lake Tourism in Mongolia

24.4.1 Eastern Mongolia

In eastern Mongolia (see Chap. 6), Lakes Buir and Khukh attract dominantly domestic travelers for beautiful sceneries, fishing, viewing birds, and swimming activities in the freshwater lake basins.

Lake Buir (see Chap. 7; Fig. 24.3) is the second most famous for fishing in the country with many types of fishes *Hemiculter warpachovskii Nikolskii*, *Microphysogobio tungtingensis*, *Hucho taimen*, *Brachymystax lenok*, *Pike* (*Esox lucius Linnaeus*), *Leuciscus waleskii*, *Horse gudgeon* (*Hemibarbus labeo*), *Pseudaspius leptocephalus*, *Mongolian culter* (*Chanodichthys mongolicus*), *White culter* (*Culter alburnus*), *Prussian carp* (*Carassius gibelio*), *Asian common carp* (*Cyprinus rubrofuscus Lacepede*), and *Silurus asotus*.

Lake Yakhi has been a protected area designated as natural reserve (Table 24.1) since 1998, a geoheritage of steppe landscape because most nomadic birds from the southeastern and southern Asia gather in the lake basin for their habitats and food resources. In addition, Lake Yakhi is the most exciting wildlife locality for steppe antelopes. Lake Ganga is a home lake for swans, and many domestic tourists tend to the lake to watch *the meeting of thousands of swans* each year. However, recently

Fig. 24.3 A tourist camp on the Lake Buir shore. Photo by the Dornod Tourism Association on August 3, 2020

the event has been cancelled due to lake's ecological problems related to shrinkage and water pollution.

Lake Khangal, one of glacial lakes in the Khentii Mountain Range, is a national monument (Table 24.1) because of its uniqueness. Lakes Khagiin Khar and Khar Zurkhnii Khukh, glacial lakes within the Siberian taiga forest in the Khentii Mountain Range, recently attract visitors with their extraordinary spectacular scenes. Lakes Avarga Toson and Gurvan are also famous for domestic tourists with mineral water and therapeutic muds.

In the region, 1,359,843 international and domestic tourists visited through seven ports of entry between 1999 and 2018 (National Statistics Office of Mongolia, 2020), and they constituted 4.1% among the total number of tourists in the country. No event linked to lakes in the region was planned in 2020 (Tourism News of Mongolia, 2020). However, this region hosts important historical places where tourists are much attracted. In recent years, the administration of the Khentii aimag, a homeland of Chinggis Khaan, plans to develop domestic tourism based on the historical places dating back to the thirteenth century, including Onon and Balj River drainage basins, Lake Khar Zurkhnii Khukh, Mt. Burkhan Khaldun,[8]

[8] Mt. Burkhan Khaldun is a cultural site registered as a World Heritage in 2015 (UNESCO, 2020). Mt. Burkhan Khaldun, the birthplace of Chinggis Khaan, his reputed burial site, and his establishment of the Mongol Empire in 1206, is located in the Khentii Mountain Range or Khan Khentii special protected area at the border between the Siberian taiga coniferous forest and the vast Central Asian steppe.

Kherlen Khuduu Aral, and others (Khentii aimag, 2014), directly and tangibly associated with the great history depicted in *The Secret History of the Mongols*. In the plan, new museums, an airport in the Binder soum, and other touristic spots will be constructed (Khentii aimag, 2014). The administration of the Dornod aimag also plans to contribute and promote local industries of handicraft productions for tourists in order to develop tourism in the aimag, a major historical monumental place for Mongolian and Russian soldiers fighting in 1939 against the Japanese and Manchurian militaries during the war in the Khalkh River and Lake Buir basins.

24.4.2 Southern Mongolia

In southern Mongolia (see Chap. 9), international and domestic tourists numbering 19,257,039 (National Statistics Office of Mongolia, 2020) have conducted journeys during the last two decades, and they constituted 58.5% among the total tourists in the country. However, almost all of them have probably been directed to the destinations for scenery of sand dunefields in the Govi[9] desert instead of journey to the Govi lakes. The Govi lakes, especially those in the Valley of the Lakes, have attracted scholars internationally for their importance in Quaternary science (see Chaps. 2, 9–11, and 21). For many lakes in the region, favorable activities are fishing and watching birds (Table 24.2).

In 2020, the *Sand Dunefield Festival* was planned to be held at Lake Gun, an aeolian small lake in the Govi Altai Mountain Range (Tourism News of Mongolia, 2020). In this region, tourists are primarily excited about passing through a vast Govi desert dotted with oasis, walking through sand dunefields, riding two-humped camels (Fig. 24.2c), visiting dinosaur boneyards of remnant outcrops (Fig. 24.1b), landing in the Valley of the Lakes, and watching stars at night. Diverse landscapes varying from the high Govi Altai Mountain Range to the low Govi desert (see Chap. 9) are present for tourists to visit.

The Govi harsh landscapes are challengeable for tourists to pass through. Europeans are aware of the famous travel of Marco Polo, and they are attracted to the appraisal of the Govi geological landscapes. Marco Polo was known as the travel extraordinaire, cultural icon, and advisor to Khubilai Khaan, the grandson of Chinggis Khaan. Marco Polo and his ensemble were a great asset to the Khaan as advisors and informants as they had brought back much useful news while on their travels around the Asian Continent. Today, Marco Polo's statue is commemorated in the Ulaanbaatar's business district as an icon of city tour, intelligence, and wonder.

[9] *Govi* has been often spelled as *Gobi* in the international literature. Govi is the correct English transliteration from Mongolian *Говь*.

24.4.3 Western Mongolia

Western Mongolia (see Chap. 12) plays the most important role in the nation's tourism because it hosts various spectacular landscapes including glaciated high mountains with numerous glacial lakes in the Mongolian Altai Mountain Range and large freshwater and saline lakes in the Great Lakes Depression, which provide scientific significances for studying their evolutions from the Little Ice Age to the present-day conditions. Lakes Uvs, Ulaagchnii Khar, Khyargas, and Khar Us are protected areas and important geoheritage sites (Table 24.1). In particular, Lake Uvs[10] (see Chap. 13), one of the strictly protected areas in Mongolia (Table 24.1), boasts an extraordinary landscape diversity ranging from cold desert to desert-steppe and steppe, conifer, deciduous, and floodplain forests to diverse wetlands and marshlands, freshwater and saltwater systems, mobile and fixed sand dunefields, and even tundra (UNESCO, 2020).

There were in total 2,361,681 international and domestic tourists who visited the region between 1999 and 2018 (National Statistics Office of Mongolia, 2020), and it amounted to 7.2% of the total number of tourists in the country. The Ulgii and Khovd airports are the major transportation ports for both international and domestic visitors. Most tourists visiting Mongolia pass by freshwater and saline lakes such as Lakes Khoton, Khurgan, Dayan, Tolbo, Khar Us, Dalai, and Khar on the way to and back from the high mountains with modern glaciers (see Chap. 12) that are their major destinations. Saline lakes such as Lakes Uvs, Khyargas, Durgun, and Uureg are famous for domestic tourists.

In this region, a great number of lake events were planned to be held in 2020, i.e., numerous tourism events with examples of ice archery in Lake Tolbo, ice fishing in Lakes Khyargas and Airag, and a bird festival in Lake Uvs (Tourism News of Mongolia, 2020). These events constitute 66.6% of the total events related to lakes in Mongolia. Most lakes in the region are freshwater, and they are most favorable for fishing, watching birds, and swimming activities (Table 24.2). For instance, Lakes Khurgan and Khongor Ulun are famous for the fish species *Oreoleuciscus humilis*, while Lake Tal hosts abundant fishes such as *Mongolian Thymallidae*, *Tinca tinca*, and *blowfish*. Moreover, Lake Khoton is abundant with the Mongolian grayling *Thymallus brevirostris* and Siberian grayling *T. Thymallus arcticus* (Slynko et al., 2010).

In this region, tourists are excited with climbing, hiking, and skiing on the high mountains of the Mongolian Altai Mountain Range where modern glaciers reside (see Chap. 12; Fig. 24.1c), riding horse, swimming in large lakes and rivers, and sunbathing on the lake beaches. One of cultural aspects expressed in the landscape of the Mongolian Altai is stone painting left by nomads.

[10] Lake Uvs, the northernmost of enclosed basins in Central Asia and a transnational World Heritage property in the heart of Asia, is important for migrating birds, waterfowl and seabirds, and is the major biomes of eastern Eurasia (UNESCO, 2020).

24.4.4 Northern Mongolia

Northern Mongolia (see Chap. 15) is the second most attractive region for international and domestic tourists. Today, Lake Khuvsgul that is a national park (Table 24.1) and is internationally well known for its unique landscape, valuable surface water resources (see Chap. 16), and the natural, academic, and cultural values. In this region, 4,630,082 international and national travelers visited from 1999 to 2018 (National Statistics Office of Mongolia, 2020), and they represented 14.1% of the total tourists in the country. The Murun airport is a major transportation port for visitors, especially foreigners.

Most lakes in the region are favorable for recreation, fishing, watching birds, swimming, and therapeutic activities (Table 24.2). For instance, domestic and international tourists visit Lake Khuvsgul for taking cruises, swimming, and fishing. Lake Khuvsgul (Fig. 24.1d) is the most famous for fishing in the country with the abundant diverse types of fishes: *Brachymystax lenok*, *Thymallus arcticus nigrescens*, *Coregonus autumnalis migratorius*, *Phoxinus phoxinus*, *Rutilus rutilus Lacustris*, *Cobitis taenia sibirica*, *Nemachilus barbatulus toni*, *Perca fluviatilis*, and *Lota lota* (Bogoyavlensky et al., 1989). Recently, the ice festival in Lake Khuvsgul (Fig. 24.4) has attracted many international and national tourists in winter. In 2020, ice festival events were planned to be held at Lake Khuvsgul and an artificial lake (Tourism News of Mongolia, 2020).

In this region, tourists are excited with swimming, fishing, and taking cruises and boats in Lake Khuvsgul and large rivers, riding reindeer while experiencing the

Fig. 24.4 Horse sleighing during the ice festival on Lake Khuvsgul. Photo by Ch. Batzaya

culture of the Tsaatan[11] ethnic group in the paleolake Darkhad basin and high mountains in the Khuvsgul Mountain Range. The Tsaatans make painstaking use of every morsel of hunted animals, skin, bones, horn, and internal organs for clothing, tools, and urts[12] as well as for food.

24.4.5 Central Mongolia

In central Mongolia (see Chap. 18), a total of 5,320,874 international and national tourists landed through the largest airport in the country, the Chinggis Khaan International Airport, during the last 20 years (National Statistics Office of Mongolia, 2020), and they represented 16.1% of the total number of tourists in the country. However, tourists entering through this airport tend to go to different destinations in other regions. For instance, the most common final destinations are Lake Khuvsgul in northern Mongolia, high mountains with modern glaciers in the Mongolian Altai Mountain Range in western Mongolia, and sand dunefields in the Govi desert in southern Mongolia, passing through specific tourist spots in each region such as Lakes Terkhiin Tsagaan and Ugii in central Mongolia (see Chap. 19 and 20).

Most lakes in this region are freshwater, and fishing and watching birds are the dominant activities at the lakes (Table 24.2). For example, international tourists visit Lake Ugii for fishing in summer, whereas domestic tourists have various purposes both in summer and winter. Tourists visit Lake Terkhiin Tsagaan, which is a national park (Table 24.1), for enjoyment of its spectacular natural scenery, fishing, and swimming.

Lakes Terkhiin Tsagaan and Ugii are famous for fishing in the country with the diverse types of fishes: *Siberian Thymallidae*, *Brachymystax lenok*, *Pike* (*Esox lucius Linnaeus*), *Perca fluviatilis*, *Rutilus rutilus Lacustris*, and *Lota lota* in Lake Terkhiin Tsagaan and *Brachymystax lenok*, *Pike* (*Esox lucius Linnaeus*), *Perca fluviatilis*, *Rutilus rutilus Lacustris*, *Lota lota*, *Leuciscus idus*, *Manchurian gudgeon* (*Gnathopogon strigatus*), and Prussian carp (*Carassius gibelio*) in Lake Ugii. Additionally, Lake Khuduu has abundant fishes such as *Perca fluviatilis* and *Pike* (*Esox lucius Linnaeus*). Lake Khuis, a natural monument (Table 24.1), attracts tourists for its beautiful landscape surrounded by high mountains in the Khangai Mountain Range.

In recent years, travelers are excited about sled dogs and mushing through the snow on the ice over the Tuul and Terelj rivers as their attractive natural landscapes

---

[11] *Tsaatan* (Цаатан) in Mongolian means *reindeers* in English. The Tsaatans, an ethnic group in northern Mongolia, have cultural similarities in tradition, religion, and lifestyle with Alaskan natives (e.g., *Eskimos*, *Athabascans*, etc.). In order to survive under the harsh condition in the frigid landscape, the Tsaatans hunt, make fire, and create warm garments.

[12] *Urts* (*Урц*) in Mongolian meaning Tsaatan's ger, and it has a cone-shaped shelter made from woods covered by reindeer skin.

within the Siberian taiga forest in the Khentii Mountain Range and as advantageous location closer to Ulaanbaatar (Orkhonselenge et al., 2021). Numerous mineral waters in this region are also famous for domestic tourists with their therapeutic treatment.

24.5 Summary

Lakes of Mongolia are one of the most attractive destination categories for scholars and tourists because of their fascinating landscapes, inspiring distinctiveness, and spectacular untouched nature prevalent in lake basins. International and domestic visitors admiring the extraordinary unique lakes in Mongolia, besides the country's great history, culture, and nomadic lifestyle, have been increasing in recent years. In particular, Lake Khuvsgul in northern Mongolia, the hot touristic spot in the country, has been booming up with tourists for fishing, swimming, and taking cruises and boats in summer and those doing with horse-, reindeer-, and dog-sledging during the ice festival in winter. In recent years, saline Lakes Uvs and Khyargas and freshwater Lakes Khar Us and Khar in the Great Lakes Depression in western Mongolia; Lakes Terkhiin Tsagaan, Ugii, and Khuis in central Mongolia; and Lakes Buir, Khukh, Khagiin Khar, and Khangal in eastern Mongolia have been attracting tourists for watching seabirds, sunbathing, fishing, swimming, and recreating. Only 6.5% of the touristic events in Mongolia were planned in relation to lakes in 2020; however, lake-related events may increase in the coming years. Today, there are considerable objective resources and opportunities to develop tourism in Mongolia, but highly qualified human resources and research-based information for tourism are yet to come.

References

Baatar, S., Dorj, T., & Gankhuyag, D. (2000). *Mongolia in brief*. 148 p. Mongolia.

Becker, J. (2008). *Mongolia: Travels in the untamed land*. Imprint I.B. Tauris, 344 p. ISBN: 978-18451-16491.

Bogoyavlensky, B. A., Batjargal, B., Martinov, V. P., Kozlov, U. P., Borobiev, V. V., & Batsuuri, D. (1989). *Atlas of Lake Khuvsgul: Mongolian People Republic*. State Printing Industry of Geodesy and Cartography, SSSR, 118 p. [In Russian with an abstract in English, German and Mongolian].

Bridge, J., & Demicco, R. (2008). *Earth surface processes, landforms and sediment deposits*. Cambridge University Press, 815 p. ISBN: 978-0-521-85780-2.

Faccini, F., Gabellieri, N., Paliaga, G., Piana, P., Angelini, S., & Coratza, P. (2018). Geoheritage map of the Portofino Natural Park (Italy). *Journal of Maps, 14*(2), 87–96.

Garn, H. S., Elder, J. F., & Robertson, D. M. (2003). *Why Study Lakes? An Overview of USGS Lake Studies in Wisconsin*. U.S. Department of the Interior U.S. Geological Survey. Fact Sheet 063–03. 8 p.

Gilmour, J. (2017). *Among the Mongols*. 228 p. ISBN: 978-1978-1484.

Government of Mongolia. (1994). *Law of Specially Protected Areas of Mongolia*. Ulaanbaatar. Retrieved from https://www.legalinfo.mn/law/details/479?lawid=479. . [In Mongolian].

Heritage of Mongols, Non-Governmental Organization. (2017). *Mongolia: Nomadic by Nature. A documentary film*. [In English, German, Russian, Japanese, Korean, and Chinese].

Holden, T., Karlin, A., Kohn, M., Skolnick, A., & O'Malley, T. (2018). *Mongolia (Travel Guide)* (1st ed.). Lonely Planet Publications Pty Ltd. Colorcraft Ltd, 296 p. ISBN-13: 978-8408-190806. [In Spanish].

Khentii aimag. (2014). *Journal of Alungoo ecology, 12 (6)*. Tepe Ltd Printing, 88 p. Retrieved from www.alungooecology.org.

Kohn, M. (2005). *Mongolia (Country Guide)*. Lonely Planet Publications Pty Ltd. Colorcraft Ltd, 296 p. ABN: 36-005-607-983.

Kohn, M. (2006). *Dateline Mongolia: An American Journalist in Nomad's Land* (1st ed.). Rdr Books Publication, 345 p. ISBN-13: 978-157-143-1554.

Kohn, M. (2008). *Mongolia (Country Travel Guide)* (5th ed.). Lonely Planet Publications Pty Ltd. Hang Tai Printing Company, 312 p. ISBN-13: 978-174-10457-8-9.

Kohn, M. (2017). *Dateline Mongolia: An American journalist in nomad's land* (2nd ed.). Blacksmith Books Publication, 344 p. ISBN: 978-988-13765-8-9.

Kohn, M., & Starnes, D. (2011). *Mongolia (Travel Guide)* (6th ed.). Lonely Planet Publications Pty Ltd, 300 p. ISBN-13: 978-174-17931-7-8.

Kohn, M., Kaminski, A., & McCrohan, D. (2014). *Mongolia* (7th ed.). Lonely Planet Publications Pty Ltd, 311 p. ISBN: 978-1-74220-299-0.

National Statistics Office of Mongolia. (2020). *Annual report of national statistical data in 2019*. [In Mongolian].

New York Times. (2020). *52 places to go in 2020*. Retrieved from https://www.nytimes.com/interactive/2020/travel/places-to-visit.html?fbclid=IwAR0s9gh1o8CO4n1PKvfkDdNThhYteuR8IDlCa0jqjl5hQrXrAze7xJ1zxC8.

Orkhonselenge, A., Uuganzaya, M., Davaagatan, T., & Enkhbayar, G. (2021). Late Holocene Peatland Evolution in Terelj and Tuul Rivers Drainage Basins in the Khentii Mountain Range of Northeastern Mongolia. *Water, 13*, 562. https://doi.org/10.3390/w13040562

Protected Areas of Mongolia. (2017). Retrieved from www.mpa.gov.mn. [In Mongolian].

Robinson, C. (2010). *Mongolia: Nomad empire of eternal blue sky. Odyssey books and guides*. Airphoto International Ltd. Tengri Group, 536 p. ISBN: 978-962-217-808-3.

Shagdar, S. (2008). *A hundred road to journey through Mongolia*. [In Mongolian].

Slynko, Y. V., Mendsaykhan, B., & Kasanov, A. N. (2010). On Intraspecies Forms of the Mongolian Grayling (*Thymallus brevirostris* Kessl.) from Hoton Nur Lake (Western Mongolia). *Journal of Ichthyology, 50*(1), 28–37.

Telegraph. (2020). *14 beautiful countries to visit without coronavirus*. Retrieved from https://www.telegraph.co.uk/travel/lists/beautiful-countries-without-coronavirus/?fbclid=IwAR3FuvR6kJkPKUOme2LBUrT4PvjsfY9TiJsC9uDkFO-KWt2c5kmpQ193nyg.

Tourism News of Mongolia. (2020). Retrieved from www.tourismnews.mn. [In Mongolian].

Tserensodnom, J. (2000). *Catalogy of Lakes in Mongolia*. Shuvuun Saaral Publishing, 141 p. [In Mongolian].

UNESCO. (2020). *World Heritage List*. Retrieved from https://whc.unesco.org/en/list/&order=country#alphaM.

University of Maine. (2019). *The Economics of Lakes—Dollars and Sense*. Retrieved from https://www.maine.gov/dep/water/lakes/research.html.

Index

A. Orkhonselenge et al., *Lakes of Mongolia*, Syntheses in Limnogeology,
https://doi.org/10.1007/978-3-030-99120-3

L